small arms survey 2002

counting the human cost

HEI

A Project
of the Graduate Institute
of International Studies,
Geneva

OXFORD
UNIVERSITY PRESS

Great Clarendon Street, Oxford OX2 6DP

Oxford University Press is a department of the University of Oxford.
It furthers the University's objective of excellence in research, scholarship,
and education by publishing worldwide in

Oxford New York

Auckland Bangkok Buenos Aires Cape Town Chennai
Dar es Salaam Delhi Hong Kong Istanbul Karachi Kolkata
Kuala Lumpur Madrid Melbourne Mexico City Mumbai Nairobi
São Paulo Shanghai Singapore Taipei Tokyo Toronto

with an associated company in Berlin

Oxford is a registered trade mark of Oxford University Press
in the UK and in certain other countries

Published in the United States
by Oxford University Press Inc., New York

First published 2002

British Library Cataloguing in Publication Data
Data available

Library of Congress Cataloging-in-Publication Data
Data available

ISBN 0–19–925173–8
ISBN 0–19–925172–X (pbk)

1 3 5 7 9 10 8 6 4 2

Typeset in Garamond Light
by Latitude
Printed in Great Britain
on acid-free paper by
Butler and Tanner Ltd.,
Frome, Somerset

Foreword

The world is awash with small arms and light weapons, now numbering more than 600 million, enough for one in every ten people on earth. While states bear most of the responsibility for controlling these weapons, no one can deny the global scope of the human tragedy that has arisen from their illicit acquisition and use. These weapons have prolonged or aggravated conflicts, produced massive flows of refugees, undermined the rule of law, and spawned a culture of violence and impunity. In short, the excessive accumulation and illicit trade of small arms is threatening international peace and security, dashing hopes for social and economic development, and jeopardizing prospects for democracy and human rights.

In such a climate, it is not at all surprising that voices both inside and outside of government are demanding new efforts to address this common peril. In July 2001, the world community took its first steps forward in building global norms to assist governments, and regional and international organizations, in rising to this challenge. The occasion was the opening at United Nations Headquarters of the UN Conference on the Illicit Trade in Small Arms and Light Weapons in All Its Aspects—the first global conference to address this issue in a common framework. After intense and difficult negotiations, participating states adopted a comprehensive programme of action to deal with this global scourge.

The challenge ahead is to convert words into deeds, an effort that must involve all levels of society—from the global community to the individual citizen. Progress will depend upon an educated citizenry, co-operation from arms manufacturers, enlightened legislation, disciplined enforcement efforts, extensive international sharing of information, and vigorous national commitments to multilateral action. The common denominator for success in all these areas is the availability of information. Indeed, when it comes to the 'good governance' of small arms, there is no substitute for reliable facts.

The first edition of the yearbook, the *Small Arms Survey 2001: Profiling the Problem* was widely praised by diplomats, government officials and activists as a valuable resource and for the part it played in raising global awareness of the many dimensions of small arms issues. I commend this current edition to the widest possible readership. Only with accurate information and sound analysis will the international community be able to formulate the policies, and carry out the measures, equal to the challenges presented by small arms.

Kofi Annan
Secretary-General of the United Nations
March 2002

Contents

Chapter 5: Reaching Consensus in New York: The UN 2001 Small Arms Conference

Chapter 6: Strengthening Controls: Small Arms Measures

Chapter 7: Changing Attitudes: Weapons Collection and Destruction

About the Small Arms Survey

The Small Arms Survey is an independent research project located at the Graduate Institute of International Studies in Geneva, Switzerland. Established in 1999 with the generous financial support of the Swiss Federal Department of Foreign Affairs, it currently receives additional funding from Belgium, Canada, Denmark, France, the Netherlands, Norway, Sweden, and the United Kingdom.

The objectives of the Survey are: to be the principal source of public information on all aspects of small arms; to serve as a resource centre for governments, policy-makers, researchers, and activists; to monitor national and international initiatives (governmental and non-governmental) on small arms; and to act as a clearing house for the sharing of information and the dissemination of best practices. The Survey also sponsors field research and information-gathering efforts, especially in affected states and regions.

The project has an international staff with expertise in security studies, political science, law, economics, development studies, and sociology. It collaborates with a worldwide network of researchers, partner institutions, non-governmental organizations and governments.

Notes to readers

Abbreviations: Topic-specific lists of abbreviations are placed at the end of each chapter.

Chapter cross-referencing: Chapter cross-references appear capitalized in brackets throughout the text. For example, in Chapter 7: 'During armed conflicts, weapons are often widely distributed, not only among those actively participating in the hostilities but also among the civilian population (HUMANITARIAN).'

Exchange rates: All monetary values are expressed in current US dollars (USD). When other currencies are additionally cited, unless otherwise indicated, they are converted to USD using the 365-day average exchange rate for the period 1 September 2000 to 31 August 2001.

Small Arms Survey: The plain text—Small Arms Survey—is used to indicate the overall project, while the italicized version—*Small Arms Survey*—refers to the publication itself. The *Survey*, appearing italicized, refers generally to past and future editions of the yearbook. The Survey, appearing as plain text, refers to the Small Arms Survey project and its activities.

Web site: For more detailed information and current developments on small arms issues, readers are invited to visit our web site at: http://www.smallarmssurvey.org

Small Arms Survey

Graduate Institute of International Studies

12, Avenue de Sécheron

CH-1202 Geneva, Switzerland

Tel.: +41 22 908 57 77

Fax: +41 22 732 27 38

Email: smallarm@hei.unige.ch

Web site: www.smallarmssurvey.org

Acknowledgements

This is the second edition of the *Small Arms Survey* yearbook. It is a collective product of the staff of the Small Arms Survey project, based at the Graduate Institute of International Studies in Geneva, Switzerland. A vast number of researchers in Geneva and elsewhere have contributed to this volume, and it has benefited from the input and advice of a number of officials, activists, experts, and colleagues from around the world.

The principal chapter authors for the *Survey* 2002 are Peter Batchelor, Maria Haug, Aaron Karp, Katherine Kramer, Keith Krause, Edward Laurance, Glenn McDonald, Robert Muggah, and Camilla Waszink. They were assisted by a large number of in-house and/or external contributors who are all mentioned by name in the relevant chapters. Detailed comments on the chapters were provided by Philip Alpers, Eric Berman, Cate Buchanan, Michael Brzoska, Robin Coupland, Spyros Demetriou, Sami Faltas, Dominic Furlong, David Gorman, Martin Griffiths, Kate Joseph, Edward Laurance, Peter Lock, Lora Lumpe, Patricia Lewis, Nicholas Marsh, Sarah Meek, Greg Puley, Michael Renner, Elisabeth Sköns, Pieter Wezeman, and Herbert Wulf.

Peter Batchelor and Keith Krause were responsible for the overall planning and organization of this volume. Aaron Karp provided valuable editorial inputs during the in-house review process. Estelle Jobson was responsible for managing the production of the *Survey*, co-ordinating with the copy editor, Michael James, and the design team. Layout and design was done by Vick Arnò and Nicoletta Forni of Latitude. The cover was designed by Rob Milliken. Dominic Byatt and Amanda Watkins from Oxford University Press provided significant support and encouragement throughout the production of the *Survey*.

Administrative and research support was provided by Delphine Zinner, Fridrich Strba, Véronique Burkhalter, Silvia Cattaneo, Anna Khakee, and Derek Miller. The project also benefited from the support of personnel of the Graduate Institute of International Studies, in particular Peter Tschopp and Daniel Warner, and staff members Wilfred Gander and Nicole Mouthon.

We are very grateful to the Swiss Government for its generous financial and overall support of the Small Arms Survey project, in particular Raimund Kunz, Peter Maurer, Stefano Toscano, Lukas Schifferle, Matthias Halter, Nadia Fischer, and René Haug. Financial support for the project was also provided by the Governments of Belgium, Canada, Denmark, France, the Netherlands, Norway, Sweden, and the United Kingdom. The project also benefits from the assistance and support of a number of other governments and international agencies, including UNIDIR, UNDDA, WHO, UNDP, and the ICRC.

In Geneva, the project has received support and expert advice from: Lucas Assunçao, David Atwood, Deborah Berlinck, Cate Buchanan, John Borrie, Christophe Carle, Randall Harbour, Martin Griffiths, Olivier Guerot, Peter Herby, Judit Kiss, Alexander Kmentt, Thymen Kouwenaar, Etienne Krug, Bennie Lombard, Patrick McCarthy, David Meddings, Camilo Reyes Rodriguez, Rakesh Sood, and Fred Tanner.

Beyond Geneva, the Project also received support from a number of colleagues. In addition to those mentioned above and in specific chapters, we would like to thank: Pete Abel, Péricles Gasparini Alves, Antonio Rangel Bandeira, Dipankar Banerjee, Ian Biddle, Loretta Bondi, Jakkie Cilliers, Wendy Cukier, Ian Davis, Paul Eavis, Rubem César Fernandes, Virginia Gamba, William Godnick, Owen Greene, Björn Hagelin, João Honwana, Don Hubert, Sally Joss, Jan Kamenju, John Kierulf, Philippa King, Steffen Kongstad, Eugenia Piza-Lopez, Andrew Mack, Lisa Misol, Geraldine O'Callaghan, Yeshua Moser-Puangsuwan, Johan Peleman, Maxim Pyadushkin, Jakkie Potgieter, Robert Scharf, Chris Smith, Gerald Steinberg, David Viveash, Lieke van de Wiel, Michel Wéry, Brian Wood, and Steve Wright. Our sincere thanks to many other individuals (who remain unnamed) for their continuing support of the project. We also express our apologies to anybody whom we have forgotten to mention.

Keith Krause
Programme Director

Peter Batchelor
Project Director

Small Arms Survey 2002

Editors
Peter Batchelor and Keith Krause

Editorial Consultant
Aaron Karp

Principal Chapter Authors
Peter Batchelor, Maria Haug, Aaron Karp, Katherine Kramer, Keith Krause, Edward Laurance, Glenn McDonald, Robert Muggah, and Camilla Waszink.

Publication Manager
Estelle Jobson

Layout and Design
Vick Arnò and Nicoletta Forni

Cover Design
Rob Milliken

Copy Editor
Michael James

Introduction

2001: A watershed year for international efforts

The year 2001 marked a watershed in the global effort to understand and address the problems created by the proliferation and use of small arms and light weapons. The *United Nations Conference on the Illicit Trade in Small Arms and Light Weapons in All its Aspects* represented an unprecedented high-level effort to achieve international agreement on the nature of the problem and the most important measures to deal with its different dimensions. Its *Programme of Action*,[1] adopted by consensus, represented the culmination of several years of effort within and outside of the UN system to raise the profile of the small arms issue on the international agenda. In the months leading up to the Conference itself, almost all governmental and non-governmental activity was directed towards influencing its outcome.

The *Small Arms Survey 2001* was also successfully launched at the 2001 UN Small Arms Conference in New York, and was widely cited and used throughout. It continues to be the single most important source of information and research on various aspects of the small arms issue. Such attention testifies to the continuing shortage of reliable and impartial information and analysis on different aspects of the problem of small arms. In this edition of the *Survey*, we break new ground by focusing on the humanitarian dimensions of small arms availability and use, as well as by providing original research on arms-producing companies. We also examine in depth the emerging field of weapons collection and destruction.

This edition of the *Survey* breaks new ground—focusing on humanitarian dimensions of small arms availability and use and providing original research on arms-producing companies.

It is of course too early to draw up a definitive balance sheet for the UN Conference. Our analysis of the conference, presented in Chapter 5, focuses on its ambitious and far-reaching goals: to define the nature of the problem; to provide a tool for policy-makers and activists; to mobilize national and multilateral resources; and to promote the development of international and national norms.

Significant progress was made towards these goals. The *Programme of Action* commits states to implement a wide range of measures at the national, regional, and global levels affecting the trade (legal and illicit), production, and holdings of small arms and light weapons. These measures include strengthening national regulation of the production and transfer of small arms; ensuring that manufacturers mark all weapons (and keep appropriate records) to allow tracing of seized weapons; improving the system of end-user certification to diminish the risk of diversion and illicit trafficking; encouraging the destruction of weapons seized in criminal investigations, collected in post-conflict disarmament programmes, or surplus to national requirements; and increasing the security of small arms stocks. Such measures, if implemented, would go a long way towards improving the situation worldwide.

On the other hand, the *Programme of Action*'s language provides states with a wide margin to exercise discretion or interpretation, via such clauses as 'where applicable', 'as appropriate', 'where needed', or 'on a voluntary basis'. Many important issues were also dropped from the *Programme of Action*, such as commitments to negotiate an international mechanism to trace illicit small arms (only a UN feasibility study was agreed), initiatives to regulate the civilian possession of weapons, increased transparency in the production, stockpiling, and trade in small arms, specific criteria (such as a code of conduct) governing authorized exports,

An Eritrean soldier. (©Panos Pictures/Heldur Netocny)

and greater regulation of arms brokering.[2] And finally, documents themselves, no matter how well crafted, do not themselves destroy surplus weapons, secure stockpiles, punish illicit traffickers, or increase the security of vulnerable groups.

The success of the *Programme of Action* will ultimately be measured by its ability to catalyse a wide range of multilateral and national measures. Compared with the multilateral climate on the issue only five years ago, a relatively wide consensus has emerged that the unconstrained availability of small arms and light weapons ought to be addressed multilaterally. Small arms are, in other words, 'independent variables' in complex (and often conflictual) situations that are susceptible to concrete policy measures.

The role of the Small Arms Survey

How will the Conference have an impact on the work of the Small Arms Survey? In addition to our ongoing research activities, over the past year the staff and collaborators of the Small Arms Survey were active in different aspects of the Conference process. Field research projects were also conducted throughout 2001 in Georgia, Yemen, the Republic of Congo (Brazzaville), and elsewhere.

The *Programme of Action* highlights a wide range of possible negative consequences associated with the proliferation and use of small arms and light weapons. These include:

- increasing the intensity of contemporary conflicts;
- diminishing the security of vulnerable groups such as women and children or refugees and internally displaced persons;
- increasing the violence associated with large-scale criminal activity (and the concomitant burden on the criminal justice system);
- eroding development gains and the prospects for socio-economic development;
- threatening humanitarian relief operations and workers; and
- increasing the public health burden associated with firearms violence.

Small arms and light weapons are implicated in complex ways with all these problems. Little work, however, has been done to assess systematically the impact of small arms availability and/or use, or to evaluate the effectiveness of particular policy measures. The Small Arms Survey project has an important role to play here, within its core mission of independent research and analysis of all aspects of the problem of small arms and light weapons. Our work is slowly shifting from collecting existing information to *generating* new data through longer-term research and information-gathering projects and partnerships.

Our focus on the small arms themselves (production, stockpiles, transfers) is also complemented by research efforts that focus on the *impact* and effects of small arms and light weapons proliferation, and on the evaluation of *measures* to deal with the problems they pose. This research is attempting to answer questions such as:

- What are the long-term costs and consequences of small arms availability, proliferation, and misuse?
- What are the costs and benefits associated with different policy measures and strategies?
- What are the costs of inaction (such as growing insecurity, stalled socio-economic development, distortions in economic allocations, and declining provision of basic services)?

This kind of research requires baseline data collection in affected states and regions, information-gathering systems for surveillance research (e.g. concerning health and policing impacts and costs), long-term research partnerships, and conceptual and practical tools for assessing the impact of small arms proliferation on humanitarian initiatives and socio-economic development. We make a start on this ambitious agenda in Chapter 4, which deals with the humanitarian impacts of small arms. New projects have already been launched in some of these areas, in collaboration with a wide range of institutional partners.

Beyond the 2001 UN Small Arms Conference

The conference process is only one piece of a much larger and more complex tapestry of efforts to tackle the problems posed by the accumulation and use of small arms and light weapons. Internationally, one of the most prominent areas of activity has been weapons collection and destruction programmes, a survey of which can be found in Chapter 7 of this yearbook. Recent initiatives by the United Nations Development Programme in the Republic of Congo (Brazzaville) and Albania have also laid the foundation for expanded collection and destruction programmes, often linked with projects for the Demobilization, Disarmament and Reintegration (DDR) of ex-combatants in post-conflict settings. Another area of activity, centred around the Organization for Security and Co-operation in Europe (OSCE), has focused on stockpile security, formal information exchanges, and the enhancement of national legislative and export control systems through bilateral and multilateral capacity-building efforts. Finally, grassroots initiatives, especially in Latin America, South Asia, and parts of sub-Saharan Africa, are also expanding. These focus on the local and human impacts of weapons proliferation, such as urban crime, communal conflicts, or large-scale insurgencies. All of these efforts involve a wide range of partners and stakeholders, from states to international organizations, armed forces, national police, local government authorities, and non-governmental organizations. They testify to the multi-dimensional nature of the problem and to the need to engage a wide range of actors depending on the particular context. In an attempt to understand better some of these dimensions, Chapter 4, covering the effects of small arms proliferation and use, focuses on one particular set of actors: those affected by the consequences of conflicts and complex humanitarian emergencies.

Small-arms control efforts involve a wide range of partners and stakeholders, from states to international organizations, armed forces, national police, local government authorities, and non-governmental organizations.

Highlights from the yearbook

The *Small Arms Survey 2002* updates and extends our previous research and tackles some new topics. Chapters on production, stockpiles, transfers, and measures build on the *Small Arms Survey 2001*, while our assessment of weapons collection and destruction and the 2001 UN Small Arms Conference represents new areas of attention. And, as noted above, the chapter on 'effects' narrows its focus in this yearbook to concentrate on the humanitarian impacts of small arms.

Chapter 1 (Products and Producers) focuses on developing baseline data on small arms **companies** instead of producer states, and attempts to identify the most important arms producing firms worldwide. Our research database now includes information on approximately 1,000 companies that produce (or have the capacity to produce) small arms and ammunition, in nearly 100 states. As a result, our total estimated annual worldwide production of small arms and light weapons and ammunition has increased to approximately USD 7.4 billion, including both commercial and military-style firearms. This increase in our estimate of the value of global production does *not* necessarily reflect an increase in the volume of production; in fact, many firms remain in a financially questionable state, and overall demand for new weapons does not appear to be increasing. In many states production is also dormant or *ad hoc,* or limited to ammunition. But the *Survey* now has a clear basis on which to build more in-depth research into specific firms, states, and regions, and this year's chapter makes a start by examining two of the world's most significant licensors of small arms: FN Herstal (Belgium) and Heckler and Koch (Germany/UK), whose weapons are produced under licence in more than 35 countries worldwide.

Chapter 2 (Stockpiles): Our estimate of total global stockpiles of small arms and light weapons has risen this year to at least 638 million, significantly higher than the 2001 estimate of at least 551 million. Why the relatively large change? The main reason is a more refined (and higher) estimate of the total civilian holdings of weapons that now includes many states and regions, including in particular China and South Asia, for which reliable estimates were previously unavailable. Calculations of government (mainly military) and insurgent-force holdings of small arms have remained relatively stable.

As the chapter points out, however, it is easy to become obsessed with these figures and to miss their wider significance. The main purpose of estimating military holdings of weapons is to encourage states to practise greater transparency in this area, one of the last bastions of military secrecy. From such transparency can come accountability and dialogue, especially concerning the development of appropriate criteria for determining excessive and destabilizing accumulations of weapons, and for identifying weapons that are surplus to national requirements.

Our estimate of total global stockpiles of small arms and light weapons has risen this year to at least 638 million, significantly higher than the 2001 estimate of at least 551 million.

Chapter 3 (Transfers): We have refined our estimate of global transfers of small arms and light weapons to USD 4 billion a year. Small arms transfers have proved to be an extremely difficult area to research. This year, however, we have attempted to analyse more systematically two areas: the overall level of transparency in the trade (and improvements in it), and the variety of ways in which small arms move from the legal to the illicit markets. These vectors include arms supplies to non-state actors, violations of arms embargoes or end-user undertakings, leakage from state arsenals or private owners, battlefield recoveries, and the 'ant trade'. Given the debate over the scope of the 2001 UN Small Arms Conference, which treated the illicit trade in 'all its aspects', it is important to note that virtually all illicit weapons begin their life cycle in legal production and trade, and that greater transparency and oversight of the legal trade is thus a crucial tool for tackling illicit transfers.

Chapter 4 (Humanitarian Impacts) draws on fieldwork conducted in various countries for the Reference Group on Small Arms of the United Nations Inter-Agency Standing Committee. To date, almost all the evidence we have is anecdotal, although it almost all points in the same direction: increased availability of small arms exacerbates humanitarian crises, and increases the difficulty (or cost) associated with humanitarian and relief efforts. The chapter identifies a series of indicators, and draws together evidence from field studies in a variety of countries to demonstrate the direct (death and injury) and indirect (displacement, reduced access to basic services) impact of small arms in humanitarian crises and complex emergencies. It also examines the frameworks of international humanitarian and human rights law within which small arms could potentially be considered, and identifies some of the limits of a humanitarian approach to the problem.

Chapter 5 (UN Conference): Our analysis of the 2001 UN Small Arms Conference provides an accessible overview of the main issues for non-experts, discussing not just the outcome of the Conference but also some of the diplomatic and political complexities of the process that led up to it. It also tries to reflect the very different views that exist concerning the process and outcomes of the Conference, including the role played by NGOs. The main points from the chapter (and the Conference) have been presented above. Ultimately, the success of the Conference and its outcome will be judged not by what it accomplished but by how the international community uses the outcome to develop concrete measures.

Chapter 6 (Measures) examines a series of regulatory regimes proposed or employed, at the national, regional, and global levels, for the purpose of controlling small arms at various points in their life cycle. While the coverage is selective, the areas the chapter examines—including marking and tracing, end-use certification, stockpile management, and the regulation of civilian possession—are widely recognized as key components of broader efforts to control small arms. These studies are complemented by an examination of a specific instrument (the recently concluded *Firearms Protocol*), a subregion (eastern Africa), and a type of actor (NGOs). The chapter highlights the contrasting stages of development of the various control regimes, with the regulation of arms brokering lagging far behind, despite growing recognition of the crucial role illicit brokering plays in small arms proliferation. The chapter also points to the limitations imposed by divergent state practices in many of these areas.

Chapter 7 (Weapons Collection and Destruction) directs a spotlight on what is becoming a growing area of international activity. It traces the origins of weapons collection programmes, and analyses some of their early successes and failures. It also highlights the variety of different approaches that have been attempted, and stresses that a number of factors shape the success or failure of any intervention. This analysis is especially important for policy-makers and advocates, given the importance that weapons collection (and broader DDR programmes) are going to play over the next few years.

The costs of small arms proliferation and use are ultimately meaningful only if measured in, or translated into, human terms.

Conclusion

The forward-looking process that was established by the 2001 UN Small Arms Conference, which includes an international meeting in 2003 to review implementation of the *Programme of Action*, ensures that political attention will remain focused on small arms issues for years to come. The expansion of specific national, bilateral, and multilateral projects in different regions also ensures that data collection, evaluation, and analysis will remain a crucial task for the Small Arms Survey project. In addition to pursuing this aim, the *Small Arms Survey 2003* will focus more specifically on the developmental dimension of small arms availability and use, and will present in-depth case studies of specific countries and regions, in order to develop insights and knowledge that is useful to all small arms stakeholders.

In all aspects of its work, and especially in the 2002 edition of the *Survey*, the Small Arms Survey project tries to draw attention to the human dimension of the small arms issue. This is not dramatic license, but rather based on our belief that the costs of small arms proliferation and use are ultimately meaningful only if measured in human terms. This approach poses several challenges. Sometimes the effects or consequences can, if approached with care, be measured, weighed, and compared (the economic burden of health and injury, for example), in order to provide an impetus to political action or innovative policy interventions. That the effects are often intangible and difficult to measure and compare makes them, however, no less important for us to grasp.

Endnotes

1 United Nations General Assembly. 2001. *Programme of Action to Prevent, Combat and Eradicate the Illicit Trade in Small Arms and Light Weapons in All Its Aspects.* 20 July. Reproduced in UN document A/CONF.192/15.

2 These were the main issues identified by major NGO groups, including the Biting the Bullet project (London), the Small Arms Working Group (Washington), and the International Action Network on Small Arms (IANSA).

SEMI AUTOMATIC
SEMI OUTOMATIESE
calibre: 7.62mm
R1
south african
calibre: 9mm
star
south african
calibre: 7.62
ak-47
calibre: 9mm
carl walther p38
calibre: 7.62mm
ppsh
calibre: 7.62mm
tokarev
calibre: 9mm
makarov

A Sick or Dying Industry?
Products and Producers

1

Introduction

In May 2001 Smith & Wesson, an icon of the American firearms industry, was sold by its British owner, Tomkins, to an American company Saf-T-Hammer. In 1987, anticipating a gun boom, Tomkins had bought the gun-maker for USD 112 million. Now facing a less promising market, it unloaded the firm for just USD 15 million (*Los Angeles Times*, 15 May 2001). At the time industry observers declared that the firearm business, with production levels at a 30-year low, was 'a dying industry' (*Associated Press*, 14 August 2001). However, a few months later, largely as a result of the 11 September 2001 terrorist attacks in the United States, government and industry figures began to show significant increases in gun sales in the US and elsewhere (*Washington Post*, 22 October 2001). Do these events give any insights into the current state of the global small arms industry and its future trends? Is small arms production a declining business, or have the events of 11 September 2001 transformed its fortunes, at least in the short term?

This chapter examines the current state of the global small arms industry. It provides an annual update of, and new information on, trends and patterns in the global small arms industry, including geographic distribution (number of companies and countries that produce small arms) and the value and volume of global production. The chapter includes a regional survey, which examines small arms production in the Middle East. It also focuses on the most important small arms-producing firms worldwide in order to provide a more detailed picture of the global industry. Licensed production is a key feature of global small arms production, particularly in developing countries, and the chapter provides an analysis of two of the world's most significant licensors of small arms—FN Herstal (Belgium) and Heckler and Koch (Germany/UK)—whose weapons are produced in more than 35 countries worldwide.

The following questions are addressed in this chapter:

- **What are the current trends in the global small arms industry?**
- **Which are the major small arms-producing companies and where are they located?**
- **Who are the major licensors of small arms and in which countries are their products produced under licence?**

All the material presented in this chapter is based on information obtained from open public sources, including official information, annual and specialized defence publications, corporate and non-governmental information services, defence exhibitions, company promotional material, and primary field research in selected countries and regions.

The following are among the key findings of the chapter:

- More than 1,000 companies worldwide are involved in some aspect of small arms production.
- At least 98 countries produce, or have the capacity to produce, small arms and/or ammunition.
- Global production of military-style small arms has declined in recent years.
- Nearly seven million commercial firearms were produced worldwide during 2000.
- Global production of military-calibre small arms ammunition has remained relatively constant in recent years.

A weapons show, South Africa. (© Gérard Klijn)

- The total value of global small arms production, including ammunition, in 2000 was at least USD 7.4 billion.
- At least 11 countries in the Middle East legally produce small arms and/or ammunition.
- 13 countries dominate the global market for small arms.
- Current domestic production of firearms in the US is still way below the peak of 1993–94.
- Licensed production is an easy way to gain market share, evade strict export controls, or facilitate exports to prohibited destinations.

A lack of detailed information about global production, particularly time series data, as well as the significant differences between the various sectors and sub-sectors of the global small arms market makes it difficult to determine the long-term trends in the global small arms industry. Unlike the global small arms industry, other sectors of the global defence industry seem to be enjoying increased sales after nearly a decade of painful down-sizing, restructuring, and consolidation. A recent report suggests that in 2000 arms transfer agreements with developing countries totalled USD 36.9 billion, the highest level since 1993, and an increase of nearly nine per cent over 1999 (Grimmett, 2001, p. 3). According to the Stockholm International Peace Research Institute (SIPRI, 2001, p. 302), 'the global defence industry, more generally, and particularly the Western defence industry, after half a decade of rapid concentration in the context of shrinking markets, has entered a new phase of reorganisation in which a smaller number of large companies face a constant if not growing level of demand for new military equipment'. Levels of global defence spending have been increasing in recent years (SIPRI, 2001; BICC, 2001) and many of the world's major defence contractors, particularly in the US, are likely to benefit from increased defence spending as the US and other governments respond to the terrorist attacks in the US on 11 September 2001 (*Defence News*, 24–30 September 2001).

However, the small arms industry is unlikely to benefit significantly from recent increases in demand for defence equipment, because it differs from other sectors of the defence industry in at least two important ways. First, its main customers, as reflected in the distribution of global stockpiles (STOCKPILES), are individuals rather than states or governments. Second, public opinion is increasingly demanding action against the many harmful effects flowing from the availability and misuse of small arms, thus prompting governments (and the international community) to consider stricter regulation of the production of these types of weapons.

BOX. 1.1 Definition of small arms and light weapons

In the *Small Arms Survey*, 'small arms and light weapons' covers both military-style weapons and commercial firearms (handguns and long guns), following the definition set out in the *Report of the Panel of Governmental Experts on Small Arms* (UNGA, 1997).

- *Small arms:* revolvers and self-loading pistols, rifles and carbines, assault rifles, sub-machine guns, and light machine guns.
- *Light weapons:* heavy machine guns, hand-held under-barrel and mounted grenade launchers, portable anti-tank and anti-aircraft guns, recoilless rifles, portable launchers of anti-tank and anti-aircraft missile systems, and mortars of less than 100mm calibre.

The *Survey* uses the terms 'small arms', 'firearms', and 'weapons' interchangeably. Unless the context dictates otherwise, no distinction is intended between commercial firearms (e.g. hunting rifles), and small arms and light weapons designed for military use (e.g. assault rifles).

The global small arms industry: Annual update

Salesman and customers at a gun shop in Moscow.

This section provides an update on the survey of the global small arms industry that was provided in the *Small Arms Survey 2001*. It focuses on the distribution of production (i.e. the number of countries and companies that produce small arms), general trends and patterns, and the size and scope of the industry (i.e. the value and volume of production).

Distribution: Bigger and smaller?

How many countries worldwide have the capacity to produce small arms? Is the number of producer countries growing? The *Small Arms Survey 2001* estimated that 95 countries have the capacity to produce small arms. However new information and research suggests that at least 98 countries worldwide have the capacity to produce small arms (see Appendix).[1] Table 1.1 shows that the largest geographic concentration of producer countries—over 40 per cent—is in Europe and the Commonwealth of Independent States (CIS). This estimate of the global distribution of small arms producing countries should be treated with caution, however. In some countries (e.g. Bangladesh, Dominican Republic, and Guinea), the lack of reliable information, both official and unofficial, makes it difficult to ascertain whether any small arms are currently being produced, and if so whether regularly or only on an *ad hoc* basis. Some countries are involved in the production only of components rather than final products; while in others, small arms production involves relatively marginal activities such as loading or filling ammunition cartridges.

At least 98 countries worldwide have the capacity to produce small arms.

TABLE 1.1 Global distribution of small arms producing countries, 2000–01

Region	Number (2000)	Percentage	Number (2001)	Percentage
Europe/CIS	39	41	41	42
North/Central America	6	6	5	5
South America	10	11	11	11
Asia-Pacific	19	20	20	21
Middle East	11	11	11	11
Sub-Saharan Africa	10	11	10	10
Total	**95**	**100**	**98**	**100**

Source: Appendix

The difficulty in distinguishing between end producers and intermediate producers makes it hard to determine the total number of companies producing small arms. In addition companies go out of business or change their names. Small arms are often produced in different divisions, subsidiary companies, or plants that are parts of larger companies. Thus, the total number of end producers is likely to be much lower than the total number of producers, including intermediate producers (Omega Foundation, 2001b).

More than 1,000 companies worldwide are involved in some aspect of small arms production.

The *Small Arms Survey 2001* estimated that small arms are produced by at least 600 companies worldwide, of which at least half are located in the US. However, new information and research suggests that more than 1,000 companies worldwide are involved in some aspect of small arms production, whether as intermediate and/or end producers.[2] Nearly half of these companies (47 per cent) are located in Europe and the CIS. The rest are concentrated in North/Central America (mainly the US), with smaller numbers in South America, sub-Saharan Africa, the Middle East, and the Asia-Pacific region. The types of companies are extremely diverse, ranging from small family-owned businesses to subsidiaries or business units of large, multinational, defence-industrial conglomerates. The size of the companies ranges from establishments with fewer than 20 to over 1,000 employees.

TABLE 1.2 Global distribution of small arms companies, 2001

Region	Number (2001)	Percentage
Europe/CIS	485	47
North/Central America	351	34
South America	37	4
Asia-Pacific	92	8
Middle East	46	4
Sub-Saharan Africa	31	3
Total	**1,042**	**100**

Source: Omega Foundation (2001b)

The apparent increase in the number of companies, and countries, involved in some aspect of small arms production does not necessarily indicate an increase in the absolute size, or production capacity, of the global small arms industry. What is clear is that the industry is more widely distributed than was previously known. However, the global small arms industry is smaller today than it was during the 1980s and 1990s in terms of the volume of production, and has less production capacity. Furthermore, even though more than half of the world's countries are currently engaged in the production of small arms, the international market for small arms is still dominated by only 13 countries (Small Arms Survey, 2001, p.15).

TABLE 1.3. Acquisitions of small arms producers (selected)

Target company	Purchaser	Date
Saco Defense (US)	General Dynamics (US)	2001
Santa Barbara (Spain)	General Dynamics (US)	2001
Primex Technologies (US)	General Dynamics (US)	2001
Smith & Wesson (US)	Saf-T-Hammer Corp (US)	2001
Heckler & Koch (Germany)	BAE Systems/Royal Ordnance (UK)	1991
Sako Ltd (Finland)	Beretta (Italy)	1999
Vursan (51%) (Turkey)	Beretta (Italy)	1999
Luigi Franchi (Italy)	Beretta (Italy)	1995
Mauser Werke (Germany)	Rheinmetall DeTec (Germany)	1996
Buck Werke (Germany)	Rheinmetall DeTec (Germany)	1998
Eurometaal (66%) (Netherlands)	Rheinmetall DeTec (Germany)	1999
Oerlikon Contraves (Switzerland)	Rheinmetall DeTec (Germany)	2000
Australia Defence Industries (Australia)	Thomson-CSF (France)	1999
Suhler Jagd-und Sportwaffen (Germany)	Steyr-Mannlicher (Austria)	1994
SK Jagd-und Sportmunitions (Germany)	Nammo Lapua Oy (Finland)	1992
Carl Walther (Germany)	Umarex Sportwaffen (Germany)	1993

Source: Company reports

Like the defence industry more generally, the small arms industry has become more concentrated in recent years through mergers and acquisitions (SIPRI, 2001) in response to factors like cuts in defence spending and the decline in demand as a result of defence downsizing, particularly in NATO countries and in the countries of the former Soviet Union. This concentration has taken place mainly at the company level. For example, some of the world's well-known small arms producers have been acquired either by larger defence companies (e.g. Saco Defense acquired by General Dynamics) or by other major small arms producers (e.g. Sako Limited acquired by Beretta).

The diversity of countries and companies producing small arms makes it difficult to generalize about the global small arms industry. The regional survey in the third section provides a more detailed picture of small arms production in the Middle East. The fourth section provides detailed case studies of selected countries and companies to help provide a more accurate picture of the state of the global small arms industry.

Estimating the value and volume of global production: Some first steps

The *Small Arms Survey 2001* provided some initial estimates for the value and volume of global production of small arms and ammunition in 1980–99 and in 2000. Most of these estimates were preliminary, based on incomplete information. For example, they excluded figures for the production of commercial firearms such as handguns, hunting and sporting rifles, and associated ammunition. However, new research and information enables us to refine and update our estimates of the value and volume of global small arms production for 2000, the latest year for which relatively comprehensive information is available.[3]

In attempting to estimate more precisely the global value and volume of small arms production, this chapter makes a distinction between military-style small arms and commercial firearms, and associated ammunition. In the last few years more detailed data on small arms production, particularly production of commercial firearms, has become available, and it is now possible to provide more reliable estimates of the value and volume of global production.

Volume of production

Military-style small arms: The *Small Arms Survey 2001* estimated that more than **815,000** military-style small arms (pistols, rifles, sub-machine guns, and machine guns) were produced worldwide during 2000 (Small Arms Survey, 2001, pp. 12–13). This figure, it was noted, was less than half the annual average production of 2.1 million during the period 1980–99. It appears that the volume of global production of military weapons has continued to decline in recent years, and it is estimated that approximately **720,000** weapons were produced worldwide during 2001 (Forecast International, 2002); but these estimates of global production have to be treated with great caution.[4] The acquisition of small arms for the US armed forces amounted to nearly 40,000 in 2000, representing five per cent of the global volume of production (STOCKPILES). No significant increase in domestic production for the US armed forces is expected during 2001.[5]

Commercial firearms: New information and research suggests that the global volume of production of commercial firearms (handguns and long guns) in 2000 was approximately **seven million**. According to new estimates compiled by the World Forum on the Future of Sport Shooting Activities (WFSA), approximately **6.9 million** commercial firearms were produced worldwide during 2000 (WFSA, 2001).[6] Nearly three-quarters of this total (74 per cent) were produced in the United States (four million) and in the European Union (1.1 million). The rest (1.7 million) were produced in countries such as Brazil, China, Canada, Japan, and the Russian Federation (WFSA, 2001). While figures for US production can be confirmed by annual production figures from the Bureau of Alcohol,

Nearly seven million commercial firearms were produced worldwide during 2000.

Tobacco and Firearms (BATF), there is no way of confirming the figures for other countries. According to the BATF, more than **one million** firearms are imported into the US every year (United States, BATF, 1999). Furthermore, the WFSA figures do not give any indication of long-term trends in the global production of commercial firearms. It is worth noting that the ratio of the volume of global production of commercial firearms to military-style small arms is approximately 9:1. In the US the ratio is 100:1.

Ammunition: The *Small Arms Survey 2001* estimated that at least **15 billion** units of military-calibre small arms ammunition, of all types and calibres, were produced worldwide during 2000 (Small Arms Survey, 2001, p. 14). This figure, it was noted, was significantly lower than the average annual production of 21 billion units during the period 1980–99. The volume of global small arms ammunition production has remained relatively constant in recent years, and it is estimated that approximately **16 billion** units were produced during 2001 (Forecast International, 2002). These estimates of global production likewise have to be treated with great caution.[7] The acquisition of small arms ammunition for the US armed forces amounted to approximately one billion units in 2000, representing six per cent of the global volume of production. No significant increase in domestic production for the US armed forces is expected during 2001. No information on the volume of production of ammunition for commercial firearms is available.

These estimates for the volume of global small arms production in 2000 are preliminary, and are not comparable with other figures. Based on the information presented above, it seems that the volume of global small arms production, particularly military-style small arms, is declining, largely because of lower demand. The trends in the volume of global production of commercial firearms are, however, more difficult to determine. The volume of the production of ammunition, particularly military-calibre ammunition, seems to be relatively constant. Overall these trends in the volume of production seem to suggest that the current production capacity of the global small arms industry is smaller than during the Cold War era.

Value of production

Military-style small arms: If we assume that at least 815,000 military-style small arms were produced in all countries during 2000, and using the average price of USD 410 paid for an M16 rifle by the US armed forces during 2000 as a rough proxy, the value of global production of military-style small arms in 2000 can be estimated at **USD 335 million**.[8]

Commercial firearms: The total value of shipments (production) of commercial firearms in the US in 1997 was USD 1.2 billion, and the total number of firearms produced in that year was 3.5 million (United States, Census Bureau, 1997a). This means that the average price of each firearm was USD 335. Using this figure, adjusted to 2000 prices (USD 358) as a rough proxy, the value of global firearms production in 2000, based on the estimate of worldwide production of seven million commercial firearms, can be estimated to be at least **USD 2.5 billion**.

Ammunition: Information on the value of global small arms ammunition production is even more difficult to obtain. It is estimated that 1.06 billion units of military-calibre small arms ammunition, of all types, were procured by the US armed forces during 2000. The total value of the US armed forces' contracts for small arms ammunition during the same year was USD 215 million (Forecast International, 2002). This means that the average price for each unit of ammunition produced in the US in 2000 was 20 US cents. Given that at least 15 billion units of military-calibre ammunition were produced in 2000, and using the figure of 20 cents as a rough proxy, the value of global production of military-calibre small arms ammunition in 2000 can be estimated to be at least **USD 3 billion**.

There are no reliable estimates for the volume of production of ammunition for commercial firearms, which makes it difficult to estimate the value of production. In 1997 domestic shipments (production) of ammunition for commercial firearms in the US were valued at USD 859 million (United States, Census Bureau, 1997b). Adjusted to 2000 prices this is equivalent to **USD 917 million**. Given that the US accounts for at least 58 per cent of the global production of commercial firearms (WFSA, 2001), it can be assumed that the value of US ammunition production for commercial firearms *vis-à-vis* the rest of the world is of a similar magnitude. Therefore it can be assumed that the value of global production of commercial firearms ammunition in 2000 was approximately **USD 1.6 billion**.[9]

Global estimates

The value of global small arms production is difficult to estimate, given the lack of detailed information about prices and the costs of production in many countries, particularly some of the major producers such as China. Therefore, it is possible to provide only very rough estimates of the value of global small arms production, based on multipliers derived from the volume of production.

The value of global production of military-style small arms, including ammunition, was estimated to be at least USD 3.3 billion in 2000. In contrast, the value of global production of commercial firearms, including ammunition, was estimated to be at least USD 4.1 billion during the same year. Thus, the commercial firearms industry appears to be the larger sector, by value, of the global small arms industry.

Based on the estimates presented above, it is clear that the total value of global small arms production, including both commercial firearms and military-style small arms, in 2000 was at least **USD 2.8 billion**. The value of the production of commercial firearms is probably more than seven times the value of the production of military-style small arms. The total value of global small arms ammunition production, for both commercial firearms and military-style small arms, in 2000 was at least **USD 4.6 billion**. Thus, the total value of global small arms production, including ammunition, in 2000 was estimated to be at least **USD 7.4 billion**.[10]

It is worth noting that the estimates for the value of global ammunition production are at least 1.5 times greater than the estimates for the value of global small arms production. This ratio of 2:3 for the value of small arms production to ammunition production is similar to the ratio presented in the *Small Arms Survey 2001*. However, the ratios for the value of production of ammunition for military-style small arms and commercial firearms are quite different.

These estimates for the total value of global small arms production in 2000 are very tentative, and not comparable with other years. It is also difficult to ascertain the trends in the value of global production, given the absence of information from some of the world's major producers. Future editions of the *Small Arms Survey* will attempt to provide more robust and updated estimates of the global value and volume of small arms production.

TABLE 1.4 Estimates of the global value and volume of small arms production, 2000

	Volume of production, 2000 (estimate)	Multiplier (USD)	Value of production, 2000 (USD, estimate)
Military small arms	815,000	410	335 million
Commercial firearms	7,000,000	358	2.5 billion
Military-calibre ammunition	15 billion units	0.20	3 billion
Commercial ammunition			1.6 billion
Total			**7.4 billion**

Sources: Company Reports; Forecast International (2002); United States, BATF (1998; 1999; 2000); United States, Census Bureau (1997a; 1997b); WFSA (2001)

Regional survey: Small arms production in the Middle East

The *Small Arms Survey 2001* provided a regional survey of small arms producers, given the fact that the global small arms industry is not homogeneous and that the production of small arms and ammunition varies considerably between countries. In this chapter, we focus specifically on one region, the Middle East[11] which is characterized by high levels of conflict (and therefore a high demand for weapons, including small arms and light weapons) and a large and growing number of domestic small arms producers.[12] Future editions of the *Small Arms Survey* will focus on small arms production, in terms of both countries and companies, in different regions.

MAP 1.1 Small arms production in the Middle East

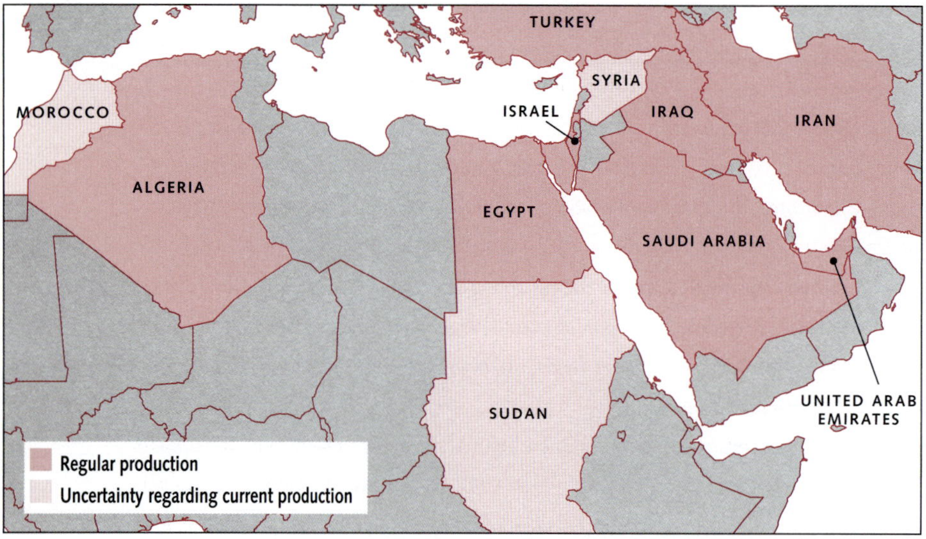

The Middle East as a region remains one of the world's largest recipients, by value, of major conventional arms. During the period 1993–2000, five of the world's top ten recipients of conventional arms—Saudi Arabia, United Arab Emirates, Egypt, Israel, and Kuwait—were in the Middle East (Grimmett, 2001). Despite the fact that the region is a major recipient of conventional arms, at least 11 countries in the Middle East produce, or have the capacity to produce, small arms and/or ammunition.[13] There is some uncertainty regarding current production in Morocco, Sudan, and Syria. None of the other countries in the region, including the Palestinian territories and Western Sahara, is known to produce, or have any capacity to produce, small arms on a regular basis (Feldman and Shapir, 2001).[14] The production of small arms and ammunition in the Middle East is summarized in Table 1.5.

States in the region have established domestic small arms production capabilities, and defence industries more generally, for a wide variety of reasons. These include: military and law-enforcement needs (e.g. strategic independence), political utility, national prestige, economic and diplomatic leverage, and/or employment creation (Miller, 2001). While anecdotal evidence suggests that many countries in the region (e.g. Iran, Saudi Arabia, Turkey) are attempting to expand their small arms production capabilities, official and unofficial information about the actual value and volume of production is difficult to obtain. There is a noticeable lack of public information about small arms production in the Middle East, and, with the exception of Israel and Turkey, it is almost impossible to assess the current state of the small arms industry in the region, particularly in terms of the value and volume of production.

Craft production

Many countries in the Middle East have a long history and culture of craft production of small arms. This type of production typically occurs in small private workshops or homes without any legal (i.e. governmental or company) authorization. It results in weapons of generally poor quality, and includes repair capabilities for damaged weapons in small workshops. Most craft production involves the manufacture of simple single-shot weapons or illicit copies of existing types of small arms. In many countries in the region it is the only form of production that occurs; in others it exists alongside an established domestic production capability for small arms.

Palestinians have long been suspected of developing and maintaining a craft production capability for various types of weapons (STOCKPILES). During 2001, Israeli security forces discovered an extensive network of workshops and factories for manufacturing various types of small arms and light weapons (e.g. mortar rounds, hand grenades and anti-tank rockets) in the Gaza Strip (*Jerusalem Post*, 23 May 2001). The discovery was made after the arrest of Sa'adi Ashi, the owner of a metal production facility. The facilities in Gaza have subsequently been destroyed, but it is unlikely that all craft production in the areas administered by the Palestinian Authority has ceased.

Regular production

The majority of small arms production in the region occurs in state-owned factories or companies, and generally involves the production of foreign weapons under licence. At least seven states in the region are involved in the regular production of small arms. Egypt, Iran, Israel, and Turkey have the most well-developed production facilities in the region, and these countries are also all engaged in the legal export of small arms. At least 11 states in the region are known to produce, or have the capacity to produce, small arms and/or ammunition on a regular basis.

Israel is the only innovative producer of small arms in the region. Countries such as Egypt, Iran, and Turkey, which are important regional producers, tend to concentrate on licensed production of foreign products. In some limited cases, minor modifications on the licensed weapons are made, and occasionally innovation is evident. In general, these countries simply manufacture the foreign products according to original specifications, usually designated with a different (local) name.

Israel is the only innovative producer of small arms in the Middle East.

Algeria produces various types of small arms under licence in state-owned arsenals (Feldman and Shapir, 2001, p. 88). It is not clear whether the country has a capability to produce small arms ammunition. No further information concerning current production is available.

Egypt is one of the oldest and largest of the region's small arms producers, and manufactures a wide range of small arms and ammunition (Feldman and Shapir, 2001, p.112), yet there is a lack of public information there, both official and unofficial, about current small arms production. Almost all the weapons produced in Egypt are based on Soviet-era designs, despite the deployment in recent years of NATO-style weapons for military use, acquired through import, generally from the US. Egypt continues to produce the AK series assault rifle, often with modest modifications for the export market, and maintains a highly active production base, all under state control (Gander and Cutshaw, 2001b).

The production of small arms in Egypt is under the control of the Ministry of Defence and Military Production (MDMP), which controls all the country's small arms and ammunition manufacturing plants; the best-known of these is Factory 54, the Maadi Company for Engineering Industries. It was founded in 1949 to produce rifles and machine guns under licence. It currently produces the Misr 7.62mm assault rifle (a copy of the Kalashnikov AKM) and a variety of pistols, rifles, sub-machine guns, and machine guns (Gander and Cutshaw, 2001b). A number of other factories produce various small arms and ammunition of various calibres, mostly under licence from the former Soviet Union. The most important factories for ammunition production include Aboukir Engineering Industries (Factory 10) and Shoubra Company for Engineering Industries (Factory 45) (Gander and Cutshaw, 2001a).

Iran has a well-developed domestic defence industrial base, and all defence production falls under the umbrella of the Defence Industries Organisation (DIO). The DIO produces a wide range of small arms including G3 rifles and MP5 sub-machine guns under a (lapsed) licence agreement with Heckler and Koch (Germany) (Abel, 2000). The Islamic Revolutionary Guard Corps has its own production facilities for the manufacture of various types of small arms and light weapons (*Jane's Sentinel Security Assessment*, 10 August 2000). Ammunition Industries, which is part of the DIO, is the main domestic producer of small arms ammunition.

Iraq has a wide range of domestic defence production capabilities, including the capacity to manufacture small arms (Feldman and Shapir, 2001, p. 151). However, many of these production capabilities were severely damaged or destroyed during the Second Gulf War. While the full extent of the damage is unknown, it is likely that many of these capabilities have been revitalized in recent years. In early 2002 it was reported that at least three types of small arms, including the Tabuk (a modified version of the AKM rifle), were in production (Forecast International, 2002).

Israel is the region's only innovative producer of small arms, thanks largely to its links with US defence producers. The country has a well-developed indigenous capacity to design various types of small arms and light weapons. Israel is also one of the world's major arms producers and exporters and it was recently ranked as the eighth largest supplier of arms to developing nations during 1997–2000 (Grimmett, 2001). The country's production of small arms has declined in recent years mainly because of a drop in foreign orders following the collapse of the Soviet Union and the end of contracts with many traditional foreign customers, along with a growing shift on the part of the Israeli Defence Force (IDF) towards US suppliers to meet its defence equipment requirements, thanks to continuing US military aid (Sadeh, 2001). The main domestic producer of small arms is the state-owned company Israel Military Industries (IMI), which was ranked as the 59th largest arms producing company in the Organisation for Economic Co-operation and Development (OECD) and developing countries by SIPRI in 1999 (SIPRI, 2001, p. 309). The company currently employs 4,100 people and in 2000 had total sales of USD 417 million, of which less than ten per cent is accounted for by sales of small arms. KSN Industries, a private company, produces various types of pistols. The main domestic producers of small arms ammunition, in addition to IMI, include Kalia Israel Cartridge Co, Reshef Technologies, and Soltam Ltd (Gander and Cutshaw, 2001b).

Morocco has a state-owned factory that manufactures small arms ammunition mainly for domestic consumption (Feldman and Shapir, 2001, p. 227). No further information concerning current production is available.

Saudi Arabia established its own domestic defence industry in the late 1970s. Production of small arms and ammunition is centred at the Al-Kharj Arsenal, part of the Prince Sultan Military City south-west of Riyadh. The country manufactures the G3 and ammunition under licence from Heckler & Koch (Germany), M1 rifles and M16 assault rifles under licence from Colt (USA), and ammunition for the M60 machine gun. It also manufactures hand grenades under licence from Plasticas Oramil (Spain) (*Jane's Defence Weekly*, 18 August 1999). During 2001, Saudi Arabia and Pakistan announced a joint venture to manufacture several types of small arms, mainly products under licence from Heckler & Koch, at the Al-Kharj Arsenal (Forecast International, 2002).

Sudan produces small arms ammunition in a state-owned company, the Military Industry Corporation, which has received support from a variety of countries in eastern Europe and the Middle East (*Jane's Sentinel Security Assessment*, 22 February 2000). There are also unconfirmed reports that Pakistan Ordnance Factories (POF) helped recommission Sudan's small arms factories a few years ago (Omega Foundation, 2001c). The Military Industry

Corporation is designed to meet the country's domestic needs but the country remains a net importer of ammunition. There are unconfirmed reports that the country manufactures small arms (*Africa News Service*, 13 August, 2001). No further information concerning current production is available.

Syria established a domestic small arms ammunition production capability in the 1980s with assistance from the former Soviet Union (Feldman and Shapir, 2001, p. 291). However, the country is currently not self-sufficient in ammunition production, and still imports from countries in eastern Europe, Egypt, and the Russian Federation (Forecast International, 2002). No further information concerning current production is available.

Turkey is one of region's largest and most export-oriented producers of small arms. Together with Israel it is also the region's most transparent arms producer, providing a fairly high degree of public information about domestic defence production. The production of small arms is carried out by the state-owned company Makina ve Kimya Endustrisi Kurumu (MKEK). The company currently employs more than 8,000 people and was ranked by SIPRI as the 60th largest arms-producing company in the OECD and developing countries in 1999 (SIPRI, 2001, p. 309). Total arms sales in 1999 were USD 480 million, up from USD 440 million in 1998. MKEK is a holding company and has

Turkey and Israel are the most transparent small arms producers in the Middle East.

TABLE 1.5 Current production of small arms and ammunition in the Middle East*

Country/Territory	Small arms production			Ammunition production	Small arms and ammunition production
	Craft**	Regular	Innovative		
Algeria		X		#	X
Bahrain	X				
Egypt		X		X	X
Iran		X		X	X
Iraq		X		X	X
Israel		X	X	X	X
Jordan	X				
Kuwait	X				
Lebanon	X				
Libya	X	#			
Morocco		#		X	X
Oman	X	#			
Palestinian territories	X				
Qatar	X				
Saudi Arabia		X		X	X
Sudan		#		X	X
Syria		#		X	X
Tunisia	X				
Turkey		X		X	X
UAE				X	X
Western Sahara	X				
Yemen	X				
Total	**11**	**7 (5)**	**1**	**10 (1)**	**11**

Notes: * Assessment of current production capabilities based on both primary and secondary sources.
** Craft production. This column is for countries that have craft production as their highest level of production. Countries with regular or innovative production sometimes also have craft production, but this is not registered here.
X Known production.
Unknown production (number of unknown producers).

Sources: Feldman and Shapir (2001); Forecast International (2002); Gander and Cutshaw (2001b); Omega Foundation (2001b); Reed (2001).

12 affiliated companies in different parts of Turkey (Reed, 2001). The company produces a wide range of small arms and light weapons including pistols, Heckler & Koch MP5 sub-machine guns, mortars, grenades (with launchers), and artillery systems. MKEK produced the G3 assault rifle under licence from Heckler & Koch for a number of years; slight changes in manufacture led to the appellation G3A7. Production of the G3A7 has now ceased. The main MKEK plant for the production of small arms is Silahsan AS (in Kirikkale). Ammunition production in Turkey takes place in a number of MKEK factories including Kapsulsan AS (in Ankara), Barutsan (Elmadag), Cansas AS (Cankiri), Celiksan AS (Kirikkale), Elroksan AS (Elmadag), Fiseksan AS (Ankara), and Muhimmatsan AS (Kirikkale) (Reed, 2001, p. 64).

The United Arab Emirates has a company in Abu Dhabi, the Adcom Company, which produces a range of small arms ammunition mainly for domestic use (Forecast International, 2002). No further information concerning current production is available.

The world's most important producers of small arms: Company and country profiles

The *Small Arms Survey 2001* provided a preliminary ranking of the world's small arms producing countries. It identified at least three countries—China, the Russian Federation, and the US—as major producers, and at least 23 countries, mainly in Europe, as medium producers. The only medium small arms producing countries outside Europe were Brazil and Israel.[15] Despite the fact that more than half the world's countries produce, or have the capacity to produce, small arms, and the emergence of 'new' producers such as India, Pakistan, Singapore, South Africa, and Taiwan, the global small arms market is still dominated by 13 countries: Austria, Belgium, Brazil, China, France, Germany, Israel, Italy, the Russian Federation, Spain, Switzerland, the UK, and the US. Of these, all except Belgium and Switzerland are also the world's most important producers of conventional weapons (Grimmett, 2001).[16] This section provides detailed case studies of the most important small arms companies in the 13 countries that dominate the global small arms market.

MAP 1.2 The world's most important small arms producers

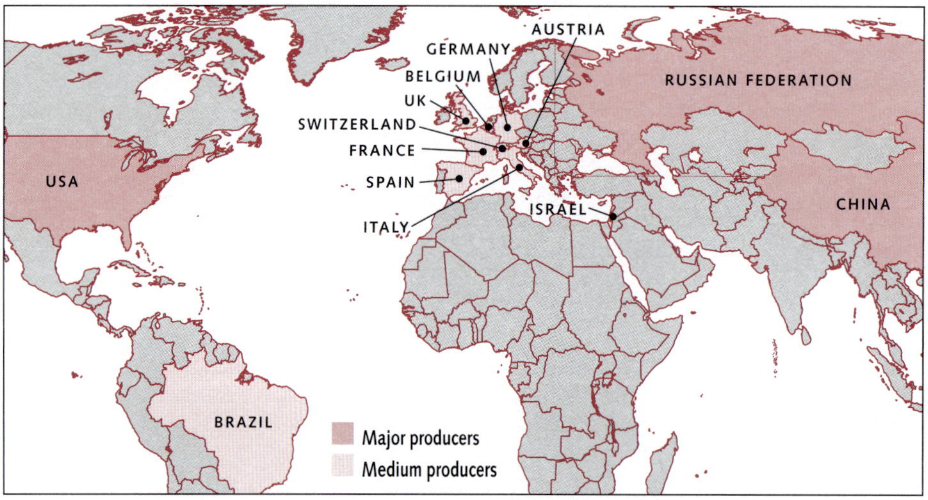

In some countries, such as Austria, Germany, Italy, and the US, the major producers of small arms are private companies, while in others, like Belgium, China, France, Israel, and Switzerland, they are state-owned. In some countries, such as Brazil, there is a mixture of private and state ownership. Among European firms there are some cases of cross-national ownership, such as FN-Herstal (France/Belgium); Heckler & Koch (UK/Germany); SIG Sauer (Germany/Switzerland/USA), and Beretta (France/Italy). In addition, many European companies, like FN Herstal, Beretta, and SIG have well-established US and/or other foreign subsidiaries.

Most of these companies manufacture a range of small arms products, including ammunition, for both military and civilian markets, although some companies are more dominant in one or the other market. Some companies (e.g. Giat Industries) also manufacture other categories of weapons like tanks and military aircraft, and in some cases small arms production makes an insignificant contribution to total sales. FN Herstal and Heckler & Koch are the world's most important licensors of small arms, and their products are manufactured, collectively, under licence in more than 35 countries worldwide. The AK series assault rifle, originally produced by Izhmash in Russia, and subsequently produced in at least 19 other countries, is the world's most widely distributed military-style small arm (Small Arms Survey, 2001, p. 20).

Most of the companies are smaller than companies in other sectors of the defence industry. With the exception of CNGN in China, and Izhmash in the Russian Federation, most companies (e.g. Heckler & Koch, Glock, Beretta) have less than

TABLE 1.6 The world's most important small arms producers, selected financial indicators, 2000

Country	Company	Total sales* (2000)**	Total employment*** (2000)
Austria	Glock	ATS 200 million (USD 14m)	380
	Steyr-Mannlicher	EUR 16 million (USD 15m)	135
	Hirtenberger	ATS 402 million (USD 28m)	260
Belgium	FN Herstal	BEF 3,7 billion (USD 86m)	929
Brazil	Forjas Taurus	USD 47.5 million	1,656
	Amadeo Rossi	USD 5.4 million	1,354
	CBC	USD 46.5 million	900
	Imbel		2,220
China	CNGN (Norinco)	CNY 13,1 billion (USD 1,5b)	456,000
France	Giat Industries	FRF 3,6 billion (USD 517m)	7,500
Germany	Rheinmetall DeTec	EUR 1,6 billion (USD 1.5b)	9,000
	Mauser-Werke	EUR 44 million (USD 42m)	235
	Heckler & Koch (UK)	DEM 159 million (USD 77m)	570
	JP Sauer & Sohn	DEM 50 million (USD 24m)	320
	Umarex	DEM 66 million (USD 32m)	190
Israel	Israel Military Industries	USD 417 million	4,100
Italy	Beretta SpA	EUR 130.3 million (USD 123m)	908
Russian Federation	JSC Izhmash	USD 170.7 million	25,400
	JV Degtyarev Plant	USD 59.7 million	15,368
	Izhevsky	USD 46.6 million	14,954
Spain	Santa Barbara	ESP 14 billion (USD 80m)	2,020
Switzerland	Swiss Ammunition Enterprise (RUAG Suisse)	CHF 270 million (USD 168m)	1,092
UK	Royal Ordnance Plc	GBP 153 million (USD 232m)	1,974
US	Alliant Techsystems	USD 1 billion	6,500
	Primex Technologies	USD 530 million	2,850
	Sturm, Ruger & Co	USD 202 million	1,778
	Smith & Wesson	USD 133 million	670
	OF Mossberg	USD 75 million	585
	Colt's Manufacturing	USD 72 million	600
	Remington Arms		2,200

Notes: * *Includes sales of small arms.*
** *Exchange rates with the USD as of 1 January 2001.*
*** *Includes those employed directly in small arms production.*

Sources: Company reports

a thousand employees, and in many cases only a few hundred.[17] State-owned companies (e.g. IMI in Israel, Santa Barbara in Spain) tend to have larger numbers of employees than the private companies (e.g. Beretta in Italy, Glock in Austria).

Many small arms companies have experienced declining sales and profitability in recent years.

Many companies have experienced declining sales and profitability in recent years, as reflected in dramatic declines in employment. In some countries (e.g. Belgium, France, Israel) this declining performance has prompted direct government financial support. In other countries (e.g. US) government support has involved the awarding of multi-year, or follow-on, contracts. The US market for both military-style and civilian small arms is critical to the survival and profitability of almost all of Europe's small arms producers. Both FN Herstal (Belgium) and Beretta (Italy), through their US subsidiaries, rely heavily on contracts from the US armed forces. Glock (US) is one of the most important suppliers of handguns to law-enforcement agencies in the US. Private companies, particularly those still owned by family interests (e.g. Glock, Beretta, Sturm, Ruger & Co) have tended to perform better than state-owned ones (e.g. FN Herstal, IMI) or those controlled by shareholder interests (e.g. Heckler & Koch).

The financial performance and employment levels of companies has an impact on productivity, and many state-owned companies have exhibited lower levels of labour productivity than their private counterparts.

Table 1.6 provides selected financial indicators for some of these companies. Much of the information on total sales and employment is not comparable. In some companies (e.g. Giat Industries, CNGN, Santa Barbara, IMI, Rheinmetall) sales of small arms account for an insignificant proportion of total sales, while in other companies (e.g. Glock, Sturm, Ruger & Co) they account for a (sometimes 100 per cent) dominant proportion of total sales.

Major producers

This subsection provides detailed case studies of the world's three major small arms producers: China, the Russian Federation, and the US. Each country study provides information on the major small arms producing companies, including selected financial information, where available.

China[18]

China is regarded as one of the world's major small arms producers and exporters, but there is almost a complete lack of official and unofficial information about the size and scope of domestic small arms production (Frankenstein, 1999, p. 191). Recent figures suggest that China was the fifth largest supplier of conventional arms to developing countries during the period 1993–2000, with sales of USD 7.2 billion: an average of USD 970 million a year (Grimmett, 2001). Between 1987 and 1994 China was a major supplier of rifles to the US market, accounting for more than 40 per cent of total US rifle imports, including military-style weapons (Diaz, 1999, p. 73). China is also a major supplier of small arms to Africa, and was the largest supplier, by volume, of civilian firearms to the South African domestic market between 1997 and 1999 (Chetty, 2000).

Defence production, including small arms production, in China is entirely controlled by the state. It occurs either in large state-owned defence-industrial companies or in companies and factories owned and controlled by the People's Liberation Army (PLA).[19] The state-owned defence industry is guided by the civilian State Council and supervised by the Commission of Science, Technology and Industry for National Defence (COSTIND). COSTIND is also the clearing house for arms production and co-ordinates military needs directly with the PLA's General Armament Department. China's defence industry today comprises large state-owned industrial corporations who also produce civilian goods. Most of these corporations used to be secret 'machine-building' industrial ministries headed by military officers. Now they are civilian-run and profit-seeking and have increasingly sought to diversify their activities beyond their core military business (Brömmelhörster & Frankenstein, 1997; Frankenstein, 1999).

The major small arms producing company in China is the state-owned **China North Industries Group Corporation** (also known as Norinco). Norinco was established in 1980 as the official arms trading company under the Fifth Ministry of Machine Building (Gill and Medeiros, 2000). In 1988 the company was formed under the Ministry of Machinery and Electronics, incorporating 160 enterprises and more than 700,000 employees. Between 1980 and 1990 Norinco earned a total of USD 12 billion from arms sales, peaking at USD 2.5 billion a year in 1986 and 1987. In the 1990s its combined sales of military and civilian products averaged about USD 2 billion annually. However, weapons account for only 20–30 per cent of overall production (*Jane's Sentinel Security Assessment*, 20 September 2000).

Under a reorganization of state-owned companies, in July 1999 Norinco was divided into China South Industries Group Corporation (CSG) and China North Industries Group Corporation (CNGN). CSG (with its 260,000 employees and 64 companies) will refocus almost entirely on civil production. CNGN will be the new weapons producer, and will produce most of China's small arms, apart from some small PLA factories.

Despite restructuring, CNGN is still large and difficult to manage. It consists of 131 companies, including 83 industrial enterprises, and 456,000 employees (Brömmelhörster, 2001). It accounts for nearly all state-controlled small arms production (besides ammunition and spare parts). State secrecy makes it impossible to provide exact figures for its small arms production. The bigger factories, however, are in Xian, Beijing, Wuhan, Dalian, Shijiazhuang, and Shenyang.

Small arms production accounts for probably less than five per cent of CNGN's overall turnover, which includes the production of tanks, armoured vehicles, artillery guns, missiles, bombs, and explosives (Reed, 2001). Prior to restructuring in 1999, most of the enterprises within Norinco made losses; although in recent years there have been significant attempts to get rid of unprofitable activities (Frankenstein, 1999).

Russian Federation[20]

The Russian Federation is one of the world's major defence producers, and it is consistently ranked as one of the most important arms exporters in the world. In the period 1993–2000, the country achieved sales of USD 30.5 billion—an average of USD 3.8 billion a year—in conventional arms to developing countries, second only to the US (Grimmett, 2001). According to unofficial figures, total revenues from arms exports amounted to USD 3.5 billion in 2000 (*Moscow Profil,* 7 May 2001). The defence industry of the Russian Federation, including the small arms sector, has been drastically downsized and restructured since the end of the Cold War and the collapse of the former Soviet Union in 1991. However, in recent years the output of the Russian arms industry has been rising (Cooper, 2001, p. 317). In addition, exports of military products now account for 34 per cent of the total output of the Russian arms industry (Cooper, 2001, p. 319).

In Russia more than 85 per cent of all small arms are produced in Izhevsk.

Russia's military industrial complex comprises about 1,700 enterprises: about 30 per cent state enterprises (GUPs), 30 per cent joint-stock companies (JSCs) with state participation, and the rest private (*Moscow Profil,* 7 May, 2001). According to the Centre for Analysis of Strategies and Technologies (CAST, 2001b) a number of trends have become evident in recent years. The main centres for the production of both military-style and civilian small arms in the Russian Federation are Izhevsk (Udmurtia), Tula (central Russia), and Kovrov (Vladimir region). More than 85 per cent of all domestically produced small arms are manufactured in Izhevsk. JSC Izhmash is regarded as the leading domestic producer and exporter of small arms. The production facilities in Tula, apart from the Instrument Design

Bureau, have lagged behind the facilities in Izhevsk, in terms of both design and manufacture for domestic and export markets. As in the past, some firms (e.g. TsKIB SOO) still concentrate on design, while others focus on production (e.g. Izhmash). Often it is unclear whether specific firms are engaged in the design or the production of small arms.

The Russian small arms industry is currently experiencing a process of product diversification. The Kalashnikov era is likely to end in the next few years, leading to a transition from the key types of small arms produced during the Soviet era. Currently, even the latest models of the Kalashnikov assault rifle are merely upgraded versions of the basic AK-47, usually with a different calibre or higher fire rates (e.g. Izhmash Nikonov assault rifle AN-94). However, many companies are developing numerous new weapons, particularly for special and law-enforcement task forces. In addition, the Russian Defence Ministry has invited tenders for new small arms for the Russian armed forces. The final choice will be made in 2002–3 and supplies of new types of small arms to Russia's armed forces will begin simultaneously.

Since the ending of centralized government orders for weapons, plants often develop and produce firearms in small batches to meet orders from security forces and paramilitary divisions of the Interior Ministry. Such a flexible approach gives plants at least some chance to win orders for military-purpose goods, but it leads to de-standardization in small arms. Besides, the commercial viability and usefulness of producing so many varieties of firearms with similar technical specifications is not always evident.

A final trend is the production of hunting and sporting guns as a way of raising total production volumes. Most of these civilian weapons have been derived from military-style small arms.

MAP 1.3 Small arms production centres in the Russian Federation

The main small-arms producers in the Russian Federation are located in Izhevsk, Tula, and Kovrov.

JSC Izhmash (Izhevsk): Izhmash is Russia's biggest manufacturer of small arms. It was founded in 1807 as the Izhevsk Arms Plant, and became the main producer of small arms in the Soviet Union era. In 1948 it launched the production of AK-47 Kalashnikov assault rifles. Today Izhmash is a major engineering facility that, in addition to small arms, makes sporting and hunting guns and various civilian products. Its large and successful civilian division moderates its dependency on arms contracts. It also uses part of the profits from its civilian production to fund its small arms Research and Development (R&D).

At the end of 2000, following an additional share issue and the restructuring of some USD 86 million in debt to the federal government, Udmurtia's stake in Izhmash shrank from 34.35 per cent to 19.8 per cent, while the share of the Ministry for Property Relations grew from 25.5 per cent to 57 per cent, giving the federal government full control over Izhmash operations. With its powerful research and design division, Izhmash currently manufactures the latest models of Russian small arms. Mikhail Kalashnikov, the world-famous designer of the AK series assault rifle, still works at the plant. In 1997 Izhmash was made the federal scientific and production centre for small arms, with Kalashnikov as its head. This status has given the company a virtual monopoly over the design and manufacture of new models of small arms in the Russian Federation.

© Associated Press/Vladimir Vyatkin

Mikhail Kalashnikov with a copy of the world-famous AK-47 rifle.

Even though Izhmash began the production of combat weapons with AK-47 assault rifles, it currently manufactures only the latest models of the Kalashnikov assault rifle family: AK-101–108 and is the Russian Federation's only producer of them. Some models (AK-101 and AK-102) are produced solely for export because they have the standard NATO calibre of 5.56mm. The plant still has the potential to manufacture the AK-74 assault rifle adopted by the Soviet Army in 1974 and its upgraded version AK-74M, though there is no information about current production.

The plant also produces some new-generation small arms such as the Nikonov assault rifle (AN-94), which has been supplied to Russia's airborne troops. Since 1964 Izhmash has been making the Dragunov sniping rifle (SVD) and its more advanced modification, the SVDS. It also manufactures the Bizon-2 sub-machine gun in small batches of several hundred pieces for special-tasks forces and the Interior Ministry. Izhmash is also Russia's biggest producer of hunting guns derived from the Kalashnikov assault rifle (CAST, 2001b).

Izhmash currently employs 25,400 people, up from 22,900 in 1999. In 1999 total sales amounted to USD 90 million; in 2000 they soared to USD 170.7 million.[21] The Izhmash Board of Directors aims to almost double total sales in 2001 to USD 290 million.[22] The company was ranked (by total sales) as the sixth largest defence company in Russia in 2000, up from eighth in 1999 (CAST, 2001a).

Izhevsky Mekhanichesky Zavod (Izhevsk): The plant was founded in 1942 to manufacture small arms for the army. It is still state-owned. Nowadays it is a multi-faceted industrial facility producing various civilian goods in addition to pistols, hunting, sporting, and air guns. Since the Second World War it has been specializing in the production of the Makarov pistol, the main handgun used by the Soviet army. It also makes a special version for law-enforcement purposes. In the past few years the plant has intensified its R&D on new types of combat pistols expected to be adopted by the Russian army, navy, and law-enforcement agencies. So far the new models have been made in test batches of several hundred pieces on order from defence and law-enforcement agencies for troop trials. Designers making the latest models have considered manufacturing them according to Western standards in order to encourage exports. In 2000 the company had total sales of USD 46.6 million (USD 40.9 million in 1999) and employed nearly 15,000 people. Exports, mainly of sporting and hunting weapons, accounted for 35 per cent of total sales (CAST, 2001a).

JSC Tulsky Oruzheiny Zavod (Tula): The plant was founded in 1712 and for a long time was the main producer of small arms in the Russian empire and later in the Soviet Union. In 1961 it launched the production of the Kalashnikov assault rifle. During the Soviet era it produced anti-tank guided missiles and under-barrel grenade launchers. In 1991–92 the share of government orders in overall output slumped from 100 per cent to between seven per cent and ten per cent, and the production of small arms was suspended. In 1993 Tula Arms Plant became a joint stock company and changed its strategy. It expanded the production of hunting and sporting guns while in the defence sector it concentrated on the manufacture of small batches of high-precision pistols and sub-machine guns for Special Forces. In recent years the plant has seen an increase in production and profits, mainly as a result of export orders, which currently account for 82 per cent of its output. The main export articles are hunting and sporting guns that are sold to over 20 countries (CAST, 2001b).

KBP-Instrument Design Bureau (Tula): The KBP-Instrument Design Bureau, which was founded in 1927, manufactures a wide spectrum of armaments, including small arms. It is the main Russian designer and manufacturer of portable anti-tank guided missiles. During the past few years the company has lived mainly on exports. A subsidiary of the KBP design bureau, the Central Research and Design Bureau of Sporting and Hunting Guns (TsKIB SOO), specializes in the design of pistols, revolvers, machine pistols, sniping rifles, and small sub-machine guns. The firearms are designed and produced in small batches on order from the Russian Interior Ministry for special law-enforcement units. The facility also produces hunting and sporting guns (CAST, 2001b).

JSC Kovrov Mechanical Plant (Kovrov): The Kovrov Mechanical Plant (KMP) was founded in 1950 for the production of small arms and light weapons. Since the 1950s it has been making various modifications of the Kalashnikov machine gun. In 1993 the plant was transformed into a joint stock company. In the late 1990s KMP tried to offer a range of new models of small arms on the market. It designed and prepared for serial production the 9A-91 assault rifle for police task forces, the AEK-971 assault rifle, and the AEK-919K Kashtan sub-machine gun. In 2000 its design office produced the first test samples of the AEK-973 assault rifle (CAST, 2001b).

Despite KMP's aggressive marketing policy to develop new models of small arms for both domestic and foreign markets, so far there is no information available about any export contracts or domestic deliveries to the Russian armed forces or domestic law-enforcement agencies. The Kashtan sub-machine gun, however, is a candidate in a tender of the Russian Defence Ministry, law-enforcement and security forces, and the final choice will be made in 2002–3 (CAST, 2001b).

JSC V.A. Degtyarev Plant (Kovrov): The plant was founded in 1917 for the production of machine guns. In the late 1950s it began to produce air defence guns, anti-tank guided missiles, and portable air-defence missile systems. Now it is a joint stock company, controlled by the MDM Group. In recent years the company has designed and produced the Kord heavy machine gun and various types of revolvers. In 2000 the company had total sales of USD 59.7 million (USD 51.9 million in 1999), and employed over 15,000 people (CAST, 2001a).

BOX 1.2 Russian small arms ammunition production

The Russian ammunition industry in general is currently in a critical condition. Annual production volumes have plummeted from five billion units in Soviet times to 50 million today.[23] Government contracts are virtually non-existent because the Russian Defence Ministry, unable to afford new cartridges, is using up its old arsenals. This situation is forcing most companies to switch to the production of cartridges for hunting and sporting guns that have a greater export potential. Current manufacturers of small arms ammunition include: Vympel State Production Association (Khabarovsk territory); Novosibirsk Low Voltage Equipment Plant (Moscow); Ulyanovsk Machinery Plant (Samara region); Pribor Federal Scientific and Production Centre (Moscow); and Bazalt State Research and Production Enterprise (Moscow).

Source: CAST (2001b)

TABLE 1.7 Russia's most important small arms producers, selected financial indicators, 1999–2000

Rank*		Company	Total sales (USD m)		Number of employees		Ownership
1999	2000		1999	2000	1999	2000	
8	6	JSC Izhmash	90	170.7	22,900	25,400	State
13	12	VA Degtyarev Plant	51.9	59.7	n/a	15,368	MDM Group
15	16	Izhevsky Mekhanichesky Zavod	40.9	46.6	n/a	14,954	State

*Note: * Based on ranking (by total sales) of top 20 Russian defence companies*
n/a not available
Source: CAST (2001a)

United States[24]

The US is recognized as one of the world's major small arms producers and exporters. During the period 1993–2000 the US was ranked as the world's major supplier of conventional arms to developing countries with sales of USD 61.5 billion, an average of USD 7.6 billion a year (Grimmett, 2001). More than 300 local companies produce small arms and/or ammunition, and the country is estimated to have one of the world's largest domestic markets for small arms (United States, Census Bureau, 1997a; 1997b). Without exception, there is more official and unofficial information about the production of small arms in the US than for any other country.

© Associated Press/EJ Flynn

Small arms for sale, California.

The small arms industry is not a significant sector of the US economy. Total employment amounted to 16,770 people in 1997, the latest year for which official figures are available. Recent unofficial information suggests that the number of people employed in small arms production has remained constant or declined since 1997. The total value of production (shipments) was USD 2.059 billion (USD 1.2 billion for small arms, and USD 859 million for ammunition) in 1997 the latest year for which official figures are available (United States Census Bureau, 1997a; 1997b). Unlike some of the other major small arms producing countries, the small arms industry in the US is dominated by privately-owned companies, and a handful of companies (including two foreign-owned ones) dominate the domestic market for both military and commercial small arms, including ammunition.

TABLE 1.8 US domestic production of firearms, 1998–1999

Type	Production 1998	Production 1999	% change
Pistols	960,365	995,446	3.7
Revolvers	324,390	335,784	3.5
Rifles	1,345,899	1,569,685	16.6
Shotguns	1,036,520	1,106,995	6.8
Machine guns	32,866	22,490	-31.6
Other/misc.	25,151	55,114	119.1
Total	**3,725,191**	**4,085,514**	**9.7**

Sources: United States, BATF (1998; 1999)

The small arms industry is not a significant sector of the US economy.

Since 1997, there have been small increases in total domestic production of small arms, and just over four million units were produced in 1999, the latest year for which official figures are available (United States, BATF, 1999). The increase between 1998 and 1999 was nearly ten per cent, with the largest increases coming from rifles (17 per cent) and other/miscellaneous firearms (119 per cent). Domestic production in 1999 (4.08 million) was slightly lower than annual average production for the period 1980–99 (4.1 million) (United States, BATF, 1998; 1999).

Despite the small increase in 1999, total domestic production of firearms is still way down from the peak in 1993–94, when total production was over five million units a year. More recent unofficial information suggests that the firearm industry, particularly the handgun business, may be a 'dying industry' (Cameron Hopkins, quoted in *Virginian Pilot*, 14 August 2001) and that production levels are at a 30-year low (*Associated Press*, 16 April 2001). Imports of handguns into the US are also lower than in recent years. The industry's decline is variously attributed to: market saturation, stiffer rules for buyers (the Handgun Violence Prevention Act of 1993), a massive decline in the number of licensed gun dealers (104,000 today, down from 284,000 in 1992), lower levels of crime, and declining public acceptance of owning handguns (*Associated Press*, 16 April 2001). The number of handguns produced for the US military has also declined dramatically in recent years, and purchases between 1993 and 2000 were 80 per cent lower than in the previous eight-year period. The drop in purchases by the US military is attributed to 'smaller armed forces and the greatly diminished threat in the post-Cold War era' (Greg Fetter, quoted in *Associated Press*, 16 April 2001).

TABLE 1.9 Top US small arms companies, selected financial indicators, 1998–2000

Company	Total sales* 1998 (USD m)	Total sales* 2000 (USD m)	Employment** 2000
Sturm, Ruger & Co	211.6	202.7	1,778
Smith & Wesson	140	133	670
OF Mossberg	20	75	585
Colt's Manufacturing	70	72	600
Saco Defense	65	31	240
Marlin Firearms	50	52.4	440
FN Manufacturing	47 (est.)	45.9 (est.)	390
Crossman Corp	45	29	310
Savage Arms	26	25	200
Beretta USA	n/a	30	200
Remington Arms	n/a	n/a	2,200
Total	**674.6**	**696**	**7,613**

Notes: * *Includes sales of small arms and other products.*
** *Includes employment of those directly involved in small arms production*
n/a not available

Sources: Ezell (2001); Gale Research Inc (2001); Reed (2001); Standard & Poor's Corporation (2001).

Against this background of a declining US handgun market, Smith & Wesson, one of the most famous small arms producers for nearly 150 years, was sold for USD 15 million to Saf-T-Hammer, having been bought by British company Tomkins for USD 112 million in 1987 (*Los Angeles Times*, 15 May 2001). Total production at Smith & Wesson in 1999 was 334,491, down from 680,717 in 1995 (*Associated Press*, 16 April 2001). The company, together with other manufacturers, has since 1998 been faced with lawsuits from a range of cities and other government bodies concerning negligence. The lawsuits sought reimbursement from gun makers for the high costs of policing violence and of treating gunshot victims. Rather than fight in court, Smith & Wesson negotiated a government settlement, agreeing to include locks on all its guns and to introduce a range of other safety features and changes in its marketing practices. Other gun makers did not follow Smith & Wesson's lead, and blamed the company for selling out (*Associated Press*, 16 April 2001). As a result of its decision, the company's sales have suffered in recent years because of boycotts from distributors, dealers, and customers (Ezell, 2001).

TABLE 1.10 Top US producers of firearms, ranked* by weapons type

Type	Company	Production 1998	Production 1999	% change	Exports/total production (%) 1999
Pistols	Sturm, Ruger & Co	161,058	213,876	32.8	3,519 (1.6%)
Revolvers	Smith & Wesson	139,583	152,724	9.4	37,857 (24.7%)
Rifles	Sturm, Ruger & Co	332,538	426,226	28.2	453 (0.1%)
Shotguns	Remington Arms Co	336,527	364,354	8.3	11,121 (3%)
Total**		**3,725,191**	**4,085,514**	**9.7**	**245,528 (6%)**

Notes: * *Ranked according to 1999 production.*
** *Total production of all types of firearms, by all US producers.*

Sources: United States, BATF (1998; 1999)

Sales by individual companies have mostly either remained unchanged or declined over the last few years. Production and total sales by the top companies declined between 1997 and 2000 (Ezell, 2001). Only one major American company—O.S. Mossberg, which manufactures mainly shotguns and rifles for both civilian and law-enforcement markets—has significantly increased its sales, from USD 20 million in 1998 to USD 75 million in 2000, largely as a result of good business planning and increasing demand for long guns, particularly shotguns (Ezell, 2001).

Sturm, Ruger & Co is the most important firearms producer in America, in terms of the value of sales, employment, and total production. It dominates domestic production of pistols and rifles for the commercial market, and is the only US-based producer that offers products in all four categories: pistols, revolvers, rifles, and shotguns. Rifles account for nearly 40 per cent of the company's total sales. It is also an important supplier of military-style small arms.

The most important US contractors (producers) of military-style small arms are **Colt's Manufacturing** (M4 carbine, M16 rifle), **FN Manufacturing** (a subsidiary of FN Herstal of Belgium: M16 rifle, M249 squad automatic weapons and M240B medium machine gun), **Saco Defense** (a subsidiary of General Dynamics: M2HB machine gun and MK19-3 grenade machine gun), and **Beretta USA**, a subsidiary of Beretta of Italy. Both FN Manufacturing and Saco Defense are completely dependent on government sales, while Colt's Manufacturing is less so because of its civilian business (Ezell, 2001).

Two companies dominate the production of military calibre ammunition for the US military: Alliant Techsystems and Primex Technologies. Alliant Techsystems had total sales of over USD 1 billion in 2000, and employed 6,500 people. Its conventional munitions group, which manufactures ammunition from small arms to tank ammunition, had sales of USD 421 million in 2000, accounting for nearly 40 per cent of the company's total sales. All the company's small arms ammunition is manufactured at the Alliant Lake City Small Calibre Ammunition Company (Hoover's Inc., 2001: 2 August).[25] In 2000 Alliant was awarded a ten-year contract, worth USD 100 million annually, to supply small arms ammunition for the US Army (Forecast International, 2002). Primex Technologies, which was acquired by General Dynamics in early 2001, had total sales of USD 530 million in 2000, and employed 2,850 people (Extel Financial Limited, 2001: 1 February).[26] Small arms account for less than ten per cent of the total sales of Primex.

Medium producers

This subsection provides detailed case studies of the world's most important medium-sized small arms producers: Austria, Belgium, Brazil, France, Germany, Israel, Italy, Spain, Switzerland, and the United Kingdom. Each country study provides information on its most important small arms producing companies, including selected financial information. Together with the world's three major producers—China, the Russian Federation, and the US—these countries dominate the global small arms market.

Austria[27]

Austria has a relatively small domestic defence industry, but the country is a major producer of small arms.

Austria has a relatively small domestic defence industry, and local defence companies, unlike in neighbouring Switzerland, have been largely excluded from the European-wide defence industry consolidation of recent years. The total turnover of the Austrian defence industry in 2000 was EUR 174 million (USD 164 million) (Austrian Defence Industry Association, 2001). However, the country is a major producer of small arms for both military and civilian markets, and a number of local companies are involved in some aspect of small arms production (Omega Foundation, 2001b). Two companies—Glock and Steyr-Mannlicher—are the most important domestic producers, for both military and civilian markets, and are also major exporters. Other local companies include Hirtenberger (ammunition), Dynamit Nobel Graz (ammunition), Andres & Dwonsky Karlstein (sporting pistols), and Voere Kufsteiner Gerätebau Handelsges (hunting and sporting weapons) (Lock, 2001).

Glock: A metal producer founded in 1963, Glock entered the small arms market only in 1981. Since 1981 the company has supplied more than 2.5 million pistols to armies and law enforcement agencies in more than 50 countries, as well as to civilians in more than 100 countries. The company produces a range of nearly 40 pistol models in different sizes and calibres for the military, law-enforcement, and civilian markets. It is estimated that more than 90 per cent of the company's production is exported. At least 5,000 police departments and agencies in the US at State, federal, and local levels are equipped with Glock pistols, and an estimated 500,000 of the company's pistols are carried daily in the US as service handguns (Gander and Cutshaw, 2001b).

In addition to its production facilities in Deutsch-Wagram (Austria), the company has production facilities in the US, France, and subsidiaries (or trading companies) in Hong Kong, Uruguay, and Madeira (Hoover's Company Profile Database, 2001). Financial details about the company, which is still privately owned by the Glock family, are difficult to obtain. However, recent reports suggest that total sales for the company amounted to ATS 220 million (USD 14 million) in 2000, and that total employment was 380 people (Lock, 2001).[28]

Steyr-Mannlicher AG: Established in 1864, Steyr-Mannlicher AG operates as an independent subsidiary of Steyr-Daimler-Puch AG and produces a wide range of products for military, law-enforcement, and civilian markets. The company's products for the military market include pistols, rifles, sub-machine guns, and a light support weapon (Gander and Cutshaw, 2001b). The company produced the 7.62mm FN FAL rifle under licence (renamed the StG-58) during 1958–70, and approximately 130,000 units were produced.[29] The Steyr 5.56mm AUG rifle, which is in service in at least 24 countries, is produced under licence in Australia and Malaysia; the Steyr 7.62mm SSG 69 sniper rifle is produced under licence in Greece. The company is also a major producer of various types of small arms ammunition.

The company owns 51 per cent of the shares in a German company, Suhler Jagd-und Sportwaffen GmbH, which produces various sporting and hunting rifles, thus complementing its existing product range of commercial firearms. In June 2001, an Austrian company, Cura Investholding GmbH, acquired 100 per cent of the shares of Steyr-Mannlicher, including its shares in Suhler Jagd-und Sportwaffen GmbH. According to company reports, Steyr-Mannlicher had total sales of EUR 16 million (USD 15 million) in 2000, and employed 135 persons.[30]

Belgium[31]

While Belgium is not a significant producer of major conventional arms, it is still one of the world's most important small arms producers. At least 15 companies in Belgium are involved in some aspect of small arms production (Omega Foundation, 2001b). However, domestic production of small arms and ammunition is dominated by one company, FN Herstal, which is part of the Groupe Herstal (Mampaey, 1998; 2000).

FN Herstal was established in 1889 (named Fabrique Nationale d'Armes de Guerre) when several small arms manufacturers from Liège joined forces to fulfil a government order of 150,000 Mauser rifles. In later years the company ventured into the civilian market with hunting rifles and bicycles (much later also trucks and cars). FN Herstal had hoped that diversification would act as a safety-net for the company when military orders were low, but it mismanaged its diversification policy (Mampaey, 2000, p. 21).

In the late 1970s FN was drastically reorganized into four divisions (Mampaey, 1997). At the same time it began to expand worldwide with the acquisition of Browning (1977) and the creation of subsidiaries in Portugal (FN Viana, 1973), Brazil (FN do Brasil, 1977), and the US (FN Manufacturing, 1981). In 1980 it had a major success when NATO adopted the FN SS-109 ammunition (5.56mm) as its standard. But not all was well with the company. In the early 1980s the demand for small arms from traditional clients fell, and the workforce in the arms production division was regularly underemployed (Mampaey, 1997, p. 5). FN was restructured again during the mid-1980s in the hope that decentralizing the industrial (i.e. non-military) activities would lower costs. In 1987 the Browning division was turned into a subsidiary of FN Herstal, and was quickly followed by FN Moteurs and FN Industry. FN Moteurs was finally sold in 1989 to the French company SNECMA (*Jane's Defence Weekly*, 24 June 1989).

At the end of 1990 FN's major shareholder, the *Société Générale de Belgique*, sold 92 per cent of its equity to the French state-owned defence group Giat Industries. The remaining eight per cent was bought by the Région Wallonne, the Walloon regional government. The FN Group was renamed Groupe Herstal, while FN Herstal became known as *Fabrique Nationale Nouvelle Herstal* (FNNH).[32] By this time employment was down to 1,200 from over 10,000 in the early 1980s (Mampaey, 2000, p. 26). But all was not well either with Giat Industries (see below). As a result of mounting losses, Giat Industries sold FN do Brasil (Brazil) and its shares in Beretta SpA (Italy) and began to search for a buyer for Groupe Herstal. One interested company was Colt's Manufacturing, FN's American rival. The sale of Groupe Herstal to Colt was vetoed by the Région Wallonne, and at the end of 1997 Giat Industries sold all its remaining equity (92 per cent) in Groupe Herstal to the Région Wallonne. Currently, the small arms activities of the Groupe Herstal are structured into two groups: (1) military products, which includes two subsidiaries: FN Herstal (Belgium) and FN Manufacturing (US); and (2) commercial products, which includes five subsidiaries: Browning International (Belgium), Browning S.A. (Belgium), Browning Inc (US), Browning Viana (Portugal), and US Repeating Arms Co Inc (US).

In recent years the Région Wallonne has been providing Groupe Herstal with substantial sums of money to keep the company afloat (Danssaert, 2001). In 2000, FN Herstal's turnover was BEF 3.7 billion (USD 86 million), with exports amounting to 98 per cent of total sales. Weapons sales accounted for 73 per cent of total sales, while ammunition accounted for 15 per cent (Danssaert, 2001). In the same year total employment was 929, down from 1,206 in 1997, when the company was sold by Giat Industries (Mampaey, 2000).

Brazil[33]

Brazil was the largest supplier of handguns to the US during 1991–95.

Brazil is one of the developing world's major defence producers and a major producer of small arms, particularly for the US market (*Jane's Defence Weekly*, 21 June 2000). At least 18 companies are involved in some aspect of small arms production (Omega Foundation, 2001b). The major domestic producers of small arms include the state-owned Industria de Material Belico do Brasil (Imbel) and a number of private companies. Imbel, which was established in 1974, is a non-quoted public company, and in 2000 employed 2,220 people (Graham & Whiteside Ltd, 2001). The company produces a wide range of small arms and ammunition, mainly for the military market. Its products include pistols, rifles, and sub-machine guns (Gander and Cutshaw, 2001b). The company started producing the 7.62 FN FAL rifle under licence in 1964 (250,000 have been produced to date), and currently produces the 5.56mm version (Danssaert, 2001). More than 90 per cent of the 2,000 .45 calibre pistols that the company exports each month are bound for the US market. Springfield Armory is the distributor of Imbel products in the US. Currently, between 40 per cent and 50 per cent of the company's total production is exported (Gazeta Mercantil, 26 October 2001).

The most important private companies in Brazil include Forjas Taurus (handguns and sub-machine guns), Amadeo Rossi (handguns), and Industria Nacional de Armas (INA) (sub-machine guns). Both Forjas Taurus and Amadeo Rossi are major suppliers of handguns to the US civilian market, and in 1996 Brazil was the second largest supplier of handguns to the US, having been the largest supplier throughout the period 1991–95 (Diaz, 1999).

Forjas Tauras is Brazil's leading producer and exporter of handguns. In 2000 it had total sales of USD 47.5 million and employed 1,656 people. Over 60 per cent of the company's sales in 2000 consisted of exports to nearly 60 countries. **Amadeo Rossi** had total sales of USD 5.4 million in 2000, and employed 1,354 people.

Brazil is also a major producer of small arms ammunition. The most important producers include Fabrica National de Cartuchos e Municoes in Sao Paulo, **Companhia Brasiliera de Cartuchos** (CBC) in Sao Paulo, and Fabrica Realengo in Rio de Janeiro. CBC, which is owned by Imbel, is the largest and oldest ammunition producer in Brazil and one of the largest ammunition producers in the developing world. In 2000 the company had total sales of USD 46.5 million and employed 900 people.

The financial performance of CBC, Taurus, and Rossi has declined in real terms in recent years, and this is reflected in the trends in Brazil's small arms exports (TRANSFERS). Taken together, the three companies had total sales of almost USD 100 million in 2000, and collectively employed nearly 4,000 people. In July 2000, the Brazilian Ministry of Defence effectively 'closed' the local small arms market to foreign competitors in an attempt to protect the local small arms industry (*São Paulo Valor*, 12 July 2000).

France[34]

While France is a significant producer and exporter of major conventional armaments, it pales by comparison with other European countries such as Austria, Belgium, Germany, and Italy in terms of small arms production. The country was ranked as the third largest supplier of conventional arms to developing countries in the period 1993–2000 with sales of USD 24.7 billion, or USD 3 billion a year (Grimmett, 2001). More than 30 companies in France are involved in some aspect of small arms production, although most production is for domestic consumption (Omega Foundation, 2001b). The state-owned company, Giat Industries, is the most important domestic producer of small arms and ammunition. The other major domestic producer is a private company PGM Precision (sniper rifles).

Giat Industries has a history dating back to the eighteenth century. The French small arms industry was established with the creation of Manufacture d'Armes de St Etienne in 1764. A number of armaments factories were subsequently established at Tulle (1777), Toulouse (1792), Rennes (1793), Roanne (1918), Le Mans (1927), Sal Bris (1933), and Issy-les-Moulineaux (1936). In 1945, all of these factories were grouped together under Direction des Études et Fabrications d'Armements (DEFA). In 1965, DEFA became Direction Technique des Armements Terrestres (DTAT) and was included in the Direction Ministérielle des Armements (DMA). These arsenals were regrouped in 1971 with the creation of Groupement Industrial des Armements Terrestres (GIAT). On 1 July 1990, GIAT became Giat Industries, SA. In the same year Giat Industries acquired Matra Manhurin, a well-known producer of small arms and light weapons.

Giat Industries is one of the world's leading producers of land systems, although its products are also sold to air forces, navies, gendarmeries, and police forces. The company is in the process of diversifying its activities into other sectors in co-operation with other industries. Luchaire, which produces ammunition and other light weapons, and Manurhin, which produces ammunition, are both wholly-owned subsidiaries of Giat Industries.

Giat Industries manufactures a wide range of small arms for the military market, including the FAMAS 5.56mm assault rifle, Apilas anti-tank weapon, FR-F2 sniper rifles, and a general-purpose machine gun. The company is also one of Europe's most important ammunition producers, manufacturing a range of small arms ammunition and grenades. Between 400,000 and 500,000 of the FAMAS F1 rifle have been produced, but it has not sold well in comparison with other assault rifles such as the AK series, G3, or FN-FAL. It is currently in service in six countries, including France (Small Arms Survey, 2001, p. 20).

Despite improving financial figures, the company is still losing money and relies heavily on government bail-outs to keep production lines moving. The company is still recovering from huge financial losses incurred in the mid-1990s as a result of currency speculations. Since 1991, the French government has poured an estimated FRF 22 billion (USD 3.1 billion) into Giat Industries (*Jane's Defence Weekly*, 9 May 2001).

Since 1991 the French government has poured an estimated USD 3.1 billion into Giat Industries.

In 2000, the company had a turnover of FRF 3.6 billion (USD 517 million), down from FRF 5.7 billion (USD 819 million) in 1999 (*Jane's Defence Weekly*, 9 May 2001), although its order book in 2000 was worth FRF 18.2 billion (USD 2.6 billion). At least 70 per cent of total sales were for the domestic market, with 30 per cent for exports. Only 20 per cent of total sales were accounted for by sales of weapons and ammunition systems, including small arms and light weapons. It is likely that sales of small arms account for less than five per cent (roughly USD 25 million) of the company's total sales. In 2000 Giat employed 7,500 people down from over 10,000 in 1998 (*Jane's Defence Weekly*, 10 May 2000). Giat Industries was ranked as the 36th largest arms producing company in the OECD and developing countries by SIPRI in 1999 (SIPRI, 2001, p. 308).

Despite its financial difficulties, the French government refuses to let Giat Industries fail, and it will therefore continue to have a strong presence in the global defence market. The downsizing of the French Army is reducing domestic requirements, forcing the company to continue to search for export markets (Reed, 2001, p. 306). There is also speculation that Giat Industries may be partly or wholly privatized in the near future.

Germany[35]

Germany is one of the world's major producers of conventional arms, ranking as the sixth largest supplier of major conventional arms to developing countries in the period 1993–2000 with sales of USD 5.8 billion, an annual average of USD 725 million (Grimmett, 2001). The country has a long history of small arms manufacturing, and is one

Germany is one of Europe's major producers of small arms for both the military and the civilian markets.

of Europe's major producers of small arms for both the military and the civilian markets. Currently more than 30 companies in Germany are involved in some aspect of small arms production (Lock, 2001; Omega Foundation, 2001b; 2001c). The most important domestic manufacturers of small arms and ammunition for military and police markets include Heckler & Koch (owned by BAE Systems of the UK), JP Sauer & Sohn, Carl Walther (owned by Umarex Sportwaffen), Rheinmetall DeTec, Diehl, and Dynamit Nobel.

Rheinmetall DeTec is the largest arms producer in Germany and was ranked by SIPRI as the 22nd largest arms producing company in the OECD and developing countries in 1999 (SIPRI, 2001, p. 308). In recent years the company has acquired a number of other small arms manufacturers including Mauser Werke, Buck Werke, Eurometall (the Netherlands), and Oerlikon-Contraves (Switzerland). In 2000 the company had total sales of EUR 1.6 billion (USD 1.5 billion) and employed 9,000 people. In the same year Mauser-Werke, which produces a range of sniper rifles, had sales of EUR 44 million (USD 42 million) and employed 235 people (Graham & Whiteside Ltd, 2001). Rheinmetall DeTec has very little interest in the small arms sector, and recently sold the trademark of Mauser pistols to JP Sauer & Sohn (Lock, 2001).

Germany's most important domestic manufacturers of pistols for the military and police markets are **JP Sauer & Sohn**, Carl Walther, and Heckler & Koch. JP Sauer was founded in the eighteenth century and started production of arms in 1871. In late 2000, the company acquired the small arms business of SIG Arms of Switzerland (see below). In 2000 the company had total sales of DEM 50 million (USD 24 million), and employed 320 people. The company's US subsidiary SIG Arms Inc had turnover of DEM 45 million (USD 22 million) in the same year, and employed 70 people. The company produces about 80,000 pistols annually, almost half for export to the US market.

Carl Walther was purchased in 1993 by two non-defence companies, and has since been streamlined and merged into Umarex Sportwaffen. The present management of Umarex has announced its intention to exit government (i.e. military) markets and concentrate on civilian markets (Lock, 2001). In 2000 Umarex had total sales of DEM 66 million (USD 32 million) and employed 190 people. Domestic production of rifles, sub-machine guns, and other crew-served small arms and light weapons (e.g. heavy machine guns) is dominated by Heckler & Koch. Mauser Werke, which is owned by Rheinmetall DeTec, produces a 7.62mm sniper rifle. Dynamit Nobel is the most important local producer of ammunition for the military market.

Heckler & Koch, which was established in 1948 and acquired by BAE Systems (UK) in 1991, dominates the production of small arms in Germany. The company produces a wide range of products for the military, police, hunting and sporting markets. In addition to production facilities in Germany, the UK, and the US, the company's products have been produced under licence in nearly 20 countries. In 2000 Heckler & Koch had total sales of DEM 159 million (USD 77 million), and employed more than 500 people. The company, together with Alliant TechSystems and Brashear LP, is involved in the development of the Objective Individual Combat Weapon (OICW) for the US armed forces.

The company **Fritz Werner**, which until 1990 was state-owned and is now a subsidiary of Ferrostaal AG (part of the MAN group), has played a key role in establishing domestic production facilities for small arms and ammunition in a wide range of countries, usually in co-operation with German small arms producers including Heckler & Koch, Dynamit Nobel, and Rheinmetall.

A large number of German companies are involved in the manufacture of high precision, and expensive, weapons for the hunting and sporting market, as well as personal weapons, mainly for the US market. Some companies also specialize in producing high-quality barrels for top of the range customized sporting and hunting weapons. In 1999 these firms supplying the hunting and sporting market had a combined turnover of DEM 325 million (USD 157 million) (Lock, 2001). The main companies supplying the civilian market include JP Sauer & Sohn, JG Anschütz, Blaser Jagdwaffen, SK Jagd-und Sportmunitions (owned by Nammo Lapua Oy of Finland), Intertex, Suhler Jagd-und Sportwaffen (51 per cent owned by Steyr-Mannlicher of Austria), and Umarex Sportwaffen (including Carl Walther). Intertex produces a range of pistols for the high end of the civilian market.

Israel[37]

Israel is one of the world's most important arms producers and exporters. The country was ranked as the seventh largest supplier of conventional weapons to developing countries in the period 1993–2000 with sales of USD 2.6 billion, an average of USD 325 million a year (Grimmett, 2001). At least ten companies are involved in some aspect of small arms production (Omega Foundation, 2001b). The state-owned company Israel Military Industries (IMI) is the most important domestic producer of small arms and ammunition. In addition to IMI, smaller private-sector firms, such as KSN Industries (pistols), Soltam (mortars), and the Kalia Israel Cartridge Company (ammunition) also manufacture small arms for domestic use and export (Gander and Cutshaw, 2001b).

© Associated Press/Nati Harnik

Israeli children with an Uzi sub-machine gun.

In the past decade the Israeli government has injected USD 1.23 billion into IMI.

Israel Military Industries (IMI), known as Taas in Hebrew, was created during the 1920s, when members of the Jewish self-defence forces built a number of factories for manufacturing explosives, small arms, and ammunition. On 15 May 1948, when Israel declared its independence, IMI was officially founded under the auspices and direction of the Ministry of Defence.

IMI produces a range of small arms and light weapons including the Uzi sub-machine gun, the Galil rifle, the Negev-Light Machine Gun (LMG), the Jericho Semi-Automatic Pistol, the Tavor Assault Rifle, the SR-99 Semi Automatic Sniper Rifle, as well as ammunition for these weapons. In 1998, weapons sales of the Negev machine gun, Jericho pistol, Desert Eagle pistol, Uzi SMG, and the Galil rifle accounted for 88 per cent of IMI's total sales of small arms, while the provision of spare parts accounted for 12 per cent of total sales.[38]

Revenues from sales of the Uzi, both to the Israeli government and abroad, have played an important role in the growth of IMI. Over the years, the profits from these sales have provided the basis for the expansion of IMI. As IMI and other defence companies expanded, they also became important factors in national industrial and economic development (Steinberg, 1983). Over 60 per cent of IMI products are exported to foreign armed forces, defence manufacturers, and law-enforcement and security agencies.[39]

In 1990, IMI employed 11,000 workers and was one of Israel's largest firms (*Jane's Defence Weekly*, 25 October 2000). In 1990, IMI became a government-owned corporation and now consists of nine individual profit centres and 12 plants. However, following years of consecutive growth, a severe downturn in sales took place during the 1990s.

The loss of sales was caused by a drop in foreign orders following the end of the Cold War and the expiry of contracts with foreign customers. In addition, the growing shift of the IDF towards American suppliers (linked to the terms of US military aid) reduced IMI's sales to the IDF (Sadeh, 2001).

Currently, IMI employs approximately 4,100 workers (*Jane's Defence Weekly*, 25 October 2000). After significant losses, the firm made profits in the late 1990s, but in 1999 total sales were USD 485 million, down from over USD 500 million the previous year. In 2000, IMI's total sales were USD 417 million. However, the company posted a USD 49 million loss, bringing its losses in the last two years to approximately USD 107 million (Steinberg *et al.*, 2001). The company stated that 2000 was a difficult year due to problems in its small arms factories. However, as part of IMI's reorganization plan, the manufacture of small arms was moved from IMI's Ramat Hasharon facility to the plant in Kiryat Shmona (Steinberg *et al.*, 2001).

As IMI expanded, small arms production played a relatively smaller role and was shifted to the periphery as part of a broad government policy to promote population movement away from the concentration in Israel's coastal plain near Tel Aviv (Steinberg, 1983, p. 292). The Small Arms Factory group (SAF) includes the Keshet factory in Kiryat Shmona. Sales from the SAF in 1999 amounted to about six per cent of IMI's total sales. The SAF currently employs 70 people, down from over 300 in 1999. The division caused IMI a USD 28 million loss in 2000, but was expected to almost break even in 2001 (*Business Arena*, 13 November 2001). About 80 per cent of the SAF's products are exported, while the remaining 20 per cent are slated for domestic use.[40] SAF's profits are primarily generated by sales of pistols, sub-machine guns, machine guns, rifles, and spare parts.

IMI was ranked as the 59th largest arms producing company in the OECD and developing countries by SIPRI in 1999 (SIPRI, 2001, p. 309). In the past decade the Israeli government has injected USD 1.23 billion into IMI (*Business Arena*, 24 October 2001). In response to the crisis and the increase in government subsidies, successive governments sought to restructure the defence industrial sector, including IMI, but this has been resisted by the trade unions (Sadeh, 2001). Recently, there has also been speculation about merging IMI with the Israel Armament Development Authority (Rafael). As a result, the future of IMI remains in doubt.

Italy

Italy was ranked as the eighth largest supplier of conventional arms to developing countries in the period 1993–2000, with sales of USD 2.5 billion, an average of USD 312 million a year (Grimmett, 2001). Nearly 40 companies in Italy are involved in some aspect of small arms production (Omega Foundation, 2001b). Beretta SpA is the country's most important domestic small arms producer, and produces a wide range of products for both military and civilian markets. Other important domestic producers of small arms for the military market include Bernardelli (rifles), Franchi (rifles), Tanfoglio (pistols) and Benelli (pistols). Both Franchi and Benelli are owned by Beretta (Cattaneo, 2001).

Beretta SpA has been selling small arms since 1526, and by the 1950s it was regarded as one of Italy's most important weapons factories.[36] The company's main factory in Brescia produces around 1,500 firearms a day, and an estimated 75 per cent of the company's total production is for the civilian market (hunting and sporting). Beretta also provides firearms to the armed forces of several countries. The company's products have been produced under licence in Indonesia (sub-machine guns), Brazil (sub-machine guns), Morocco (rifles), Egypt (pistols), US (pistols), and Austria (machine guns) (Gander and Cutshaw, 2001b).

The largest foreign market for the company's products is the US, where the company has a subsidiary, Beretta USA Corporation. In 1985 Beretta USA was awarded its first contract to supply 92F pistols (renamed M9) to the US Army, Air Force, Navy, Marines Corps, and Coast Guard. Some law-enforcement agencies have adopted the 92F as well, particularly the US Border Patrol and Immigration and Naturalization Service agents. Throughout the 1990s Beretta was awarded various contracts to supply pistols to the US armed forces. In late 2001, following the terrorist attacks on the US, the Pentagon asked Beretta to speed up delivery of its 20,000 Benelli M4 shotguns, for use by the Marines and Special Forces (*Corriere della Sera*, 4 November 2001).

Beretta owns or controls a number of other small arms companies in Italy and other countries.

Today the company is wholly owned by the Beretta family. In 1997 the family re-acquired the 36 per cent equity that had been owned by the French company, Giat Industries, for the previous 20 years. During financial year 2000, Beretta had total sales of EUR 130.3 million (USD 123 million), a 21 per cent increase over 1999. Of total sales, domestic sales accounted for 38 per cent, and exports 62 per cent. All aspects of the company's business have witnessed increases in recent years. The increase in exports in the last few years has been helped by a more favourable lira-USD exchange rate. This has lead to increased deliveries to Beretta USA and to the opening of a new distribution network through Beretta Australia. In 2000 the company sold 200,000 weapons, up from about 170,000 in 1999; of these, more than 50 per cent were sport shooting firearms, about 18,000 were AR70/90 assault rifles for the Italian Army, and the remaining items were pistols for police and civilian customers, whose sales increased by 13 per cent. The company employed 908 people in 2000, slightly down from 926 in 1999.

The Beretta Group, which includes Beretta SpA, owns or controls a number of other small arms companies in Italy and other countries. The acquisition of Sako Ltd (Finland) in 1999 allowed the group access to the Northern European/Scandinavian market. Also, the share of Sako exports to the US (about 50 per cent of its production) allowed Beretta to further consolidate its presence in the North American market. Beretta has a worldwide network of distributors, including 21 in Italy and 71 in more than 60 other countries (Cattaneo, 2001).

During fiscal year 2000 the Beretta Group's total sales were USD 291.7 million, up from USD 233 million in 1999. Since 1995, with the exception of 1996, both sales and net profit of the company have increased. Exports have increased as a share of total sales, from 64 per cent in 1995 to 80 per cent in 1999. Foreign sales currently account for about 80 per cent of total sales, and of these more than half go to the North American market.

Spain[41]

During 1996–2000 Spain was the 11th biggest exporter of arms in the world and the sixth biggest exporter in Europe, according to SIPRI (2001, p. 357). More than 20 local companies are involved in some aspect of small arms production (Omega Foundation, 2001b). During the 1990s its accumulated debts pushed the small arms industry in Spain into a serious crisis. The country's major small arms producer, the state-owned company Santa Barbara, was finally bought by the US-based company General Dynamics in March 2001, after a protracted takeover battle with Germany's Rheinmetall and Krauss-Maffei. Of the three largest private companies (Astra, Llama Gabilondo, and Star) only Llama Gabilondo has survived and is currently being transformed to a shareholder company, Fabrinor. The company had total sales of ESP 750 million (USD 4.2 million) in 2000 and manufactures about 20,000 small arms a year, mainly pistols. More than 80 per cent of the company's production is destined for the US market; the remaining 15 per cent is distributed to various countries, including a small portion of sales to the Spanish army (*El Pais,* 25 February 2001). Both Astra and Star were previously the major domestic producers of pistols for the military market.

Santa Barbara was established in 1960 when factories of the Spanish Army grouped together to develop, manufacture, and market defence material. Formally known as Empresa Nacional Santa Barbara de Industrias Militares, the state-owned company was bought by General Dynamics on 30 March 2001 for USD 4.4 million, after the company suffered huge financial losses throughout the last decade. The company is currently being privatized as a result of its acquisition by General Dynamics.

In 2000 the Spanish government admitted that Santa Barbara was 'technically bankrupt', with accumulated losses of more than USD 800 million.

Santa Barbara is the main supplier of small arms and ammunition for the Spanish armed forces. Its small arms products include the Cetme assault rifle, the C-75 Special Forces rifle, and the Ameli 5.56mm light machine gun (Gander and Cutshaw, 2001b, p. 855). In 1998 Heckler & Koch reached agreement with the Spanish armed forces to supply 115,000 new G-36 assault rifles. The initial 15,000 rifles were supplied from the HK production facilities in Germany, but the remaining 100,000 were manufactured in Spain by Santa Barbara under a licensed production agreement. Santa Barbara built a new production facility, known as FACOR, at La Coruna in north-west Spain, to produce the rifles (*Jane's Defence Weekly*, 22 July 1998).

Santa Barbara has been operating at a loss for several years. In 2000 the Spanish government acknowledged that the firm was 'technically bankrupt' and it is estimated that in the last decade the company accumulated losses of more than USD 800 million. In 2000, Santa Barbara's losses were reported at USD 22 million (*Jane's Defence Weekly*, 18 October 2000). In the same year the company's defence turnover was ESP 14 billion (USD 80 million), and the company employed 2,020 people (1,731 in 1999). Santa Barbara exports 7.1 per cent of its total production; the rest of its production is for domestic consumption. The company's products are in use in more than 47 countries worldwide.[42]

While Santa Barbara has suffered from considerable financial difficulty throughout the decade, analysts have said that the company will benefit from its recent acquisition by General Dynamics.[43] However, there have been concerns within the European Union about the takeover because it gives the US a major foothold in the European defence market. General Dynamics is planning to invest approximately EUR 60 million (USD 57 million) in capital improvements over the next five years to make Santa Barbara competitive in the world market and maintain a strong military and industrial capability for Spain (*Defence Systems Daily*, 26 July 2001). With the privatization of Santa Barbara, the company can be expected to expand its global market as well as remain the principal supplier for the Spanish armed forces.

Switzerland[44]

Until recently, Switzerland had a well-developed domestic arms industry, including a number of companies producing small arms. However, two of its most important producers of small arms, SIGARMS and Oerlikon-Contraves, are now foreign-owned (by German companies), reflecting the absence of a national industrial policy to maintain domestic defence capacities under all circumstances. But the companies are continuing the tradition of German-Swiss co-operation in small arms production, while at the same time reflecting the European-wide consolidation of defence production (Lock, 2001).

The major domestic producer of small arms for the military market is the **SIGARMS Group** (formerly SIG Arms), which was purchased from the Schweizerische Industrie Gesellschaft (Swiss Industrial Company) by German investors in late 2000 (*Agence France Press*, 4 October 2000). The 'new' SIGARMS Group, which is currently in a transition process, has subsidiary companies and production facilities in Switzerland (SAN Swiss Arms AG, Hämmerli AG), Germany (JP Sauer & Sohn GmbH and Blaser Jagdwaffen GmbH), and the US (SIG Arms Inc). The company's production facilities in Switzerland will probably be closed down and production activities rationalized in Germany (at JP Sauer & Sohn in Eckernförde) and the US (at SIG Arms Inc in Exeter, New Hampshire). The new company has about 400 employees worldwide, and produces a wide range of small arms for the military market, including pistols and rifles. It owns the trademarks for Mauser, Blaser, Hämmerli, and JP Sauer & Sohn, and will continue to use the SIG name on the P210 pistol. The SIG name will also be used in conjunction with Sauer on all SIG-Sauer handguns (e.g. P220 pistol).[45]

The SIG-Sauer 9mm P226 Pistol, of which more than 400,000 have been produced, is used by many armed forces, including the US Coast Guard, UK armed forces, and the US Federal Bureau of Investigation. The SIG 540 series rifle, which is in service in more than 20 countries (including ten in Africa), has been produced under licence in Chile, France, and Portugal (Gander and Cutshaw, 2001b). The ammunition sections of SIG Arms are now controlled by the German company Rheinmetall DeTec, reflecting the European-wide consolidation of small arms ammunition production (Lock, 2001). SIGARMS also produces a wide range of weapons for the civilian market, mainly through its subsidiaries Hämmerli AG and Blaser Jagdwaffen. The company Brügger and Thomet is mainly involved in the production of accessories (e.g. sights) for small arms for the military and police markets. It also acts as an agent for a number of foreign producers (e.g. Heckler & Koch). In 2000 it had total sales of CHF 15 million (USD 9.3 million). Other Swiss producers of small arms, mainly for the hunting and sporting market, include Grünig & Elmiger and Martin Tuma Engineering (Lock, 2001).

The state-owned **RUAG Suisse Group**, which was founded in 1863, includes both the Swiss Ammunition Enterprise (SM) and the Swiss Ordnance Enterprise (SW). It was ranked as the 50th largest arms producing company in the OECD and developing countries by SIPRI in 1999 (SIPRI, 2001, p. 309). According to company information, in 2000 RUAG had total sales of CHF 941 million (USD 584 million) and 3,843 employees. SM, which has plants in Altdorf and Thun, is the major domestic producer of small arms ammunition for both military and civilian markets, and in 2000 had total sales of CHF 270 million (USD 168 million) and 1,092 employees (Graham & Whiteside Ltd, 2001).

The UK is no longer considered a major small arms producer.

United Kingdom[46]

The UK is one of Europe's largest defence producers, and the country was ranked as the world's fourth largest supplier of conventional weapons to developing countries during the period 1993–2000 with sales of USD 8.9 billion: an average of USD 1.1 billion a year (Grimmett, 2001). Although many companies are involved in various aspects of small arms production, the UK is not in the same league as other European producers such as Belgium, Germany, or Italy in terms of the value and volume of its domestic small arms production (Omega Foundation, 2001a; 2001b)

For centuries, small arms production in the UK was concentrated at the small arms arsenal at Enfield. The arsenal was privatized in the late 1980s and transferred to a new green-field site in Nottingham in the early 1990s (Lock, 2001). The UK no longer produces pistols for the military market, and its armed forces are equipped with pistols manufactured in Belgium (FN Herstal) and Switzerland/Germany (SIG Arms).

The major domestic producer of small arms and ammunition for the military market is **Royal Ordnance** (RO), which is the small arms division of **BAE Systems Plc**. RO also owns Sterling Armaments Company (since 1988) and Heckler & Koch GmbH (since 1991). The acquisition of Heckler & Koch in 1991 gave RO access to 74 per cent of the European small arms market and a significant share of the world market (Forecast International, 2002).

BAE Systems is one of the world's largest defence firms, and was ranked as the third largest arms producing company in the OECD and developing countries by SIPRI in 1999 (SIPRI, 2001, p. 307). In 2000 the company's total sales were USD 14.3 billion, and it employed 84,900 people worldwide. Royal Ordnance, which is part of BAE Systems' RO Defence division, produces small arms and ammunition. It had total sales of GBP 153 million (USD 232 million) in 2000, and employed 1,974 people. The production of small arms is concentrated in Heckler & Koch, which includes facilities in the UK (Nottingham), Germany (Oberndorf), and the US.[47] Heckler & Koch GmbH (Germany), which is part of RO, had total sales of DEM 159 million (USD 77 million) in 2000, and employed 570 people. The production facility in Nottingham will close during 2002 because a contract to modify 200,000 faulty SA-80 rifles was awarded to the Heckler & Koch plant in Germany in 2000 (*Daily Mail*, 5 October 2000).

Other important domestic producers include Accuracy International (sniper rifles), Armalon (sniper rifles), BMS Trading (rifles), Conjay Arms (small arms ammunition), Manroy Engineering (machine guns), FR Ordnance (sub-machine gun), and Parker Hale (sniper rifles) (Omega Foundation, 2001a; 2001b; Forecast International, 2002).[48] Most of these firms have fewer than 100 employees and concentrate on a few niche products. Accuracy International, which manufactures precision rifles, opened a US manufacturing plant in 1997 to assist with sales to the US military. Manroy Engineering manufactures the M2HB machine gun, under licence, from FN Herstal. Since the closing of the Enfield small arms factory, the company also has the manufacturing rights for the L7 General Purpose Machine Gun, which is a modified version of the MAG (FN Herstal).

Licensed production

Licensed production is an easy way to evade strict export controls.

Licensed production, an important feature of the global small arms industry, has been operating since the late nineteenth century.[49] Licensed production can have a number of political, strategic, and economic benefits for the licensor. Allowing weapons to be produced under licence increases political influence or strengthens political and military co-operation, and can help to standardize military equipment among allies. In economic terms, licensed production may allow companies to achieve returns on their R&D costs and increase their global market share. This is particularly important for countries that are competing with established defence producers. Licensed production can also be an easy way to evade strict export legislation and controls or to facilitate exports to prohibited destinations. Many developing countries, for a range of strategic, political, and economic reasons, want to be self-sufficient in the production of small arms and associated ammunition. Licensed production agreements are a cost-effective and less risky way for licensee countries to obtain tried and tested foreign defence technology. Thus, in most developing countries, legal production of small arms almost inevitably includes licensed production of foreign weapons.

The globalization of small arms production, and the increase in the number of countries with the capacity to produce small arms, has been primarily facilitated by licensed production agreements, and a number of studies have documented the increase in licensed production of weapons in developing countries.[50] A 1995 report estimated that licensed production was taking place in at least 21 developing countries, 16 of which were also exporting the small arms they manufactured (Klare, 1995). Since the 1960s at least 14 countries—Austria, Belgium, the Czech Republic, France, Germany, Israel, Italy, Portugal, Singapore, South Africa, Sweden, Switzerland, the United Kingdom, and the US—have established small arms licensed production agreements with some 46 countries (Abel, 2000, p. 88).

This section defines licensed production, and examines some of the conceptual aspects of licensed production and joint ventures. It provides case studies of two of the world's most significant licensors of small arms—FN Herstal (Belgium) and Heckler & Koch (Germany/UK)—and focuses on some of the countries in which their products are produced under licence. Neither Belgium nor Germany in the early Cold War period was a major defence producer, and both used licensed production to compete with some of the major producers such as France, the UK, the US, and the former Soviet Union.

FN Herstal small arms products.

FN Herstal and Heckler & Koch are two of the world's most important licensors of small arms.

These two case studies are used to highlight some of the key features of licensed production of small arms. However, details of licensed production agreements are almost impossible to obtain because of commercial confidentiality, thus making it difficult to compare the two cases presented in this chapter. What is clear is that licensed production of small arms normally involves (1) a licensed production agreement, which includes provisions for some kind of royalty payment; (2) a technology transfer transaction (including plant and infrastructure and some technical assistance); and (3) some provision for ongoing technical assistance. Interestingly, licensed production has not saved either FN Herstal or Heckler & Koch from the financial difficulties that they are currently experiencing.

It is worth noting that the German company Fritz Werner, previously state-owned and since 1990 a subsidiary of Ferrostaal AG (part of the MAN Group), has played a key role in helping establish arms production facilities in a number of developing countries, including Ethiopia, India, Indonesia, Iran, Iraq, Malaysia, Myanmar, Nigeria, Saudi Arabia, Sudan, and Thailand. A number of these facilities have been established for the licensed manufacture of FN Herstal or Heckler & Koch weapons (Lock, 2001; Omega Foundation, 2001c). In most cases, the production facilities (and technical know-how) were set up as part of a turnkey project, and were regarded by the recipient countries as more important than the licences themselves.

Definitions: Licensed production

When defence companies want to enter a foreign market, they have several methods at their disposal, such as:

- direct export;
- licence agreements;
- joint ventures;
- direct foreign investment (establishing a wholly owned domestic subsidiary); and
- buying into an existing foreign corporation.

Several basic concepts are essential to understanding licence agreements. The World Intellectual Property Organization (WIPO) defines a **licence** as a right conferred by a patent, or by a protected utility model, industrial design, new plant variety, or trademark ('industrial property'); a licence is given by the owner of that right ('licensor') to another person ('licensee'), to perform certain acts which are covered by that right in exchange for a payment ('royalty') (WIPO, 1977).

Licensed production can be defined thus: 'A company in country A contracts with a company in country B to undertake the legal production of its products. In terms of a licensed production agreement, the licensing company in country A usually provides technical data or copies of the products to be produced in country B, and sometimes provides machine tools or assists in the setting up of production facilities' (Small Arms Survey, 2001, p. 9). The latter may include technical consultation and training or other relevant know-how as part of the licence agreement, depending on the technological sophistication of the acquirer. But normally the licensing of relevant know-how is part of a separate know-how agreement. Licensed production agreements are also commonly referred to as licensed manufacturing agreements, co-production agreements, technology-transfer agreements, and are sometimes classified within the general term of 'offsets'.[51]

Sometimes a joint venture is more desirable for the licensor and licensee. A joint venture involves the co-operation of two or more otherwise independent companies, which are linked, through the venture, in the pursuit of a common commercial, financial, or technical activity (Muchlinski, 1995). In more practical terms the activity may entail the production of a product, the supply of services, or R&D (Blakeney, 1989).

According to Muchlinski (1995) the exact form of the joint venture depends on the wishes of the parties to the joint venture and on national law, and may include either one of the following types:

- a contractual joint venture: A contract defines the respective obligations of the parties. Usually established for a limited duration. Example: the international consortium;
- an equity joint venture: the parties establish a legal entity, which is responsible for the daily management of the joint enterprise. Usually established for an indefinite duration. Examples: the partnership or limited liability company.

Licensed production can include co-production, technology transfer agreements, and offsets.

Hart and Bereskin (1996, p. 2) identify three types of licences:

- non-exclusive licence: the licensor retains the right to use the licensed property and the right to grant additional licences to third parties;
- sole licence: the licensor agrees not to grant licences to third parties, but the licensor retains the right to use the licensed property;
- exclusive licence: the licensor cannot grant licences to third parties, nor can he use the licensed property.

When a licence is granted the **scope of the licence** must be defined. The licensor can place restrictions on the licensee's right to use the licensed property.

Subject matter: the owner of intellectual property has several exclusionary rights, including 'the right to make, have made, use and sell the patented invention' (Hart and Bereskin, 1996, p. 3). The licensor may grant the licensee all or only some of these rights. The licensor may, for instance, place restrictions on production quantity, but may also concern itself with production quality. The scope of the rights granted determines the size of the royalty.

Territory: the licensor may wish to grant an exclusive licence to different foreign licensees in each country. A problem arises when the licensor wants to limit the right to export. Can the licence agreement limit the licensees' right to export? According to Muchlinski (1995), reasonable territorial restrictions have been accepted under the competition laws of developed countries. Developing countries, on the other hand, are less happy with export restrictions, arguing that such restrictions simply inhibit their ability to earn foreign exchange through exports.

Improvements: since improvements can be developed by the licensor and the licensee, licence agreements: (a) include terms in the agreement providing for prompt disclosure of such improvements to the other party; (b) provide for the ownership of any rights in the improvement of technology; (c) provide obligations with respect to seeking patent protection; and (d) provide obligations with respect to granting further licences or granting back licences (Hart and Bereskin, 1996, p. 4). A big obstacle is defining the scope of new developments and the inclusion of grant-back provisions. These provisions 'oblige a licensee to transfer to the licensor any improvements in the technology represented by the licensed industrial property rights' (Blakeney, 1989, p. 35). Such provisions are met with hostility when these obligations are not reciprocal. Know-how is in most countries excluded from these grant-back obligations (Blakeney, 1989, p. 36).

In most cases the licensee pays the licensor royalties in exchange for the use of the intellectual property. The size of the royalty varies with the scope of the rights granted.

Payment of royalties can include one or more of the following:

- lump sum payment (used primarily for the transfer of know-how);
- annual fixed sum payments;
- fixed sum per unit produced;
- a percentage of the selling price of the product.

The licensee is obliged to pay these royalties for the duration of the licence agreement. However, a problem can occur when the intellectual property rights expire. Sometimes the termination date of the licence agreement extends well beyond the expiry date of the intellectual property rights. Most countries accept that when the intellectual property rights expire all royalty payments can be terminated (Blakeney, 1989).

Case study 1: FN Herstal[52] and licensed production in Sub-Saharan Africa

FN Herstal is one of the world's oldest and most important producers of small arms and the most important domestic producer of small arms and ammunition in Belgium. FN Herstal's various small arms products, including the FN-FAL assault rifle and the FN MAG machine gun, are in service in more than 90 countries worldwide (Small Arms Survey, 2001, p. 20). It is estimated that five million to seven million FN-FAL rifles and 150,000–200,000 FN MAG have been produced worldwide (Small Arms Survey, 2001, p. 20).

In addition to the manufacture of small arms (both military-style and civilian) in its own production facilities in Belgium, Portugal, and the US, FN Herstal's small arms products and/or ammunition have been manufactured under licence in more than 20 countries worldwide (see Table 1.11). In addition to licensing its products, the company, through its engineering division, has often provided technical assistance to licensees with respect to the production infrastructure needed to manufacture small arms and ammunition.[53]

The following case studies of Kenya, Nigeria, and South Africa provide examples of how FN Herstal, through licensed production agreements and technology transfer transactions, has played a key role in helping various countries in sub-Saharan Africa to establish their own domestic small arms production capabilities. In Kenya and Nigeria, the company provided technical assistance through a technology transfer to help establish the necessary infrastructure to produce weapons, in addition to licensing the manufacture of its products. The licensed production of FN Herstal's products also provided the foundation for the development of South Africa's domestic small arms industry, including the indigenous design and export of various types of small arms products.

MAP 1.4 FN Herstal licensed production

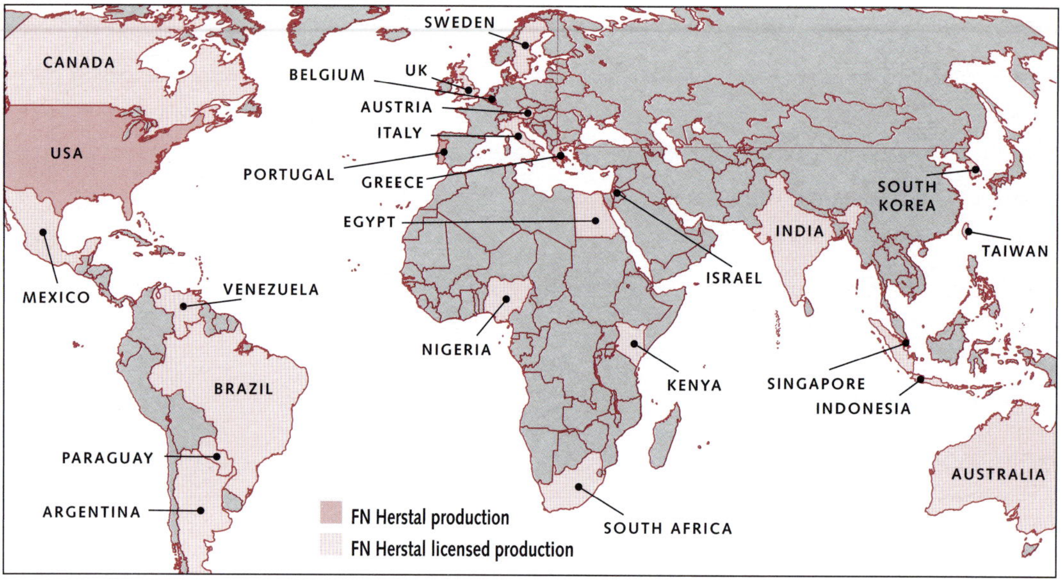

1. Kenya: Kenya Ordnance Factories Corporation, Eldoret[54]

The decision to establish a domestic arms production capability in Kenya (code-named Ngano Project) was mooted in the mid-1970s, when the then President Amin of Uganda made a number of territorial claims against Kenya. The Ngano Project remained on paper for many years, until in 1989 the government allocated KSH 300 million (approximately USD 6 million) to the construction of an ammunition factory at Eldoret.[55] Construction continued sporadically during the next few years, and was finally finished in mid-1996. The first batch of ammunition was produced in November 1996.

KOFC has a production capacity of 20 million rounds of ammunition per year.

The factory was initially a project of the Department of Defence, but was later converted into a State Corporation, the Kenya Ordnance Factories Corporation (KOFC). The company, which is wholly owned by the Kenyan government, has a board of directors including top military officials, and is subject to the provisions of the Corporations Act. President Daniel arap Moi and other senior politicians reportedly have interests in the company (*Jane's Intelligence Review*, 1 February 1998).

FN Herstal played a key role in the establishment of KOFC. The company was chosen, according to normal tender procedures, to supply the production infrastructure (equipment and machinery) for the factory, but because of cash flow problems was also called upon to construct the buildings, though in a reorganized funding arrangement. The deal between FN Herstal and the Kenya government, which was signed in 1988, was reportedly worth USD 80 million and also made provision for FN Herstal to buy any surplus production (*Jane's Intelligence Review*, 1 November 1996). At least three Belgian companies were involved in the construction of the plant: FN Herstal, M.A.D., and New Baron & Levoque International. The latter two companies worked as subcontractors for FN Herstal and/or the Kenyan Department of Defence, and built most of the infrastructure (Danssaert, 2001).

The factory has two production lines with a capacity of 40,000–60,000 rounds of ammunition a day, depending on the combinations to be manufactured (approximately 20 million a year). The factory makes three calibre types of ammunition in seven varieties:

TABLE 1.11 FN Herstal: Worldwide licensed production

Country	Company/ factory	9mm HP Browning	7.62mm FN FAL	7.62mm FN FALO	5.56mm FNC	7.62mm FN MAG	5.56mm Minimi	5.56mm FAL	Ammo
Argentina	DGFM	X	X			X		X	X
Australia	ADI		X	X			X		
Austria	Steyr-Mannlicher		X						
Brazil	Imbel/Fabrica de Itajuba		X					X	
Canada	Canadian Arsenals		X	X					
	John Inglis	X							
	Diemaco						X		
Egypt	Maadi Company					X ?			
Greece	Hellenic Arms Industry						X		
India	Government Rifle Factory, Ishapore		X			X			
Indonesia	PT PINDAD	X			X				
Israel	IMI		X						
	KSN Industries	X ?							
Italy	Beretta SpA						X		
Kenya	Kenya Ordnance Factory Corporation								X
Korea (South)	Daewoo Precision Industries				X?		X?		
Mexico	Fabrica de Armas		X ?						
Nigeria	DICON	X	X						X
Paraguay	DINDUSMIL								X
Singapore	Singapore Technologies					X			
South Africa	Armscor/Denel		X						
Sweden	FFV Ordnance				X	X			
Taiwan	Combined Service Forces Arsenals					X ?	X?		
UK	RSAF Enfield		X			X			
	RO Birmingham		X			X			
US	FN Manf. Inc					X	X		
	DS Arms		X						
Venezuela	CAVIM		X						X

Note: X? indicates uncertainty regarding licensed production

Source: Danssaert (2001)

FN Herstal's products have been produced under licence in more than 20 countries.

- 9mm ammunition for the FN35 Browning pistol, and the various sub-machine guns (e.g. Sterling, Uzi, or H&K MP5) used by the Kenya armed forces;
- 7.62mm x 51mm ammunition (in ball, blank, and tracer) for the FN-FAL and G3, which are the main rifles used by the Kenya armed forces; and
- 5.56mm ammunition (in ball, black, and tracer) for weapons used by the Kenya police.

The ammunition is stamped with a KOFC mark and a batch number that gives an indication of the year of manufacture. The factory has a quality control laboratory and three support workshops, including one for machining and repairing tools used in the main factory, and one for making wooden packing boxes. There are currently no facilities for the manufacture of small arms.

Current employment at KOFC is 230 people (60 per cent civilian, 40 per cent military), and the company is run by a retired Major-General. The main clients of the company are the Kenya armed forces, Kenya police, and Kenya Wildlife Services. Production far exceeds domestic demand and a recent newspaper article asked where the 20 million bullets which are ostensibly produced every year by KOFC go. It also noted that there are unconfirmed reports that the type of ammunition produced by the company has been supplied to armed groups in the region (*The Nation*, 13 August 2001). However, the most popular weapon in the region is the AK series assault rifle, which uses 7.62mm x 39mm (short) ammunition, which is not produced by KOFC. Thus the identity of the recipients of KOFC's excess production remains unclear.

2. Nigeria: Defence Industries Corporation of Nigeria (DICON)[56]

The Defence Industries Corporation of Nigeria (DICON) was established in 1964 under an act of parliament to produce small arms and small arms ammunition for the Nigerian armed forces. At the time, the Federal Government of Nigeria contracted the German company Fritz Werner GmbH, in a turnkey project, to provide the necessary production infrastructure and to train local staff.[57] On completion of the project, DICON was producing such items as the BM59 (rifle), shotguns for hunting, and M36 hand grenades.

Nigeria acquired licences for the production of small arms from a number of foreign sources.

Over the years, the Nigerian government has acquired technical assistance and licences for the production of small arms and ammunition from a number of foreign sources, including Beretta (Italy) and FN Herstal (Belgium). Nigeria reportedly acquired a licence to produce the FN-FAL assault rifle in 1977, and the rifle was designated NR1 (Nigerian Rifle 1). However, this is disputed by the Director-General of DICON, who claims that the licence to produce the FN-FAL assault rifle, the FN 9mm pistol, and 7.62mm ammunition was acquired in 1983 (Ebo, 2001). In addition, FN Herstal provided the technology and infrastructure required for the manufacture of these products, having replaced Fritz Werner as DICON's technical partner in the early 1980s. Despite this lack of clarity, it is known that between 1982 and 1984 FN technicians were present in the Kaduna factory providing technical assistance for the upgrade of existing equipment. Since 1970 the Ordnance Factory at Kaduna has been reportedly manufacturing 7.62mm NATO and 9mm NATO cartridges (Ezell, 1988). The annual capacity is 12 million rounds, with a possible maximum of 36 million rounds a year (Ebo, 2001).

Currently, DICON's capacity for production of small arms and ammunition is at the same level as during the 1980s. It produces the following products (current 2001 prices):

- NR1 Light Automatic Rifle (NGN175,250, USD 1,586).
- High-powered pistols (NGN85,000, USD 770).
- Single-barrel shotgun (N35,000, USD 317).
- General-purpose machine gun (no current price).

Most of these products, according to DICON, are based on the licences and production infrastructure acquired from FN Herstal in 1983. No new licences or production infrastructure have been acquired since then. DICON also has the capacity to repair and recondition the above weapons. In a bid to utilize its excess production capacities, the company has in recent years diversified into non-military markets, including the production of salt and of metal parts for vehicles and furniture.

During the 1990s, DICON's arms manufacturing was almost dormant due to a lack of purchases by the Nigerian government. Successive governments seem to have been more interested in importing weapons than in manufacturing them. The late President Sani Abacha spent USD 17 million on imported rifles, despite the fact that DICON has large amounts of rifles in its inventory (BBC Worldwide Monitoring, 18 October 1999). Currently, most of DICON's revenue comes from its investment in 20 per cent of the shares of UNION-DICON Salt, the largest salt-making enterprise in Nigeria.[58] Even this diversification has not been sufficient to keep the company afloat—in June 2000 a financial crisis obliged the company to place its staff on indefinite leave without pay (*Africa News Service*, 3 June 2000). It has been estimated that USD 100 million is needed to revive the moribund company (Forecast International, 2002).

DICON's turnover in 2000 was NGN 17.4 million (USD 157,500), and the company declared no profit. DICON currently employs 651 people and has two main production facilities. The Ordnance Factory is located at Kakuri, Kaduna South, and produces small arms and associated ammunition. The other facility is located at Bauchi and has the capacity to produce armoured personnel carriers, propellants, and explosives (Ebo, 2001).

The Nigerian government is keen to revitalize DICON, and is currently planning to phase out production of 7.62mm ammunition in favour of 5.56mm ammunition (Forecast International, 2002). At the end of 2000, the Nigerian Ministry of Defence announced the inauguration of the Board of Directors of DICON (*Africa News Service*, 21 February 2001). The Ministry proclaimed that '[w]e are committed to the principle of self reliance in weapons production... The Board is mandated to devise by all means necessary, measures that would enhance our capability to produce small weapons for our military and other relevant government agencies. It is hoped that by encouraging the development of DICON, the nation would not only be self reliant but would save the colossal amount of foreign exchange that is devoted to the procurement of arms overseas' (quoted in *Africa News Service*, 21 February 2001).

In early 2001 Denel, the South African arms company, expressed interest in taking over daily management of DICON under a joint venture agreement. Both companies are hoping that, by revitalizing DICON, they would 'jointly gain the major share of any future West African re-equipment programmes' (*Jane's Defence Weekly*, 21 February 2001).

3. South Africa: Armscor/Denel

In the early 1960s, South Africa began to move away from procuring complete weapon systems from abroad and instead moved towards local assembly and licensed production of weapons. The Armaments Development and Production Corporation of South Africa (Armscor) was created in 1968 with the explicit mandate to develop South Africa's domestic defence industry and to supervise the manufacture of armaments. During the first decade of its existence Armscor expanded rapidly through the acquisition of private companies and the creation of new subsidiaries. With the reorganization of the South African arms industry in 1975–76, Armscor also became responsible for the procurement of armaments. The UN mandatory arms embargo of 1977 fuelled the expansion of the South African arms industry even further (Batchelor and Willett, 1998, pp. 30–2).

South Africa's Denel has expressed an interest in a joint venture with DICON.

In the early 1990s the domestic demand for arms declined, after the ending of apartheid and the restructuring and downsizing of the South African Defence Force (SADF). In 1992 a new state-owned commercial company, Denel, was formed, and it took over all of Armscor's manufacturing facilities. Armscor retained responsibility for arms procurement, the sale of surplus arms, and the promotion of the South African defence industry (Batchelor and Willett, 1998).

Armscor's small arms production was concentrated in two subsidiary companies: Lyttleton Engineering Works (LIW) and Musgrave. The production of small arms ammunition took place at Pretoria Metal Pressings (PMP). All three companies became part of Denel in 1992. In 1996, Musgrave was closed down and its assets were transferred to the Vektor division of LIW (Batchelor and Dunne, 1998). In 1997 Vektor became a separate division of Denel. It manufactures small arms for both the military and the civilian markets. The production of small arms ammunition for both military and civilian markets still takes place at PMP.

The UN mandatory arms embargo of 1977 fuelled the expansion of the South African arms industry.

South Africa has manufactured a number of small arms products under licence from FN Herstal since the early 1960s. These products have provided an important base for the development of South Africa's domestic small arms production capabilities.

South Africa acquired the 7.62mm FN FAL (designated the R1) for the SADF in 1960. Licensed production began in 1963 at LIW and ran until 1978,[59] when it was replaced by the R4/R5 (Ezell, 1988). A lighter version of the R1, the R2, was also acquired for the SADF. According to Etienne Hennop of the Institute for Security Studies (ISS), this rifle never went into production, but it did lead to the development of the R4/R5 rifles.[60] On the other hand, the South African Department of Defence says that the R2 was imported. Stevens and van Rutten (1981) list another model of the FN FAL rifle, the R3, which was produced under licence in South Africa. The 7.62mm FN MAG was acquired for the SADF in 1960 (Ezell, 1988). Given the South African defence industry's abilities to re-engineer and upgrade existing products, it is possible that the FN MAG formed the basis for the Vektor 7.62mm SS77 general-purpose machine gun that is currently in production, and in service with the South African National Defence Force (SANDF) and in Kuwait.[61]

Case Study 2: Heckler & Koch[62] and licensed production in the Middle East

Heckler & Koch GmbH (HK) has been manufacturing small arms since 1948. The company started operations in Oberndorf/Neckar, Germany, as a manufacturer of sewing machine parts and gauges for the machine tool industry. It is now one of the world's largest and most successful producers of small arms, and its weapons are in use in more than 90 countries worldwide (Small Arms Survey, 2001, p. 20). In recent years, HK has launched at least five new weapon systems and is also playing a key role in developing the next generation of military small arms for the US armed forces, the Objective Individual Combat Weapon (OICW). The HK G3 assault rifle, which has been the German army's standard rifle since 1959, is in service in more than 60 countries (Small Arms Survey, 2001, p. 20). At least seven million G3 rifles have been produced worldwide, and in addition to being produced in Germany it has also been produced under licence in at least 15 countries. Most of these licensed production agreements were established during the 1960s and 1970s.[63] However, not all of these agreements are still operational.

HK and Royal Ordnance had a close relationship long before RO purchased HK in 1991. HK and Britain's Secretary of State for Defence signed a licence production agreement in April 1970, with amendments in 1972 and 1977. Under the terms of the agreement, the Royal Small Arms Factory (RSAF) at Enfield manufactured various HK weapons and sold these to Heckler & Koch (UK), which was established in March 1973 and existed mainly as a sales company. The licence transferred to RO when it was established in January 1985. The manufacturing of HK weapons was undertaken at first by RSAF Enfield and later, after the privatization, at the RO Nottingham factory.

By the late 1980s, HK was facing serious financial problems. One key factor of this was the German government's decision not to order the company's new 4.7mm G11 rifle, the so-called caseless cartridge weapon, after paying more than DEM 100 million in development costs. HK had anticipated some DEM 60 million (USD 39 million) from the first production batch for the German Bundeswehr in September 1990, with a total potential value of DEM 2.7 billion (*Jane's Defence Weekly*, 26 January 1991). With financial problems mounting, HK went looking for a buyer. Despite public announcements by Giat Industries (France) in January 1991 that it had acquired HK, by March 1991 HK had become a wholly owned subsidiary of the UK defence giant British Aerospace (now BAE Systems Plc).

One of BAE's key objectives for the acquisition of HK in March 1991 was, according to the company, 'to regain a small arms R&D base' (*Jane's Defence Weekly*, 25 January 1992). RO had lost its UK small arms production facilities when its Enfield plant (RSAF) was closed in 1987. In addition, HK hoped to benefit from the greater worldwide marketing reach of BAE/RO. The acquisition of HK gave RO access to more than 70 per cent of the European small arms market and a significant share of the world market (*Jane's Defence Weekly*, 25 January 1992).

A technical co-operation agreement, signed in 1972 between RO and HK, had led to HK providing production equipment and barrels for the UK's 5.56mm SA80 rifles, while RO had assembled a wide range of HK weapons in the UK. The RO small arms plant at Nottingham was also manufacturing components for the HK plant at Oberndorf. Most of the production for Heckler & Koch (UK) was exported, as there was little domestic requirement for G3 rifles in the UK.

After 1991, when HK became a wholly owned subsidiary of Royal Ordnance Plc (itself part of the UK defence company, BAE Systems Plc), it became more difficult to obtain disaggregated information on the turnover, production, and export details of the German-based factory. Since 1998, Heckler & Koch Ltd (UK) has become a dormant company, and its operations are reported within the overall annual accounts for Royal Ordnance Plc and BAE Systems Plc. This makes it difficult to obtain disaggregated details of turnover and production.

In 2000 Heckler & Koch GmbH (in Germany) had total sales of DEM 159 (USD 77 million), and employed 570 people (Graham & Whiteside Ltd, 2001). During the same year, Royal Ordnance Plc (including Heckler & Koch GmbH) had total sales of GBP 153 million (USD 232 million), exports of GBP 38 million (USD 57 million), and total employment of 1,974 people (ICC Information Group, 2001).

MAP 1.5 Heckler & Koch licensed production

During 2000, BAE Systems tried unsuccessfully to sell HK to Colt's Manufacturing (US). No other potential buyers have been identified. In late 2000 a GBP 83 million contract to modify 200,000 faulty SA-80 rifles was awarded to HK's Oberndorf factory in Germany, in preference to the RO factory in Nottingham (*Daily Mail*, 5 October 2000). Many commentators agreed that the contract was awarded to the German factory to make HK more attractive to potential buyers. As a result, the RO facility in Nottingham will be closed during 2002.[64]

HK currently has manufacturing facilities in three countries: Oberndorf in Germany, Nottingham in the United Kingdom, and Sterling, Virginia, in the US. In addition, HK has liaison offices in four countries: Switzerland, Saudi Arabia, United Arab Emirates, and Thailand. Since 1957, HK weapons have been manufactured in at least 18 countries under licensed production agreements (see Table 1.12).[65] Currently, licensed production is taking place in at least ten countries including Iran, Greece, Mexico, Pakistan, Portugal, Saudi Arabia, Spain, Switzerland, Thailand, and Turkey. The G3 rifle has been produced under licence in 15 countries and currently ten countries—Iran, Germany, Greece, Mexico, Myanmar, Pakistan, Portugal, Saudi Arabia, Turkey and the UK—are producing the weapon. The MP5 sub-machine gun has been produced in eight countries and currently five—Germany, Greece, Mexico, Turkey, and the UK—are producing the weapon. Despite a statement by BAE Systems that the original licensed production agreement is no longer operational, Iran's Defence Industries Organisation continues to manufacture and export both the G3 rifle and a version of the MP5 (known as the Tondar).

Heckler & Koch weapons are produced in a large number of countries worldwide. This subsection provides case studies of three countries in the Middle East—Turkey, Iran, and Saudi Arabia—which produce HK weapons under licence. Like FN Herstal, HK, through its licensed production agreements and technology transfer transactions, has played a key role in helping various countries in the Middle East to establish their own domestic small arms production capabilities. It is also worth noting that Pakistan Ordnance Factories (POF), which manufactures a range of HK products under licence, has played a key role in helping various countries in the Middle East, including Iran, Oman, Saudi Arabia, and Sudan, to establish and develop their small arms production capabilities (Omega Foundation, 2001c).

1. Turkey: Makina ve Kimya Endustrisi Kurumu (MKEK)

MKEK, part of Turkey's state-owned Machine and Chemicals Industries Board, manufactures a wide range of small arms, ammunition, and other weapons systems under licensed production agreements, including a range of HK small arms. It has produced the G3 assault rifle since 1967 and the MP5 sub-machine gun since the 1980s.

In January 1998, MKEK signed a licensed production agreement with HK worth USD 18 million to produce 200,000 HK33 5.56mm assault rifles over a period of ten years (*Jane's Defence Weekly*, 14 January 1998). Some sources indicate that up to 500,000 HK33s will be produced (Omega Foundation, 2001c). The HK33 is intended to replace the 350,000 G3s which are in service with the Turkish Army, and which were produced under licence from HK. In 1998, while signing the contract for production of the HK33, it was reported that MKEK would continue to produce the G3 rifle for export (*Jane's Defence Weekly*, 14 January 1998).

MKEK has increased its exports in recent years, including many weapons made under licence. In 1995 the company reported that it had sold its products to some 38 countries. Its most popular export products include the HK MP5 machine gun (Omega Foundation, 2001c).

TABLE 1.12 Heckler & Koch: Worldwide licensed production

Products country (company)	G3	MP5	HK33	G36	Pistols	HK21	Notes/other
Brazil	X						
France (Manufacture Nationale d'armes de St Étienne)	X	X					No current production
Greece (Hellenic Arms Industry)	X	X	X		X	X	40mm grenade machine gun
Iran (Defence Industries Organisation)	X	X					
Italy (Luigi Franchi)							40mm grenade launcher
Malaysia (Munora Holdings)	X						
Mexico (Fabrica de Armas)	X	X			X	X	
Myanmar/Burma (State Arsenals)	X						
Norway (Norsk Forsvarsteknologi)	X						No current production
Pakistan (Pakistan Ordnance Factory)	X	X			X		
Philippines	X						
Portugal (INDEP)	X	X				X	
Saudi Arabia (Al Kharj Arsenal)	X						
Sweden (FFV)	X						No current production
Thailand (Army Weapons Production Centre)	X		X				
Turkey (MKEK)	X	X	X				
Spain (Santa Barbara)				X			
Switzerland (Brugger + Thomet)		X					

Source: Omega Foundation (2001c)

In 2000, the German government approved a licence for Fritz Werner to export and install an ammunition plant in Turkey.[66] The plant will produce 5.56mm calibre ammunition for use in the HK33 assault rifles. A Belgian company, New Lachausseé, part of the Georges Forrest International Group, will assist in constructing the ammunition plant.[67] The contract between MKEK and Fritz Werner is worth BEF 1.8 billion (USD 42 million), while the contract between MKEK and New Lachausseé is worth BEF 300 million (USD 7 million) (Herssens, 2001).[68]

2. Iran: Defence Industries Organization

Iran's defence industry dates back to the early 1970s. By the early 1990s the country's defence industry employed 45,000 people in 240 state-owned plants and an estimated 12,000 privately owned workshops (*International Defence Review*, 1 April 1994). The local defence industry has received technical assistance from a number of countries including the Russian Federation, North Korea, China, Pakistan, Argentina, Brazil, Taiwan, and Germany (*Jane's Defence Weekly*, 1 February 1992). In 1991 it was reported that Pakistan and Iran had jointly set up a small arms ammunition factory and that experts from POF had visited Iran for this purpose (*Inter Press Service*, 14 June 1991).

All defence production facilities in Iran fall under the umbrella of the Defence Industries Organisation (DIO). The Islamic Revolutionary Guard Corps has its own production facilities for the manufacture of various types of small arms including grenades, mortars, and ammunition (*Jane's Sentinel Security Assessment*, 10 August 2000).

In 1966, the Government of West Germany signed a deal with the Iranian government for the licensed production of the HK G3 assault rifle at the Mosalsalsasi weapons factory (Herssens, 2001). The actual management of the licensed production programme was supervised by the German company Fritz Werner (Herssens, 2001). It is through Fritz Werner that spare parts, production tooling, and raw materials are still believed to be provided to Iran and by the end of 1977 the Jangafzarsazi factories were reported to be producing a range of small arms and light weapons.[69]

The G3 production line was damaged during the early stages of the Islamic revolution of 1979, but it was rebuilt and continues to produce the G3 rifle. Before the revolution, the factory was reported to be producing about 145,000 G3s a year. Since then, the production rate has been estimated to be about 50,000 per year (Ezell, 1988). If these estimates are correct, then reports that Iran had supplied 50,000 G3 rifles to Sudan in 1991 would mean that a whole year's production had been exported in one deal (*Jane's Defence Weekly*, 9 May 1992). The DIO are also manufacturing and marketing a version of the MP5 which they designate the MPT 9mm (Tondar) (Omega Foundation, 2001c).

Pakistan and Saudi Arabia have recently established a joint venture in small arms production.

BAE Systems/Heckler & Koch has stated that the licensed production agreements for the G3 rifle and MP5 sub-machine guns with Iran are no longer valid. As the German government refuses to discuss the licensed production agreements on grounds of 'commercial confidentiality', it is not possible to ascertain the exact status of the original agreements.

The DIO continues to produce G3 rifles and variants of the MP5 sub-machine gun. At the Defence Services Asia arms exhibition in Malaysia, in April 2000, the company was marketing the MPT 9K 9mm automatic weapon, which bears a striking resemblance to the HK MP5K 9mm briefcase system.[70]

3. Saudi Arabia: Al-Kharj Arsenal

Saudi Arabia established a domestic defence industry in the late 1970s. Production of small arms and ammunition is centred at the Al-Kharj Arsenal, part of the Prince Sultan Military City south-west of Riyadh. The US, Germany, UK, and Belgium are reported to have assisted in establishing the country's domestic production capability. In 1979, the German company, Fritz Werner, was involved in setting up a plant to produce HK G3 assault rifles and ammunition under licence (*Jane's Defence Weekly*, 18 August 1999). By the mid-1980s, the Al-Kharj Arsenal was also manufacturing M1 rifles and M16 assault rifles under licence from Colt's Manufacturing (US) (*Jane's Defence Weekly*, 18 August 1999).

TABLE 1.13 Export destinations of HK licensed-produced weapons by licensees

Licensee company (country)	G3	MP5	Unspecified
Manufacture Nationale d'armes de St Étienne (France)	Lebanon Germany (Berlin) Djibouti Burkina Faso	Germany (Berlin)	
Hellenic Arms Industry SA (Greece)	Libya Burundi		
Defence Industries Organisation (Iran)	Sudan		
Munora Holdings (Malaysia)	Chile		
Pakistan Ordnance Factory (Pakistan)	Bangladesh	Kenya South Africa Philippines Burma	
INDEP (Portugal)	Zimbabwe		
Al Kharj Arsenal (Saudi Arabia)	Yemen		
MKEK (Turkey)	Cyprus Germany	Indonesia	Jordan Pakistan Norway Switzerland UK

Source: Omega Foundation (2001c)

In September 2001, it was reported that POF and Saudi Arabia had agreed their first joint venture in small arms production at the Al-Kharj Arsenal (*Jane's Defence Weekly*, 19 September 2001). It is likely that the joint venture will involve the manufacture of several types of small arms, including products manufactured under licence from Heckler & Koch.

Heckler & Koch weapons manufactured under licence have been allegedly exported, sometimes in violation of the original licensed production agreement.

Exports of licensed products

A key issue associated with licensed production is the export, by licensees, of weapons produced under licence to third countries in violation of the original licence agreement (Omega Foundation, 2001c). As mentioned earlier, the licensor may limit the right of the licensee to export to third countries through a general (or specific) territorial restriction in the licence agreement (Muchlinski, 1995).

The exports of HK weapons produced under licence by MKEK (Turkey) and by Iran's DIO would appear to be in direct contradiction of statements by HK officials that the right to re-export was excluded from the licensed production agreements signed with states such as Turkey and Iran.

In 1982, the German government placed further restrictions on the export of completed weapons. To obtain the approval of the German government, the weapons manufactured under licensed conditions are for the sole need of the country concerned and must not be re-exported. Exemptions from the re-export exclusion are possible only after prior consultation with the German government.[71] The German government still refuses to discuss whether it gave permission for MKEK to export MP5 sub-machine guns to Indonesia (Omega Foundation 2001c).

Table 1.13 provides details of transfers of HK weapons produced by licensee companies. The licensees claim export deals to many more countries but have not provided details of the weapons supplied. However, it is clear that HK weapons manufactured under licensed production agreements have allegedly been exported, sometimes in violation of the original agreements (Omega Foundation, 2001c).

Conclusion

This chapter has provided an assessment of the state of the global small arms industry. While it is clear that parts of the industry—certain companies in various countries—are 'sick', it is not possible to ascertain, with any real precision, whether the global industry is 'dying', that is, in long-term decline. The lack of both official and unofficial information about the value and volume of global small arms production makes it almost impossible to predict the industry's long-term trends. In the short term, certain factors that will determine the health of the industry can be identified, based on the information presented in this chapter. At the level of firms, factors such as market segmentation and concentration, and internal production and marketing decisions (e.g. finding niche opportunities) will have a significant impact on the health of firms. At a more general level, factors such as the economic and security environment in particular countries, negative public sentiment towards these types of weapons, and/or growing public pressure on governments to implement stricter national controls on the production of, trade in, and use of small arms are likely to have a significant impact on the health of firms and on the global industry more generally.

The information presented in this chapter demonstrates that global small arms production is more widely distributed than previously identified. Currently, more than 1,000 companies in at least 98 countries worldwide are involved in some aspect of the production of small arms and/or ammunition. The increase in the number of countries and companies that produce, or have the capacity to produce, small arms does not necessarily indicate an increase in the size or production capacities of the global small arms industry, but simply better information.

The chapter has also provided new and updated estimates on the value and volume of global small arms production, including commercial firearms. The *value* of global small arms production, including ammunition, for 2000 was estimated to be at least **USD 7.4 billion**. In terms of *volume*, it is estimated that nearly **eight million** small arms, including commercial firearms, were produced during 2000, of which more than 70 per cent were produced in the US and the European Union. The global volume of small arms production has declined in recent years, and is at much lower levels than during the Cold War era. There are also indications that the production of military-style small arms will continue to decline in the coming years. The long-term trends in the production of commercial firearms are currently unclear, but production volumes may start to increase as a result of the 2001 terrorist attacks in the US. Ammunition production for both military-style small arms and commercial firearms seems to be relatively constant at present.

Despite its global distribution, small arms production remains fairly marginal economically, both in terms of its contribution to national economic activity, and as a share of the value of global defence production. For example, in Austria the three most important small arms companies have combined total sales of USD 57 million, and collectively employ fewer than 800 people. The production of small arms is a mature industry with few barriers to entry, and is not a significant source of employment. In the US, only about 16,000 people are employed in the small arms industry.

The regional survey of small arms production in the Middle East provides new and more detailed information on the small arms production activities of countries and companies in the region. While at least 11 countries in the Middle

East have the capacity to produce small arms and/or ammunition on a regular basis, only a few countries can be regarded as significant producers. Most production takes place in state-owned companies, and involves local production under licence from foreign suppliers (e.g. Heckler & Koch). Israel, the region's largest and most innovative producer of small arms, is currently experiencing a crisis in its domestic defence industry, and it is unclear whether the country's small arms production capabilities, which are concentrated in the state-owned company Israel Military Industries, will survive the current crisis. At the same time, a number of other regional producers including Turkey, Iran, and Saudi Arabia are expanding their domestic production capabilities and aggressively targeting export markets. Recently, it has emerged that Pakistan Ordnance Factories has played a key role in supporting the development and expansion of small arms production capabilities in a number of countries in the Middle East including Iran, Oman, Saudi Arabia, and Sudan.

The chapter has also provided detailed information on the financial performance and production activities of the world's most important small arms companies in the 13 countries that dominate the global market for small arms. The overall picture is mixed: some firms are in deep crisis, while others appear to be prospering. This is largely a function of the fact that the small arms industry appears to be no longer governed by global military and/or economic trends. The industry is now so segmented that the success or failure of particular companies is often related to internal factors (e.g. marketing decisions) rather than external factors (e.g. the business cycle). Thus, the industry's long-term trends are increasingly obscure and unpredictable.

In the world's three major producers—China, Russia, and the US—the picture is difficult to assess because of a lack of detailed information, particularly with respect to China. China's most important small arms producer, the state-owned company CNGN (or Norinco), was radically restructured in 1999, but largely because of a lack of both official and unofficial information, it is not clear whether the company is still a major player in the global small arms market. In the Russian Federation, it is clear that a small number of companies, including JSC Izhmash, are re-emerging as significant players in the global small arms market, after nearly a decade of painful downsizing and restructuring in the Russian defence industry. The growing success of these Russian firms is linked to product diversification (e.g. commercial firearms) and increasing export sales. It is worth noting that more people are currently employed in JSC Izhmash (25,400) than in the entire firearms industry in the US.

The US small arms industry is at present in fairly bad shape, particularly handgun producers. However, the US still accounts for nearly 60 per cent of all commercial firearms produced worldwide. In recent years, well-known companies such as Smith & Wesson and Colt's have suffered either flat or declining sales. A few firms such as Sturm, Ruger & Co and OF Mossberg have, however, remained profitable and actually achieved increases in both domestic and export sales. While the US domestic market has declined in recent years from its peak in 1993–94, there are indications that the market might improve in the next few years as a result of the terrorist attacks on the US in September 2001. Demand for small arms from the US armed forces has also been in decline for a number of years, and this has had a negative impact on the US military's most important small arms contractors (producers) such as Colt's, FN Manufacturing, Beretta, and Saco Defense.

In the world's ten most important medium-sized producers—Austria, Belgium, Brazil, France, Germany, Israel, Italy, Spain, Switzerland, and the UK—the picture is also mixed. In some countries (e.g. Austria, Brazil, Italy) privately-owned firms like Glock, Taurus, and Beretta are doing relatively well, and witnessing either constant or increasing sales and profits. In almost all cases these companies rely heavily on sales to the US market (both military and civilian).

Total employment in most of these companies is usually fewer than a few thousand employees, and in some cases only a few hundred. In other countries such as France, Israel, and Spain, large state-owned defence companies (e.g. Giat Industries, IMI, Santa Barbara) which also produce small arms are experiencing significant financial problems. Most of these companies are currently making significant losses, have large numbers of employees, and low levels of labour productivity. Without continuing government support, it is unclear whether the small arms production activities in these firms will be maintained.

Concentration and consolidation at the company level has been increasing in recent years and is likely to continue, particularly among European and US producers. The various segments of the global small arms market (both military and civilian) also continue to be dominated by a handful of companies in these 13 countries, despite the emergence of a few 'new' players in recent years (e.g. Pakistan, Singapore). Some countries, such as Brazil, have taken significant steps to protect their domestic small arms industry from foreign competition.

Licensed production, a key feature of the global small arms industry particularly amongst developing countries, is examined in some detail in this chapter using case studies of two of the world's most significant licensors of small arms: FN Herstal (Belgium) and Heckler & Koch (Germany/UK). Altogether the products of these two firms have been produced under licence in more than 35 countries worldwide. These case studies show how licensed production can be used by licensees (e.g. Turkey) to develop and expand their domestic small arms production and export capabilities, and by licensors (Heckler & Koch) to increase their global market share, to evade strict export controls, and/or to facilitate exports to prohibited destinations.

Overall, the information presented in the chapter presents a somewhat contradictory picture. At the company level, some firms appear to be prospering while others are in crisis. Certainly some of the private firms (e.g. Beretta) appear to have more chance of survival, although of course, in some cases, state-owned companies will continue to attract government support for political and/or strategic reasons (e.g. Giat Industries). At the global level, the small arms industry seems to be shrinking (in terms of production capacity). It is also more segmented, and more concentrated than during the Cold War period, and is experiencing a significant decline in demand, particularly in the military market and certain sub-sectors of the civilian market (e.g. handguns).

Finally, what are the most important factors that will determine the industry's trends in the near future? Certainly the events of 11 September 2001 and increasing US domestic civilian demand for firearms may provide some temporary relief for the US small arms industry as a whole, and certain firms in various other countries (e.g. Austria, Brazil, Italy) that are major suppliers to the US market. Other factors, such as the expansion of licensed production or increased export sales, might also improve the short-term prospects for certain firms, depending on the stringency of domestic export controls.

However, in the longer-term, factors such as slower rates of economic growth in various regions such as North America, Europe, and Asia, negative public sentiment concerning the impact of small arms on violence, crime rates, etc., or the growing national and international efforts (e.g. within the United Nations) to regulate and tighten controls over the production, transfer, possession, and use of small arms could have a negative impact on the future prospects for the global small arms industry. In addition, shrinking demand without a corresponding decline in the number of suppliers has made the global small arms market intensely competitive, with the result that many companies have been forced to sell their products at low and often unprofitable prices. This practice does not bode well for the future of the industry.

1. Appendix

Small arms producers, 2002

	Country	Small arms*	Ammunition
1	Albania	x	x
2	Algeria	x	
3	Argentina	x	x
4	Armenia	x	x
5	Australia	x	x
6	Austria	x	x
7	***Bangladesh***	x	x
8	Belarus	x	x
9	Belgium	x	x
10	Bolivia		x
11	Bosnia & Herzegovina	x	x
12	Brazil	x	x
13	Bulgaria	x	x
14	Burkina Faso	x	x
15	***Cambodia***	x	x
16	Cameroon		x
17	Canada	x	x
18	Chile	x	x
19	China	x	x
20	Colombia	x	x
21	Croatia	x	x
22	Cuba	x	x
23	***Cyprus***		x
24	Czech Republic	x	x
25	Denmark	x	x
26	***Dominican Republic***	x	x
27	Ecuador		x
28	Egypt	x	x
29	Estonia	x	
30	***Ethiopia***	x	
31	Finland	x	x
32	France	x	x
33	Georgia	x	
34	Germany	x	x
35	Greece	x	x
36	Guatemala		x
37	***Guinea***	x	x
38	Hungary	x	x
39	India	x	x
40	Indonesia	x	x
41	Iran	x	x
42	Iraq	x	x
43	Israel	x	x
44	Italy	x	x
45	Japan	x	x
46	Kazakhstan	x	
47	Kenya		x
48	Korea, North	x	x
49	Korea, South	x	x
50	Kyrgyzstan	x	x
51	Lithuania	x	
52	Luxembourg	x	
53	Macedonia		x
54	Malaysia	x	x
55	***Malta***		x
56	Mexico	x	x
57	***Moldova***	x	
58	Monaco	x	x
59	Morocco	x	x
60	Myanmar (Burma)	x	x
61	***Nepal***	x	x
62	Netherlands		x
63	New Zealand	x	x
64	Nigeria	x	x
65	Norway	x	x
66	Pakistan	x	x
67	***Papua New Guinea***	x	
68	Paraguay		x
69	Peru	x	x
70	Philippines	x	x
71	Poland	x	x
72	Portugal	x	x
73	Romania	x	x
74	Russian Federation	x	x
75	Saudi Arabia	x	x
76	Singapore	x	x
77	Slovakia	x	x
78	Slovenia	x	x
79	South Africa	x	x
80	Spain	x	x
81	Sudan	x	x
82	Sweden	x	x
83	Switzerland	x	x
84	Syria		x
85	Taiwan	x	x
86	Tanzania		x
87	Thailand	x	x
88	Turkey	x	x
89	Uganda		x
90	Ukraine	x	x
91	United Arab Emirates		x
92	United Kingdom	x	x
93	United States	x	x
94	Uruguay		x
95	Venezuela	x	x
96	Vietnam		x
97	Former Republic of Yugoslavia	x	x
98	Zimbabwe		x

*Notes: * Small arms include firearms (handguns and long guns) and light weapons.*
Bold italics: *uncertainty regarding current production.*

Sources: BICC (2001); Feldman and Shapir (2001); Forecast International (2002); Gander and Cutshaw (2001a; 2001b); Reed (2001); Jane's Sentinel Security Assessment (various issues); OMEGA Foundation (2001b); United Nations (1999)

1. List of Abbreviations

Armscor	Armaments Development and Production Corporation of South Africa
BATF	US Bureau of Alcohol, Tobacco, and Firearms
CAST	Centre for Analysis of Strategies and Technologies
CBC	Companhia Brasiliera de Cartuchos
CIS	Commonwealth of Independent States
CNGN	China North Industries Group Corporation
COSTIND	Commission of Science, Technology and Industry for National Defence
CSG	China South Industries Group Corporation
DEFA	Direction des Études et Fabrications d'Armements
DICON	Defence Industries Corporation of Nigeria
DIO	Defence Industries Organisation
DMA	Direction Ministérielle des Armements
DTAT	Direction Technique des Armements Terrestres
FNNH	Fabrique Nationale Nouvelle Herstal
GIAT	Groupement Industrial des Armements Terrestres
GUP	State enterprise (Russian Federation)
HK	Heckler & Koch
IDF	Israeli Defence Force
IMI	Israel Military Industries
IP	Intellectual property
ISS	Institute for Security Studies
JSC	Joint-stock company (Russian Federation)
KMP	Kovrov Mechanical Plant
KOFC	Kenya Ordnance Factories Corporation
LIW	Lyttleton Engineering Works
LMG	Light Machine Gun
MDMP	Ministry of Defence and Military Production (Egypt)
MKEK	Makina ve Kimya Endustrisi Kurumu (Turkey)
Norinco	China North Industries Group Corporation
OECD	Organization for Economic Co-operation and Development
OICW	Objective Individual Combat Weapon
PLA	People's Liberation Army
PMP	Pretoria Metal Pressings
POF	Pakistan Ordnance Factories
RO	Royal Ordnance
RSAF	Royal Small Arms Factory
SADF	South African Defence Force
SAF	Small Arms Factory group
SANDF	South African National Defence Force
SIPRI	Stockholm International Peace Research Institute
SM	Swiss Ammunition Enterprise
SRIC	Security Research and Information Centre
SW	Swiss Ordnance Enterprise
UNCTAD	United Nations Conference on Trade and Development
WIPO	World Intellectual Property Organization
WFSA	World Forum on the Future of Sport Shooting Activities

1. Endnotes

1 The Small Arms Survey has in its possession verifiable information on the production of small arms and/or ammunition in 98 countries, based on research conducted by Omega Foundation.

2 The Small Arms Survey has in its possession verifiable information on the production of small arms and/or ammunition in 1,042 companies, based on research conducted by Omega Foundation.

3 It is not possible to provide estimates for 2001, as no information is available on the production of commercial firearms for this year.

4 Many of the countries and companies identified by the Small Arms Survey as current producers of small arms are not included in the Forecast International estimates.

5 Information from Forecast International (2002) suggests that the US army procured 29,000 small arms in 2000, and was likely to procure 25,000 in 2001. No figures are available for navy, air force, and marine corps procurement.

6 These figures are difficult to verify, although the figures for US production can be confirmed by annual production figures from the BATF.

7 Many of the countries and companies identified by the Small Arms Survey as current producers of military-calibre small arms ammunition are not included in the Forecast International estimates.

8 In 2000, the US armed forces procurement of small arms, based on contracts, was worth USD 98 million. In 2001 procurement was estimated to be worth USD 82 million (Forecast International, 2002).

9 The average price for units of commercial ammunition is lower than for military-calibre ammunition. The average price is likely to be 10–12 US cents a round (see <http:www.shooterz.com/ammo.html>), compared with 20 US cents for military-calibre ammunition.

10 This estimate is a significant increase on the figure of USD 4 billion presented in the *Small Arms Survey 2001*. Based on new information and research, it reflects a better estimate of the value of global small arms production for both military and commercial markets.

11 The countries of the Middle East include: Algeria, Bahrain, Egypt, Iran, Iraq, Israel, Jordan, Kuwait, Lebanon, Libya, Morocco, Oman, Palestinian territories, Qatar, Saudi Arabia, Sudan, Syria, Tunisia, Turkey, United Arab Emirates, Yemen, Western Sahara (Feldman and Shapir, 2001).

12 The information in this section is based on Miller (2001), Steinberg *et al.* (2001), Feldman and Shapir (2001), Reed (2001), and *Jane's Sentinel Security Assessment* (2001).

13 See the *Small Arms Survey 2001* pp. 38–40 for details of small arms producers in the Middle East.

14 Some countries in the region, such as Jordan, Libya, and Oman are reported to possess a domestic capacity to produce small arms and/or ammunition (Feldman and Shapir, 2001; Forecast International, 2002). However, it is difficult to verify the reliability of this information.

15 Some countries in east and central Europe, such as Bulgaria, the Czech Republic, Hungary, and Romania are significant exporters of small arms (often surplus, or used arms). However, most of these countries are not currently producing significant quantities of new weapons.

16 According to BICC (2001, p. 62), the major European producers of conventional weapons are France, UK, Germany, Italy, Spain, and Sweden.

17 In the US, only 55 out of a total of 191 producers of small arms have more than 20 employees (United States, Census Bureau, 1997a).

18 Information in this section from Brömmelhörster (2001)

19 See Mulvenon (2001) for a discussion of arms production in the PLA military enterprises.

20 Information in this section from CAST (2001a; 2001b).

21 Figure covers both military and civilian output of Izhmash. A report of the Izhmash press service, 20 January 2001 <http:www.izhmash.ru>

22 INFO-TASS database, Vega base, 18 December 2000.

23 Statement by First Deputy Director of the Russian Agency for Conventional Armaments, Valery Rychkov, 2 April 2001.

24 Information in this section from V. Ezell (2001).

25 The company was ranked by SIPRI as the world's 39th largest defence company in the OECD and developing countries in 1999 (SIPRI, p. 309).

26 The company was ranked by SIPRI as the world's 69th largest defence company in the OECD and developing countries in 1999 (SIPRI, p. 309).

27 Information on Austria's small arms producers from Lock (2001).

28 It is not clear whether these figures are merely for production activities in Austria, or for the company as a whole.

29 See Danssaert (2001) for details of Steyr's licensed production of the 7.62mm FN FAL rifle (renamed the StG-58). The blueprints and the tooling for the StG-58, together with spare parts, were sold by Steyr to the US firm DS Arms. The company currently produces the StG-58 in several versions.

30 See <http://www.steyr-mannlicher.com>

31 Information in this section from Danssaert (2001) and Mampaey (1997; 1998; 2000).

32 During 1995 the original name FN Herstal was re-installed.

33 Information on Brazil's small arms producers from Viva Rio/ISER (2001) and Omega Foundation (2001b).

34 Information in this section from Riddell (2001).

35 Information on Germany's small arms producers from Lock (2001).

36 Information on Beretta from Cattaneo (2001).

37 Information on Israel's small arms producers from Steinberg *et al.* (2001).

38 Israel (2001).

39 See <http:www.imi-israel.com>

40 It should be noted that IMI's sales of small arms to the IDF over recent years have consisted primarily of the light sub-machine gun 'Negev' and spare parts for firearms. Indeed, the majority of small arms used by the IDF are imported from the US. Some of these American-made products are American army surplus and are therefore significantly cheaper than their Israeli counterparts (Israel, 2001).

41 Information on Spain's small arms producers from Riddell (2001).

42 See <http:www.santa-barbara-sa.es>

43 General Dynamics' purchase of Santa Barbara was prompted by the company's expertise in land systems (e.g. tanks) rather than small arms and light weapons, and by its funded backlog of USD 2 billion, including a contract for the Leopard main battle tank (*Defence Systems Daily*, 26 July 2001).

44 Information on Switzerland's small arms producers from Lock (2001).

45 See <http:www.gunnery.net>

46 Information on the UK's small arms producers from Omega Foundation (2001a; 2001b).

47 The US subsidiary of Heckler & Koch was established in 1976.

48 The Omega Foundation has detailed information on all the UK's small arms producers. See also NISAT <http:www.nisat.org>

49 For a detailed account of licensed production see Abel (2000).

50 See studies by Keller (1995) and Klare and Andersen (1996).

51 For more details of these various terms see United States, Bureau of Export Administration (1988).

52 Information on FN Herstal from Danssaert (2001) and Mampaey (2000).

53 See Danssaert (2001) for a detailed description of the history of FN Herstal's Engineering division.

54 The information in this section is from SRIC (2001) and Danssaert (2001).

55 Eldoret is the tribal homeland of President Daniel arap Moi. During 1996 an international airport was built in Eldoret.

56 The information in this section is from Ebo (2001), Danssaert (2001), and Reed (2001).

57 According to Gen. Danjuma, DICON was established with the aid of Fritz Werner (Germany), Beretta (Italy), and FN Herstal (Belgium). See 'Defence Industry Corporation Reported to be in Financial Difficulty', *Radio Kudirat*, Nigeria, Voice of Democracy, 15 October 1999.

58 See <http:www.nigerianbusiness.com/nigreg/dicon.html>

59 Personal communication with Johan Jooste, South African Department of Defence, 20 June 2001.

60 Personal communication with Ettienne Hennop, 23 January 2001.
61 Personal communication with Ettienne Hennop, 23 January 2001.
62 Information on Heckler & Koch from Omega Foundation (2001c).
63 Until 1981 German legislation prohibited German companies from exporting weapons to many regions of the world, particularly conflict zones. As a result, licensed production was a way of getting around this legislative prohibition.
64 See <http.www.rofnottingham.co.uk> for more details of the closure of the RO facility in Nottingham.
65 A company in the US, Special Weapons LLC, claims to be the home of the best copies of HK weapons in the world. It is unclear whether this production of 'copies' is licensed or authorized by HK.
66 Fritz Werner won the contract in competition with FN Herstal (Belgium). See Herssens (2001).
67 Georges Forrest previously obtained Belgian licences for the construction of ammunition plants in Iran and China. See Herssens (2001).
68 The French company Manurhin received a sub-contract from Fritz Werner worth BEF 400 million (USD 9 million).
69 Information from DIO company brochure (1998).
70 DIO brochure, Defence Services Asia exhibition, Malaysia, April 2000.
71 Presentation by Georg Stevens, Bonn Bureau of Heckler & Koch, to workshop on Small Arms organised by the Swiss Government, 22–23 November 1999, Geneva.

1. Bibliography

Abel, Pete. 2000. 'Manufacturing Trends: Globalizing the Source.' In Lora Lumpe, ed. *Running Guns: The Global Black Market in Small Arms.* London: Zed Books, pp.81–104.

Austrian Defence Industry Association. 2001. *Austrian Military Suppliers.* Vienna: Austrian Defence Industry Association.

Batchelor, Peter and Susan Willett. 1998. *Disarmament and Defence Industrial Adjustment in South Africa.* Oxford: Oxford University Press.

Batchelor, Peter and Paul Dunne. 1998. 'The Restructuring of South Africa's Defence Industry.' *African Security Review.* Vol. 7, No. 6, pp. 27-43.

Battilega, John, *et al.* 2001. *Transformation in Global Defence Markets and Industries: Implications for the Future Warfare.* Washington, DC: National Intelligence Council. <http://www.odci.gov/nic/pubs/research_supported_by_nic/battilega/index.htm>

Bitzinger, Richard. 1999. 'Globalisation in the Post-Cold War Defence Industry: Challenges and Opportunities.' In Ann Markusen and Sean Costigan, eds. *Arming the Future: A Defence Industry for the 21st Century.* New York: Council on Foreign Relations.

Blakeney, Michael. 1989. *Legal Aspects of the Transfer of Technology to Developing Countries.* Oxford: ESC Publishing.

Bonn International Center for Conversion (BICC). 1998. *Conversion Survey 1998: Global Disarmament, Defence Industry Consolidation and Conversion.* Oxford: Oxford University Press.

__. 2001. *Conversion Survey 2001: Global Disarmament, Demilitarisation and Demobilisation.* Baden-Baden: Nomos Verlagsgesellschaft.

Brom, Shlomo and Yiftah Shapir, eds. 2000. *The Middle East Military Balance 1999–2000.* London: MIT Press.

Brömmelhörster, Jörn. 2001. *China's Small Arms Producers.* Background paper. Geneva: Small Arms Survey.

__. and John Frankenstein, eds. 1997. *Mixed Motives, Uncertain Outcomes–Defence Conversion in China.* Boulder, Colorado: Lynne Rienner.

Cattaneo, Silvia. 2001. *Italy's Small Arms Producers.* Background paper. Geneva: Small Arms Survey.

Centre for Analysis of Strategies and Technologies (CAST). 2001a. 'Top 20 Russian Defense Companies.' *Eksport Vooruzheniy.* Vol. 6, No. 28, pp. 51–54.

__. 2001b. *Russia's Small Arms Producers.* Background paper. Geneva: Small Arms Survey.

Chetty, Robert, ed. 2000. *Firearm Use and Distribution in South Africa.* Pretoria: National Crime Prevention Centre.

Cheung, Tai Ming. 2000. 'Storming the Capitalist Citadel: The Chinese Army's Quest for Profits In the Reform Era'. Unpublished manuscript.

Cooper, Julian. 2001. 'Russian military expenditure and arms production'. In Stockholm International Peace Research Institute. *SIPRI Yearbook 2001: Armaments, Disarmament and International Security.* Oxford: Oxford University Press, pp. 313–22.

Danssaert, Peter. 2001. *FN Herstal: Licenced production.* Background paper. Geneva: Small Arms Survey.

Diaz, Tom. 1999. *Making a Killing: The Business of Guns in America.* New York: The New Press.

Ebo, Adedeji. 2001. *Defence Industries Corporation of Nigeria.* Background paper. Geneva: Small Arms Survey.

Evron, Yosef. 1980. *The Defense Industry in Israel.* Tel Aviv: Ministry of Defense.

Extel Financial Limited. 2001. *Extel Cards Database 2001.* London: Extel Financial Limited.

Ezell, Edward. 1988. *Small Arms Today.* Second Edition. London: Arms and Armour Press.

Ezell, Virginia. 2001. *Small Arms Producers in the USA.* Background paper. Geneva: Small Arms Survey.

Feiler, Gil. 1998. 'The Military Industries of the Arab World in the 1990s.' In Efraim Inbar and Benzion Zilberfarb, eds. *The Politics and Economics of Defence Industries.* London: Frank Cass.

Feldman, Shai and Yiftah Shapir, eds. 2001. *The Middle East Military Balance 2000–2001.* London: MIT Press.

Forecast International. 2001. *Ordnance and Munitions Forecast.* Newtown, Connecticut: Forecast International. January.

__. 2002. *Ordnance and Munitions Forecast.* Newtown, Connecticut: Forecast International. January.

Frankenstein, John. 1999. 'China's Defense Industries: A New Course?' In James Mulvenon and Richard Yang, eds. *The People's Liberation Army in the Information Age.* Santa Monica, California: RAND, pp. 187–216.

Gale Research Inc. 2001. *Ward's Business Directory of US Private and Public Companies 2001.* Detroit: Gale Research.

Gander, Terry and Charles Cutshaw, eds. 2001a. *Jane's Ammunition Handbook: 2001–2002.* Coulsdon: Jane's Information Group.

__. 2001b. *Jane's Infantry Weapons: 2001–2002.* Coulsdon: Jane's Information Group.

Gill, Bates and Evan Medeiros. 2000. *Chinese Arms Exports: Policy, Players and Process.* Carlisle, Pennsylvania: Strategic Studies Institute.

Gonchar, Ksenia. 2000. *Russia's Defence Industry at the Turn of the Century.* Brief No.17. Bonn: BICC.

Graham & Whiteside Ltd. 2001. *The Major Companies Database 2001.* Leverett, Massachussetts: Rector Press.

Grimmett, Richard. 2001. 'Conventional Arms Transfers to Developing Nations 1993–2000.' *CRS Report for Congress,* 16 August.

Hart, Roger and Daniel Bereskin. 1996. *The Licensing and Commercialising of Intellectual Property.* Toronto: Bereskin and Parr. <http:www.bereskinparr.com/art-html/LicensingCommercialofIP.html>

Herssens, Hilde. 2001. *A Turkish Ammunition Plant under Belgian Licence: Will Belgium be Accessory to Human Rights Violations and Instability?* Paper produced for the Flemish Network on Small Arms, Antwerp.

__. 2002. 'Licence Agreements.' Unpublished manuscript. University of Antwerp.

Hoover's Inc. 2001. *Hoover's Company Profile Database 2001–American Private Enterprises.* Austin, Texas: Hoover's Inc.

ICC Information Group Ltd, 2001. *ICC Financial Analysis Reports 2001.* Hampton: ICC Information Group Ltd.

Institute for Research on Small Arms in International Security (IRSAIS). Various years. *Small Arms World Report.* Alexandria: IRSAIS.

International Institute for Strategic Studies (IISS). 2001. *The Military Balance 2001–2002.* London: International Institute for Strategic Studies.

Israel. 2001. *State Comptroller's Report 2000* (unofficial translation). Jerusalém.

Jane's Sentinel Security Assessment China and Northeast Asia. 2000. 'China: Defence Production and R&D'. Coulsdon: Jane's Information Group.

Jane's Sentinel Security Assessment. 2001. 'The Gulf States'. Coulsdon: Jane's Information Group.

Keller, William. 1995. *Arm in Arm: The Political Economy of the Global Arms Trade.* New York: Basic Books.

Klare, Michael. 1995. 'Light Weapons Diffusion and Global Violence in the Post-Cold War Era.' In Jasjit Singh, ed. *Light Weapons and International Security.* Delhi: Pugwash Conferences on Science and World Affairs, pp. 1–40.

__. and David Andersen. 1996. *A Scourge of Guns: The Diffusion of Small Arms and Light Weapons in Latin America.* Washington, DC: Federation of American Scientists.

Klieman, Aharon. 1998. 'Adapting to a Shrinking Market: The Israeli Case.' In Efraim Inbar and Benzion Zilberfarb, eds. *The Politics and Economics of Defence Industries.* London: Frank Cass.

Lock, Peter. 2001. *Small Arms Production in Austria, Germany and Switzerland.* Background paper. Geneva: Small Arms Survey.

Mampaey, Luc. 1997. *FN Herstal: Quel avenir pour la tradition armurière?* Les Rapports du GRIP, No. 97/2. Bruxelles: GRIP.

__. 1998. *L'Industrie belge de défense.* Les Rapports du GRIP, No.98/1. Bruxelles: GRIP.

__. 2000. *Groupe Herstal S.A: L'Heure des décisions.* Les Rapports du GRIP, No. 6. Bruxelles: GRIP.

Miller, Derek. 2001. *Small Arms Production in the Middle East.* Background paper. Geneva: Small Arms Survey.

Muchlinski, Peter. 1995. *Multinational Enterprises and the Law.* Oxford: Blackwell.

Mulvenon, James. 2001. *Soldiers of Fortune: The Rise and Fall of the Chinese Military-Business Complex,* 1978–1998. New York: M.E. Sharpe.

Omega Foundation. 2001a. *UK Small Arms Producers.* Background paper. Geneva: Small Arms Survey.

__. 2001b. *Global Distribution of Small Arms Producers.* Background paper. Geneva: Small Arms Survey.

__. 2001c. *Heckler & Koch: Licenced production.* Background paper. Geneva: Small Arms Survey.

Reed, John, ed. 2001. *Jane's World Defence Industry: Issue Nine.* Coulsdon: Jane's Information Group.

Reed Reference Publishing. 2001. *Directory of Corporate Affiliations 2001.* New Providence, New Jersey: Reed Reference Publishing.

Reiser, Stewart. 1989. *The Israeli Arms Industry.* New York: Holmes and Meir.

Riddell, Alison. 2001. *Small Arms Production in France and Spain.* Background paper. Geneva: Small Arms Survey.

Sadeh, Sharon. 2001. 'Israel's Beleaguered Defense Industry.' *Middle East Review of International Affairs.* Vol. 5, No.1. March. <http://meria.idc.ac.il/journal/2001/issue1/jv5n1a5.html>.

Security Research and Information Centre (SRIC). 2001. *Kenya Ordnance Factory Corporation.* Background paper. Geneva: Small Arms Survey.

Shichor, Yitzhak. 1998. 'Conversion and Diversion: The Politics of China's Military Industry after Mao.' In Efraim Inbar and Benzion Zilberfarb, eds. *The Politics and Economics of Defence Industries.* London: Frank Cass.

Sköns, Elisabeth, *et al.* 2001. 'Military Expenditure and Arms Production.' In Stockholm International Peace Research Institute. *SIPRI Yearbook 2001: Armaments, Disarmament and International Security.* Oxford: Oxford University Press, pp. 223–322.

Small Arms Survey. 2001. *Small Arms Survey 2001: Profiling the Problem.* Oxford: Oxford University Press.

Standard & Poor's Corporation. 2001. *Standard & Poor's Register of Companies 2001.* New York: Standard & Poor's Corporation.

Steinberg, Gerald. 1983. 'Israel.' In Nicole Ball and Milton Leitenberg, eds. *The Structure of the Defense Industry: An International Survey.* London: Croom Helm.

__. 1986. 'Israel: High Technology Roulette.' In Michael Brzoska and Thomas Ohlson, eds. *Arms Production in the Third World.* London: Taylor and Francis.

__. *et al.* 2001. *Small Arms Production in Israel and the Palestinian Authority.* Background paper. Geneva: Small Arms Survey.

Stevens, Blake and Jean van Rutten. 1981. *The Metric FAL-the Free World's Right Arm.* Cobourg, Ontario: Collector Grade Publications.

Stockholm International Peace Research Institute (SIPRI). 2001. *SIPRI Yearbook 2001: Armaments, Disarmament and International Security.* Oxford: Oxford University Press.

Stohl, Rachel. 1998. *Deadly Rounds: Ammunition and Armed Conflict.* Washington DC: British American Security Information Council. May.

United Nations. 1999. *United Nations International Study on Firearm Regulation.* New York: United Nations.

United Nations Conference on Trade and Development (UNCTAD). 1983. *United Nations Conference on an International Code of Conduct on the Transfer of Technology.* Geneva: UNCTAD. <http:www.unctad.org>

__. General Assembly (UNGA). 1997. *Report of the Panel of Governmental Experts on Small Arms.* A/52/298 of 27 August.

United States. Bureau of Alcohol, Tobacco and Firearms (BAFT). 1995. *Annual Firearms Manufacturing and Export Report.* Washington DC: Department of the Treasury.

__. 1998. *Annual Firearms Manufacturing and Export Report.* Washington, DC: Department of the Treasury.

__. 1999. *Annual Firearms Manufacturing and Export Report.* Washington, DC: Department of the Treasury.

__. 2000. *Commerce in Firearms in the United States.* Washington DC: Department of the Treasury.

United States. Bureau of Export Administration. 1988. *Offsets in Military Exports.* Washington, DC: Department of Commerce.

United States. Census Bureau. 1997a. *Small Arms Manufacturing–1997 Economic Census.* Washington, DC: Department of Commerce.

__. 1997b. *Small Arms Ammunition Manufacturing–1997 Economic Census.* Washington, DC: Department of Commerce.

Violence Policy Center. 2000. *Firearms Production in America: 1975–1997. A State-by-State Breakdown.* Washington, DC: Violence Policy Center.

__. 2001. *Firearms Production in America: 2000 Edition.* Washington, DC: Violence Policy Center.

Viva Rio/Institute for Religious Studies (ISER). 2001. *Brazil's Small Arms Producers.* Background paper. Geneva: Small Arms Survey.

World Forum on the Future of Sport Shooting Activities (WFSA). 2001. *The Facts.* Rome: WFSA.

World Intellectual Property Organisation (WIPO). 1977. *Licensing Guide for Developing Countries.* Geneva: WIPO. <http:www.wipo.int>

Acknowledgements

Main contributors

Jörn Brömmelhörster, Silvia Cattaneo, Centre for Analysis of Strategies and Technologies (CAST), Peter Danssaert, Adedeji Ebo, Virginia Ezell, Peter Lock, Derek Miller, Omega Foundation, Alison Riddell, Gerald Steinberg, Security Research and Information Centre (SRIC), and Viva Rio/Institute for Religious Studies (ISER).

Other contributors

Aharon Etengoff, Hilde Herssens, Sharon Hurwitz, Shlomit Stern, and Michal Vahab.

Red Flags and Buicks:
Global Firearm Stockpiles

2

Introduction

When Chinese police announced that they had confiscated some 600,000 illegal firearms in the spring of 2001, they probably sought to demonstrate their ability to deal with one of the country's growing social problems (*The Economist*, 2001c). They may not have appreciated the window they were opening onto one of the most mysterious aspects of the global distribution of small arms. As a result of these revelations and others like it elsewhere, the picture of global small arms ownership gradually is becoming clearer.

The size of the global small arms stockpile is large enough to guarantee that the issue of small arms proliferation will continue to gain in political importance in the coming years. Since the 2001 UN Small Arms Conference, global awareness of the importance of small arms issues has grown and many pressing questions about global stockpiles are now being sufficiently answered to activate concrete policy responses. Despite the fact that some important details are still lacking, there is widespread agreement on the scope of the problem, on the scale of the global stockpile, and on which aspects are of the greatest concern.

This chapter broadly outlines the size of the global small arms stockpile, where the world's small arms are, and who controls them. It also illuminates trends of greatest concern. Key findings of this chapter include:

- The global small arms stockpile is at least 16 per cent greater than previously estimated, with over 638,900,000 firearms.
- Although far less numerous, shoulder-fired rocket launchers and mortars merit careful consideration as well.
- China and South Asia are major centres for small arms, home to tens of millions of firearms each.
- Insurgent arsenals are growing fastest in Colombia, the Palestinian territories, Central Asia, and parts of Indonesia. They appear to be stable or diminishing elsewhere.
- Small arms stockpiles change in size and composition primarily though the re-circulation of old weapons rather than new production.
- Destruction programmes have reduced global inventories by over four million small arms in the last decade.

The global small arms stockpile is at least 16 per cent greater than previously estimated, with 639 million firearms alone.

The assessment in this chapter reveals in most cases that the global stockpile is larger than was estimated in the *Small Arms Survey 2001*. In a few situations, however, the assessment identified decreases. Most of these changes are due to a greater awareness about small arms in a growing number of countries. The actual increase from new production is a small—but not insignificant—part of this new estimate of the global stockpile (PRODUCERS). Most of the increases revealed in this chapter can also be attributed to better information and research. In fact it is likely that the total number of small arms in the world has actually grown at a slower rate in the last year, than in previous years. Further adjustments, moreover, will be inevitable. Several major countries, for example, remain virtual blank spots on the map of global small arms stockpiles.

Confiscated firearms, Liberia. (© Panos Pictures/Heldur Netocny)

Caution also is in order. Widely cited in the news media, the *Small Arms Survey 2001* estimate of the total global firearms stockpile has joined the list of key numbers used to illustrate serious issues of public policy. Comparable statistics are increasingly common. Examples include the figure of 300,000 child soldiers around the globe promoted by the Coalition to Stop the Use of Child Soldiers, 22 million refugees according to the UN High Commissioner for Refugees, and the estimate of 230 million to 245 million landmines stockpiled worldwide from the International Campaign to Ban Landmines (UNHCR, 2000; International Campaign to Ban Landmines, 2001). Like these, the global firearms estimate is as much a symbol of the small arms issue as a representation of it.

Like any overall number it is subject to misuse. As Joel Best points outs in *Damned Lies and Statistics* (2001), numbers come with agendas. Census figures, for example, matter not just as measures of population but also as a basis for political arguments over sharing resources. Crime statistics often reveal more about those who use them than they do about crime itself. Analysts of some of the statistics mentioned above are concerned about these issues as well (Bottigliero, 2000). This chapter strives to help minimize the biases typical in political numbers.

Above all, this chapter stresses the need for greater transparency and more careful consideration of small arms data. With these goals in mind, it is important that others assess the scale of the small arms problem as well. The military procurement analyst Gregory Fetter (2001c) has arrived at an alternative figure for the total number of firearms in the world today of approximately 594,000,000.[1] Of equal importance, he calls attention to the definitional problems of small arms and light arms, of deciding what to include and exclude. This is an issue that was deliberately avoided at the 2001 UN Small Arms Conference (CONFERENCE) and continues to pose obstacles to consensus in international law enforcement and small arms policy.

The concerns addressed in this chapter include:

- **How many small arms and light weapons are there in the world?**
- **How does the number of firearms compare with that of various types of light weapons?**
- **Who controls more of the world's small armaments: the military, police, or civilians?**
- **Which regions are experiencing the greatest expansion of weapons stocks?**

For practical reasons, the chapter focuses mostly on firearms, from handguns to heavy machine guns; these are the most numerous small arms and those for which data is most available. They also cause the most destruction, making them the target of most international attention. Changing patterns of distribution are revealed not only through raw numbers but also through price, a new key indicator used here. Price can be a red flag, warning when small arms dangers grow especially acute. The chapter also examines global inventories of two less commonly considered types of small arms: shoulder-fired anti-tank rockets and light mortars. The latter include weapons improvised from readily available scrap, such as car parts.

The second section of the chapter builds upon the approach pioneered in the *Small Arms Survey 2001*, surveying the major categories of small arms owners, from the police and armed forces to private civilians. It shows how important changes in military doctrines and insurgent warfare are changing not only the global stockpile but its distribution as well. Finally, the third section reviews the regions of the world where there were important changes in stockpiles or our understanding of those stockpiles. It estimates for the first time the approximate size and distribution of small arms stockpiles in South Asia and China.

New approaches to small arms and light weapons

Contemporary small arms policy is governed by a general sense that the quantity of guns in the world is becoming an independent force in global affairs, surpassing the influence of existing regulations. While this undoubtedly is the case, it represents a simplification of the small arms issue. Factors other than raw numbers of weapons also can reveal critical aspects of the evolving challenge to policy-making.

As shown here, *price* can be an equally insightful indicator of the nature of small arms problems. The preoccupation with growing firearms inventories, while certainly justified, is an oversimplification. It inadvertently draws attention away from other light weapons and issues of serious concern. To help correct this imbalance, the basic issues surrounding the stocks of *shoulder-fired anti-tank rocket launchers* and *light mortars* are introduced in this chapter. Although the growth of the global inventory of small arms has become a steady trend, it is somewhat misleading. As shown in the section on weapons destruction, stockpiles do go down, especially when major organizations—typically the police or armed services—decide to destroy surplus or excess equipment.

The price of small arms: A key indicator

In parts of the world where there are no figures on the number of small arms, the best guide to their availability often is price. Prices convey a general sense about demand and supply, demonstrating whether or not they can be easily obtained. Especially in situations where firearms are illegal, prices may be the only way of establishing whether they are commonplace or scarce, whether smuggling is effective or embargoes are working.

Prices are no panacea for overcoming the lack of information about quantities. While inventory data is subject to inaccuracies, modest fluctuations in supply can create big changes in price. Black markets for small arms appear to be exceptionally sensitive to changing political conditions; weapons worth thousands of dollars each in the midst of fighting can be virtually worthless the day peace breaks out. The resumption of hostilities may drive prices up just as quickly.

Especially where firearms are illegal, prices may be the only way of establishing their prevalence, whether smuggling is effective or embargoes are working.

Another basic problem is that every firearms market, especially every illegal market, is unique; there does not appear to be a universal relationship between prices and supply. Each situation has to be judged on its own terms. An AK-47 may be worth USD 2,000 in both the United States and Kashmir. In both cases legal barriers on importation are the basic cause of exaggerated prices. But in reality the situations could not be more different. The American market is dominated by collectors, the Kashmiri market by secessionist rebels; American buyers split hairs over makes and models, the Kashmiri trade feeds a vicious civil war. The meaning of seemingly identical prices also varies. The cost of a Kalashnikov is not an insurmountable problem for typical buyers in the United States, but identical prices pose much greater obstacles for most Kashmiris. Taste can affect price too, as demonstrated by the cost of AK-47s in Columbia compared to M16s (see Tables 2.1 and 2.2).

Information about prices is best in legal markets, where the relationship between supply and demand is easily established. Prospective buyers need only turn to the Internet or trade publications. Producers and dealers can gauge market trends and regulate supply to minimize price fluctuations. The real value of prices for insight into stockpiles is in black-market situations, especially where insurgencies and rebellions create strong demand. The study of black-market gun prices is too primitive to reveal the number of guns available on any particular market, but it can inform observers about trends in supply and demand.

Black-market prices also illustrate how the trade in second-hand equipment affects the global distribution of firearms. Newly manufactured small arms are a relatively minor factor in the overall global stockpile, contributing roughly one per cent of the global stockpile each year (PRODUCERS). Since most small arms last a long time, second-hand weapons dominate the trade. With important exceptions, it is not so much transfers of newly produced weapons as the re-circulation of second-hand equipment that cause national stockpiles to rise and fall most. The exceptions where newly produced firearms predominate include civilian and police markets in wealthier, firearms-producing countries like the United States, western Europe, parts of Brazil, South Africa, and the Middle East, and military markets for sovereign states. Elsewhere, civilian buyers tend to receive weapons that have been previously owned. Insurgencies as well appear to be heavily dependent on access to second-hand weapons. Often such weapons actually are unused, taken from long-term storage in military arsenals. But they cannot be regarded as new production.

Transfers of second-hand small arms have a major effect on the distribution of firearms, moving them from regions of high concentration to regions of sparser availability. High prices play an instrumental role in this process, attracting brokers and persuading others to disregard the legal and personal risks of selling weapons in regions of conflict.

Table 2.1 lists a sample of black-market AK-47 prices from the Small Arms Survey database. The AK-47 is especially revealing because of its ubiquity, with an estimated 70 million to 100 million in existence, and the large number of price reports. For comparison, Table 2.2 lists prices for black-market sales of the M16, which is less common, with some seven million in circulation (Small Arms Survey, 2001, p. 20). Price disparities and more rigorous export licensing by the United States and its clients largely explain the fact that reports of AK-47 sales are much more numerous, by a factor of roughly ten to one. These tables allow only general comparisons; although quantity is a major factor in pricing, for example, it is not possible to be certain that all prices listed here are for sales of a single weapon. Bulk purchases must be evaluated separately: an AK-47 which normally might cost USD 120 to 250 purchased singly can cost as little as USD 40 when bought in quantities of several thousand, even on the black market (Tamayo, 2000). The specific model, manufacturer, and condition of the weapons involved—potentially with major effects on price—usually are not known either.

TABLE 2.1 Examples of black-market prices for a single, used AK-47 rifle (various models)

Price (current USD)	Location	Year
10	Afghanistan	2001
12	Angola-Namibia border	1998
15	Mozambique	1999
25	Honduras	1997
40	Phnom Pen, Cambodia	2001
86	Uganda-Sudan border	2001
100	Nicaragua	2001
100	Warri, Nigeria	2001
120	Somalia	2000
250	Sakhkot, Pakistan	2001
400	Siberia, Russia	1998
800	Colombia	2001
1,200	Bangladesh	2000
2,400	Indian Kashmir	2000
3,000	Colombia	2000
3,000	West Bank, Palestinian territories	1999
3,800	Bihar, India	2001

Source: Press reports in the Small Arms Survey database, available on request

The most remarkable revelation here is the volatility of prices. Each price level is associated with a general set of market conditions. Since most of these prices come from journalists' reports and often represent only specific transactions or hearsay, their individual reliability is low. At the extremes, moreover, some exaggeration must be discounted. Tales of automatic rifles being traded for a chicken in Africa are a gun-trade favourite, despite their obvious absurdity. But, as a set, these prices clearly reveal major tendencies in the market.

There does seem to be a real bottom price for used, factory-made AK-47s, generally of USD 35 to 40 each. This price is found in many regions when peace arrives suddenly. In conflict situations, strong demand attracts ever-larger supplies, and with the end of fighting there simply is no demand for the weapons. This cannot be sustained. Over the ensuing months these weapons will disappear, whether bought by the formerly warring sides or exported, and prices will tend to go up somewhat.

TABLE 2.2 Black-market prices for a single, used M16 rifle (various models)

Price (current USD)	Location	Year
100	Somalia	2001
700	Moluccas, Indonesia	2001
1,100	Solomon Islands	2001
2,000	São Paulo, Brazil	2000
3,000	Colombia	2001
10,000	Palestinian territories	2001

Note: for comparison, the US Army currently pays approximately USD 387 for each new M16A4 rifle when purchased in quantity.
Source: Press reports in the Small Arms Survey database, available on request

Slightly higher prices are typical of the large-scale, state-to-state trade in second-hand models. There is a clear hierarchy in the international market for AK-47s and lookalike versions, with the Russians able to ask for the highest prices, followed by central and east European manufacturers. Chinese exporters usually charge the least of all major producers, except when North Korean versions are present. Quality seems to be comparable in all cases—this is a product of 1940s vintage, after all—so the differences are due entirely to differences in image. The only exceptions are locally made weapons from garage workshops like Pakistan's infamous Darra market, where low-quality versions typically sell new for about USD 65 each (*The Hindu*, 2001).

Legal civilian markets for individual weapons normally appear to sustain prices of USD 230 to 400 each. This is what American buyers pay for slightly modified versions only capable of semi-automatic fire, although collectable models can go for considerably more. Nothing drives up prices beyond this level like fighting. As a general rule, when black-market prices for individual small arms rise beyond the widely established prices of legal markets, social order is in jeopardy. Extremely high prices—a sign of desperation—are evidence of active fighting or organized combat. Prices for AK-47s above USD 1,000 each are a red flag, a warning sign that, market peculiarities aside, demand is so strong that normal economic considerations have been abandoned. Prices this high indicate two important aspects of the market. First, they show that border controls are having a strong effect; weapons are not getting through as quickly as prospective buyers would like. Second, they are proof that demand cannot be easily satisfied. In a better market, buyers would snap up significantly more.

Extremely high prices—a sign of desperation—are evidence of active fighting or organized combat.

Not all price increases are equally dangerous. In the United States, new Chinese-made AK-47s sold for about USD 100 each in the late 1980s. This was due to a glut on the market created when merchants overestimated the scale of public demand and over-ordered (Smothers, 1989). After the administration of President George H. Bush began to discuss banning imports in early 1989, American prices rose to roughly USD 300 each (Mathews, 1989). Although a ban was not enacted until 1994, the threat alone was enough to sustain the high prices, to the benefit of importers who sold them in unprecedented numbers anyway. Several Internet sites list second-hand 'pre-ban' fully automatic rifles selling in the United States for between USD 400 and USD 2,000. Collectable models can cost much more. This situation is highly exceptional; in most of the world prices that high are associated with black markets or active armed conflict.

Very high small arms prices should be a serious warning for outside observers. They are characteristic of situations—often very bad already—with the potential to rapidly degenerate. Although Indian authorities may believe they have stabilized the situation in Kashmir, exceptionally high gun prices there suggest otherwise. The Kashmiri market is waiting, in a sense, for a clever broker to find a new supply route. Similarly, high prices in the Palestinian territories show that there are many Palestinians without automatic rifles who desperately want them. Although less extreme, the high prices paid by guerrillas in Colombia indicate that they too could expand their forces and conduct more aggressive operations if small arms supplies were more generous.

Spiking small arms prices are a red flag, warning of a situation on the verge of degeneration.

Especially serious are rapid increases in small arms prices. Spiking prices should be a red flag to policy-makers. They announce a situation on the verge of deadly deterioration. Rapid price increases or sustained high prices, the typical result of strong demand and constrained supply, indicate that buyers are aggressively searching for sellers. Fighting may be about to start or may be set to become much worse. Authorities will be under grave pressure to prevent smuggling, aware that, if the conflict does not abate soon, the temptation of extremely high prices means that sources of supply almost invariably will emerge.

In the same way, very low prices can be a good thing, meaning that no one wants readily available guns. Low prices, created by high supply and low demand, can be linked to the end of a conflict. This situation is not without its dangers. Low prices enable less motivated gunmen—especially petty criminals—to step into greater firepower. It is through low prices that cheap thugs replace expensive guerrillas. The other danger of low prices is for outside regions. The prospect of easily stocking up attracts brokers interested in capitalizing on peace in one area to sell at higher prices in less fortunate areas, shifting the cycle of prices and violence from one region to another.

BOX 2.1 Up and down: Supply and demand in Albania and Kosovo

The images of Albanian weapons being pilfered as government offices and arsenals were ransacked in March 1997 are among the strongest images of state collapse and anarchy. In a span of days virtually the entire military arsenal of the small, impoverished country was looted. Most of the larger items, like armoured vehicles and artillery, were quickly recovered as the government restored order. But left unaccounted for were the vast majority of the country's official inventory of 643,000 firearms, 1.5 billion rounds of ammunition, and 3.5 million hand grenades, which were never recovered (United Nations, n.d.).

The effect on weapons prices was entirely predictable at first. Automatic rifles, which had not been legally traded by civilians previously and normally would have been in high demand, were worth only USD 35–40 one month later (O'Connor, 1997). But low prices invited the attention of brokers, as well as criminals and rebels. For a brief period, Albania became a major export centre in the international small arms business.

Another effect of the sudden availability of small arms was felt in neighbouring Kosovo, helping to transform the Kosovo Liberation Army (KLA) from a small rebel band into a much larger force ready to provoke the ruling Serbs. Founded in 1992, the KLA had been irrelevant for years. But with weapons pouring in it began to attract greater support, growing more confident and assertive (Wagstyl, 1999). Its first raids against Serbian police and military facilities occurred in the summer of 1997, a few months after the arsenal looting the previous March.

BOX 2.1 Up and down: Supply and demand in Albania and Kosovo (continued)

Strong demand in neighbouring regions rapidly exhausted even this large supply. Predictably, prices began to rise (see Table 2.3). One year later AK-47s were trading in Albania for some USD 300 each, considerably more than their normal market value and the gun surplus evaporated (Sagramoso, 2000/01). As the conflict in neighbouring Kosovo intensified, the price of weapons in Albania continued to rise. By the late summer of 1998, Kalashnikov rifles were trading there for USD 650 (Smith and Sagramoso, 1999).

At this point the market underwent a fundamental transformation as traders reversed the direction of the weapons flow and began to import weapons from elsewhere. For some suppliers political motives may have been primary, for others economic incentives explain everything. Many of the new weapons appeared to come through Croatia (*The Economist*, 1999). The guns released in the March crisis, in other words, helped ignite the Kosovo crisis, but were not sufficient to sustain it. With much of the funding coming from ethnic Albanians in Germany, Switzerland, and the United States, the KLA was attracting completely new supplies (Sagramoso, 2000/01).

When the Kosovo fighting ended in June 1999, the weapons path took a new twist. Although NATO commander Wesley Clark had indicated that the peacekeepers would try to fulfil the promise made at Rambouillet and disarm the KLA, without the KLA's co-operation this was never practical (Rhode, 1999). In all, over 9,000 small arms were collected from Kosovo, but reserve stocks in Albania were not touched (*Jane's Intelligence Review*, 2000). Where voluntary collection and destruction failed, market forces began to have an effect again, creating a situation in which the flow of weapons has see-sawed back and forth as political tensions rise and fall.

© Associated Press/Dimitri Messinis

An Albanian man and his son carry rifles and ammunition they have looted on 10 March 1997.

Table 2.3 Black-market AK-47 prices in Albania, 1997–1999

Period	Price (then-year USD)
Spring 1997	40
Winter 1997–98	150
Spring 1998	300
Autumn 1998	650
Autumn 1999	125

Source: Press reports in the Small Arms Survey database, available on request

With Serbia defeated and its troops withdrawn, in late 1999, Kosovo was awash with unneeded firearms. Most former soldiers appear to have kept their weapons, but some would appear with the Albanian National Liberation Army in Macedonia a little more than one year later. Others showed up in regional markets. Weapons from Kosovo, available and cheap, began to flow into the arsenals of other groups like the Real IRA in Northern Ireland (*The Economist*, 2001a). Once fighting resumed, this time in Macedonia in early 2001, the flow reversed and weapons began to flow into the region again (B92 Radio News, 2001).

The government of Albania, which had been virtually stripped of small arms, was rearmed by Western governments through NATO's Partnership for Peace. As if to make the ironies of the situation complete, allegations have since been raised charging that some of these weapons have been sold illegally by the Logistics Department of the Public Order Ministry (Feci and Hoxhaj, 2001).

BOX 2.2 Weapon prices and fashion

Although the laws of supply and demand heavily influence the global flow of small arms, it would be a mistake to think that these two factors are based on security concerns alone. In many regions, consumer taste is no less important.

In much of the Middle East and Central Asia, for example, the AK-47 is the individual weapon of choice, so much so that throughout much of this part of the world the term 'Kalashnikov culture' is immediately understood (Cohn, 2001). Buyers often will pay a premium for the automatic rifle that symbolizes rebellion in much of the world and masculine responses to social chaos elsewhere. As a result it has become a weirdly normal postmodern fashion accessory.

In Somalia, for example, a Russian-made AK-47 costs USD 250, while an American M16 left from before the country's government collapsed trades for just USD 100. 'There aren't as many American guns here', reports a trader. 'We don't think they're as good as the Russian weapons' (Coker, 2001). A reputation for reliability is important, but—more to the point—in these regions it simply would look odd for a man to be seen carrying anything other than a Kalashnikov.

If all things were equal, all automatic rifles would cost roughly the same. In Colombia this was the case in the late 1990s, where an AK-47 and an M16 both cost roughly USD 3,000 (Policia Nacional, 2001). Similarly, in the illegal markets of Brazil today, either weapon typically sells for USD 2,000. The same is true in Indonesia's Molucca Islands, where soldiers and police currently sell either weapon to rebels for USD 700 each (*O Estado de Sao Paulo*, 2000).

When choice is available, though, even guerrillas appear to be vulnerable to fashion. When Peru's intelligence chief Vladimiro Montesinos transferred 10,000 AK-47s to guerrillas of the Fuerzas Armadas Revolucionarias de Colombia (FARC), local black-market prices for AK-47s fell from USD 3,000 before to USD 600 soon after (Otero, 2000). The price of M16s, on the other hand, held steady at roughly USD 3,000 each (Policia Nacional, 2001). Apparently Colombian guerrillas are picky customers, and will pay a steep premium for their favourite.

Images of radical defiance also explain why scarce M16s feature prominently in Palestinian martial displays, where they are preferred to readily available Kalashnikovs. The latter are the standard firearms of sympathetic Islamic states, many of which are believed to assist smuggling. Suppliers also prefer them because their ubiquity makes them harder to trace when apprehended. More readily available, although still considerably scarce compared with Palestinian demand, Kalashnikovs are traded there for some USD 3,000 each. M16s are much less common. They come almost exclusively through raids on Israel. Their dangerous origins compound the inevitable appeal of using an enemy's weapon against him. Not surprisingly, their price can escalate beyond USD 10,000.

Even guerrillas appear to be vulnerable to pressures of fashion.

Light weapons: Shoulder-fired rocket launchers and mortars

The *shoulder-fired anti-tank rocket launcher* was invented, more or less simultaneously, by all the major powers in the Second World War as a portable anti-tank weapon. Simple designs like the American Bazooka, the British PIAT, and German Panzerfaust were highly effective against the thinly armoured vehicles of their day. The concept of an unguided, fin-stabilized rocket as an inexpensive anti-tank weapon had enormous military appeal and designs were continuously updated throughout the Cold War. Several types remain in production today. Although their relevance varies from country to country, they are in use with virtually all government armed forces and many insurgencies.

All of these weapons share similar strengths and weaknesses. Simple rocket launchers are unguided after launch. Most do not carry large enough warheads to be highly effective against contemporary armoured vehicles, especially main battle tanks. The anti-tank mission has shifted to the larger and more costly, fully guided systems that appeared in the 1960s. Yet unguided shoulder-fired weapons still remain highly effective against a wide range of softer targets, especially field fortifications, bunkers, and trucks.

Because it is seen as an offensive infantry weapon, not all armed forces have a doctrinal place for it. Most NATO armies, with their emphasis on fast mechanized operations, rarely stressed small groups of soldiers hunting targets of opportunity and never gave it much prominence. It was among Soviet-style forces and ex-Soviet clients that rocket launchers came into their own. Today, however, the rocket launcher has emerged as an irregular soldier's most powerful weapon. In Somalia, in October 1993, it was used to bring down American helicopters (Bowden, 1998). In Chechnya it is primarily an offensive weapon against ground positions (Thomas and Grau, 2000).

There are many versions of these weapons, as shown in Table 2.4. The image that usually comes to mind is of a Soviet RPG-7. This was introduced for service in the Red Army in 1961, was widely used in the Vietnam War, and has been used in virtually every major armed conflict since. In service with virtually every country ever trained in Soviet tactical doctrine, versions of the RPG-7 have been manufactured in Bulgaria, China, Egypt, Iran, Iraq, Pakistan, Romania, and Thailand. The cost of the weapon varies. In Chechnya, according to one source, 'The price is supposed to be $800 [for the launcher], but our men get it from the Russian soldiers for 20 bottles of vodka' (Safire, 2000). Other sources quote figures of USD 720–1,221 for the launcher, and some USD 14 for each rocket (Fetter, 2000). For evidence of its value on the battlefield, though, consider the Colombian FARC, willing to pay USD 12,000 for a single RPG-7 launcher (Policia Nacional, 2001).

© US Dept of Defense/Gus Garcia

An Ecuadorian soldier training with an M72 LAW rocket launcher.

TABLE 2.4 Examples of shoulder-fired anti-armour rockets

Type	Operational	Unit cost (USD)	Quantity made through 1999	Source	Manufacturer
APILAS	1982	1,227	124,784	France	SERAT
C-90	1985	3,122	238,728	Spain	Instalaza
M72 LAW	1963	242	6,926,483	United States	Talley
Panzerfaust-3	1987	7,065	254,774	Germany	Dynamit Nobel
RPG-7	1961	800	9,060,487	Russia, China, many others	Various
RPG-18/22	1976	783	3,044,062	Russia, Bulgaria	Various
RPG-75	1975	2,165	432,722	Czech Republic	Zeveta
Other models			1,983,941	Various	Various
Total made*			**22,065,981**		

Notes: This table presents weapons that are not strictly comparable. For example, the M72 LAW and RPG-18/22 production figures refer to rockets, since the launcher is disposable. The RPG-7 figure refers exclusively to the launchers, which are designed for reuse.

**Not all of these weapons remain in existence; unknown numbers have been expended in training and operations.*

Source: Fetter (2000)

The American equivalent is the M72 LAW (Light Armor Weapon). Easily recognized by its thin green canister, this was developed in the early 1960s as a completely disposable infantry weapon. Lighter and smaller than its Soviet counterpart, the M72 also is one of the few Western weapons actually cheaper as well, costing USD 242 each in the mid-1980s (Fetter, 2000). Various versions have been adopted by at least 34 countries. Many comparable weapons exist as well, including Russian copies like the RPG-18. In all, references list more than 20 models currently in production (Gander, 2000, pp. 374–426).

In addition, large numbers of much older rocket launchers are held in reserve. The American M20 Bazooka, a weapon from the 1950s, is obsolete in the United States, but is held in reserve elsewhere. Manufactured in China as the Type-51, it may be held in reserve there as well. Similarly, the old Soviet RPG-2 appears to be in storage in Russia, China, and many other countries. At a minimum, hundreds of thousands of these weapons still exist.

The global quantity of unguided anti-tank rocket launchers offers a revealing contrast with firearms, which outnumber them by almost 30 to one. The reasons for the overwhelming numbers of firearms are not surprising—firearms

For guerrillas, less concerned with tactical accuracy than political impact, the mortar is an ideal light weapon. The mortar is an easy weapon to improvise from scrap.

are reusable and have a vast civilian market. The military usually requires more firearms, since most soldiers and many members of the other services are issued rifles and side-arms, but distribution of specialized weapons like rocket launchers is more restricted. Yet as experiences as diverse as the Vietnam War, Somalia, and Chechnya have shown, they deserve much greater attention. Already, they have begun to appear not just among insurgencies but also in criminal hands (Mallet, 2001).

Mortars have even older origins, dating directly to the Stokes Trench Mortar invented during the First World War. The devices have changed little in subsequent years, mostly through addition of better sighting equipment. The mortar filled a niche similar to that of the rocket launcher, giving unsupported infantry 'pocket artillery', a way to lay down high-explosive suppressive fire when beyond the reach of artillery and air support. Mortars probably reached their apogee in the 1950s and 1960s, when Soviet military doctrine ensured their widespread dissemination among Soviet satellites and clients. Since then the mortar has lost popularity, especially among Western militaries, which retain it mostly for reconnaissance units and special operations forces who might find themselves in tricky situations with no one to call on for help.

In western Europe and the United States the light mortars are in decline; there is less need for their inaccurate fire on the high-tempo, casualty-adverse battlefield. The United States closed the last of its production lines in the early 1980s. Several European countries keep production lines, apparently in hope of export orders to developing countries, but the latter are usually supplied out of stock. According to Fetter (2001a), the total global stockpile of mortars is 781,894.

Insurgents long ago discovered the advantages of the mortar. For guerrillas, less concerned with tactical accuracy than political impact, the mortar is ideal. As they demonstrated long ago, it also is an easy light weapon to improvise from scrap. This was a point that Graham Greene hammered home in his classic Vietnam war novel, *The Quiet American*, in which he observed that 'Buicks make the best mortars' (Greene, 1955). Contemporary insurgencies like the Colombian FARC, the Irish Republican Army (IRA), and the Sri Lankan Liberation Tigers of Tamil Eelam (LTTE) have built mortars of their own contrivance. More recently Israel has been forced to contend with mortars being fired from Gaza, among them home-made varieties (Davies, 2001; PRODUCERS).

Adding up completely different types of small arms and light weapons, however tempting, is a deceptive and meaningless exercise. Experts regard as seriously misleading the assumption that pistols and machine guns, to take one example, are in any sense equivalent. Greater insight comes from comparing different types of small arms and light weapons, as in Table 2.5. Beyond the enormous preponderance of firearms in the global arsenal, it shows the relative quantities of the weapons discussed here.

Indian Army mortar fired against Pakistani armed forces in Kashmir on 6 September 2001.

TABLE 2.5 Comparing global inventories of major small arms and light weapons

	All firearms	Military firearms	Shoulder-fired rocket launchers	Mortars <90mm
Raw number	638,900,000	241,600,000	22,065,981	781,894
Comparative scale	1,000	377	34	1

Sources: Tables 2.4, 2.8, and 2.21; Fetter (2001a)

The impact of destruction programmes on global stockpiles

Destruction of small arms is usually associated with weapons collections programmes, typically based on voluntary turn-in or buy-back projects, encouraging individual owners to give up unwanted weapons. The focus on weapons collection distracts attention from some of the largest and most successful destruction operations, which have been undertaken by government agencies. Typically these agencies act to serve their own needs, such as reducing storage costs, liberating space, or reducing the risk of theft. In absolute terms, though, the quantities involved in organizational destruction can be much greater than individually oriented or voluntary collection programmes (WEAPONS COLLECTION).

While none but the very largest small arms destruction programmes is likely to reduce the total numbers of weapons available in a whole society by more than a few percentage points, they can significantly reduce the stockpiles of specific actors, in some cases enough to remove much of their ability to transfer small arms in large quantities. Especially when organized within an institutional setting, the effect of destruction programmes on global stockpiles becomes readily tangible. The elimination of large numbers of military rifles, for example, has a direct effect on the ability of the armed services to rapidly expand or to quickly supply foreign clients.

The focus on weapons collection draws attention away from some of the largest, most successful destruction operations undertaken by government agencies.

Table 2.6 lists the largest known small arms destruction programmes since 1990. The list includes only programmes that culminated in physical destruction of guns, as opposed to those that took civilian guns and transferred them to the police or armed forces, or released them from the armed forces to civilians or to foreign governments. Most of the larger programmes were relatively straightforward decisions within organizations. The most important exceptions were the very largest public collection and destruction programmes. Typically public small arms collection programmes attract no more than a few thousand weapons. This should come as no surprise.

Because they involve vast numbers of individual decision-makers—almost as many as there are weapons—public-oriented undertakings have to deal with much more complicated motives. They require much more organizational effort and usually must offer some form of compensation, such as money, which makes them much more costly as well. The most productive—in Australia and Britain—also were very special. They happened through legal reform in peaceful societies. Coming at a moment of moral revulsion against particular gun crimes, they also encouraged citizens to hand in their newly illegal firearms in exchange for monetary compensation.

The major small arms destruction programmes listed in Table 2.6 eliminated a total of over four million small arms during the last decade. This is equivalent to roughly 0.5 per cent of all the firearms in the global stockpile, or roughly two-thirds of one year's new production (PRODUCERS). Some of those weapons destroyed have since been replaced: roughly eight to ten per cent in one example (Faltas, McDonald, and Waszink, 2001, p. 4). While the impact on the global stockpile is not great, the impact on specific countries and organizations has been more palpable. Through destruction programmes, Australia eliminated just under 20 per cent of all firearms in the hands of its citizens. The destruction of ageing rifles by the US Army cut its total rifle inventory almost in half.

TABLE 2.6 Recent major small arms destruction programmes

Country	Organization	Weapons type	Year	Quantity
Organizational programmes				
Brazil	Rio police	Small arms	2001	100,000 *
Germany	Federal government	Firearms	1990s	500,000 *
Netherlands	Ministry of Defence	Firearms	Late 1990s	100,000 *
Russia	Ministry of Defence	Small arms	1998–2001	421,021
South Africa	SA National Defence Force	Small arms	1998–2001	262,667
South Africa	SA Police Service	Small arms	2000–2001	52,000 *
United States	Army	M-14 rifles	1993–94	479,367
United States	Army	M-16 rifles	1996	350,000 *
Individually oriented programmes				
Australia	General public	Semi-/Fully automatic long guns	1997–98	643,726
Nicaragua	Ex-guerrillas and general public	Small arms	1991–93	142,000 *
United Kingdom	General public	Firearms	1997–99	185,000 *
Others		Small arms	1990–2001	765,000 *
Total				**4,000,781**

* *Estimated*

Sources: Brazil: Galeria (2001); Faltas, McDonald, and Waszink (2001); Netherlands: author's interview with Dutch MoD official, New York, 11 July 2001; Russia: Ordzhonikidze (2001); United States: author's interviews with US Army officials; others: WEAPONS COLLECTION

Further large-scale destruction is being planned. The German government reportedly plans to destroy at least another 50,000 sub-machine guns and 400,000 rifles (DDP, 2001). In this case, most—if not all—are former East German firearms that the Berlin government simply has no use for. But even destruction of unwanted arms should not be trivialized; in the past Germany tried to solve the problem by transferring many of its unwanted East German small arms abroad.

Major changes among owners: The military, civilians, and insurgents

Patterns in ownership did not change much in 2001, but our understanding of the distribution of firearms has become more detailed and nuanced. Of greatest significance here is the role of military doctrine and strategy on small arms stockpiles. Different countries are affected in different ways. On the one hand, orthodox multipliers used to estimate small arms per soldier seriously underestimate the stockpiles among those relying on the strategy of People's War. Their stocks appear to be much larger than previously thought. On the other hand, there appears to be another trend toward smaller military establishments with fewer small arms per soldier. How the latter will affect global stockpiles in the long term is uncertain. In the short term, though, it could exacerbate the glut of second-hand equipment.

Civilian stockpiles did not change much in 2001, growing through normal purchases. The total number of insurgent small arms did not change much, but their distribution did, growing in some regions, declining in others. The fastest recent growth appears to be in Palestinian-controlled areas. But size is not everything, as the disproportionate influence of the relatively tiny stockpile of the IRA shows.

Military firearms: The link between strategy and numbers

When countries refuse to make public the size of their military small arms arsenals, the efficient way to estimate their dimensions is through comparison with countries for which accurate data is available. This approach, described in the *Small Arms Survey 2001*, facilitates comparisons based on multiplying the number of uniformed military personnel—which is generally known for all countries—by the number of small arms that countries typically have for each soldier, sailor, and airman. In the *Small Arms Survey 2001*, the small arms ratios of the Canadian armed forces were used as the basis for a conservative estimate of global military small arms inventories. The approach is useful but crude; it assumes that all countries are relatively similar in their small arms procurement practices—a major oversimplification. These assumptions and others require regular adjustment as more data becomes available.

Estimates of military small arms inventories in this chapter continue to be based on the assumption of 2.25 firearms for each member of the armed forces, as developed in the *Small Arms Survey 2001*, but this approach has important limits. While this appears to lead to a reliable, conservative estimate of global military small arms, it can lead to a distorted impression of the stockpiles of individual countries. In practice, many countries have much larger stockpiles (see Box 2.3) while a few are smaller.

Among the countries where these ratios have increased is Canada itself. New information from the Canadian government shows that in recent years the number of troops there has declined, while the military small arms inventory generally was stable. This pushed its weapon-to-troop ratio up from the figure of 2.25 used in the past to 2.5 in 2002 (see Box 2.4). Despite this change, the old multiplier continues to be used since it reflects conditions in the base year of 1987 when global military forces were their largest. But the implication of the new information should not be overlooked; actual weapon-to-troop ratios and actual military small arms inventories worldwide could be higher than previously thought.

In the US Army, the ratio of small arms per soldier has fallen from 2.3 in the early 1990s to 1.3 today. The ratio for the entire American military, including all the armed services, is even lower.

Among the countries that have allowed the size of their military small arms inventories to atrophy is the United States. For the United States Army—described at length below—the ratio of small arms per soldier has fallen from 2.3 as recently as the early 1990s to 1.3 today. The ratio for the entire American military, including all the armed services, is even lower at roughly 1.05 small arms for each uniformed airman, marine, sailor, and soldier. This decrease is the direct result of changes in national strategy.

Instead of re-evaluating basic assumptions comprehensively, this chapter starts by estimating the global stockpile total based on the multiplier of 2.25, and then adds the approximately 700,000 military weapons produced in 2001 (PRODUCERS). Other additions for specific countries are noted in the following text. Finally, the 2.1 million military weapons destroyed, as noted above, must be deducted (see Table 2.6). The net result is a total estimate of over 241 million firearms in the hands of global armed forces as of January 2002 (see Table 2.7).

Even at the height of the Cold War, when most of the military firearms in existence today were made, some countries were not satisfied with the 2.25 ratio, which they would have found too low for their national security needs. That ratio used in past and present editions of the *Small Arms Survey* is a conservative choice for estimating the scale of typical inventories during the Cold War era; some countries, on the other hand, were more aggressive in their defence planning, accumulating much larger quantities, virtually all of which are still around.

© DND Photo/Sgt Reid

A Canadian soldier with a C7 rifle.

TABLE 2.7 Revisions to the estimated global military firearms stockpile

Category	1987 base figure	Change since 2000
1987 base estimate	202,000,000	-
Global new military production	25,000,000	+ 1,000,000
Addition for People's Armies (except China)	3,200,000	+ 3,200,000
Addition for Chinese People's Liberation Army	13,500,000	+ 13,500,000
Subtraction for military weapons destroyed	– 2,100,000	– 2,100,000
Total global military firearms stocks	**241,600,000**	**+ 15,600,000**

Note: the People's Army countries estimate figure here excludes 20.8 million firearms already in the base figure.

Conservative assumptions are safe and useful rules of thumb, especially in place of additional information. But enough data is accumulating about one particular group of countries that built up much larger military small arms arsenals than conservative assumptions would lead one to believe. These mostly are former—and some current—communist-ruled states that relied on military strategies based on People's War. The latter is the well-known Maoist strategic doctrine of repelling a technically more advanced foe through a combination of guerrilla raids and massed infantry attacks, relying on sheer numbers to compensate for lack of sophisticated equipment and tactical finesse. It was a brutal approach to warfare, generally preferred by ideologically driven political leaders rather than by general officers. By the early 1970s it had fallen into widespread disfavour, a victim of its own extremely costly successes in Korea and Vietnam.

Albania was one of the last pure practitioners of People's War, and its monastic exceptionalism offers a unique insight into the small arms structure of these countries. After its military arsenals were ransacked in the nationwide looting that began in early 1997, the country became the centre of an international effort to recover its lost small arms. One result was complete reporting of Albania's lost arsenal, outlined in Table 2.8. According to United Nations reports, the country lost control of more than 643,220 small arms, although a more recent Albanian police report says the total was 549,775 (Republic of Albania, 2001). In either case, this is a very large amount of equipment for a country of 3.6 million people and a military establishment that never grew beyond 180,000 men for all armed services, active and reserve. Instead of an orthodox 2.5 small arms per soldier, Albania had something closer to 3.6.

TABLE 2.8 Arms looted from Albanian military depots, January–March 1997

Type	Number	Ratio of items/soldier
Pistols	38,000	0.21
AK-47s	226,000	1.25
Bolt rifles	351,000	1.95
Machine guns	25,000	0.14
Grenade launchers	2,450	0.01
Mortars	770	0.004
Total	**643,220**	**3.57**

Note: these figures do not include weapons from the small arms factory at Gramsh. Most of its production and stocks were intended for export
Source: United Nations (n.d.)

Albania's extra weapons were intended to arm civilians drafted into the fray of an attack on the country. Although there was little about communist Albania that was normal, the small arms force it created to fight a People's War was not exceptional. With a version of the same strategy in mind since his guerrilla successes of the Second World War, Yugoslavia's late-President Tito created an arsenal with 4.5 small arms per military member, active and reservists (Small Arms Survey, 2001, p. 75).

What of the other countries that relied on versions of People's War during the Cold War era? Cuba and Vietnam already have shown the scale of these plans. Cuba relied primarily for its defence on its 330,000-member armed forces, which it planned to support with the 1.3 million members of its Territorial Militia—this in a country of some ten million. Vietnam, with a population of roughly 70 million, supported an army of over one million, backed up with a Strategic Rear Force of three to four million. And North Korea, with a population one-third that of Vietnam (21 million), maintained an army of 1.1 million and a reserve of no fewer than six million (IISS, 1987). Since their total reserve forces are well known, standard multipliers can be used in these cases to estimate military stockpiles.

Countries in which military strategy has relied on People's War tend to have especially large military small arms stockpiles.

The Albanian military firearms multipliers are of greatest value with countries preparing for some version of People's War, though little is known about such countries. An example is Libya, with an arsenal of major weapon systems far in excess of the capabilities of its armed forces, which never totalled more than 125,000 personnel. Just as it keeps more armoured vehicles than it can hope to use, Libya almost certainly has filled its firearms arsenals very generously. Iran and Iraq also have followed the same strategy, which would lead to Albanian-style military firearms arsenals of at least some 444,000 military guns for Libya, 3.8 million for Iran, and 3.7 million for Iraq. These estimates—again, almost certainly on the low side—create a net increase of 2.4 million military guns over the otherwise anticipated minimum stockpiles of these countries.

The biggest question in the field of People's War is none other than China, where the concept originated. The People's Liberation Army remained at a regular strength of some six million personnel throughout the Cold War. How large an irregular force did Beijing plan for? Given China's population and industrial resources, the total could be virtually limitless. Applying the Albanian military firearms multipliers points to a conservative estimate of enough Chinese small arms to equip at least 27 million troops (see Table 2.7).

As in almost all other aspects of human affairs, poor information can lead to queer results in military small arms planning as well. Despite its enormous importance for security, mistakes happen. A recent example occurred in Macedonia, where several countries have helped the country build its armed forces since 1996, when the UN arms embargo was lifted. Inadvertently they helped create an army with one of the highest ratios of weapons to soldiers ever.

Macedonia started out with 5,000 Kalashnikov-type rifles inherited from the former Yugoslavia and some 40,000 bolt-action rifles. Assistance quickly began to pour in, most generously from Germany, which supplied 15,000 former East German AK-47s. This was to equip an army initially planned to reach 60,000 troops in all. But an army of such dimensions was never a serious possibility. When the goal became manifestly infeasible and politically divisive, it was reduced to 16,000. In reality, a force of just 2,500 soldiers could be created by the time civil war with ethnic Albanian rebels began (Gordon, 2001b). The weapons remain, giving the country a remarkable ratio of 24 firearms per soldier.

In Bosnia-Herzegovina, after years of NATO and bilateral military assistance, the armed forces have 450,000 small arms, ostensibly to equip an army of some 40,000 (Zivalj, 2001). This would be exceptional over-arming, except that Bosnian security planning relies on 150,000 reservists (IISS, 2001). With reservists the total weapons-to-soldier ratio is high, but not exceptionally so. For a country convinced its security was irreparably damaged by international embargoes, such hording is not difficult to understand.

BOX 2.3 Swiss military small arms

Swiss gun owners can point to a long and distinctive tradition of gun ownership, with roots in an armed neutrality. But unlike many countries with widespread firearms ownership, the Swiss are comparatively private about their firearms. The Swiss federal government does not keep comprehensive statistics on private gun ownership. Although all Swiss private gun owners must be licensed and many cantons require firearms registration, most records are maintained at the canton (provincial) level.

The major exceptions to this rule are the small arms of the Swiss Armed Forces. According to figures provided by the General Staff of the Swiss Armed Forces, the nation's military currently has a total of 695,000 rifles and pistols in its inventories. These include:

- 450,000 Stgw 90 (5.6mm) rifles*
- 175,000 Stgw 57 (7.5mm) rifles (designated for liquidation)*
- 50,000 Pist 75 (9mm) pistols*
- 20,000 Pist 49 (9mm) pistols*

This list is not complete, missing sub-machine guns, machine guns, grenade launchers, and other typical military firearms, whose numbers remain classified. Using orthodox multipliers to estimate these—but not other kinds of small arms like grenades, mortars, and shoulder-fired anti-tank rockets—the country would be expected to have a total of approximately 1,040,000 military firearms in its armed forces inventory.

This might seem high for a country with just 4,000* full-time military personnel and 29,000* conscripts in basic training every year. In reality most of these weapons are for the country's militia armed forces, which is based on 420,000* citizens (including 70,000* reservists) serving regularly as soldiers every other year parallel to their professional life who form the foundation of Swiss national security. The small arms for this total force of 453,000* personnel are stored in depots, unit arsenals, and under a special system of regulations in the homes of the country's militiamen.

With approximately 2.3 military-owned firearms for every man and woman who can be called into uniform, the Swiss Armed Forces have an arsenal proportionally larger than most NATO countries—which typically have 1.3–2.5 firearms per soldier—but lower than most other countries relying heavily on reservist forces or nation-in-arms strategies (in which category the Swiss Armed Forces belong)—who typically have 3.5 or many more firearms per soldier.

Swiss militiamen assembling.

Another distinctive aspect of the Swiss military system is the widespread distribution of Army firearms among its militiamen, who keep weapons and ammunition in locked storage at home. This is a procedure used by few other countries, of which Israel is best known. Even more distinctively, Swiss militiamen are allowed to receive former military weapons upon retirement and by proving that they execute a sport shooting activity. These are specially marked and converted for semi-automatic fire. Former military small arms in the hands of ex-conscripts are not included in the inventories of the Swiss Armed Forces, although records of such transfers are kept for ten years and shared with canton police.

The Swiss example shows how vital it is to include militiamen and reservists when evaluating the small arms of a national military establishment. The information released here offers insights into a distinctive military structure. As revealing as they are, though, these figures do not include police weapons or privately purchased firearms. Until data on those categories is collected, the total number of small arms in the country cannot be established.

*Note: * Indicates actual figures supplied by the General Staff of the Swiss Armed Forces. All other calculations are those of the Small Arms Survey.*
Sources: Correspondence with the General Staff of the Swiss Armed Forces, 5 December 2001, 21 December 2001, and 1 February 2002.

BOX 2.4 Canadian military firearms multiplier

As a conservative baseline for estimating global firearms stockpiles, the *Small Arms Survey 2001* used the example of Canada. This was not only one of the few nations for which comprehensive information on small arms inventories was available, it also led to a lower multiplier than most other examples: 2.25 firearms per every uniformed airman, sailor, and soldier. Some countries have fewer firearms per solider than Canada—Norway currently has only 1.6, Togo has 1.34, and the United States is an extreme example at 1.05 small arms for each man and woman in uniform—but most countries appear to have at least as many as Canada, if not more. Use of the Canadian-inspired multiplier probably underestimates global military stockpiles, it almost certainly does not overestimate them.

Thus it is significant that the Small Arms Survey has been informed by the Canadian government that the number of its uniformed military personnel no longer is 107,500, as it was only a few years ago. As of 2001 its authorized strength was 93,500 for all services, active and reserve. This is the maximum allowed under law; actual strength is lower. This troop level leads to a current multiplier of 2.5 firearms for every man and woman in uniform for Canada.

Source: Canadian Department of Foreign Affairs and International Trade, Ottawa, private correspondence, 24 January 2001.

Civilian ownership: Continuing to dominate the market

The most dynamic area of small arms ownership continues to be the civilian market. Civilian owners have at least 378 million firearms. This represents an increase of 25 per cent over the estimate in the *Small Arms Survey 2001*. Most of this increase is the result of better data and counting techniques.

Civilians own at least 59 per cent of the total global firearms stockpile.

Actual growth of civilian stockpiles in the year 2001 was approximately 6.9 million newly manufactured firearms. Civilian buyers, in other words, acquired roughly 85 per cent of the nearly eight million guns manufactured in 2001 (PRODUCERS). In all, civilian guns constitute over 59 per cent of the total global firearms stockpile.

While civilian sales are depressed from their historic peak a decade ago, the civilian sector of the market remains the most dynamic. Police inventories changed marginally in 2001 through normal attrition replacements and modernization. The major exception was Nepal, where police facing Maoist insurgents are in the process of converting to automatic rifles, as described below. Military forces too have generally been less affected by major new purchases, although fulfilment of old contracts and smaller purchases continue to shape their inventories. Insurgent arsenals appear to have grown slightly in 2001, but remain slight compared with civilian holdings.

The most prominent civilian market continues to be the United States, where civilians purchased more than 95 per cent of the four million firearms manufactured in the country in 1999, the last year for which official statistics are available (PRODUCERS). Combined with imports—mostly from Brazil and China—of some one million, this means the American civilian firearms stockpile grew by roughly five million that year. Production for civilian purchase in other countries added another two million. Thus new production increased the global civilian stockpile by almost seven million in the most recent year known, a level that appears to have been sustained in the years since (PRODUCERS).

Of nearly eight million firearms manufactured in 2001, some 6.9 million went to civilians.

Better reporting from governments also has made possible re-calculation of total civilian stockpiles in several countries, most notably in the Czech Republic and in the Philippines, where it was possible to estimate even illegal civilian firearms inventories for the first time. These are discussed in the regional sections below.

Several countries are gradually developing estimates for their civilian weapons inventories. In Switzerland, for example, a picture is emerging through a combination of official disclosures and public polling. Swiss federalism makes total national firearms ownership difficult to ascertain. As noted above, the Swiss military controls 695,000 automatic rifles and service pistols. Many of these are stored in the homes of the nation's seven million people (see

Box 2.3; Clasmann, 2001). The scale of total private gun ownership in the country was suggested by a recent survey. This reported that 35 per cent of all Swiss households have at least one gun (Swiss Crime Surveys, 2001). While this is not sufficient to establish the total number of civilian firearms in the country, it ranks Swiss public gun ownership higher than other well-armed countries like Australia and Canada, in which one out of four households has a gun, but below the United States, where over 40 per cent of households have at least one gun.

With fewer conflicts, insurgent demand for small arms has decreased.

TABLE 2.9 Additions to estimated civilian stockpiles

Country	Category	Total	Change
Czech Republic	Legal civilian	534,000	+ 251,000
India	All civilian	40,000,000	+ 40,000,000
Nepal and Sri Lanka	All civilian	3,000,000	+ 3,000,000
Pakistan	Legal civilian	2,000,000	+ 2,000,000
Pakistan	Illegal civilian	18,000,000	+ 18,000,000
Philippines	All civilian	5,300,000	+ 4,900,000
New production	All civilian, 2001	7,000,000	+ 7,000,000
Subtractions for disposals, 1990–2001		–1,836,000	– 1,836,000
Net change for estimated civilian stockpiles			+ 73,300,000
Approximate total civilian firearms			**378,300,000**

These and other changes, especially the addition of estimated civilian gun ownership in the countries of South Asia (below), make it possible to recognize an increase of more than 73 million previously unidentified firearms in the global total of civilian-owned weapons, which stands at a total of approximately 378 million firearms.

Despite the additions, the overall picture of the global civilian small arms stockpile remains far from complete. Very little is known about the total number of civilian small arms in many major countries including Brazil, France, Indonesia, Iran, Mexico, Thailand, Turkey, and Vietnam. Other major countries have been able to provide only partial or unreliable reports, including Germany, Poland, and Russia. With only a handful of exceptions, there currently is no way to estimate the dimensions of civilian small arms ownership for almost the entire continent of Africa and the countries of the Arabian peninsula.

© Associated Press/Ujir Magar

A Maoist rebel leader addresses Nepalese villagers.

Insurgent stockpiles: Deeds more than numbers

While states arm themselves in good times and bad, in peace and in war, rebel movements are a response to grievance, provocation, and opportunity. Their arsenals generally rise and decline as they are needed. The decline of ethnic and internal conflict is of more than academic interest for the spread of small arms. With fewer conflicts, insurgent demand for small arms has decreased. Although some conflicts clearly got worse in 2001, even more subsided. The overall impression is that there are fewer rebel soldiers in the world, and that they are not procuring as much lethal equipment. The total global insurgent stockpile thus appears to be declining very slightly. But, of greater importance, the focus of ethnic and internal warfare is shifting regionally, taking insurgent arms buying along with it and shifting the relative ranking of various regions.

As a prominent report notes, 'The number and magnitude of armed conflicts within and among states have lessened since the early 1990s by nearly half' (Gurr *et al.*, 2000). No state-to-state wars have been fought since March–June 1999, when NATO attacked Serbia over the status of Kosovo, and June 2000, when the war between Eritrea and Ethiopia was settled. With the important exception of the American-led attacks on Afghanistan, the only wars in progress at the time of writing are internal conflicts, mostly over religious or ethnic secessionism. Despite the horrors of 11 September 2001 and the patterns of insurgent and terrorist violence in places like Israel and Sri Lanka, the total number of international terrorist incidents also appears to be down sharply from the 1980s as well (Johnson, 2001). Most of the increase in individual incidents since 2000 occurred in Colombia, where guerrillas specialize in attacking the international oil industry (US, Department of State, 2000).

There are fewer rebel soldiers in the world, and they are procuring less lethal equipment.

Internal conflict is heavily concentrated in Central Africa and the broad arc of Asia, from the Iraq–Iran border, through Afghanistan and Central Asia, through South Asia and Southeast Asia, ending in the Philippines and some of the Pacific Islands. These are the regions where the illicit small arms trade is most active, sustaining insurgent arsenals in their battles with governments and each other (TRANSFERS).

In much of Africa the scale of the fighting and chaos often exaggerates the scale of the forces involved. Except for countries like Angola and Sudan, the size of rebel movements can be deceptively small. The importance of insurgent stockpiles on the order of 2,000–15,000 weapons—the size of even relatively large insurgencies in most of Central and West Africa—becomes clearer when one considers how small the government armed forces in much of the region are.

This was demonstrated by a disclosure of great importance by the Republic of Togo in July 2001, when it submitted to the United Nations Register of Conventional Arms (UNRCA) the complete small arms inventory of its armed forces. According to Togolese diplomatic officials, the submission was intended to dispel regional anxieties aroused by allegations in the Fowler report (Fowler, 2000) of Togo's complicity in arming Angolan rebels, concerns which alarmed some West African governments and aid donors.[2] The endemic guerrilla warfare afflicting much of West Africa and concerns with its own internal stability led to the growth of its armed forces, from roughly 6,500 in the mid-1990s to some 9,500 troops today (IISS, 2001). Even so, according to its report, Togo's military small arms arsenal includes just 12,850 items. With a weapon-to-troop ratio of 1.34, this is significantly lower than the usual conservative estimate of 2.25 for most countries (see Table 2.10). Since government armed forces usually need overwhelming numerical superiority to prevail over a rebel insurgency, Togo's armed forces would appear to be poorly equipped to match even a relatively small rebel movement.

© Associated Press/Rafiq Maqbool

Bodies of separatist militants from Jaish-e-Mohammed in Indian-controlled Kashmir.

TABLE 2.10 Small arms and light weapons of the armed forces of Togo

Category*	Subcategory	Number	Type
Light weapons	rocket launchers	140	40mm, 89mm
	Heavy machine guns	70	50 cal., 12.7mm
	Light machine guns	360	Various
	Mortars	76	60mm, 82mm, 100mm
Small arms	Sub-machine guns	1,700	Various
	Pistols	1,184	Various
	Rifles	9,320	FAL, G3, MAS 36, etc.
Total		**12,850**	

**Categories used by the Report of the Republic of Togo to the UNRCA.*

Source: Republic of Togo report to the UN Arms Register. Submitted 3 July 2001. Information based on the year 2000.

In Africa the decline of fighting in 2001 appears to have reduced the immediate demand for small arms. Skirmishing between rival military factions and raiding by bandit gangs continues to be a massive problem along a broad line from Sierra Leone to Kenya and Somalia, but there have been fewer reports of large military factions undertaking major operations (Achieng', 2001). In the Democratic Republic of Congo, Rwanda and Uganda appear to have withdrawn most of their own troops and reduced support for rebel clients (Fisher, 2001a). West African leaders in particular increasingly see the region's small arms problem in terms not of worsening warfare but of the large quantities of remaining firearms. The director of the Institute of Economic Affairs in Ghana estimated that there are eight million small arms left over from West African wars, now in illegal hands (Acheampong, 2001). This is the first overall estimate of the West African small arms problem.

Fresh insight into the arsenals of African insurgencies came through a report carried out by the Small Arms Survey for the International Organization for Migration (IOM) and the United Nations Development Programme (UNDP) (Demetriou, Muggah, and Biddle 2002). Based on field research in the Republic of Congo (Brazzaville), where fighting occurred from 1993 to 1999, this report concludes that the major rebel factions, with a total membership of roughly 31,000, currently have some 41,000 small arms under their control. Thousands of other weapons were acquired during the fighting and destroyed, mostly through normal wear and tear. The current Congo ratio of roughly 1.3 weapons per insurgent is close to the 1.4 ratio used in the *Small Arms Survey 2001*. While this is only one study, it reaffirms the need to treat all small arms estimates with caution.

Outside of Central Africa and Asia, active internal conflicts are important exceptions on an increasingly peaceful world map. The most visible is the war in Afghanistan, renewed after the disasters of 11 September 2001 and discussed in the South Asia section below. In Latin America the biggest exception remains Colombia, where the only current Latin American war continues. Similarly, Europe is preoccupied trying to contain fighting in Macedonia, while two other long-running conflicts refuse to end: a large one in Chechnya and a small one in Spain. In the Middle East, public and official attention is dominated by the Palestinian–Israeli conflict, while the conflict between Iran and anti-revolutionary guerrillas based in Iraq continues less noticed. These conflicts are the major drivers of illicit small arms transfers in their regions. Other, smaller, and less continuous conflicts have an effect on small arms accumulations, such as fighting in Algeria, but without comparable intensity or quantity.

Another worsening conflict is Colombia's war with rebels of the FARC and the Ejército de Liberación Nacional (ELN). It is a measure of the intensity of the fighting that, even after receiving 10,000 rifles from Peru in 1999, the rebels appear to rely on support from the government of President Hugo Chavez in Venezuela (Webb-Vidal, 2001).

Money from drugs and kidnappings does not automatically translate into battlefield successes since the rebels still have difficulty accessing global firearms markets, a fact illustrated by the high prices they must pay (see Table 2.2). Much of the pressure on them comes from the union of paramilitary groups known as the United Self-Defence Forces of Colombia (AUC). Having nearly doubled in size to 11,000 soldiers in the year 2000, it appears to rely on connections with the Army for most of its weapons (Forero, 2000).

The rise of secessionist and sectarian violence, wherever it occurs, is a crucial challenge to global peace and stability. Many devastating sub-state and internal wars continue around the world. But in terms of the sheer numbers of insurgent combatants, only the new Afghan war that began after the terrorist attacks of 11 September 2001 begins to equal the scale of the major Kurdish, Palestinian, and Afghan rebel movements of the early to mid-1990s. Reports from Afghanistan in the autumn and winter of 2001, discussed below, indicate that additional small arms were flowing into the country, but not on a scale comparable with the years of fighting against the Soviet Union (Gordon and Schmitt, 2001).

Consequently, the combined scale of global insurgent small arms arsenals appears to be in a state of gradual decline. The numbers of weapons insurgents need to wreak havoc, though, is not great. Measured by their results, even small rebel arsenals are of disproportionate importance.

Regional stockpiles: Learning more

As journalists and researchers devote more attention to small arms, information on the scale of regional stockpiles is accumulating, pushing particular national and regional totals upward. Before examining regional dynamics in detail, two caveats are in order though. First, much of the apparent growth is due exclusively to better information and analysis; the actual quantity of weapons is not changing, only our awareness of them. Second, the logic of this building-block approach dictates that most of this new information can be used only for a more complete global picture. Portraits of regional small arms distribution often remain impossible because of lingering uncertainty about particular countries. There are important exceptions, however, of which the most important to emerge was South Asia, as assessed below.

North America: Armed forces decline, civilians register

Because it has by far the largest civilian weapons market in the world, it is only natural to assume that the United States also has a massive military market as well. Its armed forces are the world's best financed with a budget of USD 328 billion for fiscal year 2002 (Sherman and Svitak, 2001). This is equal to the total defence spending of the next nine largest military powers together (IISS, 2001). With a total of more than 2.5 million men and women in uniform—active duty and reserve—it also has the fourth largest forces in terms of personnel, after China, Russia, and Vietnam. American military small arms procurement has considerable influence on global patterns, moreover, helping set strategic and tactical precedents that affect decision-makers elsewhere. The significant decline in America's military small arms inventories since the end of the Cold War has broad implications.

American military small arms procurement has considerable influence on global patterns.

During the Cold War, the United States military maintained large reserves of small arms for its armed forces as well as for allies. Assuming that it might have to fight two major wars simultaneously—in Europe and Korea—as well as be called upon to support allies in the Middle East and possibly elsewhere too, its military establishment had a policy of procuring additional weaponry when feasible and preserving older equipment when possible. This stockpile, or floating reserve, was a key element of a national strategy that stressed the ability to mobilize to fight lengthy conflicts.

In the past, the Pentagon deliberately preserved enough equipment to maintain a small arms inventory equivalent to roughly 2.3 small arms for each man and women in uniform. This was the practice from the Second World War to the 1980s, a natural part of a military strategy based on mass mobilization—through conscription—to fight two major wars. Since then, long-term mobilization, involving the training of hundreds of thousands of additional troops, ceased to be relevant. With it disappeared long-term requirements for a large floating reserve of small arms.

TABLE 2.11 Small arms inventory of the United States Army, 11 April 2001

Type	Deployed	Non-deployed	Total
Machine guns			
M2 flexible	19,999	3,785	23,784
M2 fixed	3,107	634	3,741
M60	21,552	2,560	24,112
M85	66	249	315
M249	68,640	1,737	70,377
M240	9,746	2,450	12,196
M240B	5,358	806	6,164
M240C	4,359	1,059	5,418
Grenade launchers			
M79	350	283	633
M203/M203A1	57,829	782	58,611
MK19	20,369	1,567	21,936
Current rifles and carbines			
M16A1	144,275	74,418	218,693
M16A2	640,520	1,443	641,963
M16A4	4,787	11,290	16,077
M4 carbine	77,671	787	78,458
M4A1	16,494	3	16,497
M231	6,745	9,636	16,381
Obsolescent rifles and carbines			
M1	458	3,716	4,174
M14	12,597	104,657	117,254
.22 cal. rifle	5,862	8,012	13,874
.30 cal. rifle	77	952	1,029
Sniper rifles			
M21	447	0	447
M24	2,153	110	2,263
Pistols			
M9	158,711	21,621	180,332
M11	2,789	38	2,827
.22 cal. pistol	1,830	14,196	16,026
.38 cal. pistol	541	69	610
45 cal.	4,591	28,179	32,770
Other categories			
Shotguns	5,581	1,610	7,191
M3 sub-machine gun	135	12,502	12,637
Pyrotechnic pistol	501	481	982
Total	**1,298,140**	**309,632**	**1,607,772**

Source: US Army Executive Agent for Small Arms (private correspondence, April–July 2001)

With the end of the Cold War, American strategy changed to emphasize rapid responses to smaller contingencies. Pentagon planning emphasized smaller regional crises. It was the high-tech Bosnian interventions in 1995 and Kosovo in 1999 that established the precedents for the future, not the relatively old-fashioned 1991 war against Iraq. Instead of lengthy warning followed by protracted mobilization, in the early twenty-first century wars will be fought by forces in being, 'come as you are'. Sheer numbers of troops matter less as well, especially in an environment where extensive training is essential and casualties must be rigorously minimized. Reserves are important as an affordable way of maintaining personnel that are not normally needed—like artillery units, field hospitals, and civil affairs—and preserving specific specialties, not as a basis for force expansion.

In the United States, strategic transformation led to a massive cut-back in military small arms.

The implications of this strategic transformation for small arms inventories are clearest in the US Army. With a total of over 1.6 million firearms of all types as of April 2001, it has the largest small arms arsenal of the American armed services.[3] These range from virtual-museum artefacts like .45 calibre pistols to state-of-the-art Objective Individual Combat Weapons. This stockpile is enough to equip a combined active and reserve force of 1.2 million personnel. It gives the Army the equivalent of only 1.3 firearms per soldier (see Table 2.11). This is just enough to fully equip its combat forces with a small central reserve to cover breakage and other losses. Most of the reserve holdings listed in Table 2.11 consist of newly arrived equipment that has not been passed on to its designated units—such as M9 pistols and M16A4 rifles—or obsolescent equipment the Army has not disposed of yet—such as M14 and M16A1 rifles or .45 calibre pistols. The only floating reserve of any consequence today is literally floating, kept on pre-positioned ships for emergency re-supply.

The low ratio of troops to small arms is the result of strategic choice. Since they will not be fighting extended wars, the chance of soldiers losing weapons in combat is low. Most of the small arms reserve is more important today to replace weapons being overhauled rather than weapons that might be lost in the field. The evolution of force structure also is an important consideration shaping the small arms arsenal. With an ever-greater proportion of personnel allocated behind the lines to logistic and support duties, fewer troops need firearms. Although the number is not readily calculated, a surprisingly large proportion of America's soldiers never carry a gun after basic training. For the other services, the tooth-to-tail ratios—the number of combatants per support personnel—are even lower and the proportion carrying firearms lower still (see Table 2.12).

TABLE 2.12 Total US military firearms, by service, 2001

Air Force	260,000 *
Army	1,608,000
Coast Guard	20,268
Navy (includes Marines)	800,000 *
Total	**2,688,000**
Total firearms per uniformed member	**1.05**

** Estimated*

Sources: US Army Executive Agent for Small Arms, US Coast Guard Cutter Training Division (private correspondence)

With these changes in mind, the US Army dramatically reduced its small arms inventories in the 1990s. Initially it 'demilled' large quantities of small arms through contracts for their physical destruction. Among the largest were contracts that led to the destruction of 479,367 M14 rifles in 1993–94 and roughly 350,000 M16A1 rifles in 1996. This brought a reaction from gun advocates in the Congress, who wanted excess weapons made available to the public. Since the sale of new automatic rifles to the American public is prohibited under a presidential order from 1994, they could not be dumped on the market. Instead, the Congress passed legislation amending the annual Defense Authorization Act in 1996 to prohibit the Army from destroying further 'collectable' guns. The amendment has been passed every year since. Supporters hope these firearms will be made available for sale someday.

With destruction increasingly controversial, the Army changed course. In 1995 it began to transfer many of its surplus small arms to foreign governments. Between 1995 and early 1998, the export of 321,905 older Army firearms was arranged through aid packages to foreign militaries under the Excess Defense Articles programme. The leading recipients were the Baltic states—Estonia, Latvia, and Lithuania—and Israel, all of which received the equipment free of charge, and the Philippines and Taiwan, which paid varying amounts (Lumpe, 1997; 1999). By the end of 1999, most of this equipment had been shipped. This approach had the advantage of strengthening vulnerable friends and allies while getting the guns out of the Army's own inventories and making someone else responsible for them. The security of the weapons should not be a problem since foreign governments cannot re-transfer American-supplied defence equipment without permission from Washington.

Today the US Army has its firearms inventory very close to what it believes is ideal. It still buys additional small arms for modernization, but the total inventory is changing more slowly. In 2000–01 it acquired a total of 37,224 small arms (see Table 2.13), equal to some two per cent of its holdings. It still sells some weapons to the general public. M1 semi-automatic rifles of Second World War vintage are released for sale to the public through the Civilian Marksmanship Program, a programme very popular with collectors. Currently 17,827 M1 rifles recently returned from Denmark are being disposed of this way. Other obsolescent firearms will be kept when the law requires, transferred abroad, or destroyed.

TABLE 2.13 US Army small arms acquisitions in fiscal year 2000

M2 heavy machines guns	952
M4 carbines	9,428
M11 pistols	311
M16A4 rifles	14,895
MK19 grenade launchers	1,517
M240B medium machine guns	4,623
M249 light machine guns	5,498
Total	**37,224**

Source: US Army Executive Agent for Small Arms, Rock Island Arsenal, Illinois (correspondence, May–June 2001)

There has been wide-spread satisfaction with implementation of Canada's new gun laws.

Among civilian owners, the major change in North America occurred in Canada, where comprehensive firearm licensing and registration is beginning. The initial filing period for public gun owners' licenses ended on 31 December 2000. By this date, almost two million of the nation's estimated 2.3 million firearms owners had received their licences (Canadian Firearms Centre, 2001). Owner licensing was the first part of the process mandated under the 1995 Firearms Act, widely known in Canada as Bill C-68. The second stage—registration of individual firearms—is to be completed in 2003. While there has been widespread satisfaction with the implementation of the law, there also has been resistance among some gun owners, especially in the Western provinces (Brooke, 2000; Cassels, 2000).

One of the goals of the 1995 Canadian Firearms Act was to find out how many private guns and gun owners there are in the country. This remains ambiguous, though. Early estimates of three million gun owners were scaled down to 2.3 million, a revision that critics charged was intended to make licensing appear more complete than it really was. Similarly, previously accepted estimates of 7.1 million guns in public hands have come under political pressure. Gun control supporters have suggested that the actual total of individual guns is closer to five million, with the implication that registration is going well so far, as 1.8 million firearms have been registered. Opponents charge otherwise, maintaining that only a small proportion of a much larger Canadian arsenal has been affected (MacQueen, 2001).

Europe: Civilian trends and guerilla warfare

The number of firearms in public hands in Europe is not easily estimated. Several large European countries, including France and Germany, do not release data on civilian ownership. Among those that do report on licensed civilian ownership, statistics often are suspect.

A prime example is Russia, where over 3.6 million civilian weapons—mostly hunting rifles and shotguns—were registered as of 1996 (United Nations, 1998). Because of high licensing fees and infamously poor controls over trade, it is likely that actual ownership is considerably greater. Ownership of handguns generally is illegal in Russia, but, as careful observers in Moscow testify, they are not an uncommon sight. While the total number of civilian firearms has been speculated to be 30 million or more, Russia remains a firearms enigma.

Today, the Czech Republic has many more gun owners and guns than it did a decade ago.

Even seemingly accurate statistics do not eliminate such ambiguities. This was demonstrated in a recent report from the Czech government (Czech Republic, 2001), which revealed that since major changes in the licensing law in 1991 and 1995 the number of registered firearms held by licensed weapons permit holders has risen from 90,295 in 1990 to 534,144 in 2000 (see Figure 2.1). The total number of registered firearms has increased by as much as 20 to 33 per cent from year to year. A large but unknown number of firearms still remain unlicensed.

Not only are there many more guns in the Czech Republic than there were a decade ago, there also are many more gun owners. One of the most striking aspects of the trends in gun numbers is how it is matched by an increase in the number of citizens with weapons permits, which rose from 74,604 to 303,904 during the same time period (Czech Republic, 2001). Some of this change may be a natural reaction to the collapse of communism and the celebration of personal liberty. It also appears to be accelerating due to a reform of the nation's Firearms Act in 1995, which made registration both easier and more readily enforced. As a result, a greater proportion of guns probably are being registered than before. Of equal importance are the implications for the rest of Europe; if one country in the region is experiencing unprecedented civilian ownership, others may be in the same situation. This appears to be most likely among other post-communist societies. Already Romania has reported an upsurge of private citizens buying AKM automatic rifles (*Ziau*, 2001).

Some governments have tried to release additional details on national firearms stockpiles, but problems with reporting systems, jurisdiction, and confidentiality often make their reports difficult to use. These problems were illustrated in information released by Denmark in 2001.

FIGURE 2.1 Number of registered civilian firearms in the Czech Republic, 1990–2000

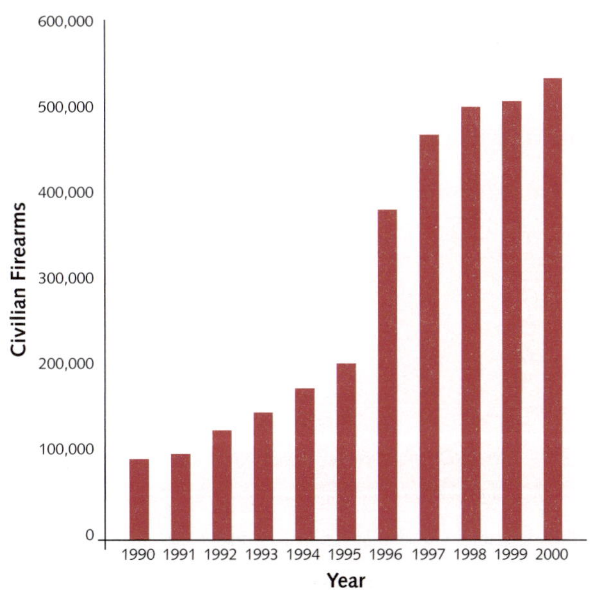

Source: Czech Republic (2001)

This confirmed existing estimates of civilian ownership of rifles and shotguns—200,000 and 650,000 respectively—but added nothing about other categories like handgun ownership or unlicensed firearms. As a result, it remains impossible to assess the general level of public small arms ownership in the country. More revealing was information about the Danish police. With 10,400 sworn—and armed—officers, Danish police forces have a total of 12,700 small arms: 11,000 pistols and 1,700 sub-machine guns. Although nothing was revealed about police special weapons—automatic rifles, grenade launchers, and suchlike—this confirms that the Danish forces are very typical of most police in the world, with little more than one firearm per officer (Denmark, 2001).

The other major stockpile issue for Europe continues to be secessionist rebel movements, especially in Chechnya, Macedonia, Northern Ireland, and Spain. Typical of insurgencies, their political and military importance is highly disproportionate to their actual size.

© Reuters/Paul McErlane

Mural honoring dead IRA members in Belfast, Northern Ireland.

The announcement by the Irish Republican Army on 23 October 2001 that it would begin to 'put its weapons completely and verifiably beyond use' gave hope that this stockpile issue might begin to ease (Brown, 2001). The small arms stockpile of the IRA has been at the centre of a fundamental dispute. The refusal of the IRA to yield its weapons to the independent commission chaired by the Canadian General John de Chastelain led to the collapse of home rule in July 2001, plunging Northern Irish politics into turmoil. The scale of the stockpile in question probably is not great, as estimated in Table 2.14. But it reportedly has grown in recent years, notably through receipt of Russian-designed RPG-22 light anti-tank rockets, believed to have been shipped from the Balkans (*Financial Times*, 2001). The first 'decommissioning'—as disarmament is called there—probably will trim this very little, but the political impact could be enormous.

TABLE 2.14 Estimates of the IRA stockpile

	Estimate 1	Estimate 2
Handguns	–	600
Automatic rifles	600	1,000
Machine guns	50–70	–
Grenades	100+	–
Explosives	2.5 tons	2 tons
Anti-tank rockets	24+	45

Note: left unresolved here is the question of IRA possession of man-portable surface-to-air missiles like the SA-7 and SA-16, which is widely suspected.
Sources: Estimate 1 from Brown (1999; 2000); Estimate 2 from Ananova (2001)

Although the IRA has enjoyed greater success raising funds in recent years, largely through smuggling and drug dealing, little of this appears to go for additional arms (Brown and Burns, 2001). Tighter British border controls undoubtedly explain part of this. One also gets the impression that the IRA may have concluded that its small arsenal is sufficient to wreak havoc, especially in a fragile society. The virtual evaporation of support for terrorist methods after 11 September 2001 apparently led it to formally abandon control over its small but politically potent arsenal.

The limited requirements of urban terror also explain the significance of a series of thefts in France by the Basque separatist movement Euskadi Ta Askatasuna (ETA). In September 1999 it stole eight tons of explosives from a firm in Brittany. Most was recovered, but enough remained to facilitate the ETA's return to violence in December 1999. It kept the small movement—with fewer than 100 combatants—active until recently. With its stockpiles depleted, the ETA raided another office of the same firm in March 2001, taking another 1.6 tons of explosives, enough to continue bombing for several years (Graham, 2001).

The rise of a new insurgency in Macedonia, the ethnic-Albanian National Liberation Army (NLA), created renewed demands for small arms in the Balkans. This began as an outgrowth of fighting by Kosovo Albanian nationalists in the neighbouring Presevo Valley, part of Serbia, in November 2000. At that time there were roughly 200 fighters (Gall, 2000). When this was resolved by permitting Serbian police to re-enter the delicate border region, the ethnic Albanians re-formed and shifted their attacks to Macedonia. When fighting began there in early 2001, rebel strength was estimated to number some 800 armed combatants (Erlanger, 2001). Flush with early successes, this grew to roughly 1,200 to 1,500 in a few weeks (Gordon, 2001a; R. Smith, 2001). By late summer, estimates of 2,000 to 2,500 full-time combatants were widely accepted (Kusovac, 2001).

Under agreements brokered by NATO in August 2001, the NLA pledged to end its military operations, disband, and turn over 3,300 weapons to NATO in exchange for constitutional changes by the Macedonian government. This deal created a major controversy over the size of the NLA arsenal. Estimates of its total size ranged from 6,000 small arms, according to sources sympathetic to the NLA, to 50,000 to 80,000, according to the Macedonian Ministry of the Interior (Fisher, 2001b).

The differences arose mostly, of course, from political expediency. But serious definitional issues are involved as well, issues aggravated when the NLA turned over some 3,800 weapons (WEAPONS COLLECTION). Does one count all firearms owned by Macedonian-Albanians, for example, or just those belonging to NLA activists? The scale of this problem is shown in a prominently published estimate, reprinted here as Table 2.15. What is beyond dispute is the essentially symbolic nature of the disarmament scheme.

TABLE 2.15 An estimate of weapons available to the Albanian nationalist National Liberation Army, summer 2001

Type	Quantity
Rifles	5,000–8,000
Obsolete rifles held by individuals	15,000–30,000
Handguns (mostly held by individuals)	40,000–55,000
Heavy machine guns	150–200
Sniping rifles	100–300
Man-portable surface-to-air missiles	20–50
Shoulder-launched anti-tank rockets	200–350
Mortars	100–200
Landmines	5,000+
Hand grenades	Thousands

Source: Kusovac (2001)

The impact of the fighting in Macedonia for regional stockpiles is obscure. The NLA is not a completely new force; many or most of its members are veterans of the KLA's struggle for Kosovo. The Macedonian Army has received large transfers of small arms, but these are not newly manufactured (TRANSFERS). This leads to the conclusion that the fighting in Macedonia is not increasing global stockpiles. Instead, previously existing weapons are being redistributed and re-circulated.

The same appears to be true in Chechnya, where fighting continues despite Russian President Vladimir Putin's declaration in October 2000 that the war was over. Official Kremlin sources maintain there are 1,500 Chechen fighters. According to the director of the Russian FSB, though, there are 5,000 active insurgents. This is far fewer than the 25,000 rebels at the beginning of the Second Chechen War in October 1999. But it helps confirm the impression that the Chechens remain capable of major operations and will continue to fight for some time to come (Galeotti, 2001).

The Russian Ministry of Defence has routinely displayed captured rebel weapons, but the number taken remains far lower than the suspected total number of Chechen combatants. The most important sources of rebel supply remain deals and extortion of Russian troops, as well as weapons imported from Russia and through Georgia (Jack, 2001).

The Middle East: Accommodating a new actor

With two vital exceptions, territorial conflict and secessionist violence in the Middle East have not grown appreciably in 2001, creating few new pressures for small arms. This is a blessing for a region only recently scourged by Kurdish secessionism, Islamic rebellion, and border tensions between states.

Of the exceptions, the least well-known is the anti-clerical Mojahideen al-Khalq, an Iranian counter-revolutionary movement based in Iraq and sponsored by Saddam Hussein. With somewhere from 15,000 to 50,000 armed followers, it is among the very largest and best-armed insurgencies active today. As it became more aggressive in 2001, mostly through cross-border raiding, it antagonized the Islamic Revolutionary Government in Tehran. This led to Iran's massive missile strikes against Khalq bases in April 2001, with 44–66 ballistic missiles—reports vary—hitting Iraqi soil, the largest single ballistic missile attack ever (*al-Jazirah*, 2001; Mojahideen al-Khalq, 2001). While there is only limited information on its weapons, Mojahideen al-Khalq almost certainly is being heavily re-armed by Iraq.

The Mojahideen al-Khalq, an Iranian counter-revolutionary movement based in Iraq, with between 15,000 and 50,000 armed followers, is one of the largest, best armed insurgencies active today.

The other exception was the continuous and barely restrained violence in the West Bank and Gaza. There are two crucial differences between the first Palestinian Intifada of 1987–91 and the uprising that began on 28 September 2000: territory and guns (Kifner, 2000). With a secure base of operations and formal legal status, Palestinian organizers can stockpile weapons legitimately and acquire them more easily. Even before fighting started between Israelis and Palestinians, the proliferation of firearms on the West Bank and Gaza had become a serious social problem, as feuding between Palestinians became increasingly deadly (Sontag, 1999).

Under the 1994 Cairo Agreement, the Palestinian Authority was able to create a 9,000-man police force, which was to be expanded to 30,000 under the 1995 Oslo II Agreement (Sayigh, 2001). The police were recruited almost exclusively out of Yassir Arafat's Fatah organization, and brought their own weapons. The new Palestine Security Services soon grew beyond the size specified in the accords to 40,000 officers or more, but arming was not a problem (Fishman and Leshem, 2001). Long before the violence began, Israeli analysts noted that weapons were trickling in, allowing the Palestinians to accumulate a total of between 30,000 and 40,000 small arms in the late 1990s. So long as tensions remained low, prices for guns on the West Bank actually went down, suggesting a bit of a glut. The price of a stolen Israeli M16—USD 3,000 in the mid-1990s—fell to USD 2,000 in 1999, while an AK-47 that had cost USD 2,400 could be found for USD 500 (Sontag, 1999). But since the fighting started, demand and prices have gone up rapidly. The most recent estimate of the total Palestinian inventory of automatic rifles is some 70,000 (Goldenberg, 2000).

The Palestinian Authority is suspected to be transforming its police into an infantry force that soon will number 60,000 (Opall-Rome, 2001). Weapons have been stolen from Israel, like the 60 rifles stolen from a kibbutz armoury on 15 March 2001 (Rudge, 2001). Smuggling became more important in 2001. Sea shipments appear to come mostly from Lebanon and Egypt to Gaza. A prominent case involved the fishing boat *Santorini*, intercepted on 7 May 2001, on its fourth trip ferrying small arms for the Palestinian Authority (Katzenell, 2001). Although the weapons were captured—including four SA-7 surface-to-air missiles, Sagger anti-tank missiles, Katyusha rockets, 30 AK-47s, and 13,000 rounds of ammunition—similar shipments apparently have got through. The *Santorini* affair gave a foretaste of the much greater controversy surrounding the arms freighter *Karine A*—loaded with light arms and major weapon systems, apparently from Iran—captured on 3 January 2002 (see Box 2.6).

© Associated Press

Israeli Defense Forces offload a variety of weapons and ammunition from the fishing boat, Santorini.

BOX 2.5 Yemen: The most heavily armed place on earth?

Yemen has a reputation as the most heavily armed society on earth. After the 1994 civil war, when the country was reunited, unsubstantiated estimates began to circulate that Yemen had 50 million or more small arms in circulation. With a population of roughly 17.5 million, this would make Yemen the world's most heavily armed country per capita. The 50 million figure has been widely repeated, even—with evident pride—by Yemenis themselves.

Field research in Yemen conducted by the Small Arms Survey, however, suggests that this commonly cited figure is a gross exaggeration. Realistic estimates point to a national total of roughly five million to eight million firearms.

TABLE 2.16 Estimated individual gun ownership in Yemen

Governate	Approximate population (2001)	Estimated males aged 15 and over	Regional firearms multiplier	Approximate individual firearms
Mareb	220,000	55,000	3	165,000
Al Jowf	203,000	51,000	3	153,000
Sad'ah	580,000	145,000	2	290,000
Sana'a	1,121,000	280,000	2	560,000
Al Dhalee	600,000	150,000	2	300,000
Amran	500,000	125,000	2	250,000
Damar	1,259,000	315,000	1.6	504,000
Ebb	2,357,000	589,000	1.5	884,000
Abyan	500,000	125,000	1.5	187,000
Taiz	2,639,000	660,000	1.2	792,000
Hadja	1,519,000	380,000	1.5	570,000
Mahweet	484,000	121,000	0.5	60,000
Hodahda	2,105,000	526,000	0.5	263,000
Shabwa	451,000	112,000	1.4	158,000
Al Baydah	607,000	152,000	0.2	30,000
Maharah	135,000	34,000	0.2	7,000
Lahj	759,000	190,000	1	190,000
Hadramou	1,045,000	261,000	0.5	131,000
Aden	677,000	169,000	0.5	85,000
Total Yemen	**17,761,000**	**4,440,000**	**1.37**	**5,579,000**

Note: figures may not agree due to rounding

The assertion that Yemen has 50 million or more firearms in public hands is a myth.

BOX 2.5 Yemen: The most heavily armed place on earth? (continued)

A more reasonable estimate is that Yemen has between five and eight million guns.

The new estimate was reached through demographic analysis and field research in 2001. Meetings with Yemeni tribesmen, professional and Western specialists permitted preparation of ordered lists of governates (provinces) showing typical gun ownership per man aged 15 and above. To avoid under-counting, deliberately high estimates were used. It was agreed that individual ownership varies from 0.2 guns per man—or one for every five men—in the capital of Sana'a, to three guns per man in Al Jowf, and Mareb (see Table 2.16).

Tribal stockpiles, under the personal control of sheikhs, also had to be included. These vary with the size of the tribe and the status of the sheikh. Interviewees agreed that Yemen has 100 major sheikhs controlling an average of 1,000 firearms each, 1,000 minor sheikhs with about 40 guns each, and some 5,500 marginal sheikhs with a stock of some eight firearms each.

Another sub-set is the country's markets and stores. The major regional markets in Abyan, Al Baydah, Jehannah, Al Jowf, and Sad'ah mostly serve the wholesale trade. There are roughly 300 shops selling firearms in Yemen, averaging roughly 100 guns each, although their actual inventories vary greatly in size.

The most elusive part of Yemeni stockpiles are the small arms of the Ministry of Defence. Although the government of Yemen does not make its firearms inventories public, it is known that there are some 100,000 personnel in its armed forces. On this basis, orthodox Small Arms Survey firearms multipliers can be used, suggesting 220,000 military firearms. Since 1994, moreover, the government has been a major buyer of small arms (Rybarova, 2001). Shopkeepers claim that the Ministry of Defence forces merchants to turn over one-third of the weapons they import. Although the exact size of the total government stockpile is unknown, it probably is not more than one-third the total in public hands, or 1.9 million.

TABLE 2.17 Estimated total firearms in Yemen

Location	Number
Individual firearms	5,579,000
Sheikhs' stockpiles	184,000
Commercial stockpiles	30,000
Military arsenals	220,000
Government inventories	1,930,000
Total	**7,943,000**

FIGURE 2.2 Estimated distribution of firearms in Yemen

This analysis shows that the 50 million figure is a myth. More reasonably, Yemen is home to a maximum of 7.9 million firearms. Given the lack of precision in this deduction, though, it is more reasonable to estimate the total at between five million and eight million (see Table 2.17). With one gun for every three people, Yemeni ownership is somewhat higher than in other well-armed countries like Canada, but far less than in the United States, where there is almost one gun per person. The most distinctive factor in Yemen is the very high proportion of fully automatic weapons. Rather than being locked away, moreover, these guns are on the street for all to see.

Source: unpublished research by Derek Miller for the Small Arms Survey, June 2001

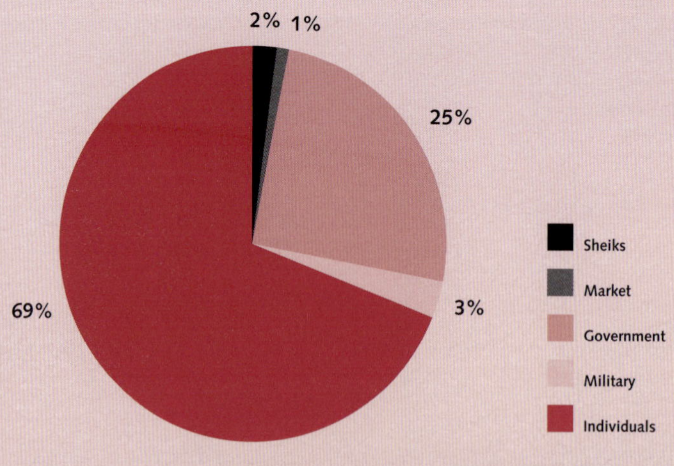

BOX 2.6 Significance beyond numbers: The Karine A affair

The political and military impact of small arms can far surpass their numerical significance. On 3 January 2002 Israeli commandos captured the freighter *Karine A* in the Red Sea. The ship's captain later testified that the ship was bound for the Palestinian enclave of Gaza, where its cargo of 50 tons of military equipment would be dropped in watertight containers to be picked up by the Palestinian Authority. The numbers of weapons involved would barely be enough to be noticed in most parts of the world (see Table 2.18), but they could have had a significant impact on the nature of the al-Aqsa Intifada.

The Israeli operation reportedly began with intelligence revelations in August or September 2001, when the Spanish-built ship was allegedly purchased by Palestinian agents and registered under the flag of the Pacific Island state of Tonga. Establishing the ownership and route of the ship has been difficult—with connections to Bulgaria, Dubai, Iraq, Lebanon, Sudan, and Yemen—leading some critics to question Israeli truthfulness about the affair (Whitaker, 2002a, 2002b).

Israeli officials maintain that the Iranian government sold the weapons to the Palestinians for about USD 10 million in an effort to build influence with the Palestinian Authority. Most of the equipment is, however, virtually untraceable: generic Soviet-style items, mostly of 1950s and 1960s design, manufactured in dozens of countries and stockpiled in scores of others. According to the Israeli Defense Forces, the landmines are clearly of Iranian manufacturer, although this does not exclude the possibility of unauthorized sales.

Initial reactions

Palestinian officials denied any involvement, criticizing the Israeli publicity as a 'propaganda campaign'. President Yassir Arafat denied that the Palestinian National Authority was acquiring arms from Iran. He arrested the official in charge of finances for the Palestinian security forces, Fuad al-Shubaki, and said he would form a committee to investigate Israeli claims (Palestinian News Agency, 2002a; 2002b; 2002c). Subsequently, Arafat has implied that the affair was an Israeli ruse (*Tehran Times*, 20 January 2002). Iranian foreign minister Dr. Kamal Kharrazi dismissed Israeli claims against Iran as 'unfounded', stating that 'the issue of the ship carrying weapons has nothing to do with Iran' (Kharrazi, 2002).

In addition to the potential political significance of an Iranian arms sale and the military impact of the weapons, the affair posed a major challenge to the authority of Yassir Arafat. To many observers, it suggested either that he lacked control over the Palestinian Authority or that he was lying. Such concerns led US President Bush to pressure Arafat to clarify his role in the controversy (Bush, 2002).

Military impact

Only part of the 50 tons of military equipment found on the ship were small arms or light weapons. Because of the sensitivity of the situation, though, these could have had major effects on the current conflict. The 212 AK-47 rifles would have represented a tiny addition to a Palestinian firearms inventory previously estimated to hold 70,000 automatic rifles. Potentially more disruptive was the small quantity of sniper rifles (30) and machine guns (18). Although few in number, these would have created much greater risks for Israeli civilian residents and forces operating in Gaza and the West Bank. The 29 light mortars would have facilitated attacks on surrounding communities.

If in fact destined for Palestinian forces, the heavier military equipment on the *Karine A* would have permitted far greater escalation, with weapons with enough range to reach more distant Israeli cities and destroy Israeli armoured vehicles. In the words of a veteran Israeli defence analyst:

> The Katyusha rockets would have extended the scope of Palestinian attacks, placing the outskirts of Ashkelon and other large Israeli population centres within range. Had some of these Katyushas and mortars been smuggled into the West Bank, Palestinian missile attacks would have been able to reach Ben-Gurion International Airport and major Israeli cities. Sources connected to the IDF General Staff say such a situation would have necessitated the re-occupation of a wide swath of territory in the West Bank and Gaza Strip—in other words, wide-scale war. (Schiff, 2002)

Serious as these implications are, future shipments could bring even greater capabilities. Unlike the arms ship *Santorini*, intercepted eight months earlier, the *Karine A* did not carry SA-7 anti-aircraft missiles. Whoever supplied the equipment apparently demonstrated some restraint.

BOX 2.6 Significance beyond numbers: The Karine A affair (continued)

TABLE 2.18 Weapons seized from the ship Karine A, 3 January 2002

Type	Quantity	Capability
Unguided rockets		
122mm Katyusha launchers	4	
122mm Katyusha rockets	62	Range 20km
107mm rocket launchers	6	
107mm rockets	283	Range 8.5km
RPG-7 launchers	51	
RPG-7 rockets	328	
RPG-18 rocket launchers	346	
Mortars		
120mm mortar tubes	10	
120mm mortar bombs	700	Range 6km
81mm mortar tubes	19	
81mm mortar bombs	686	Range 5.2km
60mm mortar tubes	10	
60mm mortar bombs	159	Range 1.7km
guided missiles		
AT-3 Sagger anti-tank launchers	6	
Sagger missiles	10	Range 3.5km
Mines and explosives		
Anti-tank mines	522	
C-4 explosives	2,200 kg	
Fragmentation grenades	735	
Firearms		
Dragunov sniper rifles	30	
PK machine guns	18	
AK-47 automatic rifles	212	
Firearms ammunition	699,200 rounds	

Source: adapted from Israeli Defense Forces, 10 January 2002

© Reuters

Fifty tons of weapons were offloaded from the Karine A ship in the Israeli harbour of Elait.

Waking giant: Small arms in the People's Republic of China

The world's most populous nation has an exceptionally opaque weapons culture. While many governments refuse to make national firearms statistics available to the public, few succeed as completely as Beijing. This review does not reveal the number of small arms in the People's Republic of China. Instead, it conveys a general sense of the number in the country. It concludes that the actual total probably is much greater than commonly assumed, especially among the People's Liberation Army (PLA) and civilian owners.

It comes as no surprise that the most populous nation has the world's largest police force. The best armed are the Chinese People Armed Police. Formerly part of the Ministry of Defence, they have a total strength of 1.1 million armed officers (Federation of American Scientists, 2001). China's civil police force is much larger, but its members normally go unarmed in most of the country. Whether or not firearms are available for all civil police is unclear. According to the 1982 national census—the latest available at the time of this writing—China's civil police numbered 3.4 million. On the assumption that only half of the police are armed, the total number of armed police is roughly 2.8 million. If it is like most countries—such as Denmark (above)—and maintains only a small reserve stockpile of police small arms, the total number of firearms for both Chinese police forces is approximately three million.

Chinese police prepare to destroy some of the 5,000 guns confiscated in Huaian county, Jiangsu province, on 2 January 2002.

The small arms inventories of the PLA are more difficult to estimate. At its peak during the Cold War, the PLA had a total strength of four million active-duty personnel and some two million reservists. In the 1980s active troop strength fell by one million. Another half-million were released in 1997–2000, mostly as military-owned factories and businesses became independent. This leaves a total strength of 2.5 million active troops (Reuters, 2001a).

Although the strength of the PLA has been cut significantly, few of the cuts involved front-line combat units, which supports the conclusion that weapons inventories have changed much less than the number of personnel. Even if one deducts reductions in PLA stockpiles by firms like PolyTech—which specialize in selling second-hand Chinese military hardware (PRODUCERS)—and allows for the possibility of small arms destruction, it appears that most of the small arms present at the peak of the Chinese armed forces in the early 1980s remain in the country today. Rather than shrinking, the military stockpile may have grown due to continued procurement.

On the basis of the armed forces multiplier of 2.25 firearms for every person in uniform, the PLA was previously thought to control approximately 13.5 million small arms. This assumed that China followed the practices typical of Western nations, based exclusively on mobilizing fully trained military personnel for war. There is compelling reason to conclude, however, that this assumption minimized actual PLA planning. Above all, it fails to account for the implications of Chinese strategic planning, based for the first 25 years after the revolution on the Maoist doctrine of People's War. In case of invasion, China expected to arm broad swaths of its population, starting with military veterans, Communist Party members, factory workers, university students, and other politically trustworthy groups. Insights from other countries relying on people's armies, especially Albania and the former Yugoslavia, point to a suitable small arms multiplier of 4.5 firearms for each active and reserve troop, to permit massive mobilization. To arm millions of likely combatants, at its peak the Chinese military inventory probably totalled at least 27 million firearms, probably the biggest in the world.

At its peak, the Chinese military inventory probably totaled at least 27 million firearms, probably the biggest in the world.

Chinese civilian gun ownership is especially difficult to evaluate. Many Chinese people, such as those from major cities and coastal regions, are under the impression that guns are scarce in their society, strictly licensed, and confined to the criminal underworld and the remote regions. Certainly this appears to have been the case until recently. A growing body of evidence from Chinese police suggests that, to the contrary, firearms are rapidly becoming more plentiful. Despite the apparent contradiction, both perspectives may be right; the vastness of Chinese society makes it possible for a relatively low proportion of people to be gun owners while their absolute numbers are relatively high.

Recent revelations of police seizures of illegal firearms point to the actual scale of firearms ownership (see Table 2.19). Reports of police seizures and destructions of civilian guns are increasingly commonplace. According to a Chinese diplomat, for example, in 1998 China confiscated and destroyed 300,000 illicit guns. While this figure seems large, he cautioned that 'every year, a big amount of illicit guns is confiscated and destroyed' (Xiang, (2000/01)). This statement received credence from many similar reports, such as the announcement that police seized another 600,000 guns from March to June 2001 (Reuters, 2001b; *The Economist*, 2001c; WEAPONS COLLECTION).

Such figures on criminal confiscations emerge mostly through periodic 'Strike Hard' campaigns. Launched by the government since 1983 to eliminate a changing list of crimes, these tend to focus on previously overlooked problems. The year 2000–01 was exceptional for mass arrests along this pattern (Craig Smith, 2001). It is possible that these reports from Strike Hard campaigns are exaggerated to curry favour and for propaganda reasons. The problem is similar to recently discovered exaggerations by Chinese fisherman, who systematically magnify their hauls, just as communist officials in the former Soviet Bloc lied about overfulfilled quotas for decades (*The Economist*, 2001d). But the consistency of firearms reports and collaborating evidence—such as photographs of Chinese firearms being destroyed—leave little doubt that gun proliferation is seen as a growing social problem.

But where do these illegal civilian firearms come from? In China, officially permitted production of small arms is dominated by one company, Norinco (PRODUCERS). With more than 130 separate companies employing more than 450,000 people, Norinco has the resources to produce weapons at such a rate. In the 1960s, for example, it routinely produced hundreds of thousands of small arms annually for foreign clients. Among the largest recipients were North Korea (670,000 small arms in the 1960s), Albania (752,000), and especially Vietnam (2.2 million in the 1960s and early 1970s) (Kong, 1994). While the state-owned Norinco has the *facilities*, though, it is hard to believe it has the *incentives* to supply the illegal domestic gun trade. A second possible source is trading firms like PolyTech, which have a reputation for being less scrupulous.

A third major source is unlicensed production. Firearms production also is reported to be a growing problem in the western provinces of Gansu, Qinghai and Sichuan, and Guangxi in the south, and to a lesser extent in Tibet (see Map 2.1). Specific cities have even acquired reputations for gun production, much like Darra in Pakistan. In Hualong County, part of the province of Qinghai, manufacturers are proud enough to stamp the firearms they build with 'Made in Hualong'. In the winter of 2000–01, police in the province of Guangxi raided 60 illegal firearms factories. In all, 240,000 illicit 'guns, moulds and tools' were reported seized in this one province. (*South China Morning Post*, 2001).

The scale of Chinese confiscations of illegally trafficked firearms can be compared to the same phenomenon in the United States. The American government confiscates an average of 34,000 illicitly traded firearms each year (US, BATF 2000).[4] The *entire* United States, in other words, with some 240 million publicly owned firearms, confiscates from illegal trafficking only 15 per cent as many guns as a *single* Chinese province. In all, China has 34 provinces, independent municipalities, and autonomous regions. It is hard to resist the conclusion that China's gun market is growing at a rate virtually without precedent elsewhere.

Even if Strike Hard reports are greatly exaggerated, the total numbers of small arms in public hands for all of China must reach into the tens of millions. At the same time, there are limits to what can be concluded from reported confiscations alone. At this point, the data is not sufficient to determine the proportion of China's total civilian stockpile that is being confiscated. But it seems that the reported confiscations account for no more than a small percentage of the total public stock of small arms.

In light of the enormous numbers of firearms confiscated by the police, China must have more publicly owned firearms than almost any other country in the world. In terms of general magnitude, it probably is surpassed only by the United States with some 240 million civilian firearms, and possibly India, home to some 40 million. While the total number of privately owned guns in China still cannot be given a single numerical value, it appears to approach and possibly surpass the latter figure.

MAP 2.1 China's firearms centres

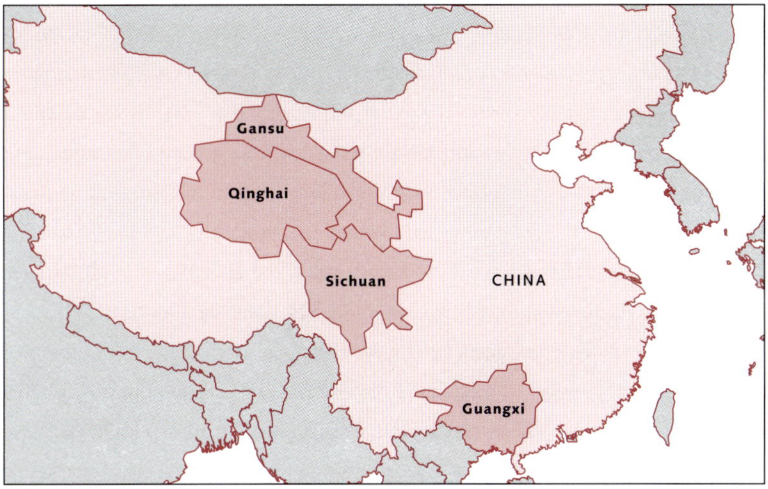

TABLE 2.19 Strike Hard gun seizures in China

Region	Date	Total seized	Report
National	1993	90,000	Police
National	1994	120,000	Police
National	1998	300,000	Diplomatic
Guangxi	Winter 2000	240,000	Police
National	Spring 2001	600,000	Police
National	Spring 2001	330,000	Police

Sources: Asiaweek (1995); The Economist (2001c); Reuters (2001a, 2001b); Xiang (2000/01)

In light of the enormous numbers of firearms confiscated by the police, China appears to have more publicly owned firearms than almost any other country in the world.

Southeast Asia: The rising prominence of insurgent forces

In other parts of Asia, such as Southeast Asia, small arms trends are more stable. The major exception is the rise of secessionism in Indonesia and the Philippines. Growing demand has fuelled increased clandestine trafficking, enabling insurgents and ethnic combatants in these countries to accumulate bigger reserves. What is less clear is whether these growing arsenals represent new weapons, increasing the total global arms supply, or whether they are redistributing weapons already in the region.

The evidence on this point is mixed. Some accounts point to quantities of new small arms being imported, often through highly circuitous routes and mostly originating in China. Other accounts suggests that most of these transfers involve local ant-trade and diversion from national armed forces, as weapons from more peaceful parts of East Asia circulate to regions caught in active warfare (TRANSFERS).

Indonesia saw the region's worst communal violence in 2001, with several parts of the country threatening to leave the union. Not all breakaway factions are well-armed. In West Papua (formerly Irian Jaya) the largest militia, the Satgas Papua, has between 7,000 and 20,000 followers. So far it has been held in check, in part because of its inability to acquire firearms. Geographically isolated and lacking a sizeable community of expatriates abroad or the support of a sympathetic government, it has no reliable way to acquire small arms. The most likely source at the moment probably is capturing armaments from the growing Indonesian military and police presence, up from around 3,000 personnel one year ago to 12,000 today (Bostock, 2001).

Fighting is even more serious at the opposite end of the archipelago in Aceh, where the Free Aceh Movement, with some 5,000 combatants, faces a government security force estimated at 32,000. Less isolated geographically, Aceh rebels have found it much easier to acquire weaponry. Their arsenal is estimated at 2,000–3,000 firearms, including M16 rifles captured or bought from government forces and AK-47s smuggled from abroad (Gunaratna, 2000/01; Harris, 2001). Many of the latter appear to originate in Thailand. Some apparently come from Thai Army arsenals, but most probably originate with new production from China channelled through Thai brokers (*Jane's Intelligence Review*, 2001).

The other major scene of separatist violence in Southeast Asia is the Philippines. In 2000–01 headlines were dominated by the Abu Sayyaf Group based in the island chain of Basilan, Sulu, and Tawitawi. This Islamic separatist faction won global press coverage by holding foreigners for ransom. Having resisted several campaigns by the Philippine army, Abu Sayyaf has enormous local prestige. Its wealth and success led to expansion from perhaps 200 armed followers in 2000-01 to roughly 1,100 combatants and several thousand followers (*New York Times*, 2001). Perceived as a money-making operation as much as a secessionist group, it can afford to turn away would-be recruits (*The Economist*, 2000). Unlike most rebel factions, Abu Sayyaf also appears to be extremely well armed. The Muslim nationalist armies on the island of Mindinao are much larger, dominated by the 15,000 strong Moro Islamic Liberation Front (Landingin, 2001). But the situation there, although fragile, has been more stable, suggesting no significant changes in small arms holdings.

Insights into the total number of civilian firearms in the Philippines came through a new Philippine government report on firearms licensing and gun ownership (Philippine National Police, 2001). This establishes that at the end of 1999 there were 706,148 privately owned firearms registered in the country, a figure growing at a rate of 10–15 per cent annually. In addition there were 329,985 'loose' or unregistered guns. The remarkable precision of the latter figure reflects a narrow definition of unregistered weapons, covering only weapons purchased from licensed dealers but not subsequently registered with the police. The combined total of 1,026,133 known firearms in public hands does not include all the unregistered guns in the country, especially in view of the large quantities reportedly sold by unlicensed dealers or manufactured in illegal factories.

FIGURE 2.3 Approximate small arms distribution in the Philippines

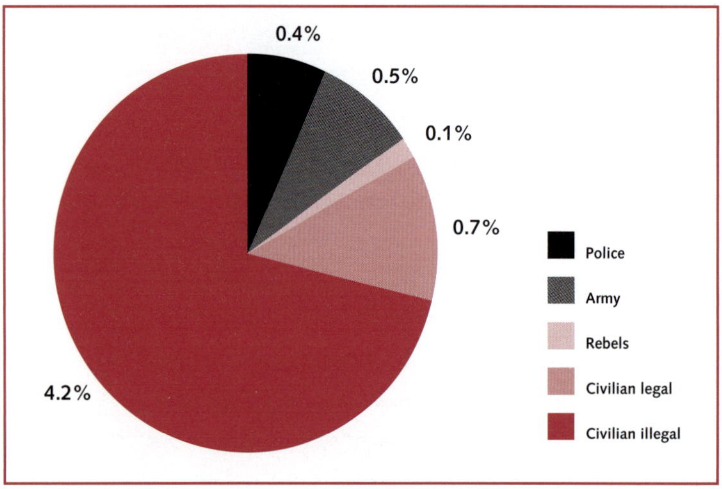

In reality the total of unlicensed firearms in the Philippines appears to be substantially greater than the confirmed figure. The real picture is suggested by the apparent fact that unlicensed firearms, although

BOX 2.7 Cambodia's leftover arsenal

MAP 2.2 The flow of weapons out of Cambodia

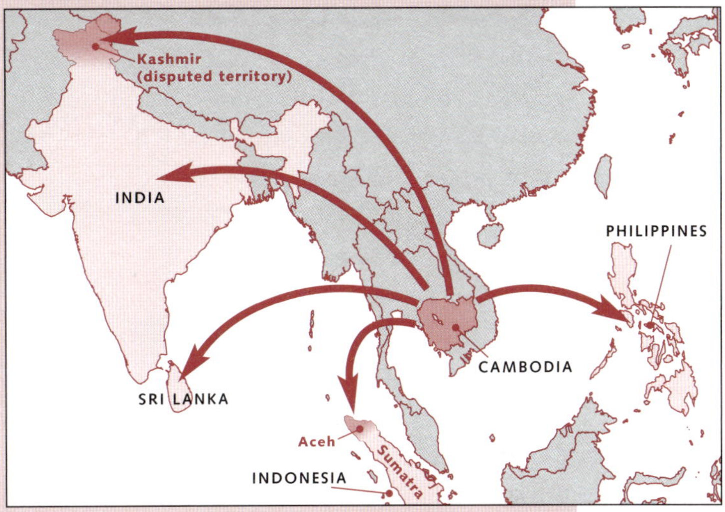

Over a decade ago, Cambodia ended a hellish era of some 20 years of warfare and mass murder. The remnants of the killing live on. The dangers of hidden landmines are well known. Of growing importance is the legacy of small arms proliferation, contributing to conflicts in other parts of Southeast Asia, including Aceh, the Philippines, Sri Lanka, north-east India, and even distant Kashmir (Chalk, 2001; Fawthrop, 2001).

Estimates of the total number of small arms in the country vary. Since mid-1988, according to the Cambodian government, a total of 112,562 small arms have been collected, and some 63,000 destroyed (WEAPONS COLLECTION).[5] The total number of small arms remaining in the country is much greater, although efforts to establish the total have been inconclusive. A prominent estimate maintains that there are around 500,000 small arms in Cambodia, half controlled by the government—both the military and the police—and half by militiamen, demobilized soldiers, and other individuals. Another estimate maintains that there are 900,000 unregistered small arms in the country (Fawthrop, 2001).

The lower, 500,000, figure corresponds to the total number of small arms to be expected based on the maximum number of combatants active in the country at the height of fighting in the mid-1980s, when there were roughly 220,000 government troops, village militiamen, Khmer Rouge, and other factions (IISS, 1988). Using conventional multipliers, one would expect that there would have been some 550,000 small arms in the country at the close of hostilities. But one should not dismiss the higher estimate of 900,000, since the lower figure is the result of a deliberately conservative approach.

Unneeded by Cambodian authorities, these weapons are haemorrhaging out of the country. Just as the leftover South Vietnamese arsenal of American-supplied weaponry became a pantry for Soviet-supported Third World clients and insurgencies in the late 1970s and early 1980s, the unneeded Cambodian arsenal is a source of supply today. In this case economic opportunity is encouraging their dissemination, not global political agendas.

supposedly less common than legally owned guns, are involved in over 92 per cent of resolved criminal firearms cases (Philippine National Police, 2001). This disparity leads to the conclusion that the total number of unlicensed firearms must be much greater than registration statistics suggest.

If any gun has an equal chance of being used in a crime, the total number of unlicensed firearms would have to be over 20 times greater than currently known by the police to account for the disparity. But in reality not all firearms are equally likely to be used in crime. Assuming that unlicensed firearms are twice as likely to be used in crime, their total probably is closer to 4.2 million. When added to the total of licensed weapons, this indicates the existence of approximately 5.3 million privately owned guns in the Philippines (see Figure 2.3).

South Asia: New data and new interpretations

Just as in most other regions of the world, in South Asia only the military firearms inventories were reasonably well understood in the past. The weapons trade appears to be accelerating in the region due to a combination of inter-state conflict—the American-led campaign against Afghanistan and border strife between India and Pakistan—and internal tension—worst in Sri Lanka, but serious in Kashmir, Nepal, and elsewhere. Previous information was good enough to permit standard estimations of the scale of armed forces and insurgent weapons inventories, but not police or privately owned weapons. From 2000–01 this problem began to ease, due largely to events in Pakistan.

In response to rising social violence and the easy availability of automatic weapons, Pakistan's Minister of the Interior, Lieutenant General Moinuddin Haider, initiated a national Arms Control Campaign in the autumn of 2000. With sectarian battles and feuding increasingly common on the streets of Pakistan's major cities and much of the countryside, small arms proliferation has become one of the country's leading social problems. The main goal of the campaign was to facilitate 'de-weaponization' through confiscation and prosecution of illegally held arms (Rizvi, 2000; WEAPONS COLLECTION). Initially, owners were encouraged to register legal firearms. During the second stage owners could surrender illegal firearms over a two-week amnesty in June 2001. Although the number of weapons received was not expected to be great, the hope was that these measures would break the culture of freely carrying Kalashnikov rifles in public and facilitate future police intervention (AP Worldstream, 2001).

MAP 2.3 South Asia

According to the 1998 census, Pakistanis owned around two million licensed firearms (*Business Recorder*, 2001). In addition, officials of the Ministry of the Interior believe there are roughly 18 million more illegally held (Bokhari, 2001). While there is no reason to doubt the scale of these numbers, they may give a misleading impression of the *distribution* of small arms in Pakistani society. Much like their counterparts in the United States and other gun-proud cultures, Pakistani firearms owners typically have more than one gun, often as many as a half-dozen. Consequently, firearms ownership is concentrated in a section of the population. Geographically, gun ownership is more common in regions like the Northwest Frontier and Baluchistan than in Sindh and Punjab, although this is changing slowly as small arms become more common everywhere (Cohn, 2001). Out of a total population of some 140 million, firearms ownership appears to extend to some four million to six million individuals.

Compared with the 20 million firearms believed to be in public hands, the inventories of the armed forces appear to be much smaller. The Pakistan Armed Forces, with a total active and reserve strength of 1.3 million—including the National Guard, as well as Frontier Corps and Pakistan Rangers under the Ministry of the Interior—probably have at least three million small arms of all types. This figure does not include the often repeated but unverified tale that the Pakistani government has three million AK-47s left over from the Afghan war of the 1980s, 'packed in grease' (Kartha, 2000).

The Pakistan figures also help clarify the distribution of firearms in India. As in Pakistan, the general size of military inventories in India was understood previously. The armed forces and especially the Indian Ministry of the Interior have been major arms buyers in recent years. Since 1998, the Army has been acquiring an average of roughly 80,000 domestically manufactured INSAS rifles annually to re-equip its combat units (Press Trust of India, 2000; *The Pioneer*, 2000). Ministry of Interior forces, like the Rajestan Rifles and other units deployed in Kashmir and other regions of violent conflict, also have been major buyers, receiving several hundred thousand AK-47s imported from eastern Europe (Raghuvanshi, 2000).

In a single deal, India can purchase more firearms than most South Asian countries have in their entire armed forces stockpile.

Insurgent arsenals in India appear to be relatively stable, as the size of major secessionist groups has not grown significantly. Active fighting in Kashmir, Nagaland, and Assam ensures a continuing demand for additional rebel small arms. In Kashmir weapons probably continue to flow in at the rate of several thousand annually, while other conflicts attract several hundred new small arms per year (*Jane's Islamic Affairs Analyst*, 2000; *Jane's Terrorism and Security Monitor*, 2000).

The best estimate of the scale of civilian ownership in South Asia was the widely cited figure of at least 73 million firearms.[6] Previously, though, this number was not very useful. Not only was it an estimate, it combined all seven countries of South Asia. With a sense of the number of civilian firearms in Pakistan, however, the larger figure can be disaggregated somewhat, although in view of currently available information it still appears high. If Pakistani guns are subtracted, this leaves roughly 50 million firearms in public hands in the rest of South Asia. India, with a population 50 times bigger than either of the next two largest South Asian countries—Nepal and Sri Lanka—is home to the great majority of these weapons, approximately 48 million, public and private. Indian small arms appear to be divided approximately seven million military, 0.6 million for the police, and some 40 million civilian.

In all, South Asia appears to be home to roughly 75 million firearms, of which some 63 million are owned by civilians (see Figure 2.4 and Table 2.20). The dominance of India and Pakistan in these figures is almost overwhelming. In a single deal, India can purchase more firearms than most South Asian countries have in their entire armed forces stockpile. An Army requirement for new rifles originally was set at 528,000 in 1993 (*Times of India*, 2001). Since then several additional orders have been placed, including a purchase from Bulgaria for 200,000 AK-47 versions (Raghuvanshi, 2000). The smaller countries of South Asia also have seen an increase in small arms inventories, largely as a consequence of continuing civil war. The 18-year war in Sri Lanka continues unabated. Although the size of the LTTE and government armies did not change significantly in 2001, both sides continue to replace wastage (Gunaratna, 2000/01).

The situation in Nepal is more dynamic, due to the rising strength of the revolutionary Nepal Communist Party, a Maoist insurgency. Since the fighting began in February 1996, the Maoists have gradually been gaining strength, and reportedly field some 1,500 armed combatants. More than 2,000 people have died in the fighting—two-thirds insurgents, one-third police and civilians—and there is a growing sense that the Maoists are gaining momentum (Bhattarai, 2001a). On 26 November 2001, escalation in the fighting led King Gyanendra to declare a state of emergency

FIGURE 2.4 Approximate distribution of small arms in South Asia

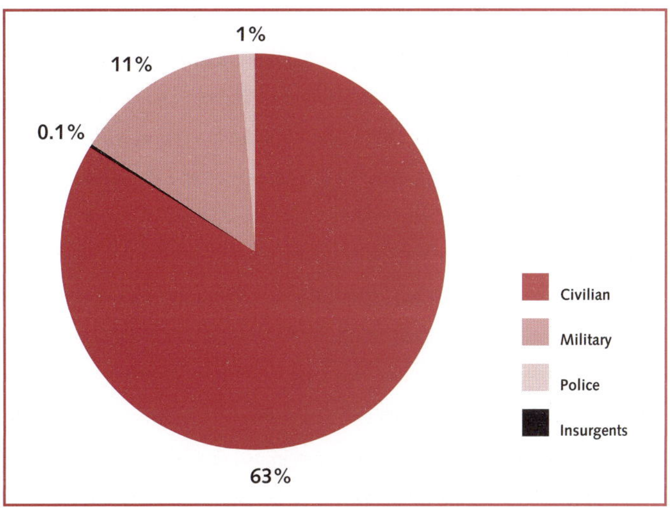

(Dugger, 2001). Nepal's civilian police, armed exclusively with handguns and Second World War-vintage Enfield rifles, are increasingly outclassed. The government in Kathmandu has requested bids to purchase 50,000 automatic rifles and machine guns to make them more competitive (Agence France Presse, 2001; Bhattarai 2001b).

TABLE 2.20 Approximate distribution of firearms in South Asia, by country (millions of firearms)

Country	Civilian	Military	Police	Insurgent
Bangladesh	0.2	0.4	0.1	0.01
India	40.0	7.0	0.6	0.1
Nepal	1.0	0.1	0.05	0.01
Pakistan	20.0	3.0	0.4	?
Sri Lanka	2.0	0.3	0.05	0.01

In Bangladesh sensitivity to the small arms issue became markedly more pronounced in recent years. As organized crime and student political factions became more violent, political leaders in Dhakka began to take the issue seriously. Greater awareness of the subversive effects of Bangladesh's role in the region as a transit country also is mobilizing concern (TRANSFERS). A consortium of analytic and non-government organizations, the National Small Arms Forum, estimates that there are some 200,000 civilian small arms in the country, with a total population of some 120 million. A total of 50,000 or so of these arms are believed to be in criminal hands (*Bangladesh Independent*, 2001; *Nation*, 2001). Other observers maintain that there are 250,000 illegal weapons alone in the country (Ahmed, 2001). The numbers involved may seem small for a country with a population of 120 million, and they are small compared with the problems of its neighbours, but Bangladesh has reacted sharply to rising gun violence. The problem has become highly charged, provoking a sharp national debate on how to respond.

Central Asia

On the borders of South Asia, the very nature of international security was altered by the attacks of 11 September 2001 and the subsequent war in Afghanistan. While small arms have been the dominant weapon in the ground fighting, second only to American bombing, the effect on stockpiles is less clear. One surprising discovery of the war was the poor quality of the armaments of the opposition Northern Alliance. Despite the region's reputation for mammoth small arms stockpiles, the Northern Alliance had to be almost completely rearmed by Russia and the United States—even with several thousand additional AK-47s—before it could begin strategic operations (Filkins and Schmitt, 2001). Pakistan continued to permit arms shipments to the Taliban until November 2001 (Frantz, 2001b). Disarming Afghanistan is increasingly recognized as essential to the nation's reconstruction (Bhutto, 2001).

In Central Asia the Islamic Movement of Uzbekistan has begun mounting deadly raids into the politically sensitive Fergana Valley of Kyrgyzstan and Uzbekistan. The movement came to international notice in August 1999 when it captured Japanese geologists for ransom (Frantz, 2001a). Since then it has extorted enough wealth and acquired enough small arms to threaten the stability of the region.

As tensions mount in both South and Central Asia, small arms procurement almost certainly will accelerate. Events since 11 September 2001 have proven fertile for craft armourers in towns like Darra in Pakistan, where prices appear to be increasing (*The Hindu*, 2001). Prices for imported weapons are accelerating as well (McGirk, 2001).

Conclusion

The overwhelming characteristic of the global small arms stockpile is continuity. Most of the weapons identified in the *Small Arms Survey 2001* remain as before. Police inventories changed very little in most countries, while the total global military stockpile grew by approximately 700,000 as a result of new production (PRODUCERS). Insurgent stockpiles also appear to have grown marginally, by less than 100,000. Although insurgents undoubtedly received considerably more arms in absolute terms, most of these weapons went to replace lost equipment, leaving overall figures almost the same as before. The greatest exception for the total stockpile was civilian purchases of new firearms, estimated to have grown by some 6.9 million in 2001 as a result of new production (PRODUCERS).

The re-transfer of second-hand weapons has the greatest effect on the distribution of the global small arms stockpile.

The biggest changes revealed in this chapter come not from new production but from the slow chipping away at unknown categories of small arms, adding previously missing information to create a more complete picture. The largest change came through improvement in the statistics for civilian ownership, especially in South and East Asia. As more information becomes available, the world's 'unknown' categories are gradually declining. The *Small Arms Survey 2002* estimate of all known firearms, at least 638 million, is significantly more complete than former estimates. Like the former figure, this is a conservative estimate; the actual global stockpile undoubtedly is greater, by several tens of millions at least, if not more.

When reassessing the number of firearms, it is important to keep in mind that higher totals for many countries, such as China, India, and Pakistan, cannot simply be added to the estimates presented in the *Small Arms Survey 2001*. Most or all of their police and military inventories were already included in the estimates for 2001. The net total global figure is given in Table 2.21. The numbers that have been added do not represent newly produced firearms. With the exception of some eight million newly manufactured firearms—both military-style and civilian firearms—the rest of the 87 million firearms being counted in the global stockpile for the first time here have existed for some time, but their existence could not be reliably demonstrated. They are included here with confidence in their reliability as estimates, as accurate as possible, and conservative where accuracy is in some doubt.

Only a small proportion of the global small arms stockpile comes from new production, less than one per cent a year. It is rather the re-transfer of second-hand weapons that has the greatest effect on the global distribution of guns. Re-circulation of old weapons, rather than purchases of new ones, accounts for the greatest proportion of global firearm movement. This has important implications for policy options like weapons marking and tracing, which must deal with existing guns to be effective. There are other options for second-hand small arms too. Weapons collection and destruction programmes have significantly reduced the inventories of several states and organizations, eliminating over four million small arms in the last few years.

FIGURE 2.5 Distribution of known global firearms, 31 December 2001

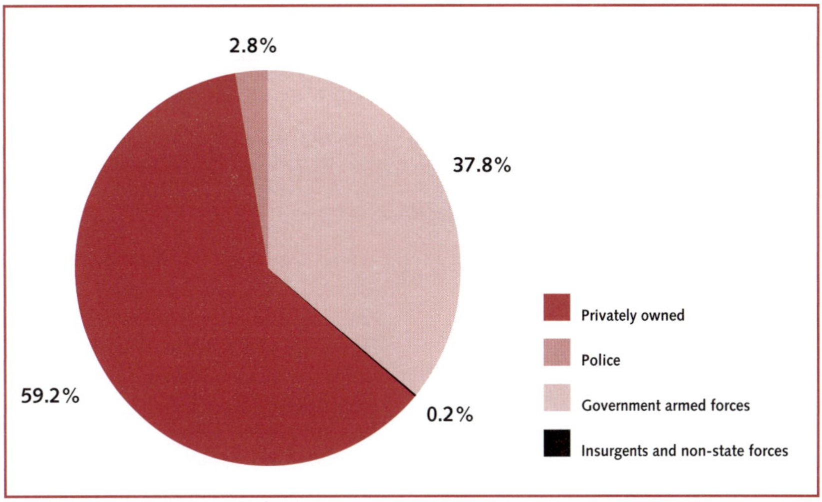

This chapter also draws attention to two other types of small arms and light weapons. Although far less numerous than firearms, shoulder-fired rocket launchers and mortars are extremely important in global conflict, adding to the destructiveness of terrorism, guerrilla warfare, and state-to-state conflict. Easily overlooked in the deliberations of both governments and global civil society, they warrant much more careful consideration in future discussions of small arms issues.

TABLE 2.21 Distribution of known global firearms, 31 December 2001

Group	Estimated firearms	Percentage
Police	18,000,000	2.8
Government armed forces	241,600,000	37.8
Privately owned	378,300,000	59.2
Insurgents and non-state forces	1,000,000	0.2
Total	**638,900,000**	**100.00**

While military small arms may be beginning to decline, civilian ownership appears to be increasing.

Not all national small arms arsenals are growing uniformly. The United States allowed its military small arms stockpile to contract significantly in the last decade. Other governments, possibly Germany and Russia, appear to be following the same path through organizational firearms destruction. Are we witnessing the start of a trend? Will other governments pursue unilateral reductions? Some governments, such as Canada, have seen their weapons-to-troops ratio rise as troop numbers were cut while arsenals remained untouched. All in all, however, even aggressive military cuts are unlikely to affect overall global stockpiles. Trends in civilian ownership are likely to offset any declines in military arsenals. In this chapter, the examples of the United States and the Czech Republic both illustrate the tendency for private gun ownership to climb quickly, irrespective of military cuts.

Other regions appear to be rapidly acquiring more firearms too. China and South Asia, long left out of the global firearms picture, are among the world's major centres for small arms ownership, each providing a home to tens of millions of civilian, police, military, and insurgent firearms.

Despite the additions, the overall picture of global small arms remains far from complete. Very little is known about the total number of civilian small arms in several major countries including Brazil, France, Indonesia, Iran, Mexico, Thailand, Turkey, and Vietnam. Others have been able to provide only partial or unreliable reports, including Germany, Poland, and Russia. With only a handful of exceptions, there currently is no way to estimate the dimensions of civilian small arms ownership for most of the entire continent of Africa and the Arabian peninsula. Illegal ownership remains almost impossible to quantify in most parts of the world.

Some of these problems can be resolved through more systematic national polling and public surveys on gun ownership. However, a comprehensive picture of the global firearms stockpile ultimately can be obtained only through greater official transparency and information-gathering efforts.

2. List of Abbreviations

AUC	United Self-Defence Forces of Colombia
ELN	Ejército de Liberación Nacional
ETA	Euskadi Ta Askatasuna (Basque Homeland and Liberty)
FARC	Fuerzas Armadas Revolucionarias de Colombia
INSAS	Indian Small Arms System (7.62mm automatic rifle)
IOM	International Organization for Migration
IRA	Irish Republican Army
KLA	Kosovo Liberation Army
LAW	Light Armour Weapon
LTTE	Liberation Tigers of Tamil Eelam
NLA	National Liberation Army
PIAT	Projector, Infantry, Anti-Tank
PLA	People's Liberation Army
UNDP	United Nations Development Programme

UNHCR	United Nations High Commissioner for Refugees
UNRCA	United Nations Register of Conventional Arms
UNTAC	United Nations Transitional Authority in Cambodia

2. Endnotes

1 Originally Fetter concluded the total number of firearms was 612 million. This was quickly amended to 594 million (Fetter, 2001b; 2001c).

2 Correspondence and discussions, November–December 2001, with official of the Togo Embassy, Washington, DC.

3 The information in the following section was furnished by US Army officials interviewed at Rock Island Arsenal, Illinois, 9 May 2001, and in subsequent communications. The author would like to express appreciation to the US Army for its co-operation.

4 In addition both countries confiscate 'crime guns' as part of normal criminal investigations. In the United States this amounts to approximately another 200,000 guns annually (US, Department of Justice, 2000). The figures for Chinese crime guns are not available.

5 Reports that the United Nations Transitional Authority in Cambodia (UNTAC) collected a total of 320,443 small arms and light weapons (Chalk, 2001) appear to be based on misunderstanding, confusing an estimate of the total number of small arms and light weapons controlled by all four Cambodian armies (Fawthrop, 2001).

6 The estimate of 73 million civilian firearms in South Asia was publicized in a conference on small arms proliferation hosted by Network for Child Welfare, inaugurated by then-Defence Minister George Fernandes (*The Hindustan*, 2000).

2. Bibliography

Acheampong, Marilyn Opoku. 2001. 'Ghana: millions of small arms said circulating in West Africa.' GHC Radio. 29 March, BBC Worldwide Monitoring.

Achieng', Judith. 2001. 'Rights-Kenya: government accused of laxity in probing cattle thefts.' Nairobi Inter Press Service. 26 March.

Agence France Presse. 2001. 'Nepal to upgrade arms in face of deadly Maoist rebellion.' 18 May.

Ahmed, Anis. 2001. 'Bangladesh leaders seek urgent move on guns.' Reuters. 17 July.

Albania, Republic of. 2001. 'Regarding some information on weapons collected and problematic areas.' Private correspondence with the Small Arms Survey. Tirana: General Directory of the Police, Ministry of Public Order. 13 November.

al-Jazirah (Doha). 2001. 'Iraq: damage caused by Iranian missile attack reportedly "not grave".' 18 April, FBIS translation.

Ananova. 2001. 'One thousand Kalashnikovs thought to be in Northern Ireland.' 23 October.

AP Worldstream. 2001. 'Pakistan begins major media campaign to seize illegal weapons.' 30 May.

Asiaweek. 1995. 'Law and order: a growing torrent of guns and narcotics overwhelms China.' 25 August, p. 36.

Associated Press. 2001. 'Cambodia seizes war weapons.' 9 July.

B92 Radio News. 2001. 'Lorry-load of arms seized in Kosovo.' 15 June.

Bangladesh Independent. 2001. '50,000 arms in hands of criminals, godfathers.' 7 July.

Best, Joel. 2001. *Damned Lies and Statistics: Untangling Numbers from the Media, Politicians and Activists.* Berkeley: University of California Press.

Bhattarai, Binod. 2001a. 'Nepal's Maoists urged to lay down arms.' Nepali Times. 17 April.

__. 2001b. 'USD 70,000,000, that's how much.' *Nepali Times.* 23 November.

Bhutto, Benazir. 2001. 'Bhutto calls for destruction of small arms in Afghanistan.' New York: Eminent Persons Group. 4 December.

Bokhari, Farhan. 2001. 'Pakistan tries to persuade gunmen to ease off the trigger.' *Financial Times.* 2–3 June, p. 3.

Bostock, Ian. 2001. 'Jakarta raises stakes in West Papua.' *Jane's Intelligence Review.* April.

Bottigliero, Ilaria. 2000. *120 Million Landmines Deployed Worldwide: Fact or Fiction?* London: Pen and Sword.

Bowden, Mark. 1998. *Blackhawk Down.* Boston: Atlantic Press.

Brooke, James. 2000. 'Canadians said to obey gun law, after all.' *New York Times.* 26 December, p. A11.

Brown, John Murray. 1999. 'Unionists signal deal over arms.' *Financial Times.* 18 November, p. 10.

__. 2000. 'Resurrection of a peace deal in Ulster.' *Financial Times.* 8 May, p. 15.

__. 2001. 'IRA says it has started to disarm.' *Financial Times.* 24 October, p. 1.

__. and Jimmy Burns. 2001. '"Ghost herds" and smuggled cigarettes fuel terror.' *Financial Times.* 26 April, p. 12.

Bush, George W. 2002. 'Remarks of the President while touring United States Coast Guard Cutter Tahoma.' Office of the Press Secretary, The White House. 25 January.

Business Recorder. 2001. 'Arms Control Campaign starts Friday.' 1 June.

Canadian Firearms Centre. 2001. 'Independent Survey Provides Updated Estimates on Number of Firearm Owners in Canada.' Ottawa. 25 January.

Cassels, David. 2000. 'Gun law aids in protecting Canadians.' *Edmonton Journal.* 19 November.

Chalk, Peter. 2001. 'Light arms trading in SE Asia.' *Jane's Intelligence Review.* March.

Clasmann, Anne-Beatrice. 2001. 'Why are the Swiss armed to the teeth?' *Deutsche Presse-Agentur.* 25 August.

Cohn, Martin Regg. 2001. 'Kalashnikov culture.' *Toronto Star.* 11 April.

Coker, Margaret. 2001. 'Somalia: amid war, famine, selling guns.' *Atlanta Constitution.* 9 July.

Czech Republic. Ministry of Foreign Affairs. 2001. *The Czech Republic and Small Arms and Light Weapons.* Prague: Ministry of Foreign Affairs.

Davies, Roger. 2001. 'Improvised mortar systems: an evolving political weapon.' *Jane's Intelligence Review.* May.

DDP. 2001. 'Germany playing pioneering role in destruction of small weapons.' DDP. 13 July. FBIS translation.

Demetriou, Spyros, Robert Muggah, and Ian Biddle. 2002. *Small Arms Availability and Trade in the Republic of Congo.* Geneva: Small Arms Survey.

Denmark. Ministry of Foreign Affairs. 2001. *Questionnaire Concerning Ownership of Small Arms in Denmark by Civilians and Police.* Copenhagen: Royal Danish Ministry of Foreign Affairs. 3 August.

Dugger, Celia W. 2001. 'Nepal curbs rights after Maoist attacks.' *New York Times.* 27 November, p. A6.

The Economist. 1999. 'Beginner's Luck.' 16 January, p. 48.

__. 2000. 'Fears of the hostage-takers.' 26 August, pp. 31–32.

__. 2001a. 'The criminal crucible that is Kosovo.' 17 February.

__. 2001b. 'Holy war in the Spice Islands.' 15 March, p. 42.

__. 2001c. 'Guns in China: the wild East.' 10 November, pp. 39–40.

__. 2001d. 'Fishy figures.' 1 December.
Erlanger, Steven. 2001. 'NATO to let Yugoslav troops into Kosovo buffer zone.' *New York Times.* 13 March, p. A8.
O Estado de Sao Paulo. 2000. 'Brazilian arms trafficking.' 16 July. FBIS translation.
Ezell, Virginia. 2000. 'Arms guru b-day offers policy forum.' *National Defense.* January.
Faltas, Sami and Joseph Di Chiaro, III, eds 2001. *Managing the Remnants of War: Micro-disarmament as an Element of Peace-building.* Baden-Baden: Nomos.
__. Glenn McDonald, and Camilla Waszink. 2001. *Removing Small Arms From Society.* Occasional Paper No. 2. Geneva: Small Arms Survey, July.
Fawthrop, Tom. 2001. '900,000 small arms plague Cambodia.' *Jane's Intelligence Review,* April.
Feci, Zilie and Anila Hoxhaj. 2001. 'Scandal in Public Order Ministry.' *Tirana Republika.* 20 July, p. 3, FBIS translation.
Federation of American Scientists. 2001. 'China'. In *World Intelligence and Security Agencies.* Washington, DC: Federation of American Scientists.
Fetter, Gregory. 2000. *Ordnance and Munitions Forecast.* Newtown, Connecticut: Forecast International, September.
__. 2001a. *Ordnance and Munitions Forecast.* Newtown, Connecticut: Forecast International, January.
__. 2001b. 'Again I say – good luck!' *Forecast International/DMS Market Alert.* 23 July.
__. 2001c. 'Just how many small arms are there in the world?' *Forecast International/DMS Market Alert.* 10 July.
Filkins, Dexter and Eric Schmitt. 2001. 'Afghan rebels claim progress and credit U.S.' *New York Times.* 7 November, p. A1.
Financial Times. 2001. 'Army still keeps grip on "bandit country".' 15 March, p. 10.
Fisher, Ian. 2001a. 'Rwanda and Uganda begin pulling troops from Congo.' *New York Times.* 1 March, p. A3.
__. 2001b. 'NATO's plan for Macedonia: splendid little disarmament.' *New York Times.* 27 August, p. A1.
Fishman, Alex and Ron Leshem. 2001. 'We gave them guns, the rest they got themselves.' *Yedi'ot Ahraronot* (Tel Aviv). 16 February, pp. 12–18, 90 FBIS translation.
Forero, Juan. 2000. 'Rightist squads in Colombia beating the rebels.' *New York Times.* 5 December, p. A12.
Fowler, Robert. 2000. Letter dated 10 March from the Chairman of the Security Council Committee Established Pursuant to Resolution 864 (1993) Concerning the Situation in Angola Addressed to the President of the Security Council. S/2000/203. New York: United Nations Security Council.
Frantz, Douglas. 2001a. 'Central Asia braces to fight Islamic rebels.' *New York Times.* 3 May, p. A1.
__. 2001b. 'Pakistan ended aid to Taliban only hesitantly.' *New York Times.* 8 December, p. A1.
Galeotti, Mark. 2001. 'No respite in Chechnya.' *Jane's Intelligence Review.* February.
Galeria, Jessica. 2001. 'Rio bulldozes 100,000 guns.' *San Francisco Chronicle.* 3 July.
Gall, Carlotta. 2000. 'Serb police, Albanian rebels and cowering villagers.' *New York Times.* 30 November, p. A3.
Gander, Terry J. 2000. *Jane's Infantry Weapons 2000–2001.* Coulsdon, Surrey: Jane's Information Group.
Goldenberg, Susan. 2000. 'Guns for sale – how stolen Israeli weapons arm Fatah's fighters.' *Guardian.* 16 December.
Gordon, Michael. 2001a. 'NATO is planning Macedonia force.' *New York Times.* 21 June, p. A1.
__. 2001b. 'A hobbled army casts a cloud in Macedonia.' *New York Times.* 6 July, p. A8.
__. and Eric Schmitt. 2001. 'U.S. seeks Afghan coalition against Taliban.' *New York Times.* 24 September, p. A1.
Graham, Robert. 2001. 'Eta in Grenoble explosives raid.' *Financial Times.* 9 March, p. 2.
Greene, Graham. 1955. *The Quiet American.* London: Heinemann.
Gunaratna, Rohan. 2000/01. 'The state-of-the-art in illicit trafficking of conventional weapons.' PfP Workshop on Industrial Aspects of Small Arms Limitations. Baden, Switzerland, 28–30 June 1999. General Staff of the Swiss Armed Forces, (Winter 2000/2001). Bern, pp. 10–22.
__. 2001. 'The structure of the GAM.' *Jane's Intelligence Review,* April.
Gurr, Ted Robert *et al.* 2000. *Peace and Conflict 2001: Global Survey.* College Park, Maryland: Center for International Development and Conflict Management, University of Maryland.
Harris, Paul. 2001. 'Indonesia prepares for assault on GAM.' *Jane's Intelligence Review.* April.
The Hindu (Madras). 2001. 'Pakistan's wild west a gun-lover's dream.' 3 October.
The Hindustan (New Delhi). 2000. 'Frightful web of small weapons.' 23 April, p. 6, FBIS translation.
International Campaign to Ban Landmines. 2001. *Landmine Monitor Report 2001: Towards a Mine-Free World.* New York: Human Rights Watch.
International Institute for Strategic Studies (IISS). 1987. *The Military Balance 1987–1988.* London: Brassey's.
__. 1988. *The Military Balance 1988–1989.* London: Brassey's.
__. 2000. *The Military Balance 2000–2001.* Oxford: Oxford University Press.
__. 2001. *The Military Balance 2001–2002.* Oxford: Oxford University Press.
Jack, Andrew. 2001. 'Chechnya human rights abuses have become worse.' *Financial Times.* 17 January, p. 3.
Jane's Intelligence Review. 2000. 'The UCK's arsenal.' November.
__. 2001. 'Thailand cracks down on arms in Aceh.' June.
Jane's Islamic Affairs Analyst. 2000. 'Internecine struggles among Kashmiri militants.' July.
Jane's Terrorism and Security Monitor. 2000. 'The killing fields of North-East India.' March.
Johnson, Larry. 2001. 'The declining terrorism threat.' *New York Times.* 10 July, p. A19.
Kartha, Tara. 2000. 'A rising spiral of proliferation.' Unpublished manuscript. Geneva: Small Arms Survey.
Katzenell, Jack. 2001. 'Israeli navy captures arms shipment.' *National Post.* 8 May.
Kharrazi, Kamal. 2002. 'Iranian Foreign Minister terms as unfounded the Zionist Regime's accusations against Iran.' Islamic Republic of Iran, Ministry of Foreign Affairs. 12 January.
Kifner, John. 2000. 'Tales of two uprisings: this time, the Palestinians have territory and guns.' *New York Times.* 18 November, p. A6.
Kong, Yan. 1994. 'China's arms trade bureaucracy.' *Jane's Intelligence Review.* February, p. 80.
Kusovac, Zoran. 2001. 'How many weapons in Macedonia?' *Jane's Defence Weekly.* 29 August, p. 3.
Landingin, Roel. 2001. 'Macapagal tries to tackle separatist rebels on two fronts.' *Financial Times.* 20 June, p. 6.
Lumpe, Lora. 1999. 'The legal side of a dirty business.' *Washington Post.* 24 January.
__. 1997. 'U.S. policy on small/light arms exports.' Unpublished paper. 11 December.
McGirk, Tim. 2001. 'Waiting games.' *Time Asia.* 10 October.
MacQueen, Ken. 2001. 'Armed, angry and defiant.' *Maccleans.* 19 February.
Mallet, Victor. 2001. 'Violent crime becomes routine in French cities.' *Financial Times.* 11 November, p. 6.
Mathews, Jay. 1989. 'Chinese sales wars flood U.S. with AK-47 rifles.' *International Herald Tribune.* 3 February, p. 3.
Mojahideen al-Khalq. 2001. 'Statement No. 4 – 66 missiles fired by mullahs.' *Iran Mojahedin.* 18 April, FBIS Translation.
Muggah, Robert and Eric Berman. 2001. *Humanitarianism Under Threat.* Geneva: Small Arms Survey, March.

Nation (Dhakka). 2001. '80,000 terrorists own 60,000 guns.' 4 May.
New York Times. 2001. 'Filipino rebel announces beheading of American.' 12 June, p. A11.
O'Connor, Mike. 1997. 'Albanian village finds boom in gun-running.' *New York Times*. 24 April, p. A8.
Opall-Rome, Barbara. 2001. 'Israeli experts differ on Mideast war prospects.' *Defense News*. 25 June, p. 11.
Ordzhonikidze, S. A. 2001. Opening statement of the Russian Federation at the 2001 UN Small Arms Conference. 10 July.
Otero, Jose. 2000. 'Seized weapons in Colon were brought from Contras.' *La Prensa* (Panama City). 19 September, FBIS translation.
Palestinian News Agency. 2002a. 'An official Palestinian statement.' 7 January.
__. 2002b. 'President Arafat in a joint press conference.' 8 January.
__. 2002c. 'President Arafat denied any involvement by PNA in the arms shipment.' 13 January.
Philippine National Police. 2001. 'Corollary functions.' Unpublished report. Quezon City: Firearms and Explosives Division, Camp Crane.
The Pioneer. 2000. 'Kenya may purchase Indian assault rifles.' 11 July.
Policía Nacional. 2001. Dirección Central de Policia Judicial, Grupo Antiterrorismo y Armados Ilegales. Bogota, 1 February.
Press Trust of India. 2000. 'India: Bulgarian defence delegation arrives; weapons deals anticipated.' 29 October.
Raghuvanshi, Vivek. 2000. 'Bulgaria to supply AK-47 to Indian government.' *Defense News*. 20 November, p. 40.
Reuters. 2001a. 'China puts military at 2.5 million.' 2 April.
__. 2001b. 'China's "strike hard" yields guns and explosives.' 12 June.
Rhode, David. 1999. 'Guerrilla pledge not to disarm contradicts goal of the allies.' *New York Times*. 7 May, p. A13.
Rizvi, Shamm Ahmed. 2000. 'Government "arms control" campaign.' *Pakistan and Gulf Economist*. 18 September.
Rudge, David. 2001. 'Jew, 7 Arabs held for kibbutz murder, arms theft.' *Jerusalem Post*. 2 April.
Rybarova, Nadia. 2001. 'Government defends rising Czech arms sales to Yemen.' Associated Press. 27 November.
Safire, William. 2000. 'There's a war on.' *New York Times*. 17 January.
Sagramoso, Domitilla. (2000/01). 'Illicit weapons in Albania.' PfP Workshop on Industrial Aspects of Small Arms Limitations. Baden, Switzerland, 28–30 June 1999. General Staff of the Swiss Armed Forces. (Winter 2000/2001). Bern, pp. 45–46.
Sayigh, Yezid. 2001. 'The Palestinian paradox: statehood, security and institutional reform.' *Conflict, Security and Development*. Vol. 1, No. 1, pp. 101–8.
Schiff, Ze'ev. 2002. 'A ship laden with strategic impact.' *Ha'aretz*. 12 January.
Sherman, Jason and Amy Svitak. 2001. 'DoD supports proposed USD 32.3 billion boost.' *Defense News*. 5 November, p. 6.
Small Arms Survey. 2001. *Small Arms Survey 2001: Profiling the Problem*. Oxford: Oxford University Press.
Smith, Chris and Domitilla Sagramoso. 1999. 'Small arms trafficking may export Albania's anarchy.' *Jane's Intelligence Review*. January, p. 27.
Smith, Craig. 2001. 'China justice: swift passage to execution.' *New York Times*. 19 June, p. A1.
Smith, R. Jeffrey. 2001. 'The making of an insurgent army in Macedonia.' *International Herald Tribune*. 2 April, p. 2.
Smothers, Ronald. 1989. 'Spurred by Bush move, gun sales boom in U.S.' *International Herald Tribune*. 18–19 March, p. 1.
Sontag, Deborah. 1999. 'West Bank shooting does what accords could not.' *New York Times*. 11 October, p. A1.
South China Morning Post. 2001. 'SCMP cites report on increase in illegal arms trading in China.' 10 February.
Swiss Crime Surveys. 2001. *Possession d'armes a feu en Suisse; évolution et rapport de situation*. Lausanne: Swiss Crime Surveys.
Tamayo, Juan. 2000. 'Montesinos suspected of role in AK-47 shipment to guerrillas.' *Miami Herald*. 20 September.
Tehran Times. 2002. 'Arafat says Karine A was used by Israeli company.' 20 January.
Thomas, Timothy and Lester Grau. 2000. 'Russian lessons learned from the battles for Grozny.' *Marine Corps Gazette*. April, pp. 45–48.
Times of India. 2001. 'Systems are not performing to optimum level.' 12 August.
United Nations. 1998. *United Nations International Study on Firearm Regulation*. New York: United Nations.
__. n.d. *Report of the Evaluation Mission to Albania*, 11–14 June 1998.
United Nations High Commissioner for Refugees (UNHCR). 2000. *The State of the World's Refugees*. Oxford: Oxford University Press.
United States. Bureau of Alcohol, Tobacco, and Firearms (BATF). 2000. *Following the Gun: Enforcing Federal Laws against Firearms Traffickers*. Washington, DC.
__. 2001. *Annual Firearms Manufacturing and Exportation Reports*. Washington, DC: US Department of the Treasury.
United States. Department of Justice. 1995. '23 per cent of those arrested for weapons offenses are younger than 18 years.' Washington, DC. 12 November.
__. Department of State. 2000. *Patterns of Global Terrorism 1999*. Washington, DC.
Wagstyl, Stefan. 1999. 'Serb aggression helps KLA to win hearts and minds.' *Financial Times*. 10–11 April, p. 2.
Webb-Vidal, Andy. 2001. 'Venezuelan guns aiding guerrillas.' *Financial Times*. 25 January, p. 16.
Whitaker, Brian. 2002a. 'Voyage of the arms ship.' *Guardian*. 14 January.
__. 2002b. 'The strange affair of Karine A.' *Guardian*. 21 January.
Wines, Michael. 2001. 'Putin scaling down Chechen war despite new fighting.' *New York Times*. 23 January, p. A3.
Xiang, Jiagu. (2000/01). 'Models of possible solutions on the national level: the situation in China.' PfP Workshop on Industrial Aspects of Small Arms Limitations. Baden, Switzerland, 28–30 June 1999. General Staff of the Swiss Armed Forces (Winter 2000/2001). Bern, p. 88.
Yangcheng Wanbao (Guangzhou, China). 2001. 'Where do illegal guns come from?' 11 February, FBIS translation.
Ziau (Bucharest). 2001. 'Forest Kalashnikovs.' 19 July, FBIS translation.
Zivalj, Husein. 2001. Statement by H.E. Husein Zivalj, Ambassador, Head of Delegation of Bosnia and Herzegovina, to the 2001 UN Small Arms Conference. New York, July.

Acknowledgements

Contributor

Derek Miller

کامران آرمز
ڈیلرز اسلحہ و کارتوس

The Legal-Illicit Link: Global Small Arms Transfers

3

Introduction

In April 2001, a group of Thai soldiers raided an airport warehouse near Bangkok. They stole 30 Austrian-made Glock pistols and fled with the booty, apparently unapprehended. To their ensuing dismay, however, their every move had been filmed. The official inquiry that followed exposed that both the shipment and the raid were integral to a scheme to sell guns on the black market. This particular incident is symptomatic of the larger, international scenario. The trade in small arms has become increasingly difficult to control and it leaves a growing trail of corruption and insecurity in its wake.

According to Article 51 of the UN Charter, states do have the right to arm themselves, and small arms are one of the universal weapons of choice. Indeed, members of staff from a wide range of government agencies carry small arms, from armed forces to government militias, local police to wildlife agents and customs authorities to airport security. Millions of ordinary civilians in virtually all countries also keep arms as their rightful property, whether for personal protection or recreation. In fact, roughly 80–90 per cent of the global trade in small arms, worth approximately USD 4 billion annually, is legal trade. Such trade supplies those states and individuals who arm themselves rightfully for defence, security, or leisure.

Yet small arms are also used in acts of violence, crime, and human rights abuses that hinder development and democracy while contributing to regional instability and insecurity. The negative effects are most keenly felt in regions at war or in post-conflict situations. Few countries, however, escape such negative impacts completely. The illicit trade, which probably accounts for no more than USD 1 billion annually, is one way in which various actors can obtain small arms for misuse. Often the same countries and individuals are involved in both aspects of the small arms trade. As the line between legal and illegal transfers blurs, the two merge into one seemingly indistinguishable problem.

This chapter examines several aspects of the linkages between legal and illicit trade in small arms. Here are some of the key findings:

- There has been no dramatic change in the overall scale of the legal and illicit trade on the previous year, although the legal trade appears to be in gradual decline.
- Through analysis of available data, it is estimated that the global legal trade in small arms is worth USD 4 billion annually. This is a revision of the USD 4–6 billion estimate in the *Small Arms Survey 2001*.
- Through official customs data, only approximately 20 per cent of the suspected global, legal small arms trade of roughly USD 4 billion can be definitively documented.
- By adding up all the other country specific declarations of the small arms trade, an additional 20–25 per cent of the global legal trade can be accounted for. Thus, public official documentation exists for approximately 50 per cent of the global legal trade in small arms.
- Specific sectors of the small arms trade appear to be declining. According to customs data, the global export of pistols and revolvers decreased by 49 per cent from 1995 to 1999.

An arms dealer in Quetta, Pakistan. (©Associated Press/Jerome Delay)

- Small arms enter the illicit realm mainly through the diversion of legal stockpiles.
- Government patronage is the leading source of arms for most non-state actors.
- Preliminary research shows that at least 54 countries have been linked to activities that allegedly violated international small arms embargoes.
- The shipment of illicit small arms by sea deserves much greater international scrutiny.

While the international community grapples with how to define the illicit trade (see Box 3.1), this chapter will examine all types of small arms transfers, showing how individuals, groups, and governments obtain the weapons responsible for a significant proportion of the death and destruction in the world today. By reviewing different types of small arms transfers, the links between the legal and illicit trade become clearer. This chapter will address the following questions:

- **What are recent trends in the legal international small arms trade?**
- **Is the state-authorized small arms trade becoming more transparent?**
- **Who are the main suppliers and recipients of small arms, both legal and illicit?**
- **How do small arms move from legal to illicit circulation?**

The first part of the chapter assesses the quantitative data that has been made publicly available. Transparency in the arms trade is key to understanding how arms get into the hands of those who misuse them. Thus, the chapter begins by examining the data available on the international legal trade in small arms. Using statistics provided by the United Nations and a number of governments, it highlights those aspects of the small arms trade that have been made public. While this picture is far from complete, it provides the only reliable and comparable information available on the subject.

The second part of the chapter examines recent trends in the illicit trade. It pays close attention to the ways in which legally owned small arms are diverted into illegal markets, examining a variety of routes, from wilful violations of embargoes to inadvertent battlefield seizures. Also considered are processes that facilitate the illicit small arms trade, especially the trade in conflict goods and sea shipping. Careful examination of events in Thailand and Bangladesh illustrate the differences between source countries and transit countries in the illicit trade, and the different risks the two face.

FIGURE 3.1 Arms transfer chain of possibilities

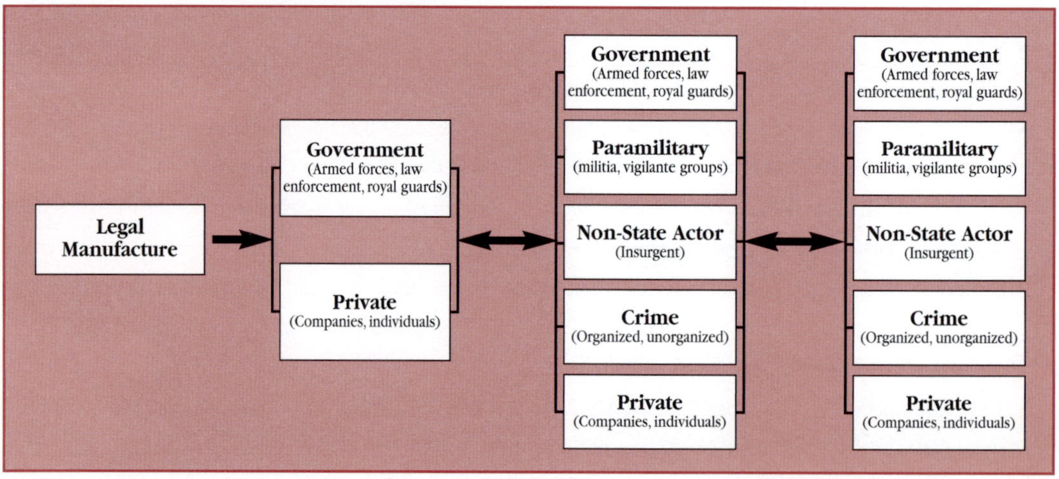

Small arms can last a very long time and pass through many different hands. The legal status of a gun can change with each new owner and in ways that seemingly mock the best intentions of policy-makers and activists. Figure 3.1 and Box 3.1 only hints at the complexity of the process. The preoccupation with legal status is somewhat academic. Ultimately, efforts to deal with small arms issues should focus on the weapons themselves.

BOX 3.1 Classifying transfers

The definition of an arms transfer is relatively simple. A *transfer* is the reallocation of small arms from the possession, either de facto or *de jure*, of one actor to another. There are always *at least* two principal actors involved, the *originator* and the *recipient*. These actors can be individuals, groups such as firms or rebel bands, or states. Facilitating actors, such as brokers and transport agents, are also often involved in the logistics of the transfer.

Other terms commonly used are much harder to define. One of the most interesting lacunae of the 2001 UN Small Arms Conference was the failure to define both small arms and light weapons, as well as illicit trade (CONFERENCE). The broader small arms community is no nearer to an agreed definition as to what constitutes a legal transfer of small arms, much less what constitutes an illicit one.

Experts on international law have made convincing arguments as to what is legal and therefore what is not (Gillard, 2000). However, these arguments are far from universally accepted and even further from being legally applicable. Attempting to determine issues of legality and accountability in regards to arms transfers is a complicated endeavour. It also ignores the fact that it is not the 'type' of transfer in and of itself that is important, but the end use of the weapons. Of course, illicit transfers may lead to more instances of misuse. While the definitional debates cannot be resolved here, the following guidelines were used in the *Small Arms Survey 2001*.

- *Legal vs illicit*. A transfer generally is legal if it fully conforms to international law and the national laws of both the exporting and importing states. An illicit transfer breaks either international or national laws. This simple division often blurs in practice, especially in ambiguous cases where legal or policy exceptions are exploited. Another source of controversy arises from transfers that are legal by these criteria—for example, which do not violate international embargoes or sanctions—but not in terms of international humanitarian or human rights law (HUMANITARIAN).
- *Authorized vs unauthorized*. Transfers can be *authorized* by both the originating and receiving governments or their agents in compliance with international and national laws, or *unauthorized* by either or both of the sending and receiving governments. This distinction is useful in attempts to attribute accountability for a transfer, but only with caution. A transfer may be authorized but, as is often the case, it never culminates in actual arms delivery for one reason or another. In a growing number of countries, moreover, governments are too weak to make authoritative decisions; and any transfers might better be described as *undefined*. This is clearest where the government structure has broken down or where the country is ruled undemocratically.
- *Intentional vs unintentional*. Transfers can also be classified as being *intentional* or *unintentional*. The transfer is intended either by the supplier or the recipient. There might be a few extraneous cases where a transfer is unintended by both parties—for example, a weapon is lost and then found—but this represents a minute portion of transfers. Determining the intention of a transfer transcends the issue of legality. It provides a basis on which to investigate matters of accountability and ultimately responsibility for any given transfer (Miller, 2001b).

The legal trade in small arms

As long as the trade in small arms occurs in accordance with national and international laws, and is authorized by the respective authorities of the exporting and importing state—that is, it is state-authorized—it is legal for the purposes of this edition of the *Survey*. Legal trade makes up the bulk of the international trade in small arms, but legal arms transfers can be diverted into the illicit market. Moreover by examining legal transfers one can see where major shipments of arms are going and when regions become saturated in arms.

Exhibitors and visitors shake hands over small arms at an aerospace technology trade show open to the public. The vast majority of small arms sales is legal.

It is clear that transparency is needed in the legal trade in small arms. Transparency allows governments and the general public to scrutinize transfers, to determine whether or not they conform to national and international regulations, and hence to better enforce said regulations. Many states see the benefits of transparency. As a result, over the last decade, official transparency in the small arms trade has definitely increased. Based on the increase in available data over 2001, it is possible to more accurately estimate the global annual value of small arms exports to USD 4 billion, down from the USD 4–6 billion range cited in the *Small Arms Survey 2001*.

Two main data sources for the international trade in small arms are examined here: *customs data* and *national reports* on arms exports and imports. Each of these sources has strengths and weaknesses. Customs data, in general, conforms to a global standard when it comes to definitions and categories of small arms. While there are limitations to customs data, as will be outlined below, it remains the only source on small arms exports and imports that is comparable from country to country. National arms export reports often go above and beyond the level of detail found in customs data. While national reports are not comparable, due to their differing methods and definitions, they often provide a clearer picture of a state's international trade in small arms than customs data does.

It is by observing this legal trade that we can clarify what constitutes the illicit or illegal trade, as well as how legally held weapons move from legal to illicit markets. Without studying the legal trade, it is impossible to understand the dynamics of the illicit trade. The illicit trade is believed to contribute to high levels of regional instability and insecurity, crime, war, and human rights abuses, and it usually is of greater concern to governments. However, illicit phenomena cannot be appreciated without a clear understanding of the legal trade in small arms from which they originate. Data available on the legal trade, looking at customs figures, national arms export reports, and other sources is discussed below.

An incomplete snapshot: Customs data on small arms transfers

In the *Small Arms Survey 2001* customs data for global exports and imports of 'military firearms' and 'pistols and revolvers' was examined.[1] This year, the *Survey* also looks at the global export and import of non-military firearms. In 1999, the latest year available, over USD 413 million in hunting and sporting firearms was reported exported through customs worldwide. For the same year, more than USD 111 million of civilian pistols and revolvers was reported exported through customs. As for military firearms, the total value covered in international customs reports was nearly USD 410 million, or about 43 per cent of the total of small arms exported through customs.

There are several problems with customs data. Above all, many countries fail to supply information on relevant categories to the UN Statistics Division. Others do so, but inconsistently. Second, customs data does not encompass the entire trade in small arms, as many legal transfers—such as military aid—do not pass through customs. Third, despite an agreed universal customs system, differences persist in reporting from country to country. For example, sometimes weapons that are only in transit and are not being imported into the country are reported as an export. The Netherlands, which does not produce small arms, but is a major transit country for shipments that pass through customs, is an example. In addition, shipments usually are reported in US dollars, yet dollar values are not always the best indicator of the small arms trade. Only a few countries report actual quantities of weapons. Many report volume by weight, which is less useful in assessing the global trade in small arms. Finally, some items included in the relevant customs categories are not what are usually considered to be small arms and light weapons, further clouding the issue.

Despite these shortcomings, customs data is the only official data on the global arms trade that allows direct comparisons between countries. While far from comprehensive, it is the most reliable indicator of the overall trade. The analysis here uses data compiled by the International Trade Center based on UN Customs Database and Commercial Trade (COMTRADE) figures as submitted by states to the UN Statistics Division.

Non-military firearms

There are several subcategories of small arms that fall under the customs code 9303 covering hunting, sporting, and target-shooting firearms. These include muzzle-loading firearms, shotguns for sporting, hunting, or target-shooting, and rifles for sporting, hunting, and target shooting. Pistols and revolvers fall under another customs category 9302. In general, COMTRADE customs data shows that:

- many countries have reported regularly for the years 1995–99; 15 of 29 countries supplied data on *pistols* and *revolvers* for all five years, with 22 of these reporting three or more years. For *hunting and sporting rifles*, 22 of 35 countries supplied data for all five years, with 30 of these reporting three or more years. For *hunting and sporting shotguns*, 25 of 35 countries supplied data for all five years, with over 90 of these reporting three or more years;
- exports of three of the four customs categories have declined over the last five years, the exception being exports of shotguns for sport, which have remained fairly constant (see Figure 3.2);[2] and
- there is a higher level of countries submitting data for these four categories of weapons than for the category of 'military weapons'.

Hunting and sporting rifles

In 1999 there were just over USD 170.8 million of hunting and sporting rifles exported—through customs—worldwide.[3] The top exporters reported were Brazil, Germany, Japan, Finland, the United States, Portugal, and the Czech Republic. There was a steady decline in the value of global exports of sporting and hunting rifles over the 1996–99 period, declining each year from the high of over USD 206 million in 1996 (see Figure 3.2 and Appendix 3.2).

Customs data for this category is the most complete of any category of small arms examined. Of those countries reporting, the top 16 provided data for all four years. With such complete data, it can be concluded that global exports of hunting and sporting rifles did indeed decline over the 1996–99 time period. The total value of hunting and sporting rifles reported exported globally over the five-year period (1995-99) was USD 947.1 million.

TABLE 3.1 Customs data for top non-military firearms exporters by export category (1999)

Pistols & revolvers	Value (USD million)	Sporting rifles	Value (USD million)	Non-military shotguns	Value (USD million)	Muzzle-loading firearms	Value (USD million)
Germany	32.2	Brazil	35.6	Italy	108.2	Spain	10.7
US	26.6	Germany	23.9	Japan	15.2	Italy	8.4
Italy	19.3	Japan	15.1	UK	13.7	US	1.2
Czech Republic	10.3	Finland	12.0	Belgium	13.4	Germany	0.2
Canada	6.6	US	11.9	US	12.3	China	0.2
Switzerland	4.2	Portugal	11.5	Russia	12.0	Czech Republic	0.05
Spain	3.6	Czech Republic	10.6	Spain	10.6	France	0.03
Argentina	3.4	Italy	8.7	Germany	10.3	Norway	0.01
Indonesia	0.9	Austria	7.8	Brazil	6.2	Switzerland	0.01
Portugal	0.8	Belgium	7.6	Turkey	5.1	Malaysia	0.01

Source: COMTRADE customs data (2001)

Non-military shotguns

In 1999 there were USD 221.7 million of non-military shotguns exported worldwide. The top exporters in 1999 were Italy, Japan, the United Kingdom, Belgium,[4] the United States, and the Russian Federation. Thirty-five countries provided export data for at least one year over the period 1995–99, but the data is less complete than that for hunting and sporting rifles (see above) in that a major exporter, the Russian Federation, did not submit data for 1995 or 1997. There is no discernible trend over the five-year period examined in shotgun exports (see Figure 3.2). The five-year total (1995–99) for reported global exports of non-military shotguns was USD 1.1 billion.

Muzzle-loading firearms

Often overlooked but important in the non-military market, there were USD 20.8 million of muzzle-loading firearms exported through customs worldwide in 1999. The top exporters were, in descending order, Spain, Italy, the United States, Germany, China, and the Czech Republic. Nineteen countries are reported exporters of muzzle-loading firearms. The global trade in these weapons decreased steadily over the period 1995–99 (see Figure 3.2 and Appendix 3.2). The five-year total value of muzzle-loading firearms reported exported through customs was USD 118.2 million.

Pistols and revolvers

According to customs data, exports of pistols and revolvers worldwide decreased substantially from 1995 to 1999.

There is also a discernible decrease in the worldwide export of pistols and revolvers reported through customs from 1995 to 1999. From a high of USD 216.5 in 1995, the trade reportedly fell to USD 111.2 million of pistols and revolvers exported globally in 1999. The 1999 top exporters were Germany, the United States, Italy, the Czech Republic, and Canada (see Table 3.1).

All of these countries have seen a substantial decrease in exports of pistols and revolvers over previous years, most by half (see Figure 3.2 and Appendix 3.2). Some export declines were even more dramatic. Portugal, for example, reported a decline from USD 10.7 million of exports of pistols and revolvers in 1995 to USD 762,000 in 1999.

In terms of data on reported imports from 1995 to 1999, the top importers—by value—of pistols and revolvers were, in descending order, the US, Germany, Thailand, Turkey, Switzerland, Mexico, the Philippines, Canada, and Venezuela.[5] Of those few countries that report on the 'quantity' of weapons imported, the United States imported 4.1 million, Thailand imported 523,000, and Canada imported 83,000 pistols and revolvers in total from 1995 to 1999.

FIGURE 3.2 Trends in worldwide small arms exports, non-military, 1995–1999

Source: COMTRADE customs data (2001); Appendix 3.2

Military firearms

Customs data on military weapons is much less complete. Recalling that this customs category also includes larger weapons systems, the data show that:

- Due to inconsistent reporting practices on the part of exporting countries, it is exeedingly difficult to discern any trends in the trade. Only ten of 29 countries supplied data every year for the five-year period in question, with only 59 per cent of the countries supplying data for three or more years.
- Several known exporters of weapons of this category do not submit data for this category of weapons, but do submit data for other categories: for example, the Russian Federation and Israel. It is clear that many countries still consider exports of this category of weapons as secret.

Customs data on exports of military weapons was reported more sporadically than those for any other category of weapons.[6] In 1999, USD 409.9 million in military weapons were reported as exported globally. The top exporters were the United States, the United Kingdom, South Africa,[7] Switzerland, and Canada. However, certain countries are noticeably absent from the 1999 data. The Netherlands, a major trans-shipment country but not an important exporter, did not submit export data in this category for 1999, nor did China, which exported USD 22.8 million in this category in 1998. Significant exporters of these weapons, such as the Russian Federation and Germany, provided no export data in this category. Therefore, customs data in the category of military weapons—again, a category that also includes arms other than those falling under the UN definition of small arms and light weapons—does not really represent the reality of this sector of the small arms trade.

TABLE 3.2 Top exporters and importers of military weapons through customs, 1999

Military arms exporters	Value 1999 (USD million)	Military arms importers	Value 1999 (USD million)
US	322.8	Cyprus	83.0
UK	32.9	South Korea	57.8
South Africa	16.3	US	25.6
Switzerland	16.3	Turkey	8.8
Canada	8.9	UK	8.8
Poland	5.3	Australia	8.6
Czech Republic	3.0	Bolivia	6.6
Croatia	1.7	Estonia	6.5
Republic of Korea	0.7	Switzerland	4.5
Turkey	0.4	Mexico	4.5

Source: COMTRADE customs data (2001)

In 1999, data on imports of military weapons was provided on an equally sporadic basis; hence, such data does not reflect the actual trade in these weapons worldwide. Nevertheless, in 1999 the top importers of military weapons who reported customs data were Cyprus, South Korea, the United States, Turkey, the United Kingdom, and Australia (see Table 3.2). Former major importers that did not provide data for this category in 1999 were the Netherlands—again, a major trans-shipment country and not a major importer; Saudi Arabia—USD 37 million in imports in 1998; and Thailand—USD 10.9 million in imports in 1998.

Other sources of customs data

Commercially available customs data can be prohibitively expensive and does not always include information on categories of interest to the small arms community. There are at least three countries whose customs data, including data on small arms transfers, is available on the web free of charge: Thailand, Chile, and the United States.[8] Customs data

BOX 3.2 Man portable air defence systems

One of the most worrisome light weapons is the 'man portable air defence system' (MANPAD) or shoulder-fired surface-to-air missiles. The best known examples are the American-designed FIM-92 Stinger and the Russian SA-16 Igla. While many governments maintain that proliferation of rifles and pistols does not constitute a grave threat to their own national security, they certainly recognize the danger of MANPADs in the hands of terrorists or insurgents.

One of the greatest fears of aviation authorities was the risk that these weapons could be used to bring down a commercial airliner in an act of terrorism. In early 2001, a witness testified in a New York courtroom that the Al-Qaida tried to acquire US-made Stinger missiles in 1993 and bring them to Sudan (Loeb, 2001).

A growing number of countries produce and export MANPADs. Available data points to the United States and the Russian Federation as the leading exporters. From 1991 to 2000, for example, the United States exported Stinger missiles to Bahrain, Egypt, Israel, and Saudi Arabia (US, GAO, 2001). In 2001 Raytheon was awarded a deal worth USD 89 million to build 1,007 Stinger missiles for Greece, Italy, and the United Kingdom (*Armada International*, 2001).

In 1997 the Russian Federation sold 440 Iglas to Singapore and licensed assembly of up to 3,000 more (CAST, 2001). In 1999 the Russian Federation sold 70 Iglas to Malaysia (Brooke, 2000). The Russian Federation also delivered several hundred Iglas worth USD 32 million to India in July 2001 (CAST, 2001).

At such prices, it is clear that sales of MANPADs, though only a tiny proportion of the total number of individual items in the global small arms trade, can make up a large chunk of the overall value of light weapon sales.

on Chile's exports and imports of small arms, available for 1997 to 1999, is useful for those studying arms transfers as it lists data by exporting and importing company. US data is available from 1994 to 1998 on the Internet. More recent data can be purchased on CD-ROM. Thailand also sets a good example when it comes to transparency in customs data, with exports and imports, including small arms categories, updated monthly on the Internet.

Through commissioned research, the Small Arms Survey has obtained some customs import data for certain countries. Data for Paraguay and Uruguay was reported in the *Small Arms Survey 2001*. For Guatemala, the customs office reported that, in 2000, Guatemala imported 31,143 pistols and revolvers and 5,675 assault rifles, hunting rifles, and shotguns (Godnick, 2001a).

Toward transparency: National export reporting

Some 22 countries now make official data publicly available in annual or other regular reports on their arms exports and/or export licences.[9] The number of countries releasing such data has grown dramatically, from only a couple just ten years ago. The majority of these reports categorize small arms and light weapons exports; however, some make no such distinction and lump all arms exports together. A number of the documents are available on the Internet, setting examples for transparency in the arms trade. Those reports available on the Internet can be accessed through the Small Arms Survey web site at <http://www.smallarmssurvey.org>

There still is no international norm governing national arms export reports. The presentation of data can vary widely from country to country, making direct comparisons virtually impossible. Sometimes data from these reports is in contradiction with other official data sources, notably customs data and reports in the open press. Ideally, reports should include country-by-country breakdowns of exports of small arms, with figures for values and quantities of weapons exported by weapon type. The same information should also be provided for approved export licences, as well as denials, deliveries, and unfulfilled orders. No country provides all of this information.

While there is enormous room for improving transparency in many of these reports, the countries producing them should be commended for providing information that was deemed secret just a decade ago. A short overview of national export reports with respect to information on small arms and light weapons follows. While the *Small Arms Survey 2001* focused on the value of small arms exports presented in each report, the *Small Arms Survey 2002* looks at how long a country has been producing public reports, the usefulness of information concerning small arms, and how easy it is for the public to access the reports. Table 3.3 presents a comparison of information provided in the various national reports.

Australia's national arms export report provides no data on small arms and light weapons.

Australia: Australian reports cover exports of 'defence and strategic goods' back to the 1994–95 time period, but do not make a distinction for small arms. The monetary value of deliveries of military goods, disaggregated by importing country, are given for the following categories: military goods, non-military lethal goods, all defence-related goods, and dual-use goods. While the report includes information on current Australian legislation on defence exports, export-licensing procedures, and export control regimes, there is no data specific to small arms and light weapons.

Austria: The Ministries of the Interior and Economic Affairs release a joint annual report on arms export licences granted and refused. The Ministry of Economic Affairs lists the total value and number of licences granted for export. Data from the Ministry of the Interior is more detailed, listing quantities exported by geographic region.

Unfortunately, categories are ill-defined and obscure. The report is available in English and German, but it is extremely difficult to obtain. A more detailed report listing export licences granted by importing country has been distributed confidentially to European Union (EU) member states.

Belgium: Since 1994 Belgium has published reports on its arms exports. These show values of actual deliveries divided by weapons category, including categories for small arms. The 1999 report was more transparent than previous reports in that it also contained a table of the quantity of export licences granted by country and weapons category. Belgium also reports arms imports.

Canada: In 1997 Canada published its first report on arms exports, with data from 1996. Canada's report is quite transparent with respect to small arms exports as they list values of exports by country and weapons category, including the category 'firearms' which covers arms falling under the 1997 UN definition of small arms. It does not, however, detail licences for exports, nor does it list exports to the United States, a major importer of Canadian arms.

Czech Republic: In 2001 the Czech Foreign Ministry published a report titled *The Czech Republic and Small Arms and Light Weapons.* This document covers the value of annual exports of small arms and light weapons as a percentage of total arms exports from 1996 to 2000. It lists the producers of small arms and light weapons in the Czech Republic, as well as numbers of civilian weapons permit holders, the number and type of registered weapons in the country, and the quantity of 'non-commercial' weapons exported. While the report is not very clear on small arms exports, it is an impressive review of other aspects of firearms in the Czech Republic. The report is published in English and is available from the Foreign Ministry.

Denmark: After intense pressure from the public and NGOs, Denmark issued its first report on arms exports in December 2000, covering the prior year. The report covers only the number of licences granted by country, which is not at all helpful in determining actual exports in terms of quantity or value.

Finland: The Finnish report on arms exports, published since 1998, is one of the better examples of transparency in national arms export reports. The 2000 edition lists the value of exports by countries and detailed weapons categories, such as 'sniper rifles and accessories'. It also provides data on licences granted by country and weapons category.

France's 2000 national report on arms exports provides some of the most transparent data on small arms transfers to date.

France: In March and December 2000, France published its first two reports on arms exports, for 1998 and 1999 respectively. In the appendices there is data covering the quantity and value of government-negotiated sales by country of small arms and light weapons, as well as those countries which received free transfers of small arms and light weapons. In addition, the report lists the number of export authorizations of small arms by country, broken down into sub-categories. France's national export report for the year 2000, published in December 2001, showed an even greater degree of transparency in small arms transfers, including several sections on small arms transfers that list authorizations for commercial exports and actual exports of military small arms and light weapons listed by quantity, type, country of destination, and year exported/authorized for the years 1997 to 2000.

In addition, Annex 9 of the report lists transfers of small arms, including free grants, carried out by the Ministry of Defence disaggregated by type, quantity, and country of destination. The 2000 French report stands as one of the most transparent national export reports issued to date by any country in terms of small arms exports.

Germany: The first report on arms exports of the Federal Ministry of the Economy and Technology in Germany was published in September 2000 for licences granted and refused in 1999. The report is quite specific in that it lists the total value of licences issued for each importing country, then lists the categories of weapons licensed, followed by the percentage value of that category as part of the total of exports authorized to that country. However, with respect to EU and NATO countries, only partial information is given. Germany's export report 2000, published in December 2001, contains a special section on the value of authorizations for military small arms and ammunition transfers for the years 1996 to 2000 broken down into the destination categories of EU countries, NATO countries (minus EU countries), and other countries. Germany does not publish any data on actual exports.

India: The Indian government provides data on the sales of the Ordnance Factories Organization (OFO), which produces military small arms and ammunition as well as larger weapons systems, vehicles, clothing, and other equipment. The value of these sales is subdivided between the Indian armed forces and other defence departments, and sales to paramilitary forces and exports. The data goes back to the 1996–97 fiscal period. The report is not very useful, as there is no indication of destination countries or of the small arms share of exports.

Ireland: The Export Licensing Unit of Ireland's Department of Enterprise, Trade and Employment provides statistics on military export licences issued. While there is no indication of monetary values of the licences or of quantity, the data is disaggregated by country and weapons category, including a specific category for small arms. One unique aspect of the Irish data is that it is updated monthly on the Internet.

Italy: The Italian Parliament has received an annual report on the import, export, and transit of arms and military equipment since 1991. The report is extremely detailed, providing information on quantity and value of licences and actual exports of equipment, with descriptions of the items. Until 1993 the Italian report also listed destination countries, but this ceased on requests from Italian defence firms claiming that it was detrimental to their competitiveness abroad (Mariani and Urquhart, 2000). The profuse data in the Italian report makes it one of the more transparent reports; however, it also is at times overwhelming and difficult to analyse. The report is available in Italian from the Chamber of Deputies' Stationery Office in Rome.

The Netherlands and Spain publish data on arms exports semi-annually. Ireland provides monthly updates.

Netherlands: The Netherlands' first arms export report, on activity in 1997, was published in 1998. The Dutch report lists the value of export licences by weapons category, and then the value of export licences by country broken down into two categories: arms and ammunition, and other military goods. With respect to actual exports, information is less useful in that it contains only a five-year total—for 1994–98—of exports attributable to 'military production'. The document stands out by providing the most information on licence refusals of any report released so far. Another distinguishing feature is that the Dutch issue semi-annual and annual reports.

TABLE 3.3 National export reports and analogous official data

Country	Data on value of total small arms exports?	Value of small arms exports disaggregated by country?	Quantity of small arms exported disaggregated by country?	Language	Available on Internet?
Australia	no	no	no	English	yes
Austria	yes	by region	no	English, German	no
Belgium	yes	no	no	French, Flemish	yes
Canada	yes	yes	no	English, French	yes
Czech Republic	yes	no	no	English	no
Denmark	no*	no	no	Danish	yes
Finland	yes	yes	no	English	yes
France	yes	yes	yes	French	yes
Germany	yes	yes	no	German	yes
India	partial	no	no	English	yes
Ireland	no*	no	no	English	yes
Italy	yes	no	no	Italian	no
Netherlands	yes	no	no	English, Dutch	yes
Norway	yes	no	no	Norwegian	yes
Portugal	yes	yes	no	Portuguese	no
Slovak Republic	yes	no	no	English	no
South Africa	yes	yes	no	English	yes
South Korea	yes	no	no	English	yes
Spain	yes	no	no	Spanish	no
Sweden	yes	no	no	English	yes
Switzerland	yes	yes	no	French, German	yes
United Kingdom	yes	no	yes	English	yes
United States	yes	yes	yes	English	yes

* *Data given on quantity of export licences issued for small arms.*

Norway: The Norwegian Ministry of Foreign Affairs issued its first annual report on weapons and related material exports in 1997, covering exports for 1996. With respect to small arms, value of total small arms exports is listed, as well as a list of recipient countries. There is no information on value of small arms exports by country or on volume.

Portugal: Since 1998 Portugal has produced an annual report on arms exports. The report lists the value of exports by weapons type and destination country, as well as the exporting company. Weapons descriptions are fairly detailed: for example, '9mm pistols' and 'ammunition <12.7mm'. The report is available in Portuguese from the Ministry of Defence.

Slovak Republic: Since 1997, the Slovak Republic has compiled data on its arms exports based on customs data. The report lists the annual value of exports by customs category, as well as total arms exports by destination country. There is no information on volume or quantity of arms exported. The report is available on request from the Slovak Ministry of Economy.

South Africa: The first year for which official data on South Africa's arms exports is available is 1996. South Africa lists value of exports by country, broken down into weapons categories. Category B covers 'all types of infantry hand-held and portable assault weapons and associated ammunition for calibre smaller than 12.7mm'. The data does not cover civilian firearms exports or data on licences.

South Korea: The Republic of Korea has made statistics on its arms exports public only once, in its Defence White Paper of 1998. Data on annual exports for 1990–97 inclusive was listed. This single report was hardly a model of transparency, but it did list the value of exports broken down by weapons category.

Spain: The Spanish government has produced an annual report on arms exports since 1998. The report lists the total value of arms exports for each of six categories, including the category 'small arms and their ammunition'. The report also lists the value of total arms exports by country of destination. Like the Dutch report, data is provided on a semi-annual basis. The report is available in Spanish.

Sweden: A pioneer in the field of transparency in arms exports, Sweden published its first report in 1985. Since 1993 the reports have included the value of arms transfers by weapons category, including a category for 'small calibre barrel weapons'. Values are also given for total exports to each country, broken down into two categories: military equipment for combat purposes, and other military equipment. Its most recent report of April 2001 included information on licences granted as well as the types of equipment that were actually exported to each country: a level of transparency that is a substantial improvement on previous years.

Switzerland: In January 2001 Switzerland published its most transparent report on arms exports to date. The report lists the value of arms exports by category and importing country for 2000. There is also a graphic showing the percentage of the value of arms export by weapons category: for example, handguns accounted for six per cent of Swiss arms exports; small arms and light weapons other than handguns accounted for over 24 per cent; and ammunition for 22 per cent. The Swiss report provides no information on licences, nor does it cover arms exported by the Swiss Ministry of Defence or other government bodies.

United Kingdom: In March 1999 the United Kingdom published its first Annual Report on Strategic Exports, covering exports for 1997. Every report since has been marked by a better format and greater transparency. The 2000 edition includes data on licences granted and refused by country, as well as data on quantities of small arms licensed for export. It also includes a table of actual exports of small arms based on customs data, that lists weapons by quantity and destination. It also lists transfers of arms, giving quantities and destination for those arms transfers that do not pass through customs, such as the government-to-government transfer of '4,550 self-loading rifles' to Sierra Leone in 2000 (see Appendix 3.1). The UK report is one of the better models of transparency, listing actual quantities of small arms exported as well as those licensed for export, by country.

United States: The United States has produced a highly detailed and transparent report on arms exports since 1997, known as the '655 Report'. The report is compiled by the State Department and the Department of Defense.

© Associated Press/Alan Mothner

The United Parcel Service no longer ships handguns through its regular ground service, as they are too easily stolen. Rifles and shotguns, less easily stolen, are still shipped over land.

The State Department portion reports export licences granted through commercial sales, with a breakdown by country of weapon type, quantity, and value. After legislation passed in 2000, future reports must include actual exports also. The Department of Defense portion lists actual exports of arms transfers negotiated by the Department of Defense, again broken down by country, weapon type, quantity, and value. This section also includes similarly specific data on transfers of surplus weapons donated through the Excess Defense Articles Program.

Best national reporting practices

While no country has yet achieved total transparency with respect to arms transfers, several reports contain elements that greatly facilitate oversight by parliamentarians and the public. Those countries that provide quantity, destination, value, and a detailed description of the weapon are the most transparent, of which the United States and France offer the best examples. Countries which list small arms exports by exporting company or agent, such as Italy and Portugal, should also be mentioned, as this gives an even clearer picture of all of the players involved in the arms transfer.

The United States, France and the United Kingdom have the most transparent arms export reports with respect to small arms, although there is definitely room for improvement.

Those countries that report actual quantities of weapons exported—the United Kingdom, France, the United States—are in the minority, but it should be emphasized that such quantitative values are much more important that monetary figures when it comes to arms exports. Another interesting case is that of Ireland, where licences granted are updated on the Internet on a monthly basis. This practice is extremely important as it allows for reaction on the part of the public or parliament to a specific licensed transfer.

Many of the reports take so long to compile that the licences may have been issued more than a year before the public learns of it. The semi-annual reports of the Netherlands and Spain are an improvement in this respect. In addition, Belgium should be mentioned as it lists imports as well as exports. While arms producers argue that secrecy is

BOX 3.3 Small arms transfers to Macedonia

Since the UN arms embargo was lifted on Macedonia in 1996, a number of states have assisted the country in building up its conventional armed forces. When the Macedonian Army was created, it started with a stockpile of 40,000 obsolescent repeating rifles and 5,000 Kalashnikovs that it had inherited from the Yugoslav People's Army. The Macedonian Army's strength when it was admitted into NATO's Partnership for Peace (PfP) programme in 1995 was 24,000 active troops and 60,000 reservists. Shortly after, NATO convinced Macedonia to cut its active troop strength to 15,000 troops. This still was much more than could be provided with modern arms (Kusovac, 2001a).

Fully aware of the delicate tensions in the region, donors wanted to strengthen the Macedonian government without igniting civil war. Reports in the German and Macedonian press that Germany donated 100,000 assault rifles and 10,000 sub-machine guns to Macedonia (*Deutsche Presse Agentur*, 2000a) in early 2000 are false, and the German government has confirmed that no such transfers have taken place.

In 1998–99, the United States provided Macedonia with 707 ageing M3 sub-machine guns. In 1999, France donated 12 'Milan'anti-tank systems (France, 2000). Ukraine, Bulgaria, Greece, and Italy donated larger systems, including armoured personnel carriers (APCs), tanks, and artillery (Varouhakis, 2001). Reports also allege that Macedonia has purchased assault rifles and mortars from RH Alan of Croatia (Kusovac, 2001b). Other weapons systems, including helicopters and APCs equipped with small arms, have been purchased from Russia and Ukraine (Clark, 2001). In July 2001, though, Ukraine suspended arms shipments to Macedonia (Pravda, 2001). Macedonia has further built up its arsenals with small arms confiscated from ethnic Albanians brought into the country during the Kosovo conflict and since the fighting began in Macedonia in 2001.

needed to protect business interests, it must be acknowledged that the most transparent country when it comes to arms exports is the United States, the largest and the most competitive arms exporter in the world.

Other official data on the global small arms trade

For many countries, the only source of data on legal small arms exports or imports is the press. Government officials often make statements to the press or to research institutions with respect to small arms exports and/or imports. Also, there is information in the form of company data from production firms, including exports. Further, a number of NGOs have published numerous reports detailing the small arms trade (see HRW 2001a, 2001b; Global Witness 1998).

The *Small Arms Survey 2001* identified 18 known exporters of small arms for which no data at all was available on the value or quantity of their annual small arms exports. Now, the value of annual small arms exports for some of these countries can be approximated. Further, more precise figures can be given for other countries.

One of the most significant exporters of small arms is the Russian Federation (PRODUCERS). The various data combined show that exports of small arms and light weapons from the Russian Federation in 2000 amounted to USD 177.5 million (CAST, 2001). This includes the approximately USD 80 million exported through the state exporting agency, Rosoboronexport, which listed a quantity of 130,000–150,000 weapons and 150 million–200 million rounds of ammunition exported.[10] The main producers of Russian small arms and ammunition, the Tula Arms Plant and Izhmash, export all their products through Rosoboronexport. The additional USD 97.5 million was exported by the Tula-based KBP Instrument Design Bureau, which has the right to conclude foreign contracts independent of Rosoboronexport. It produces man portable anti-tank missiles and infantry rocket launchers. The Russian Defence Ministry is also a minor exporter of small arms, exporting about USD 1.6 million worth in 1999 (CAST, 2001).

The Russian Federation exported approximately USD 177.5 million worth of small arms and light weapons in 2000.

Since the *Small Arms Survey 2001*, researchers commissioned by the Survey have been able to obtain figures for the small arms exports of the Ukraine. According to Ukrspetsexport, the Ukraine's state exporting agency, small arms exports make up less than three per cent of its total arms exports. Official figures for Ukraine's arms export for 2000 amounted to USD 500 million, making the value of Ukraine's annual exports in small arms in the USD 10 million–15 million range. As Ukraine produces virtually only the Fort pistol and ammunition for export, this figure seems plausible. It is likely that much of the excess small arms that the Ukraine inherited from the Soviet Union has been sold off by now.

Data is more difficult to obtain on Israel's small arms exports. However, whereas the *Small Arms Survey 2001* noted that no data at all was available, this status has since changed. There are some statistics for Israel Military Industries (IMI), which is the largest producer of small arms and ammunition (PRODUCERS). With respect to small arms sales—that is, total sales, not just exports—in 1998 IMI sold about USD 32.4 million in small arms. Of this amount, USD 19 million, or 60 per cent of the total, was accounted for by sales of the Galil assault rifle. The Negev machine gun, which has replaced the Uzi as the primary military small arm, accounted for USD 1.8 million of 1998 sales (5.5 per cent), while the Uzi accounted for only USD 0.8 million (2.5 per cent). Pistol sales (Jericho and Desert Eagle) accounted for USD 6.4 million of 1998 sales, or 20 per cent of the total. The remaining 12 per cent of sales for 1998 were for spare parts for the above weapons. According to other data, the Small Arms Factory Group (SAF), a subsidiary of IMI, exported 80 per cent of its small arms produced in 1999; the remaining 20 per cent were sold domestically. In 1999 SAF forecast sales of USD 30.2 million. Israel exported about USD 25 million worth of firearms annually in 1998 and 1999, plus an unknown value of ammunition and other items.

© Associated Press

Internal transfers of weapons are rarely tracked. Here, Russian troops supply anti-Chechen volunteers with AK-47s.

With respect to hunting and sporting firearms, the World Forum on the Future of Sport Shooting Activities (WSFA) published data in a 2001 study. It cited customs data and other official data as its sources. Although no analysis of the quality of this data is included in the study, nor any explanation of which customs categories were used, it gives a clearer picture of the world trade in hunting and sporting firearms. The WSFA found that the United States and the EU accounted for 49 per cent of the global total of exports of sport shooting firearms. It also found that United States and the EU were major importers, with some 1.6 million guns imported in 2000. According to the report, the US and EU were importers for 72 per cent of the firearms exported by the rest of the world in 2000 (see Table 3.4).

When compared with UN customs data from 1999, several inconsistencies emerge in the WSFA data. Brazil, the Russian Federation, and Japan provided data in weight (in tons), not individual guns, as did many countries in the EU. Although this measure is widely used, there is no standard definition of how many items are included in a typical ton of small arms or how much a typical ton is worth.

Another major problem concerns Chinese exports. According to official statistics, China's exports of sporting guns for 1999 totalled 65,012 in 1999, one-third of the total reported here for 2000. It is possible that Chinese exports nearly tripled in one year, but it is more likely that different things are being counted. It is to be hoped that future editions of this worthwhile report will deal with these problems.

The major players: Exporters and importers

The data examined in this chapter confirm the findings of the *Small Arms Survey 2001* that the top exporters of small arms are the United States and the Russian Federation. A careful review of the US 655 Report shows that in 1999 the US exported or authorized for export at least USD 355.7 million of small arms, ammunition, and parts. This figure is considerably lower than the number cited in the *Small Arms Survey 2001*, but this is partly because it does not include light weapons exports. While Germany and Brazil were listed as top exporters in the *Small Arms Survey 2001*, the most recent data does not confirm this. However, the data for 2000 on Brazilian and German small arms exports is not comparable to that of 1999. For example, Brazil's 2000 exports

BOX 3.4 Brazil's small arms exports

In November 2001, the NGO Viva Rio published the most comprehensive and accurate report on Brazil's small arms exports available to date. The *Small Arms Survey 2001* estimated that Brazil exports USD 100 million–150 million worth of small arms annually. While this estimate was fairly accurate for Brazilian exports in the early 1990s, exports since then have dropped substantially. Viva Rio used data from Brazil's Ministry of Development Industry and Exterior Commerce and analysed exports of small arms and ammunition from 1989 to 2000. In 2000, Brazil exported USD 69.7 million of small arms, ammunition, and accessories. This is a decrease of almost 65 per cent from the 1992 high of USD 196.4 million in exports (see Figure 3.3).

FIGURE 3.3 Brazilian exports of small arms and light weapons, ammunition, parts, and accessories, 1989–2000

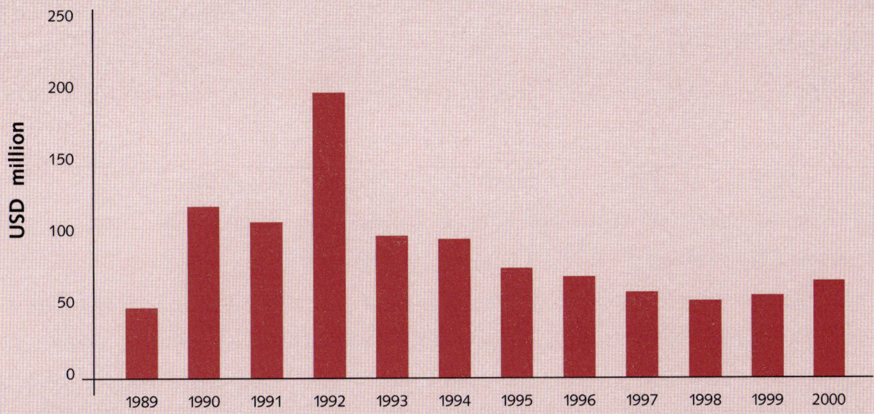

Source: Viva Rio (2001)

of USD 70 million do not include pistols and revolvers, a significant portion of the Brazilian small arms industry (see Box 3.4). Likewise, Germany's military small arms and ammunition exports of 2000, according to its national report, were only USD 14 million, but this does not include light weapons and hunting and sporting rifles, which would increase this value significantly. However, there was a significant drop in German exports of military small arms and ammunition, from USD 37 million in 1999 to USD 14 million in 2000. Table 3.5 shows the dollar value of small arms exports by country for the most recent year available.

The top exporters of small arms are the USA and the Russian Federation.

TABLE 3.4 World Forum on the Future of Sport Shooting Activities (WSFA) data on sporting firearms exports

Exporter	Sporting firearms exported in 2000 (number of weapons)
European Union	839,163
Brazil	314,000 *
United States	267,902
China	185,000 *
Russia	99,000 *
Canada	96,000 *
Japan	87,000 *
Total	**1,888,000 ***

* *Estimated*

Source: WSFA (2001)

BOX 3.5 The gun lobby vs Vanuatu

The archipelago nation of Vanuatu, some 2,000km north-east of Australia, is an unlikely place for a major confrontation over arms export policy. It is a nation of fewer than 200,000 people—mostly poor—spread out among 66 islands. Remote as it is, it could not escape from one of the United States' strongest domestic lobbying groups over small arms export controls.

Negotiating to join the World Trade Organization (WTO), Vanuatu became embroiled in a heated debate with the US that extended far beyond its original parameter. Clearly influenced by the National Rifle Association (NRA), Washington insisted that the tiny nation abandon discriminatory tariffs against small arms exporters in order to join the WTO.

'It has become one of the clichés of accession negotiations', wrote two analysts, 'that in the end [they] always come down to booze and cigarettes; but now the United States, clearly under pressure from its own gun lobby, is putting pressure on acceding countries to liberalize the trade in weapons'. Though commercial interests may have played a nominal role in the accession negotiations, it appears that the US position was conditioned by principle, rather than commmercial advantage.

Source: Grynber and Joy (2000)

As for the international trade in ammunition exports, even less data is available than for firearms. While many national reports include exports of small arms ammunition, meaningful figures are not available from customs data. The role of exports of small arms ammunition in the global small arms trade will be covered in a subsequent edition of the *Small Arms Survey*.

Comparing the top ten exporters for 1999 according to weapons category using customs data alone produces a slightly different picture (see Tables 3.1 and 3.2). Again, it should be emphasized that many countries did not provide values for all export categories. China did not supply values for military arms and pistols and revolvers in 1999, although it exported USD 22.7 million and USD 4.2 million worth respectively in 1998.

As for the major importers, we must rely on a combination of customs data and those national export reports that disaggregate their data by country. For weapons for hunting and sport, the United States is clearly the top importer. As mentioned earlier, the top importers of military firearms for 1999 according to customs data were Cyprus, South Korea, the United States, the United Kingdom, and Australia. Data from national arms export reports shows other emerging trends. The top importers of South African small arms in terms of value from 1997 to 1999 were Colombia, Singapore, Israel, and Taiwan. The top importers of Canadian small arms in terms of value in 1999, excluding the United States, were Germany, France, Thailand, and Australia. The top importers of small arms in terms of quantity from the UK in 2000 were Sierra Leone, the United States, Uruguay, and Singapore.

According to the US 655 Report, in terms of value the top countries authorized to receive commercial sales of small arms from the US in 1999 were Belgium, Germany, and the United Arab Emirates; the top recipient of actual transfers from the US Department of Defence was Greece. According to US Customs Data, in 1999 the top recipients of military rifles from the United States in terms of quantity were the United Arab Emirates, Thailand, Kuwait, Brazil, and Israel. The top recipients in terms of value of Brazilian small arms from 1999 to 2000 were the United States, Venezuela, Germany, and Argentina (Viva Rio, 2001).

When countries appear to be top importers of small arms from more than one country, it can be assumed that they are likely to be major importers. In this respect, the United States, Thailand, Israel, and Germany are countries that appear as top importers through various data sources. Other factors must also be considered in interpretation of this data, such as population and national gun laws. For example, the United States has a large population that has the right to own and carry arms. On the other hand, Singapore stands out as importing large quantities of small arms relative to its population.

TABLE 3.5 Annual small arms exports, by country with most recent year data available

Country	Year data	USD value	Data source	Remarks
United States	1999	356 m	National Report	Includes authorizations and actual transfers; includes ammunition but not light weapons
Russian Federation	2000	178 m	Official data	Actual transfers
Italy	1999	145 m	Customs data	Customs codes 9302, 930310, 930320, and 930330 (majority—USD 108 million—were sporting shotguns)
Switzerland	2000	71 m	National Report	Includes ammunition and some systems larger than light weapons
Brazil	2000	70 m	Viva Rio	Includes ammunition and accessories
Austria	1994	60 m	Customs data	Customs codes 9301 and 9302
United Kingdom	1999	53 m	Customs data	Customs codes 9301, 9302, 930310, 930320, and 930330
South Korea	1997	43 m	National Report	'Guns and Ammunition exports' (no definitions of categories given)
Poland	1999	40 m	Press report	
Belgium	1999	33 m	National Report	
Pakistan	1999	30 m	Press report	
Japan	1999	30 m	Customs data	Customs codes 930320, 930330
Czech Republic	1999	28 m	Customs data	Customs codes 9301, 9302, 930310, 930320, 930330
China	1998	27 m	Customs data	
Spain	1999	26 m	Customs data	Customs codes 9302, 930310, 930320, and 930330 (majority sporting arms)
Israel	1998	25 m	Company trade data	
Portugal	1999	16 m	Customs data	Customs codes 9302, 930320, 930330
Romania	1996	15 m	Customs data	Customs code 9301
Germany	2000	14 m	National Report	Authorizations only. Does not include hunting and sporting firearms and light weapons. Includes ammunition.
Canada	1999	13 m	National Report, US Customs import data	Firearms of 12.7mm and less (National Report), plus US customs data for pistols and sporting firearms (USD 12 million to US)
Ukraine	2000	10 m	Official data	Estimate based on official statement that 3 per cent of Ukraine's defence exports are small arms
South Africa	1999	9 m	National Report	
Turkey	1999	6 m	Customs data	Customs codes 9301, 9302, 930320, 930330
France	2000	5 m	Customs data	Customs codes 930310, 930320, 930330. National Report lists military weapons exported in quantities of weapons.
Argentina	1999	3 m	Customs data	Customs codes 9302, 930320, 930330
Croatia	1999	2 m	Customs data	Customs codes 9301, 9302
Philippines	1999	2 m	Customs data	Customs Codes 9302, 930320, 930330
Finland	2000	1 m	National Report	Includes ammunition
Denmark	1999	1 m	Customs data	Customs codes 9301, 9302, 930320, 930330
Indonesia	1999	1 m	Customs data	Customs codes 9301, 9302
Yugoslavia	1998	874,000	Customs data	Customs codes 930320, 930330
Netherlands	2000	855,000	National Report	
Greece	1999	678,000	Customs data	Customs code 930320
Slovak Republic	1999	576,000	Customs data	Customs codes 9301, 9302, 930330
Nicaragua	1999	375,000	Customs data	Customs code 9301
Senegal	1999	354,000	Customs data	Customs code 9301
Malaysia	1999	346,000	Customs data	Customs codes 9301, 9302, 930310, 930320
Australia	1999	257,000	Customs data	Customs codes 9301, 930320
New Zealand	1999	205,000	Customs data	Customs codes 9302, 930320, 930330
Venezuela	1997	122,000	Customs data	Customs code 930320
India	1998	97,000	Customs data	Customs codes 9302, 930330
Singapore	1999	75,000	Customs data	Customs code 930320
Norway	1999	57,000	National Report	
Hungary	1998	36,000	Customs data	Customs code 930330
Mexico	1998	25,000	Customs data	Customs code 9302
Panama	1997	20,000	Customs data	Customs code 930310
Chile	1999	19,000	Customs data	Customs code 9301
Ireland	1998	9,000	Customs data	Customs code 930320
Thailand	1999	8,000	Customs data	Customs code 9302
Latvia	1999	4,000	Customs data	Customs code 9301

Now the question is: are the countries that are importing arms in large quantities a source for the illicit market? That question will be answered with evidence presented in the following section.

The illicit trade in small arms

Grey market transfers are those that are legally questionable, typically authorized by either the supplying or recipient government, but not by both.

Global concern with small arms now focuses above all on how various actors—whether individuals, groups, or governments—obtain small arms and then misuse them, despite current efforts to prevent this from happening.

What exactly is an illicit transfer of arms? The 2001 UN Small Arms Conference was unable to resolve this issue (CONFERENCE). Is it a transfer that occurs outside government control? This is true in part, but this does not take into account government participation in the illicit trade, which often makes prosecution impossible. Nor does it take into account *grey market* transfers, those that are legally questionable, typically authorized by either the supplying or the recipient government, but not by both. The *Small Arms Survey 2001* defines an illicit transfer as one that violates international and/or national laws of the sending and/or receiving country. It includes both grey- and black-market transfers.

It is increasingly acknowledged that illicit transfers contribute to crime, conflict, and violence worldwide. What is less apparent is the role that legal transfers play in these same problems. Controls over the legal trade and efforts to control illicit arms trafficking are therefore linked. The *Small Arms Survey 2001* provided a regional overview of grey- and black-market transfers. In this chapter we attempt to expose some of the dynamics of the illicit trade, looking at the source of arms—both legal and illegal—and trends within the trade. Exploring in further detail the sources of arms for the illicit trade helps answer the following questions:

- How are small arms obtained for misuse?
- How many arms are coming from legal sources?
- How many arms are coming from 'illicit' sources?

Same guns, different implications: Legal to illicit

Black market transfers are those where suppliers and receivers knowingly violate international and/or national laws of both the supplying and receiving country.

An increasingly large body of evidence indicates that the primary method by which actors obtain arms for misuse is through *diversion*. Diversion is the process by which small arms move from legal government control into the illicit realm. This can be authorized or unauthorized, intentional or unintentional, since in the broadest sense diversion is simply the movement of a weapon from legal origins to the illicit realm. Diversion can occur domestically as well as internationally, though this survey focuses primarily on the international aspects.

The diversion of arms can happen at any point along the transfer chain. It occurs when governments or authorized private owners intentionally or unintentionally transfer arms to embargoed governments, unauthorized paramilitary groups, non-state actors (NSAs) such as rebel insurgents or mercenaries, criminals, or unauthorized private users (see Figure 3.4). Frequently, the process of diversion involves an intricate network of actors adept at finding loopholes in existing legal regimes, escaping detection and legal accountability. Six methods of diversion are discussed below:

- government supplies to NSAs;
- violations of arms embargoes;
- violations of end-user undertakings;
- the ant trade;

- losses from state arsenals/authorized private owners; and
- battlefield seizures and war booty.

These methods can be classified into two groups by reference to the intention of the supplier. The first four are intended by the supplier, although they may not be legally authorized. The last two, on the other hand, obviously do not reflect the intentions of the supplier. Re-transfers, as well, may not be consistent with the intentions of the original supplier.

FIGURE 3.4 Possible transfers from legal to illicit

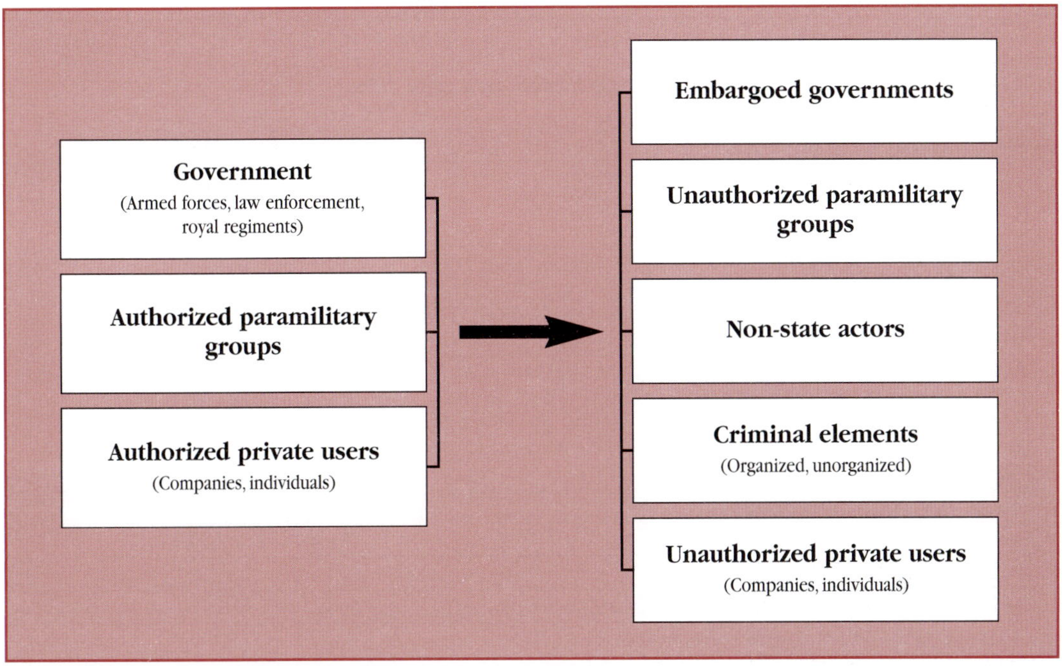

The primary method by which actors obtain arms for misuse is through diversion.

Government supplies to non-state actors

Government patronage appears to be the leading source of arms, funds, and training for the vast majority of non-state actors, though by no means their only source of supply. Government support can be politically, ideologically, or economically motivated. Nearly every region of the world has experienced this phenomenon.

There have been numerous examples of covert state support and arms transfers to NSAs during and since the Cold War. For example: Soviet Union and US support of Central American NSAs during the Cold War (Godnick, 2001a); American sponsorship of Nicaragua's Contras between 1981 and 1984; South Africa's financial and arms support of Mozambique's Renamo in the 1980s (Claiborne, 1987); Libya's supplies of arms to the Irish Republican Army (IRA) between the early 1970s and the late 1980s (Boyne, 1996); India's support of Sri Lanka's Liberation Tigers of Tamil Eelam (LTTE) between 1983 and 1987 (Gunaratna, 2000) and Bangladesh's Shanti Bahini between 1976 and 1989 (Ibrahim, 2001); Chinese arms supplies to Cambodia's Khmer Rouge until 1991 (*Bangkok Post*, 1993); US arms shipments to Cambodia's Sihanouk and Son San prior to 1991 (Phongpaichit, Piriyarangsan, and Treerat, 1998); and Thailand's support of Burmese NSAs operating along the Thai-Myanmar border from the late 1980s to the late 1990s (*Bangkok Post*, 1989).

A recent Small Arms Survey study of the Middle East shows that six governments have supported more than 19 NSAs in the region over the last ten years (see Table 3.6). An examination of apparent transfers to NSAs by Middle Eastern states—many denied by them—shows that the most common reasons are: to foster goodwill with the recipient organization for political purposes; to symbolize their political support of the recipient organization; and in some cases to support proxy warfare in order to destabilize or topple a regime. Economic motives appear not to be a factor, as the provision of arms to NSAs is not known to be profitable for states, though certain individuals might benefit. It appears that state support of NSAs is a common feature of international relations in the region, where it may be considered a standard tool of statecraft (Miller, 2001a).

Government patronage is the leading source of arms for the majority of non-state actors.

TABLE 3.6 Support for selected Middle Eastern non-state actors

Non-State Actor	Alleged government sponsor(s)
Al-Dawa al-Islamiyya (Iraq)	Iran
Al-Gama'a al-Islamiyya (Egypt)	Iran, Sudan
Groupe Islamique Armé (Algeria)	Iran, Sudan
Hizbullah (Lebanon)	Iran (through Lebanon), Sudan, and Syria
Hizbullah (Turkey)	Iran
Movement for Islamic Change, Tiger of the Gulf, Legion of the Martyr Addullah al-Huzaifi, Hizbullah-Gulf, Brethren (Battalions) of the Faithful, Islamic Peninsula Movement for Change—Jihad Wing, Jamaat al-Adala al-Alamiya and the Fighting Ansar of Allah (Saudi Arabia)	Iran, Lebanon, and Syria
Polisario Front (Western Sahara)	Algeria
The Abu Nidal Organisation (Israel)	Libya, Syria, Iran, and Iraq
The Amal group (Lebanon)	Syria
The Democratic Front for the Liberation of Palestine (Syria)	Syria
The Democratic Party of Iranian Kurdistan and the Kurdish Communist Party of Iran (Iran)	Iraq
The Democratic Revolutionary Front for the Liberation of Arabistan and the Al-Harakan al-Islamiya (Saudi Arabia)	Iraq
The Egyptian Islamic Jihad Group (Egypt)	Iran
The Hamas and Islamic Jihad (West Bank)	Syria (through Lebanon) and Iran
The Islamic Front for the Liberation of Bahrain, the Movement for the Liberation of Bahrain, and Hizbullah-Gulf/Bahrain (Bahrain)	Iran
The Mujahideen-e-Khalq (Iran)	Iraq
The Palestine Liberation Front (West Bank and Gaza)	Syria, Iraq, and Iran
The Popular Front for the Liberation of Palestine and the Popular Front for the Liberation of Palestine-Special Command (Israel)	Syria, Libya, Iran, and Iraq
The Popular Struggle Front (West Bank)	Syria, Libya, and Lebanon

Sources: Agence France Presse (2001); BBC News (1999); Jane's Information Group (2001); Miller (2001a); UK (2000a); Wilkinson (2000).

Sometimes, government support of NSAs need not be covert, especially when the group is fighting an unpopular government or when the scale of assistance makes it impossible to conceal (see Box 3.6). A prominent example of the former is the joint support of Russia, Uzbekistan, Tajikistan, Iran, and India for the Northern Alliance fighting in Afghanistan against the Taliban (Davis, 2001; Rashid, 2001b). Though never covert in nature, transfers of military equipment to the Northern Alliance became a major public issue after the terrorist attacks in New York and Washington, DC on 11 September 2001. The terrorist attacks led to an increase in material support, and the Northern Alliance emerged as perhaps the best-supplied NSA in all of 2001 (Pakistan News Service, 2001; *Toronto Star*, 2001).

However, many post-conflict countries whose NSAs received covert state support are now among the largest arm suppliers to the illicit market. Yet, despite the risks such support involves, at the 2001 UN Small Arms Conference there was strong opposition to a ban on assistance to non-state actors from the United States, the Arab League, and others (CONFERENCE).

The limits of good intentions: Violations of arms embargoes

Arms embargoes represent a unique combination of *Realpolitik*, carrot-and-stick diplomacy, and ethical foreign policy motivated above all by the hope of ending conflict or ending oppression. Traditionally, they have been used against states; however, increasingly they are also being used against NSAs, such as União Nacional para a Independência

BOX 3.6 Illegal arms in Palestinian territories

There are a number of armed militias active in the self-ruled areas of the West Bank and Gaza Strip. These include militias connected to the Palestine Liberation Organisation (PLO), such as Force 17—controlled directly by Yassir Arafat—and the Tanzim—local armed gangs loosely connected to the Fatah organization—as well as 'non-official' groups such as Hamas and Islamic Jihad.

One unofficial estimate suggested that the Tanzin groups alone hold at least 70,000 guns, primarily M16s and AK-47s, as well as other automatic rifles and some heavy machine guns. Many of these weapons were allegedly stolen from the Israeli army (Goldenberg, 2000). Another estimate from 1998 suggests that over 40,000 weapons were illegally distributed within the Palestinian Authority (Conteras, 1996; Luft 1999). Arms that can be found within the Palestinian Authority include handguns, rifles, semi-automatic weapons, machine guns, hand grenades, rocket propelled grenades, grenade launchers, anti-tank missiles, and shoulder fired surface-to-air missiles (Luft, 2000).

Palestinian groups allegedly obtain arms from the governments of Iran, Iraq, Libya, and Syria, and even from Israel via theft from Israeli army depots or from Israeli criminal elements (Miller, 2001a; Goldenberg, 2000). Other groups in the region, including Hizbullah in Lebanon, also supply arms to Palestinians (Dudkevitch, 2001). A large majority of these arms are transited through Jordan and Egypt before being smuggled into the West Bank and Gaza Strip (Rodan and O'Sullivan, 1998; Sobelman, 2001). Smuggling routes include all methods available by land, sea, and air.

A prominent example was a shipment intercepted in January 2001. This included 40 barrels containing weapons and ammunition, which were thrown from an Egyptian boat along the Gaza coast. The barrels contained 50 rocket propelled grenades, 1,000 rockets, and a large number of mortar rounds (Harel, 2001a). Another typical shipment, this one over land from Jordan, was intercepted in 2001. This included four Kalashnikov assault rifles, two M16s, a Beretta 9mm submachine gun, and 20 handguns (Dudkevitch, 2001). Foiled smuggling attempts suggest that earlier estimates on the number of illegal weapons circulating within the West Bank and Gaza Strip have to be increased.

Source: Gerald Steinberg and Human Security Research Group (2001)

MAP 3.1 Israel and Palestinian territories

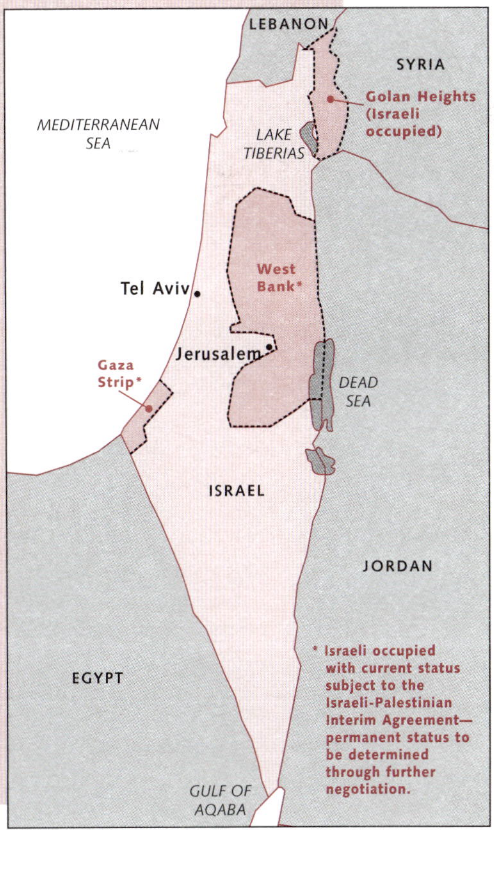

TABLE 3.7 Multilateral arms embargoes effective in 2001

Country	Effective date	Alleged violations		
Mandatory UN embargoes				
Afghanistan	19 Dec. 2000	✧Pakistan (2001)[1]		
Angola	15 Sep. 1993	▲Belgium[2]	* Israel[8]	✦ Togo[14]
(UNITA)		✧▲Bulgaria[3]	* Lebanon[9]	✧ Uganda[15]
		✦Burkina Faso[4]	✧▲ Namibia[10]	✧* Ukraine[16]
		✦Congo-Brazzaville[5]	✧ Romania[11]	✦ Zambia[17]
		✦Côte D'Ivoire[6]	✧▲*Russia[12]	
		✧▲DRC/Zaire[7]	✧▲*South Africa[13]	
Eritrea	17 May 2000 –	✧Russia[18]		
	16 May 2001			
Ethiopia	17 May 2000 –	✧France[19]		
	16 May 2001	✧Russia[20]		
Iraq	6 Aug. 1990	✧✦Belarus[21]	✧ Romania[24]	✧ United Kingdom[26]
		✧China[22]	✧ Russia[25]	✧ Yugoslavia[27]
		✦Iran[23]	✧ Ukraine	
Liberia	19 Nov. 1992	✧✦Burkina Faso[28]	* Guinea[33]	Portugal[39]
		▲Central African Republic[29]	✧* Kyrgystan[34]	▲* Russia[40]
		▲Congo-Brazzaville[30]	▲ Liberia[35]	✧* Slovakia[41]
		✦Côte D'Ivoire[31]	✧▲ Libya[36]	✧ Uganda[42]
		✦* Egypt[32]	▲* Moldova[37]	✧▲*Ukraine[43]
			✧ Niger[38]	▲ United Kingdom[44]
Rwanda	16 Aug. 1995	✧Albania[45]	✧ Italy[55]	Spain[65]
(ex-RAF)		✦* Belgium[46]	✦ Kenya[56]	✦ Tanzania[66]
		✧▲Bulgaria[47]	▲ Lebanon[57]	✦ Uganda[67]
		✧China[48]	▲ Liberia[58]	▲* Ukraine[68]
		✧Czech Republic[49]	✧ Libya[59]	▲* United Kingdom[69]
		✦▲DRC/Zaire[50]	✦ Malta[60]	* United States[70]
		✦Egypt[51]	▲ Niger[61]	Yugoslavia[71]
		✧France[52]	✧ Russia[62]	Zambia[72]
		▲Ghana[53]	✧ Seychelles[63]	✦ Zimbabwe[73]
		✧Israel[54]	✧* South Africa[64]	
Sierra Leone	5 Jun. 1998	*Belgium[74]	Ghana[78]	* Russia[82]
(RUF)		✧Bulgaria[75]	* Lebanon[79]	✧ Slovakia[83]
		▲✦ Burkina Faso[76]	✧▲✦Liberia[80]	✧▲*Ukraine[84]
		✦Gambia[77]	✦ Niger[81]	▲* United Kingdom[85]
Somalia	23 Jan. 1992	✧Djibouti[86]	✧ Ethiopia[88]	✧ Ukraine[90]
		✧Eritrea[87]	▲ Russia[89]	
Yugoslavia	31 Mar. 1998 –	✦Albania[91]	✧ Greece[93]	
(FRY)	10 Sep. 2001	✧Bosnia & Herzegovina[92]	✧ Macedonia[94]	
			✦ Montenegro[95]	

TABLE 3.7 Multilateral arms embargoes effective in 2001 (continued)

Country	Effective date	Alleged violations	
EU embargoes (non-mandatory)			
Afghanistan (Taliban)	17 Dec. 1996	Unknown	
Bosnia & Herzegovina	5 Jul. 1991	▲* Austria[96] * Belgium[97]	✧ Germany[98] ✧ Italy[99]
China	27 Jun. 1989	✧ France[100]	
DRC (formerly Zaire)	7 Apr. 1993	* Belgium[101] ✧ France[102]	
Eritrea	15 Mar. 1999 – 31 May 2001	Unknown	
Ethiopia	15 Mar. 1999 – 31 May 2001	✧ France[103]	
Iraq	4 Aug. 1990	✧ United Kingdom[104]	
Libya	27 Jan. 1986	Unknown	
Myanmar (Burma)	29 Jul. 1991	✧ Portugal[105]	
Sierra Leone	8 Dec. 1997	* Belgium[106] ▲* United Kingdom[107]	
Sudan	16 Mar. 1994	Belgium[108]	
Yugoslavia (FRY)	5 Jul. 1991 – 8 Oct. 2001	* France[109] ✧ Greece[110]	
OSCE (non-mandatory)			
Nagorno-Karbakh (i.e. Armenia & Azerbaijan)	28 Feb. 1992	✧ Russia[111] ✧ Turkey[112]	✧ Ukraine[113]

✧ *Country from which weapons originate*

* *Country nationals breaking embargoes*

▲ *Country in which flag vessels and aircraft are breaking embargoes*

✦ *Transit countries*

Source: Appendix 3.3

Total de Angola (UNITA) in Angola and the Revolutionary United Front (RUF) of Sierra Leone. Multilateral embargoes have been initiated by the UN, the EU, and the Organization for Security and Co-operation in Europe (OSCE). Some are binding and some are not. Table 3.7 presents a list of multilateral embargoes that were in effect during 2001. Embargoes can also be implemented unilaterally, as shown by the United States, which maintains a policy 'to deny licences, other approvals, exports and imports of defence services destined for' 20 countries, in addition to those countries on which the UN maintains a standing arms embargo (US Department of State, 2001).[11]

As Table 3.7 and Appendix 3.3 demonstrate, there have been allegations of violations of nearly every arms embargo currently in place.[12] Preliminary research shows that 54 countries have been linked to shipments of small arms in violation of international arms embargoes effective in 2001. The nature of non-compliance varies. Many of the states listed here deny any involvement in embargo-busting activities. Some countries such as Bulgaria and Romania claim to have sold arms in good faith to non-embargoed countries and only later discovered that the end-user certificates used to justify the purchases were forged. In other cases, arms may have been sent in good faith to recipients such as Burkina Faso, a country not under arms embargo, only to be re-transferred to an embargoed party. Of special interest is the involvement of brokers and transport agents in embargo-busting activities. The UN has taken this into account, and, since 1992, embargoes specifically hold member states responsible for the activities of their nationals and transport agents registered in their country with respect to embargo-busting activities.

Just a slip of paper: Violation of end-user undertakings

End-use certificates and related assurances are the exporting country's way of obtaining assurances about where its arms exports are going and retaining a measure of control over them once they leave the originating country's jurisdiction (Joseph, 2001). Even so, there are no guarantees. Typical end-user violations include re-transferring shipments in violation of a non-transfer pledge, covering up the identity of the real end-users—which might be countries under embargo or rebel groups—or simply forging end-user certificates.

End-use violations have been well documented in UN reports on embargoes against UNITA, the RUF in Sierra Leone, and Liberia. Most often a violation of end-user undertakings requires the assistance of a sympathetic government, or, at the very least, the collaboration of a corrupt, high-ranking government official. Such ruses are particularly successful if the assistance comes from a neighbouring country. For example:

- UNITA apparently received one official end-user certificate in July 1997 from Togo signed by the then Army Chief of Staff. Though the original document never surfaced, 18 forged versions surfaced in Bulgaria during 1997–98 of which nine were executed either partially or in full, amounting to a total of USD 14 million. Another two forged certificates were used successfully in Romania in 1996 and 1999 amounting to a total of USD 646,780 in arms. All shipments transited through Togo, where the government allegedly kept up to 20 per cent of the equipment (United Nations Security Council, UNSC, 2000c).
- In March 1999, the RUF received a shipment of 68 tons of small arms that originated in Ukraine. Investigations by the UN showed that Ukraine had authorized the export of arms to Burkina Faso, from where the shipment was flown to Liberia before being transported to the RUF. According to Ukraine, the arms had been transferred with the understanding that they would not be sent to a third party (UNSC, 1999a).
- In July and August 2000, ammunition was transported from Abidjan to Monrovia in direct violation of the UN arms embargo against Liberia. The ammunition—five million 7.62mm cartridges, weighing 113 tons—originated in Ukraine. Investigations revealed that the Côte D'Ivoire head of state, General Robert Gueï, signed the end-user certificate for the ammunition as well as for a long list of additional weapons. The certificate was authenticated by the ambassador of Côte D'Ivoire in Moscow, and on that basis the Ukrainian government issued an export permit and authorized the flight. A Ukrainian military officer accompanied the flight to guarantee delivery. Further investigation revealed that upon assuming office after a *coup d'état* General Gueï had asked for assistance from

several African heads of state in procuring ammunition. In payment for assistance in procuring the arms on behalf of Côte D'Ivoire, Liberia requested a portion of the armaments. In an interview, General Gueï admitted that most of the ammunition obtained on behalf of Côte D'Ivoire was re-transferred to Liberia (UNSC, 2000c).

- Former Argentinian President Carlos Menem was put under house arrest in July 2001 for allegedly heading a conspiracy that smuggled small arms to Croatia and Ecuador in 1991–95. According to reports, Menem signed official decrees authorizing the state weapons factory, Fabricaciones Militaries, to sell arms nominally to Panama and Venezuela. In reality, the equipment went to Croatia and Ecuador. The combined value of the 6,575 tons shipped amounted to USD 100 million (Faiola, 2001). At the time, Croatia was under a UN arms embargo, and Ecuador was fighting a border war with Peru in which Argentina was a guarantor in the peace process. And, though in November 2001 Argentina's federal court ordered Menem's release due to lack of sufficient evidence, a number of other former high-level officials remain in prison, including the former Army Chief and Minister of Defence (*Financial Times*, 2001).

The ant trade

In some regions the 'ant trade'—a term coined by the Mexican authorities to describe smuggling from the US to Mexico—accounts for a significant portion of diverted illicit arms. The ant trade is the process by which arms are bought legally in one country and then smuggled in small increments, sometimes one at a time, into another country. Typically traders are taking advantage of controls that are less strict in the country where they buy firearms than in the receiving country. The same phenomenon has been seen on the Canadian-American border and between Paraguay and Brazil (Small Arms Survey, 2001, p. 186).

Mexican officials estimate that 80 per cent of the illegal weapons in the country are illegally imported from the United States.

Mexican officials estimate that 80 per cent of the illegal weapons in the country are illegally imported from the United States (Weiner and Thompson, 2001). Arms are smuggled into Mexico both by sea and land, though it is the 3,000km land border where most of the ant trade occurs. Often the firearms originate in gun shops and pawn brokers in California, Texas, Arizona, and New Mexico. Florida is increasingly becoming a source for weapons smuggled into Mexico by air and by sea (Weiner and Thompson, 2001). Straw purchasers, legally eligible American residents, acquire firearms from a licensed dealer, concealing the identity of the intended recipient (US, Bureau of Alcohol, Tobacco and Firearms, BATF, 2001). The guns are passed on to smugglers, usually petty criminals, who take them across the border. Smugglers can repeat this process hundreds of times a year. The recipients of the small arms include drug cartels and criminals (Lumpe, 1997).

A study by the US BATF (2000) showed that 46.3 per cent of 1,530 criminal cases involving arms trafficking investigated from July 1996 to December 1998 involved a straw purchaser. The same study found that 11.1 per cent of the cases examined involved shipments across international borders. Mexico is not the only country experiencing a flow of illegal arms from the US. It is estimated that more than half of the handguns recovered from crime scenes in major Canadian cities like Toronto were smuggled into the country from the United States. These illegal firearms were mainly acquired at gun shows or via straw purchases (Canada, Department of Justice, 1995; 1997). Japan has found that 30 per cent of firearms recovered each year are illegally imported from the United States, while 21 per cent come from China (United Nations, 1999). Arms smuggled from the United States have found their way into Latin America and Northern Ireland.

© Associated Press/Jockel Finck

A half a million weapons were looted from Albanian military depots in 1997. Some turn up at markets such as this one, in northern Albania.

BOX 3.7 An attempted violation?

The difficulty of distinguishing legal and illegal deals, or even simply understanding what is going on, was illustrated by an arms affair involving actors from the Czech Republic, Bulgaria, Georgia, Israel, Ukraine, and Eritrea. Although the deal might at first appear complicated, it is surprisingly typical of questionable arms deals in the early twenty-first century.

The affair became public on 26 April 2001, when a Ukrainian plane carrying arms from the Czech Republic was detained at Bulgaria's Burgas airport. It was loaded with howitzer parts consigned to Gamma Centre, a Georgian company. Both the Czech Republic and Georgia confirmed that the arms shipment was authorized and 'completely legal' (see ①). Everything seemed in order. So why the fuss?

When the plane landed in Bulgaria, its next destination was stated as Asmara, Eritrea, a country under a UN arms embargo. However, documentation for the military cargo clearly noted that the arms were destined for the Ministry of Defence of Georgia. When confronted, the pilot repeatedly insisted that the flight was to Asmara, as was also noted in the plane's flight journal (see ②). Due to the contradiction, the authorities detained the plane pending clarification on the real final end-user.[13]

Newspaper accounts suggest that the original flight plan was to Georgia's Aspara airport, reportedly with the approval of the Czech Foreign Ministry and the Czech Ministry of Industry and Trade (*Czech News Agency*, 2001c). The pilot and crew are thought to have changed the flight plan from *Aspara*, Georgia to *Asmara*, Eritrea. The change was claimed to have been made in order to pick up additional cargo in Eritrea at the request of the client before flying to Georgia (see ③). The client in question, the Israeli company Unitag, was responsible for arranging the transportation of the arms from the Czech Republic to Georgia (*Africa News*, 2001; *Czech News Agency*, 2001d). A change in flight plans is ordinarily not a problem if authorized by the relevant governments. In this case it was not. Given that Eritrea was still under a UN arms embargo, the transport of arms from Bulgaria would have constituted a violation of international law.

Bulgarian authorities detained the flight for over a month to ask all the countries and parties involved why a request was made to fly to Eritrea with a cargo of arms in violation of the UN embargo. According to both the Georgian Deputy Defence Minister Gela Bezhuashvili and Thomas CZ—the Czech company exporting the armaments to Georgia—the shipment contained only howitzers. The total cargo was reported as weighing 30 tons and amounting to over USD 250,000 in arms (*Czech News Agency, 2001*a). Newspaper reports state that an unlisted cargo of Czech-made Kalashnikov rifles was found aboard the plane. Bulgarian authorities denied the allegation.[14]

The drama ended on 6 June 2001 when Bulgarian authorities finally authorized the plane to depart for Kopitnari, Georgia (*Czech News Agency*, 2001d; *ITAR-TASS*, 2001 (see ④). Ominously, all the companies involved have been accused of illicit activities in the past (*Czech News Agency*, 2001b).

MAP 3.2 A snapshot of an alleged embargo violation

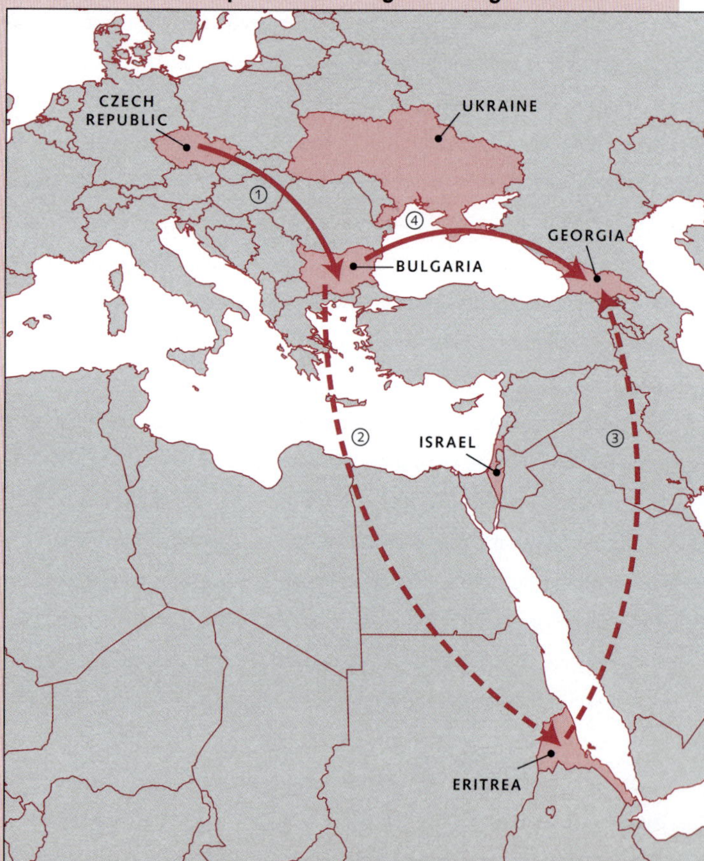

Isn't that government property? Theft from state arsenals

Unlike the previous examples of diversion, theft from state arsenals or authorized private owners occurs without the intent or authorization of the supplier. Leakage from state arsenals can occur either in large quantities or in small increments over time. Theft or pilfering from military or police stockpiles is especially serious when security collapses or corrupt officials sell or rent arms without authorization. In some cases, the arms in question had been previously confiscated by authorities or were being held as criminal evidence (WEAPONS COLLECTION). In other cases, arms have been acquired by military officers under the auspices of the ministry of defence for the purpose of reselling them on the black market.

In early April 2001, troops raided an airport warehouse in Bangkok to steal a shipment of 30 Glock pistols. The soldiers escaped with the pistols but were caught on film. An inquiry revealed that a senior air force officer authorized the shipment for the express purpose of reselling the arms on the black market. When warehouse officials refused to release the shipment to anyone other than the officer, he organized the raid. In a news interview, Deputy Defence Minister Yuthasak Sasiprapha admitted that there had been instances of corruption in the military over arms procurement. Typically, this occurs when arms importers under contract to the military are allowed to import more guns than needed by the military. These are then sold on the black market (*The Nation*, 2001a; 2001b; Tang, 2001).

Most serious are the losses resulting from a breakdown of state security, as when 549,775 military small arms were taken from Albanian arms depots by the general population in 1997 (Republic of Albania, 2001); from a collapse of the state, as in the case of Somalia; or from inadequate stockpile management, as in the early days of the newly independent states of the former USSR which suddenly found themselves in control of stockpiles whose records were all kept in Moscow (Coker, 2001a).

Theft from government stockpiles is frequently linked to corrupt military or police officers. Poor record keeping and stockpile management systems facilitate insider jobs (MEASURES). This was illustrated on 10 May 2001 when Thai police intercepted two trucks carrying 23 landmines, 60 hand grenades, 720 bars of detonating powder, and 15,000 rounds of ammunition for rifles and machine guns. The weapons, classified as non-inventoried items, had been stolen from several arms depots and were destined for Indonesian insurgents in Aceh (*The Nation*, 2001c; Tang, 2001). There are also reports of individual soldiers or officers bartering arms for food or other items in conflict zones when rations or pay has been delayed. At a road junction in Chechnya, a reporter observed Russian soldiers trying to exchange a grenade for a pack of Prima cigarettes. Another report recalled soldiers exchanging two rocket-propelled grenades for two bottles of vodka and a small box of hashish (Sventsitskiy, 2001).

Western countries, which might be expected to manage their stockpiles adequately, experience losses due to theft by both insiders and outsiders (MEASURES). In a well-publicized attack in April 2000, a British Royal Artillery armoury in Larkhill, Wiltshire, was raided by thieves who took 19 pistols, 13 shotguns, and three starter pistols. Authorities suspect that the arms were stolen to supply Britain's underworld (Evans, Bayley, and Tendler, 2000). As of August 2001, only three of the arms had been recovered (Carrell, 2001).

Theft, of course, is even more serious for private owners and dealers, who cannot afford as much protection. Valuable and useful, guns are a tempting target for thieves. According to the South African Police Service, 70 per cent of all illegal arms within circulation in the country were either lost or stolen. Between 1994 and 1998, a total of 76,797 firearms were reported as lost or stolen in South Africa. Even this figure appears to be far from complete, since, during the same time period, 84,557 illegal firearms were confiscated by police (Hennop, 2000). By way of comparison, in Guatemala, during 1996–98 a total of 3,458 registered firearms were reported as stolen (Table 3.8; Godnick, 2001a).

In the United States, of 1,530 cases of firearms trafficking investigated, 13.7 per cent involved arms stolen from licensed gun dealers, accounting for a total of 6,084 firearms. The same study found that 10.3 per cent of illegally trafficked firearms had been stolen from residences, accounting for a total of 3,306 firearms (US, BATF, 2000).

TABLE 3.8 Cumulative firearms reported stolen (or lost) in Guatemala and South Africa, 1994–1998

Year	South Africa		Guatemala	
	Number	% increase over previous year	Number	% increase over previous year
1994	7,285	n/a	n/a	n/a
1995	9,109	20.0	n/a	n/a
1996	13,746	50.9	1,003	12.4
1997	16,963	23.4	1,117	11.4
1998	29,694	75.0	1,338	19.8

n/a not available

Sources: Godnick (2001a); Hennop (2000)

TABLE 3.9 Some recent small arms seizures and government recoveries in the Philippines

Date	New People's Army seizures
12 Jun. 2001	2 M14, 1 M16 automatic rifle, and 1 grenade launcher
23 Dec. 2000	7 M16 rifles and two 9mm pistols
20 Aug. 2000	11 M1 rifles, 3 M16 rifles, 2 M14 rifles, and a cache of assorted ammunition
3 Nov. 1999	5 Armalite rifles, 1 M14 rifle, 4 .38 calibre revolvers, and 2 9mm pistols
11 Jun. 1999	High-powered rifles and ammunition
7 May 1999	Several firearms, including an M-16 rifle and ammunition
2 Apr. 1999	13 military-issued high-powered rifles and ammunition
18 Feb. 1998	Two M16 rifles and an M1A carbine rifle
3 Jan. 1998	5 M14 assault rifles, 17 M1 rifles, and an M60 7.62mm light machine gun

Sources: BBC Summary of World Broadcasts (1997; 1999b); Business World (1998; 1999; 2000; 2001); Deocadiz (1998); Deutsche Presse-Agentur (2000b); Garcia (1999); Lugo and Garcia (1999); Manila Times (2000); Philippine Star (2001); Villanueva (2001); Xinhua (2001)

Date	Government recoveries
23 Jun. 2001	3 firearms
12 Feb. 2001	Several firearms recovered
8 Jan. 2001	7 M16 rifles, 1 grenade rifle, 1 Russian AK-47 rifle, 1 US Thompson submachine gun, 1 automatic carbine, and thousands of rounds of ammunition
9 Feb. 2000	2 M16 rifles, 2 M14 rifles, 1 M76 grenade launcher, and 1 M1 rifle
3 Aug. 1997	51 high-powered firearms, including 3 M16 machine guns, 39 Armalite rifles, 5 AK-47s, and some M1 rifles

Battlefield seizures and war booty

Rebel movements have long staged attacks on police headquarters, military cantonments, and patrols specifically to steal equipment. The practice is extremely dangerous, but for desperate, isolated, or weak rebel movements it can be the only way to obtain equipment. For some NSAs, over half of the weapons held are derived from operations against the state. This would include arms seized during battles, theft from government arsenals, arms that are abandoned by government forces in conflict situations, and arms caches that are acquired by gaining territory.

It is important to note that arms are transferred in each direction between government forces and NSAs in conflict situations. During armed conflict, for example, areas of control are traded between factions. The territory gained or lost may include weapons caches. Effective control over these caches depends on who controls the territory at any given time. This is illustrated in the Philippines, where press reports frequently note the amount of arms acquired by local insurgencies during a particular raid, only to be retaken by government forces in counter-attacks (see Table 3.9). While the situation in the Philippines may be extreme, there seems to be some fluidity in the movement of arms between parties in most conflicts (HUMANITARIAN).

Other sources for the illicit trade: Illicit craft production

The most significant proportion of arms available for the illicit trade is drawn from illicit markets, including weapons that have already been diverted or which fuelled one conflict and are now being moved to another. The smaller number is derived from illicit craft production (Small Arms Survey, 2001, pp. 45–47; PRODUCERS).

Illicit craft production accounts for only a small percentage of illicit arms in circulation.

Illicit craft production—unlicensed work carried out in small workshops—accounts for only a small portion of small arms involved in illicit transfers. Buyers who can afford to be selective seem to prefer legally manufactured arms. Craft production is characterized by inferior quality, unreliability, and low cost. Craft-produced arms are often favoured by criminal elements and unlicensed individuals as well as by a few insurgency groups. Most often, craft-produced small arms appear to be manufactured and transferred domestically. There are also cases where these weapons cross international borders. This can happen in small shipments, such as the case of small arms manufactured illegally in India's Bihar State that are smuggled into Bangladesh, or in larger shipments, such as small arms manufactured illegally in the Philippines that are supplied to the Yakuza and other criminal gangs in Japan (Parreño, 1995). A well-known illicit production site is Darra Adam Khel, in Pakistan. Darra has allegedly supplied Pakistan's troubled provinces of Punjab and Sind, as well as the Taliban in Afghanistan, the IRA, and insurgency groups in Kashmir and the Middle East (Sullivan, 1998; Lallemand, 2001).

Conflict trade and illicit arms acquisitions[15]

The trade in 'conflict goods' is an increasingly salient feature of war economies of contemporary conflicts, often providing the finances and means for procurement of arms and continuation of war (HUMANITARIAN). The best known conflict goods are diamonds, which have fuelled conflicts in Angola, Sierra Leone, Liberia, and the

© ICRC/Charles Page

An arms dealer in Darra, Pakistan, a town notorious for its illicit production of small arms. Darra's arms are believed to reach non-state actors in India, the Taliban, the IRA, and insurgency groups in Kashmir and the Middle East.

Democratic Republic of Congo. Other goods, such as timber, oil, and coltan—a mineral used in a variety of high-tech products, most notably mobile phones—have also perpetuated conflict. Conflict goods can be defined as:

> non-military materials, knowledge, animals or humans whose trade, taxation or protection is exploited to finance or otherwise maintain the war economies of contemporary conflicts. Trade can take place by direct import or export from the conflict zone or on behalf of military factions (both government and non-government) by outside supporters. Conflict goods do not include arms, military aid or the services of mercenaries, although these may be paid for in kind with goods or concessions to trade in products which would then become conflict goods (Cooper, 2001b).

The profits from the trade in conflict goods are used not only to purchase arms but to supplement wages for soldiers and to pay for food, oil, and medical supplies. Such trade also generates personal wealth, particularly for leaders and key supporters. Whilst the profits from the trade in conflict goods are occasionally directed to the purchase of major weapons, small arms have been the weapon of choice in 90 per cent of the conflicts since 1990 (Small Arms Survey, 2001, p. 210). There is a dynamic nexus between the trade in conflict goods and that in small arms. There are a variety of ways in which the trade in conflict goods can facilitate arms acquisition:

Profits from conflict trade are not only used to purchase arms, but to fund wages for soldiers and to pay for food, oil, and medical supplies.

Conflict goods bartered for small arms: This is not a phenomenon confined either to the small arms market or to illicit arms sales. Perfectly legitimate major arms deals are often paid for in whole or in part with oil or other products from the purchasing country. Yet it is also a path frequently used by NSAs. This has apparently been the preferred purchasing method of UNITA in Angola, which has reason to be wary of having its funds frozen in national and foreign bank accounts. In such situations, agents who act as arms brokers for equipment going into a conflict will act as commodity brokers for goods coming out.

According to a report prepared for the UN Security Council, key brokering agents for UNITA in the early 1990s, selling about one third of UNITA's diamonds, were Fred Rindel—who allegedly went by the name Watson—and the De Decker brothers of South Africa. Typically, they would fly to Andulo in Angola where Rindel would negotiate the arms component of the deal that might include a variety of small arms and light weapons, mostly supplied from Eastern Europe. Once the arms component of the deal was agreed, Joe De Decker allegedly would sit with UNITA's own diamond experts to assess packages of diamonds, with each parcel being worth some USD 4–5 million. The brothers and Rindel would take the packages to Antwerp or Tel Aviv, where they were sold to the De Beers Diamdel buying offices. They would then return to Angola with the money and sales documents (UNSC, 2000c).

Supplementing military budgets: The proceeds from the trade in conflict goods also may be used by states to supplement defence budgets. Alternatively they can be banked, often after laundering, by insurgent groups thus presenting an important opening for the international community to intervene. Indeed, in some cases, as with the Revolutionary Armed Forces of Colombia (FARC), proceeds from conflict trade may be invested in the stock market or other perfectly legitimate enterprises for laundering to provide long-term funding. One example of trade revenues supplementing formal defence budgets is the Rwandan army, which, through the companies Rwandan Metals and Grand Lacs Metals, made an estimated USD 250 million from the export of coltan from the Congo—formerly Zaire—

in an 18-month period between 1999 and late 2000. This is much larger than the official annual defence budget of USD 70 million (UNSC, 2000b).

Paying for an indigenous defence manufacturing capability: The proceeds from trade in conflict goods may be used to fund the development of a industrial defence capacity. This has been most striking in the case of Sudan, where oil-related revenues have enabled the government to double the defence budget to USD 327 million a year and to develop two defence industrial complexes manufacturing light weapons, machine guns, ammunition, rocket propelled grenades, mortars, and ammunition (Christian Aid, 2001; CLW, 2002; HRW, 1998).

Providing infrastructure for military logistics: The roads or airstrips built by companies to facilitate the movement of goods also provide a dual-use infrastructure that can be utilized by combatants to transport both troops and arms. In Liberia, timber exports amounting to at least USD 187 million annually not only provide Charles Taylor with funds by which to train, arm, and supply his security forces and those of the RUF in Sierra Leone, but the infrastructure of the timber industry itself can facilitate the transfer of arms.

In May 2001, a cargo ship owned by the Belize-registered firm Alpha Paramount, arrived from Dakar, Senegal, and allegedly unloaded a cargo of armaments at Harper Port, Liberia. The port is controlled by the logging company MWPI through its representative, Hussein Fawaz. The ship was then loaded with logs for export from Liberia. Arms are apparently also frequently transported inside Liberia using trucks owned by the timber companies that travel on roads built by these same companies (Global Witness, 2001; HRW, 2001b).

Purchasing allies: Commodities—usually precious stones—may be used instead of cash to buy or reward favours from allies. UNITA used diamonds to win and maintain regional allies who would facilitate the provision of arms or training to UNITA forces. These included diamond payments allegedly made to ex-President Mobutu in the former Zaire, as well as to the presidents of both Burkina Faso and Togo (UNSC, 2000a). Engagement in trade of conflict goods serves as a perk for military officers and ordinary soldiers who participate in conflict. This can cement loyalty, although arguments over such perks can lead to disputes between allies as well.

Infrastructure built to accommodate legitimate trade also can be used by combatants to transport troops and arms.

Since much of the trade in conflict goods is illicit, it is difficult to quantify the value of this trade as a whole or the extent of it within any industrial sector. As the figures in Table 3.10 show, the revenue-raising potential of such trade can be quite substantial, particularly in comparison with the defence budgets of many states in the developing world or the weapons procurement budgets of armed insurgent groups.

Black-market sourcing and transiting: Thailand and Bangladesh

Once small arms enter into illicit trade, they can continue to circulate for years, before they are either confiscated and/or destroyed. Covert arms supplies, arms left over from civil wars, diverted arms, and arms produced illegally all continue to circulate in the black market. For the most part, small arms are durable, and if kept in good condition can move through many different hands and across many international borders before they become unusable.

When the trade becomes international, three principal kinds of countries are involved: *source* countries, *transit* countries, and *destination* countries. In a source country weapons first enter illicit markets; they must physically pass through transit countries to reach destination countries. In the case of black-market transfers, source countries for the most part are not producing states. Instead, they are often countries that have recently experienced war and have an abundance of leftover weapons or large stockpiles ripe for diversion. Transit countries tend to border regions of conflict. Destination countries provide the demand for illicit arms.

TABLE 3.10 Putting a price on conflict goods

Organization	Product	Period	Value of trade (USD)
UNITA (Angola)	Diamonds	1992–2000	4.1 billion
Charles Taylor (Liberia)	Timber, diamonds, iron ore	1992–6	400 million annually
	Timber	Current	Minimum 187 million
RUF (Sierra Leone)	Diamonds	Late 1990s	30–125 million annually
Taliban (Afghanistan)	Taxing opium trade	Late 1990s/2000	30–40 million
	Taxing border smuggling		75 million
Afghan opposition	Lapis lazuli	Late 1990s/2000	30–40 million
	Taxes on opium and consumer goods		12–20 million
FARC (Colombia)	Taxing cocaine trade	Late 1990s	140 million annually
	Total income from drugs, kidnapping, extortion, etc.		600 million annually
All Colombian rebel groups	Kidnapping	1996–2001	900 million
Rwandan government	Coltan from the Democratic Republic of Congo	1999–2000	250 million over 18 months
Sudan government	Oil	Current	400 million annually
Khmer Rouge (Cambodia)	Timber	Mid-'90s	120 million annually
Cambodian government	Licensing of timber exports by Defence Ministry	Mid-'90s	150 million annually

Sources: Global Witness (1998); UNSC (2000a); Alao and Olonisakin (2000); Berman (2000); Rashid (2001a); Human Rights Watch (2001a); Rangel Suárez (2000); Guardian (2001); Christian Aid (2001); Berdal and Keen (1997)

Thailand: These three categories are not always clearly distinguishable, as the case of Thailand illustrates. Though Thailand is technically considered a transit country, it is here that Southeast Asia's principal black market in arms is found. Illicit arms are trafficked through the country to be redistributed around the region. Its contemporary role in arms trafficking began when it was used as a preferred transit route for covert arms shipments from China and the United States to Cambodia's anti-Vietnamese factions beginning in the late 1970s. From the Cambodian 'pipeline', small arms were diverted by Thailand's military to ethnic groups along the Thai–Myanmar border. At the time, this fell in line with Thai policy of creating a buffer zone between Thailand and perceived external threats. When the Cambodian war ended in the early 1990s, arms began flowing back into Thailand. At first, these arms continued to flow primarily to Burmese insurgents, but as time passed, policies changed, and the trade was

It is estimated that 80 per cent of all illegal weapon consignments from Cambodia pass through Thailand.

FIGURE 3.5 Overview of Thailand's black marke

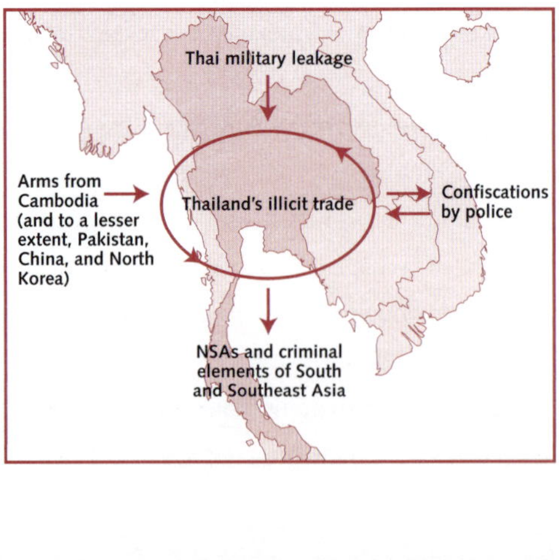

motivated less by politics and more by economics. Thus, Thailand became the primary source of arms for insurgents operating in north-east India, the Philippines, Indonesia, and Sri Lanka in addition to criminal groups operating in Southeast and East Asia (Kramer, 2001).

Today, it is estimated that 80 per cent of all illegal weapon consignments from Cambodia pass through Thailand. It is also believed that illegal arms coming from Pakistan, China, and North Korea also pass through Thailand for redistribution (Chalk, 2001). Other sources for Thailand's illicit arms market include thefts from military arsenals, illicit deals by licensed gun dealers, and confiscations made by the police that are then offloaded onto the black market by corrupt officers (Kramer, 2001).

There are several reasons why arms trafficking continues in Thailand. There is a well-established network of brokers and transport agents. Many of those involved are influential within their locality; and, though the actors have changed, traffickers continue to be members of the military and police, businessmen, gun dealers, politicians, and gangsters.[16] The residual Thai socio-political structure of patronage and the ease with which traffickers can buy co-operation and silence have allowed the trafficking to continue.[17]

MAP 3.3 Thailand

Bangladesh: Bangladesh illustrates the dangers of being a transit country. Until recently, regional hot spots in Pakistan, India, and Sri Lanka overshadowed armed violence in Bangladesh. Bangladesh attracted attention only because of its position as a conduit of arms for insurgency groups operating in north-eastern India and western Myanmar. Yet a recent study undertaken by the Small Arms Survey reveals that Bangladesh is itself a destination for illicit arms rather than solely a transit state.[18] Today arms enter the country increasingly to feed an ever-growing domestic demand. The main end-users of illicit arms are petty criminals or political cadres, armed groups affiliated with a criminal syndicate or one of Bangladesh's many political parties (*Independent*, 2001). Other recipients include smugglers and some businessmen, specifically large landowners along the coastal belt and managers of illicit logging ventures.

Bangladesh, once perceived as primarily a transit country, is increasingly becoming a destination country.

Bangladesh is the transit route for small arms deliveries to north-east Indian—Assamese and Naga—and Burmese—Rakhyan, Rohingya, and China—insurgent groups operating along Bangladesh's borders. Most of the shipments come from Thailand's black market or from Pakistan. Most of these arms shipments occurred between 1991 and 1998. Though some of them have been pilfered to feed domestic demand, or in some rare cases sold outright by insurgent groups or individual insurgents, they are not a significant source of illicit arms within the country (Ahmed, 2000; Chakma, 2000; *Daily Janakantha*, 2000).

A small percentage of the illicit trade was supplied by Bangladesh's war for independence. After independence, a large number of guns remained in the hands of both the pro- and anti-independence forces. A significant number of former

BOX 3.8 Arms deliveries

Although illicit small arms deliveries by air receive more international attention (Small Arms Survey, 2001, pp. 111–23; Johnson-Thomas, 2000), shipments by sea are no less important. Reports of illicit small arms transfers routinely mention transport by sea.

Over 60 per cent of the world's total commercial trade is moved by sea. The large volume of cargo passing through major ports each year around the world makes checking cargo very difficult. This is compounded by the fact that, with limited resources available, more attention is paid to incoming shipments than to transhipment cargoes. Thus, it is an ideal mode of transport for illicit small arms.

One of the most active regions for maritime illicit small arms trafficking is Asia. Patrolling the vast waterways and islands has proved a daunting task for the region's navies and coastguards. Insurgency groups in north-west India, western Myanmar, Sri Lanka, Indonesia, and the Philippines use anything from commercial ocean freighters to fishing boats to deliver small arms procurements. Speedboats are frequently used to unload cargo at sea and ferry the shipments across coastguard patrol lines to bring them to shore.

The Liberation Tigers of Tamil Eelam (LTTE) have a highly sophisticated sea transport network. The LTTE has set up at least 12 foreign-based shipping companies, mainly in Southeast Asia, and has a fleet of at least 11 deep-seagoing vessels at its disposal travelling mostly under Panamanian, Honduran, or Liberian flags (*Jane's Terrorism & Security Monitor*, 2000; Chalk, 1999). Ninety-five per cent of the time, the vessels transport legitimate commercial goods. It is the other five per cent of the time that they play a vital role in supplying small arms, ammunition, explosives, and other war-related material to the LTTE (Chalk, 1999).

The LTTE's maritime wing, the Sea Tigers, 'meet incoming LTTE supply vessels at pre-arranged rendezvous points and the supplies are then transferred to clandestine landing points in the many creeks and inlets in the northeast of the country' (*Jane's Terrorism & Security Monitor*, 2000). In June 2000 it was reported that the Burmese navy seized a suspected LTTE speedboat reportedly on its way from southern Thailand to Sri Lanka. In addition to 33 pistols, 42 M16 assault rifles, nine Chinese machine guns, and a large amount of ammunition, two Arakanese insurgents and one Mon insurgent were also found on board the boat (*Irrawaddy*, 2000), suggesting that the LTTE smuggles arms in collaboration with other insurgency groups in the region.

pro-independence forces later joined Maoist and Marxist extremist groups, taking their arms with them. The guns in the hands of anti-independence forces like East Pakistan Civil Armed Forces helped form the base of the Shanti Bahini armed insurgency, though they later received guns from India as well. Most of the guns supplied by the Indians were from the stocks received from the Pakistani army which had surrendered to Indians in 1971. After giving up their own struggle in 1998, some members of the Shanti Bahini began selling arms on the black market (Hosain, 1999; Ibrahim, 2001).

MAP 3.4 Bangladesh

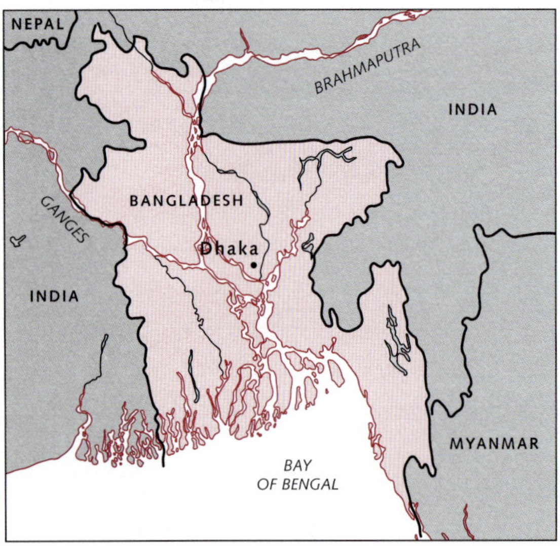

Another source of small arms is those smuggled into Bangladesh via unaccompanied air cargo, foreign-posted parcels, and, allegedly, diplomatic consignments, which are exempt from customs inspections (*Daily Manabjamin*, 2001; *Independent*, 2000). Such smuggling is carried out in collusion with customs, postal, or intelligence officers. The smuggling goes on unhindered as long as the officials along the way are paid.

The largest source of illicit arms in Bangladesh is the neighbouring Indian State of Bihar. Both illicit craft-made guns and foreign-made guns, including arms made in France, Italy, Germany, and the United States, are smuggled across Bangladesh's border with West Bengal (*Daily Janakantha*, 1999). Guns originating in Pakistan also transited through India to arrive in Bangladesh. Pistols and revolvers are sought-after mostly by criminal and political groups operating in Dhaka, while actors operating in Chittagong and Cox's Bazaar reportedly prefer assault rifles.

Thailand and Bangladesh both illustrate the possible dangers that can face small arms transit countries, not to mention the destabilizing effects for the regions in question. Much work still needs to be done to determine the risk factor of transit countries becoming either source countries or destination countries themselves, because, either way, the social, political, and economic effects can be severe.

© Associated Press/Osman Hassan

Kalashnikov assault rifles and heavy machine guns for sale in open-air markets in Mogadishu, Somalia.

Conclusion

In this chapter, we have come closer to documenting half of the value of the global small arms trade. Approximately USD 2 billion of the estimated USD 4 billion can be accounted for. While the legal trade in handguns and certain sporting firearms seems to be declining, no such trend is apparent in the trade of military small arms. It is impossible to determine trends within the illicit trade, as the data is limited and fragmented. Anecdotal evidence, however, suggests that the downward trend in the legal trade has little impact on the illicit trade.

As the legal trade becomes more transparent, and greater emphasis is placed on uncovering the illicit trade, it will become easier to estimate the value and volume of the illicit trade. Several obstacles need to be overcome. First, there is no base figure available to judge the stock of weapons currently circulating within the illicit market. Second, the market is not stagnant. Small arms flow continuously into the illicit market via diversion and illicit craft production, and out as a result of confiscation and destruction (PRODUCERS; WEAPONS COLLECTION). Only once these inflows and outflows are quantified, can we judge whether governments who are confiscating and destroying large caches are actually gaining the upper hand.

This chapter poses the question: Is there a link between the legitimate and the illicit trade in small arms and light weapons? The answer is a qualified yes. There is clear evidence in certain cases subjected to UN investigation that legal transfers do risk disintegrating into illicit ones. Consider the case of Ukrainian arms sent to Burkina Faso that were eventually re-transferred to the RUF in Sierre Leone. Further transfers are simply illegal right from the start, such as the instance when several top Argentinian government officials transferred arms to Croatia and Ecuador. Finally, there are cases in which stockpile security ultimately determines whether or not arms feed into the illicit market. Direct evidence of the links between the legal and illicit trade may not always exist. For example, the United States is one of the largest manufacturers and importers of small arms and light weapons, of both military and sporting firearms. It is no coincidence too that the United States is also a major source of illicit arms circulating in Mexico. Thailand is yet another example, as it is both a major importer of arms on the legal market and a major source of illicit arms for criminals and NSAs in Southeast Asia.

Legal and grey market transfers may place arms in the hands of those who misuse them. The key role of governments in this phenomenon needs to be scrutinized. Those states that choose to supply small arms to states or NSAs should assess whether the arms will be appropriately controlled, to prevent resulting misuse or re-transfer. If not, supplying states may well contribute to the global small arms problem.

The increasing transparency in the legal trade of small arms is a step in the right direction. By observing the national export reports for a number of countries—and the same top destinations come up over and over again—

major legal weapons flows can easily be identified. If these importing countries are in turn found to be sources of supply to the illicit trade, exporters should rethink the implications of their sales. Countries claiming to support the commercial arms trade can no longer turn a blind eye to the implications of their actions. Furthermore, the legal trade in small arms makes an insignificant contribution to the profitability of the arms industry as a whole. The key question is: How closely are governments and the international community willing to monitor and control the legal trade in small arms and light weapons in order to combat its knock-on illicit trade?

3. List of Abbreviations

APC	Armoured personnel carrier
BATF	US Bureau of Alcohol, Tobacco and Firearms
COMTRADE	UN Customs Database (Commercial Trade)
FARC	Revolutionary Armed Forces of Colombia
IMI	Israel Military Industries
IRA	Irish Republican Army
LTTE	Liberation Tigers of Tamil Eelam
MANPAD	Man portable air defence system
NRA	National Rifle Association
NSA	Non-state actor
OFO	Ordnance Factories Organization
OSCE	Organization for Security and Co-operation in Europe
PfP	Partnership for Peace
PLO	Palestine Liberation Organization
RUF	Revolutionary United Front
SAF	Small Arms Factory Group
UNCTAD	United Nations Conference on Trade and Development
UNITA	União Nacional para a Independência Total de Angola
UNSC	United Nations Security Council
WSFA	World Forum on the Future of Sport Shooting Activities
WTO	World Trade Organization

3. Endnotes

1 All the data in this section is based on the United Nations Statistics Division's COMTRADE Database (2001), and was obtained from the CD-ROM compiled by the International Trade Centre (UNCTAD/WTO). The International Harmonized System customs code for 'military firearms' is 9301, for 'pistols and revolvers' it is 9302.

2 However, this decline can be a result of other countries that do not submit their customs data on arms exports gaining a larger share of the export market. For example, Ukraine is a relative newcomer in the export of pistols, yet its exports are not included in the COMTRADE data.

3 The USD figures in this section are rounded off to the nearest USD 100,000.

4 For customs purposes this also includes exports from Luxembourg.

5 The Philippines submitted data for only three years, but still made this top importer list over the five-year period. All the other countries listed submitted data for all five years.

6 Again, it must be emphasized that this category of weapons also includes larger systems that do not fall under the category of small arms, such as missile launchers, howitzers, mortars, and guns.

7 South Africa submitted its customs data on arms exports to the UN for the first time in 1999.

8 Chile's data is available at <http://www.mercantil.com>
Thailand's data is available at <http://www.customs.go.th>
US customs data to 1999 is available at
<http://govinfo.kerr.orst.edu/impexp.html>

9 For an in-depth look at transparency in national arms export reporting, see Haug *et al.* (2002).

10 This actually covers the exports from Rosvoorouzhenie and Promexport, the two former state exporting agencies that merged in November 2000 to become Rosoboronexport.

11 The 20 countries are Angola, Armenia, Azerbaijan, Belarus, China, Cuba, Cyprus, Haiti, India, Indonesia, Iran, Libya, Myanmar, North Korea, Pakistan, Sudan, Syria, Vietnam, Yemen and Zaire.

12 For a more complete table of all multilateral embargoes since 1990, including excerpted text, please see the Small Arms Survey web site <http://www.smallarmssurvey.org>.

13 Correspondence with the Permanent Mission of Bulgaria to the United Nations, Geneva, 12 September 2001.

14 Correspondence with the Permanent Mission of Bulgaria to the United Nations, Geneva, 12 September 2001.

15 This section on the conflict trade was provided by Neil Cooper.
16 Interview with informed observer, June 2001.
17 Discussion with Yeshua Moser-Puangsuwan, Nonviolence International South East Asia, Bangkok, Thailand, June 2001; see also Chalk (2001).
18 Research for this section was conducted by Rafique al Islam during January–June 2001. Interviews were conducted with political party leaders, journalists, leaders of insurgency groups, insurgents, intelligence officers, gun dealers, and end-users.

3. Appendices

Appendix 3.1 Selected legal military small arms transfers, 1999–2001

The following sample of small arms transfers is presented to give the reader a better view of the international small arms trade. It is by no means a comprehensive list, but includes some significant military small arms transfers over three years. The terminology used for the type of weapon is as listed in the source documents.

Exporting country	Importing country	Quantity/type of weapons	Value of transfer	Date authorized or delivered
Bulgaria	India	Approximately 200,000 AK-47 assault rifles and its derivatives	n/a	Contract signed 31 October 2000
Russian Federation	Indonesia	4,000 AK-101 and AK-102 assault rifles	n/a	Delivered August 2000
	Indonesia	5,000 AK-101 and AK-102 assault rifles	n/a	Contract reported May 2001
	Bhutan	Over 1,000 AK-103 and AK-104 assault rifles	Approximately USD 1 million	Reported sale December 2000
	Mongolia	Small arms and ammunition	n/a	Reported sale February 2001
	Kyrgyzstan	2 train loads of small arms and ammunition	n/a	Delivered late 1999
	Uzbekistan	Small arms	Exchanged for cotton, gas, fruit, and vegetables	Contract signed 4 May 2001
United Kingdom	Sierra Leone	10,000 self-loading rifles	n/a	Delivered 1999
	Sierra Leone	4,550 self-loading rifles	n/a	Delivered 2000
United States	East Timor	1,200 M16 A2 assault rifles, 75 M203 grenade launchers	n/a	Announced June 2001
	Estonia	40,500 M14 rifles	USD 2.4 million (grant)	Authorized 1999 (delivery complete)
	Israel	10,000 M16 A1 rifles	USD 892,000 (grant)	Authorized 1999 (delivery complete)
	Latvia	30,500 M14 rifles	USD 2.1 million (grant)	Authorized 2000
	Lithuania	40,000 M14 rifles	USD 2.4 million (grant)	Authorized 2000 (delivery complete)

Sources: CAST (2001); Defense Security Cooperation Agency (DSCA); Raghuvanshi (2000); United Kingdom (2001); United States.

Appendix 3.2 Trends in worldwide small arms exports, non-military, 1995–1999 (Figure 3.2)

	1995	1996	1997	1998	1999
Pistols & revolvers	216,533	197,072	156,984	149,788	111,202
Hunting & sporting rifles	203,524	206,180	194,941	171,705	170,808
Shotguns & combination shotgun-rifles	197,485	245,198	209,594	237,356	221,657
Muzzle-loading firearms	29,423	25,549	21,541	20,888	20,829
Military firearms	249,768	631,521	599,337	709,268	409,933

Source: COMTRADE customs data (2001)

Appendix 3.3 Multilateral arms embargoes effective in 2001 (Table 3.7)

1 Human Rights Watch (2001a); The Economist (2001a); Lallemand (2001); West (2001).
2 Venter (1998).
3 UNSC (2000a; 2000c; 2001a); Africa Confidential (1999); Tel Aviv Ha'aretz (2001).
4 UNSC (2000a; 2000c; 2001a).
5 UNSC (2000a); Human Rights Watch (1996; 1999).
6 Human Rights Watch (1999).
7 UNSC (2000a); Human Rights Watch (1996; 1994).
8 Tel Aviv Ha'aretz (2001).
9 UNSC (2000a; 2001a).
10 Human Rights Watch (1996; 1994).
11 UNSC (2000c); Tel Aviv Ha'aretz (2001).
12 UNSC (2000a); Human Rights Watch (1994; 1999); Wood and Peleman (1999).
13 UNSC (2000a); Africa Confidential (1999); Human Rights Watch (1994, 1996; 1999); Venter (1998); Wood and Peleman (1999).
14 UNSC (2000a; 2000b; 2001a).
15 Human Rights Watch (1999).
16 UNSC (2000a); Africa Confidential (1999); Human Rights Watch (1999).
17 Human Rights Watch (1999).
18 SIPRI (2001).
19 Bouveret and Elomari (2001).
20 SIPRI (2001).
21 BBC Summary of World Broadcasts (1999a).
22 Berry (2001).
23 Berry (2001).
24 Comaroni (2001).
25 Berry (2001); Blank (2001).
26 Foreign Report (1993).
27 Berry (2001).
28 UNSC (2000b); Human Rights Watch (2000; 2001b); UNSC (2000c).
29 UNSC (2000c).
30 UNSC (2000c).
31 Farah (2001); UNSC (2000c).
32 UNSC (2000c).
33 UNSC (2000c).
34 UNSC (2000c).
35 UNSC (2000c).
36 Wood (2000); UNSC (2000c).
37 UNSC (2000c).
38 Jane's Terrorism & Security Monitor (2001).
39 UNSC (2000c).
40 Jane's Terrorism & Security Monitor (2001).
41 UNSC (2000c).
42 UNSC (2000c).
43 Jane's Terrorism & Security Monitor (2001); UNSC (2000c); Human Rights Watch (2001b).
44 Jane's Terrorism & Security Monitor (2001).
45 Watt and Norton-Taylor (1999); Wood; Peleman (1999).
46 UNSC (1998b); Berghezan (1997).
47 UNSC (1997; 1998b); Wood and Peleman (1999).
48 Foreign Report (1998); Wood and Peleman (1999).
49 CTI News Agency (1996).
50 UNSC (1996, 1997; 1998a); Human Rights Watch (1995); Wood and Peleman (1999).
51 UNSC (1997).
52 Human Rights Watch (1995); Wood and Peleman (1999).
53 Wood and Peleman (1999).
54 Watt and Norton-Taylor (1999); Wood and Peleman (1999).
55 UNSC (1997).
56 UNSC (1998b).
57 Human Rights Watch (1995).
58 Human Rights Watch (1995).
59 Human Rights Watch (1995).
60 UNSC (1997).
61 UNSC (1997).
62 Interfax News Agency (1995).
63 UNSC (1996); Human Rights Watch (1995); Indian Ocean Newsletter (1994); Wood and Peleman (1999).
64 UNSC (1996); Human Rights Watch (1995); Wood and Peleman (1999).
65 UNSC (1997).
66 UNSC (1998b).
67 Indian Ocean Newsletter (1996; 1998).
68 UNSC (1997); Interfax News Agency (1995).
69 UNSC (1998b); Evans (1996); Bevins (1997); Norton-Taylor (1995); Wood and Peleman (1999).
70 Human Rights Watch (1995).
71 UNSC (1997).
72 UNSC (1997; 1998a).
73 UNSC (1998b).
74 Wood (2000).
75 Africa Confidential (2001).
76 UNSC (2000b); Beeston (2000); Wood (2000); Human Rights Watch (2000).
77 Wood (2000).
78 Wood (2000).
79 UNSC (2000b).
80 UNSC (2000b); Farah (2000); The Economist (2001); Beeston (2000); BBC News (2000); Wood (2000); Africa Confidential (2001; 2000a; 2000b); Human Rights Watch (2001a).
81 UNSC (2000b).
82 UNSC (2000b).
83 Wood (2000); Leppard *et al.* (1999).
84 UNSC (2000b); Beeston (2000); Wood (2000); Human Rights Watch (2000).
85 UNSC (2000b); Wood (2000); Leppard *et al.* (1999); Legg and Ibbs (1998).
86 BBC Summary of World Broadcasts (2001); Agence France Presse (2001).
87 Bloch (2000); Xinhua (1999); Jane's Islamic Affairs Analyst (2000).
88 Indian Ocean Newsletter (2001); Coker (2001b); Xinhua (1999).
89 Xinhua (1999).
90 Indian Ocean Newsletter (2001).
91 UNSC (1999b); SIPRI (1999); Smith and Sagramoso (1999).
92 UNSC (1999b).
93 Berghezan (1997).
94 Mitevski (2000); Smith and Sagramoso (1999).
95 Josar (2001).
96 Berghezan (1997).
97 Fuzes (2001); Berghezan (1997).
98 Berghezan (1997).
99 Berghezan (1997).
100 Bouveret and Elomari (2001).
101 Venter (1998).
102 Venter (1998).
103 Bouveret and Elomari (2001).
104 Foreign Report (1993).
105 Far Eastern Economic Review (1992); Selth (1996).
106 Wood (2000).
107 UNSC (2000b); Wood (2000); Leppard *et al.* (1999); Legg and Ibbs (1998).
108 Human Rights Watch (1998).
109 *Neue Zürcher Zeitung* (2000).
110 Berghezan (1997).
111 Rogers (1997); Howard (1997); Hadzhizade (1999); ITAR-TASS (1997); Iberia (1997); Interfax (1997); Banks (1996); Jane's Sentinel Security Assessment (1999).
112 Banks (1996).
113 Foreign Report (1995).

3. Bibliography

Africa Confidential. 1999. 'Deadly Diamonds.' Vol. 40, No. 8. 16 April.

__. 2000a. 'Liberia/Sierra Leone: Godfather to the rebels.' Vol. 41, No. 13. 23 June.

__. 2000b. 'Liberia: Cross-border crisis.' Vol. 41, No. 15. 21 July.

__. 2001. 'Minimal Contracts.' Vol. 42, No. 13. 29 June.

Africa News. 2001. 'Eritrea; Ukraine Denies Arms Shipment.' 16 May.

Agence France Presse. 2001. 'Fledgling Somali government accused of illegally importing arms.' 19 February.

Ahmed, Tofail. 2000. 'Cox's Bazar a new route of arms trafficking.' *Daily Janakantha.* 16 January.

Alao, Abiodun and Funmi Olonisakin. 2000. 'Economic Fragility and Political Fluidity: Explaining Natural Resources and Conflicts.' *International Peacekeeping*, Vol. 7, No. 4. Winter, pp. 23–36.

Albania, Republic of. 2001. 'Regarding Some Information on Weapons Collected and Problematic Areas.' Tirana: Ministry of Public Order, General Directory of Police. 13 November.

Armada International. 2001. 'News.' March, p. 10.

Australia. Department of Defence. 2001. *Annual Report Exports of Defence and Strategic Goods from Australia 1999–2000.* Canberra: Directorate of Strategic Trade Policy and Operations, Industry Division, Department of Defence. May. <http://www.dmo.defence.gov.au/id/export/ar2000.pdf.>

Austria. Bundesministerium für Inneres. 2000. *Report on the Export of Arms by the Republic of Austria for the year 1999, pursuant to paragraph 8 of the operative provision of the EU code of Conduct on Arms Exports.* Vienna: Bundesministerium für Inneres.

Bangkok Post. 1989. 'Why the Weapons War May be Unwinnable.' 1 July.

__. 1993. 'Wimol: Arms Belong to "a Third Country".' 9 December.

Banks, Tony. 1996. 'Azeris want new look at Tashkent accord.' *Jane's Defence Weekly.* 18 December, p. 15.

BBC *News.* 1999. 'World: Middle East Egypt Jails Arms Smugglers.' 30 September. <http://news.bbc.co.uk/hi/english/world>

__. 2000. 'Liberia selling arms for diamonds.' 6 July. <http://news6.thdo.bbc.co.uk/hi/english/world/africa/newsid_821000/821726.stm>

BBC Summary of World Broadcasts. 1997. 'Communist rebels retreat as army takes camp.' 5 August.

__. 1998. 'Russian aircraft delivers Macedonia's arms purchases.' 4 May. Original source: MIC news bulletin, Skopje, in English, 29 April 1998.

__. 1999a. 'Belarus president denies supplying arms to Iraq.' 20 February.

__. 1999b. 'President orders "sustained all-out war" against communists'. 5 November.

__. 2001. 'Government denies delivering arms to Somalia.' 8 February.

Beeston, Richard. 2000. 'Britons named in UN report on "blood diamonds".' *The Times.* 22 December, p. 14.

Belgium. 2001. *Rapport du Gouvernement au Parlement sur l'application de la loi du 5 août 1991 sur les importations, les exportations et les transferts d'armes, de munitions et de matériel pouvant servir à un usage militaire ainsi que les technologies y afférentes.* July. < http://www.diplobel.org/news/rt/index.htm>

Berdal, Mats and David Keen. 1997. 'Violence and Economic Agendas in Civil Wars: Some Policy Implications.' *Millennium: Journal of International Studies.* Vol. 26, No. 3. Special Issue: War Endings: Reasons, Strategies and Implications, pp. 795–818.

Berghezan, Georges. 1997. *Ex-Yugoslavie: L'embargo sur les armes et le réarmement actuel.* Les Rapports du GRIP. Brussels: GRIP.

Berman, Eric. 2000. *Re-armament in Sierra Leone: One Year After the Lomé Peace Agreement.* Occasional Paper No. 1. Geneva: Small Arms Survey.

Berry, Jessica. 2001. 'Iraquis Step Up Secret Russian Weapons Trade.' *London Sunday Telegraph.* 25 February.

Bevins, Anthony. 1997. 'Foreign Office let arms through.' *Independent.* 22 January.

Blank, Stephen. 2001. 'Russia seeks to profit from Iranian rearmament.' *Jane's Intelligence Review.* 1 April.

Bloch, George. 2000. 'No compromise for Eritrea and Ethiopia.' *Jane's Intelligence Review.* 1 January.

Bouveret, Patrice and Belkcem Elomari. 2001. *Ventes d'armes de la France: Tour du monde des pays acquéreurs, rapport 2001.* Lyon: Observatoire des transferts d'armements.

Boyne, Sean. 1996. 'Uncovering the Irish Republican Army.' *Jane's Intelligence Review.* 1 August.

Brooke, Micool. 2000. 'South-East Asian targets for Russia's Igla and Kilo.' *Asia-Pacific Defence Reporter.* June/July, p. 34.

Business World. 1998. 'Guerrillas kill army officer.' 20 February.

__. 1999. 'Rebels hostage Agusan mayor.' 11 May.

__. 2000. 'NPA rebels attack post.' 22 August.

__. 2001. 'Seven NPAs killed.' 26 June.

Canada. Department of Foreign Affairs and International Trade. 2000. *Export of Military Goods from Canada Annual Report 1999.* Ottawa: Export and Import Controls Bureau, Department of Foreign Affairs and International Trade. September. <http://www.dfait-maeci.gc.ca/~eicb/export/milit_tech-e.htm>

Canada. Department of Justice. 1995. *Report of the Smuggling Work Group.* Ottawa: Department of Justice.

__. 1997. *Firearms Recovered By Police: A Multi-Site Study.* Daniel Antonowicz Consulting. Ottawa: Department of Justice. July.

Carrell, Sverin. 2001. 'Stolen Army Weapons Used by London Yardie Drug Gang.' *Independent.* 5 August.

Centre for Analysis of Strategies and Technologies, Moscow (CAST). 2001. *Deliveries of Russian Small Arms and Light Weapons.* Background paper. Geneva: Small Arms Survey. August.

Chakma, Budhajit. 2000. 'The rebels of Myanmar are active along Naikongchari and Thansi border. Weak groups are selling out arms.' *Daily Prothom Alo.* 25 November.

Chalk, Peter. 1999. 'The Tamil Tiger Insurgency in Sri Lanka.' In Abdel-Fatau Musah and Niobe Thompson, eds. *Over a Barrel: Light Weapons & Human Rights in the Commonwealth.* New Delhi: Commonwealth Human Rights Initiative, pp. 65–91.

__. 2001. 'Light arms trading in SE Asia.' *Jane's Intelligence Review.* 1 March.
Christian Aid. 2001. *The Scorched Earth: Oil and War in Sudan.* March. <http://www.christian-aid.org.uk/indepth/0103suda/sudanoil.htm>
Claiborne, William. 1987. 'South Africa, Mozambique Form Commission to Probe Massacre.' *Washington Post.* 7 August.
Clark, Dave. 2001. 'Arms Dealers Hover as Macedonia Struggles with Rebels.' *Agence France Presse.* 10 May.
Coker, Margaret. 2001a. 'Illegal Soviet Weapons Fuel Wars Around World.' *AJC Newspaper Online.* 8 July.
__. 2001b. 'Somalia: Amid war, famine, selling guns "guarantees my family will be fed".' *Access Atlanta.* 9 July.
Comaroni, Bogdan. 2001. 'Romanian Daily Says Weapons-Sales-to-Iraq Allegations Reemerge as NATO Expands.' *Ziua.* FBIS Translated text. 19 June.
Conteras, Joseph. 1996. 'Locked in a Box.' *Newsweek.* 14 October.
Cooper, Neil. 2001a. *Conflict Trade and Arms Acquisitions.* Background paper. Geneva: Small Arms Survey.
__. 2001b. 'Conflict Goods: The Challenges for Peacekeeping and Conflict Prevention.' *International Peacekeeping.* Vol. 8, No. 3, p. 27.
Council for a Livable World. 'Russia sold Mic-29s to Sudan.' Arms Trade Newswire. 4 January. <http://www.clw.org>
CTI News Agency. 1996. 'Company denies exporting arms to Rwanda and Zaire.' BBC Summary of World Broadcasts. 18 November.
Czech News Agency. 2001a. 'Weapons detained in Bulgaria exported by Thomas CZ-BTV.' 1 May.
__. 2001b. 'Czechs attempted to smuggle weapons to fighting in Eritrea.' 9 May.
__. 2001c. 'UN Investigating Case of Detained Plane with Czech Weapons.' 30 May.
__. 2001d. 'Mysterious plane carrying Czech weapons leaves Bulgaria.' 6 June.
Czech Republic. Ministry of Foreign Affairs. 2001. *The Czech Republic and Small Arms and Light Weapons.* Prague: Ministry of Foreign Affairs.
The Daily Janakantha. 1999. 'Illegal Fire-arms Sneaked Through the Northern Border.' 29 April.
__. 2000. 'Safe shelter of fundamentalist in deep forest: Hot trading of arms and drugs going on.' 25 August.
The Daily Manabjamin. 2001. 'Brisk trading of arms in Dhaka.' 15 January.
Davis, Anthony. 2001a. 'Pakistan in quandary over new sanctions against the Taliban.' *Jane's Intelligence Review.* 1 February.
Deocadiz, Christina. 1998. 'Guerrillas raid armoury.' *Business World.* 7 January.
Deutsche Presse-Agentur. 2000a. 'Deutschland begann mit Rüstungshilfe für Mazedonien.' 13 January.
__. 2000b. 'At least six communists rebels killed in southern Philippine clash.' 9 February.
Dudkevitch, Margot. 2001. 'Jordanian Held for Arms Smuggling.' *Jerusalem Post.* 23 April.
The Economist. 2001. 'Sierra Leone and Guinea: Fuelling Africa's Wars.' 13 January.
Evans, Andrew. 1996. 'Government "knew of Rwanda arms claims 11 months ago".' Press Association Newsfile. 2 December.
Evans, Michael, Matthew Bayley, and Stewart Tendler. 2000. 'Raiders Steel 35 Guns From Army Barracks.' *The Times.* 13 April.
Faiola, Anthony. 2001. 'Menem Under House Arrest for Weapons Sales.' *International Herald Tribune.* 8 June.
Farah, Douglas. 2000. 'Liberia Bolsters Support for Rebels in Sierra Leone.' *International Herald Tribune.* 21 June.
__. 2001. 'Arms Flow Into Liberia/Fighting Threatens to Draw in Ivory Coast: In West Africa's Borderlands, Wars Start to Flare Up Again.' *International Herald Tribune.* 7 August.
Far Eastern Economic Review. 1992. 'Portuguese Men-of-War.' 12 November.
Financial Times. 2001. 'Argentine Court Releases Menem.' 21 November.
Finland. Ministry of Defence. 2001. *Export of Defence Materiel.* Helsinki: Ministry of Defence. <http://www.vn.fi/plm/ekvas.htm#puolu>
Foreign Report. 1993. 'Britain's Iraq-gate "Scandal".' 11 November.
__. 1995. 'The New Arms-Trader: Ukraine.' 6 April.
__. 1998. 'Small is nasty.' 15 October.
France. Ministère de la Défense. 2000. *Rapport au Parlement sur les exportations d'armement de la France en 1999.* Paris: Ministère de la Défense. December. <http://www.defense.gouv.fr/actualites/dossier/d82/index.html>
__. 2001. *Rapport au Parlement sur les exportations d'armement de la France en 2000.* Paris: Ministère de la Défense. December. <http://www.defense.gouv.fr/actualites/dossier/d118/accueil.htm>
Fuzes, Oszkar. 2001. 'East-Bloc Connections Fuel War.' World Press Review On-Line. 28 June.
Garcia, Cathy Rose A. 1999. 'Rebels increasing in Central Visayas.' *Business World.* 9 July.
Germany. Bundesregierung. 2000. *Bericht der Bundesregierung über ihre Exportpolitik für konventionelle Rüstungsgüter im Jahr 1999 (Rüstungsexportbericht 1999).* Berlin: Bundesministerium für Wirtschaft und Technologie. <http://www.bmwi.de/Homepage/Politikfelder/Au%dfenwirtschaft%20%26%20Europa/Exportkontrolle/Exportkontrolle.jsp>
__. 2001. *Bericht der Bundesregierung über ihre Exportpolitik für konventionelle Rüstungsgüter im Jahr 2000 (Rüstungsexportbericht 2000).* Berlin: Bundesministerium für Wirtschaft und Technologie. <http://www.bmwi.de/Homepage/Politikfelder/Au%dfenwirtschaft%20%26%20Europa/Exportkontrolle/Exportkontrolle.jsp>
Gillard, Emanuela-Chiara. 2000. 'What's Legal? What's Illegal?' In Lora Lumpe, ed. *Running Guns: The Global Black Market in Small Arms.* New York: Zed Books Limited, pp. 27–52.
Global Witness. 1998. *A Rough Trade: The Role of Companies and Governments in the Angolan Conflict.* London: Global Witness. <http://www.oneworld.org/globalwitness/reports/Angola/title.htm>
__. 2001. 'The Role of Liberia's Logging Industry on National and Regional Insecurity.' Briefing to the United Nations Security Council. January. <http://www.oneworld.org/globalwitness/press/gwliberia.htm>
Godnick, William. 2001a. *Small Arms Transfers in Central America.* Background paper. Geneva: Small Arms Survey.
__. 2001b. *Stockpiles/Inventories.* Background paper. Geneva: Small Arms Survey.
Goldenberg, Suzanne. 2000. 'Guns for sale—how stolen Israeli weapons arm Fatah's fighters.' *The Guardian.* 16 December.
Grynber, Roman and Roy Mickey Joy. 2000. 'The Accession of Vanuatu to the WTO.' *Journal of World Trade.* Vol. 34, No. 6, pp. 159–73.
The Guardian. 2001. 'Scot held hostage in Colombia tells of will to survive.' 26 April.

Gunaratna, Rohan. 2000. 'Sources of Arms Supplies to the LTTE: Successes and Failures of the Sri Lankan State.' In Dipankar Banerjee, ed. *South Asia at Gun Point.* Colombo: Regional Centre for Strategic Studies.

Hadzhizade, Asya. 1999. 'The Truce is Being Violated.' *Defence and Security.* 21 June.

Harel, Amos. 2001a. 'PA Smuggled Arms Via Sea, IDF Suspects.' *Ha'aretz.* 29 January.

__. 2001b. '1,000 Rockets Caught on the Way to Palestinians.' *IMRA.* 4 February.

Haug, Maria, Lora Lumpe, Nicholas Marsh, and Martin Langvandslien. 2002. *Shining a Light on Small Arms Exports: The Record of State Transparency.* Occasional Paper. Geneva: Small Arms Survey.

Hennop, Ettienne. 2000. 'Illegal Firearms in Circulation in South Africa.' In Virginia Gamba, ed. *Society Under Siege: Managing Arms in Southern Africa.* Towards Collaborative Peace Series, Vol. 3. Pretoria: Institute for Strategic Studies, pp. 61–76.

Hosain, Syed Anwar. 1999. *War and Peace in the Chittagong Hill Tracts—Retrospect and Prospect.* Dhaka: Osman Gani.

Howard, Glen E. 1997. '"Scuds" Threaten Shaky Stability.' *Jane's Intelligence Review.* 1 June.

Human Rights Watch. 1994. *Angola: Arms Trade and Violations of the Laws of War Since the 1992 Elections.* New York: Human Rights Watch. November.

__. 1995. *Rwanda/Zaire: Rearming with Impunity, International Support for the Perpetrators of the Rwandan Genocide.* Vol. 7, No. 4, May.

__. 1996. *Angola: Between War and Peace: Arms Trade and Human Rights Abuses Since the Lusaka Protocol.* Vol. 8, No. 1. February.

__. 1998. *Sudan: Global Trade, Local Impact: Arms Transfers to all Sides in the Civil War in Sudan.* Vol. 10, No. 4. August.

__. 1999. *Angola Unravels: The Rise and Fall of the Lusaka Peace Process.* New York: Human Rights Watch. September.

__. 2000. *Neglected Arms Embargo on Sierra Leone Rebels.* Briefing paper. New York: Human Rights Watch. 15 May.

__. 2001a. *Afghanistan: Crisis of Impunity: The Role of Pakistan, Russia, and Iran in Fueling the Civil War.* Vol. 13, No. 3(C). July.

__. 2001b. *No Questions Asked: The Eastern Europe Arms Pipeline to Liberia.* Briefing paper. New York: Human Rights Watch. 15 November.

Iberia. 1997. 'Defence Minister Said Involved in Arms Smuggling.' 28 January.

Ibrahim, Syed Md. 2001. *Parbattya Chattagram: Shantiprakria-o-Poribesh-Paristhiti Mullayan.* Dhaka: Ahmed Mahmudul Hoque of Mowla Brothers.

The Independent (Bangladesh). 2000. 'Smugglers using names of foreign mission!' 1 October.

__. 2001. 'Arms and Politics.' 21 March.

India. Ministry of Defence. *Defence Production and Supply.* <http://www.mod.nic.in/product&supp/welcome.html>

The Indian Ocean Newsletter. 1994. 'The Seychelles: Caught red-handed.' 16 July.

__. 1996. 'Uganda: The Banyamulengue File.' 21 December.

__. 1998. 'Private arms trade.' 28 February.

__. 2001. 'Ethiopia: War moves from north to south.' 10 February.

Interfax. 1997. 'Yeltsin Sacks Officials Involved in Arms Sales to Armenia.' 23 October.

Interfax News Agency. 1995. 'Foreign Relations: Foreign Ministry denies reports of arms sales to Rwanda.' BBC Summary of World Broadcasts. 22 June.

Ireland. Department of Enterprise, Trade & Employment's Export Licensing Web Site. Military Statistics. <http://www.entemp.ie/export/statistics.htm>

Irrawaddy. 2000. 'Burmese sink Tamil boat.' June.

Italy. Camera dei Deputati. 2000. *Relazione sulle operazione autorizzate e svolte per il controllo dell'esportazione, importazione e transito dei materiali di armamento nonché dell'esportazione e del transito dei prodotti ad alta tecnologia, Anno 1999.* Rome: Chamber of Deputies' Stationery Office.

ITAR-TASS. 1997. 'Yerevan Doubts Rodionov Statement on Illegal Arms Shipments.' 14 March.

__. 2001. 'Plane carrying Czech-made weapons arrives in Georgia, Bulgaria apologizes.' 7 June. <FBIS-SOV-2001-0607>

Jane's Information Group. 2001. *Jane's World Insurgency and Terrorism.* January–April.

Jane's Islamic Affairs Analyst. 2000. 'Uneasiness about Somalia's destabilising role.' 1 April.

Jane's Sentinel Security Assessment. 1999. 'Azerbaijan.' 16 July.

Jane's Terrorism & Security Monitor. 2000. 'Sri Lankan guerrillas campaign at sea as well as on land.' 1 September.

__. 2001. 'United Nations exposes illegal arms sales.' 1 January.

Johnson-Thomas, Brian. 2000. 'Anatomy of a Shady Deal.' In Lora Lumpe, ed. *Running Guns: The Global Black Market in Small Arms.* London: Zed Books.

Josar, David. 2001. 'U.N. Police Intercept Weapons Shipment at Border Post in Western Kosovo.' *European Stars and Stripes.* 18 June.

Joseph, Kate. 2001. *End-User Certificates.* Background paper. Geneva: Small Arms Survey.

Kramer, Katherine. 2001. 'Legal to Illegal: Southeast Asia's Illegal Arms Trade.' *Kasarinlan: A Philippine Journal of Third World Studies.* Vol. 16, No. 2.

Kusovac, Zoran. 2001a. 'Macedonia Teeters on the Brink of Disaster.' *Jane's Intelligence Review.* Vol. 13, No. 6. 1 June.

__. 2001b. 'Macedonian Arms Steps Up Arms Procurement.' *Jane's Defence Weekly.* Vol. 35, No. 18. 2 May.

Lallemand, Alain. 2001. 'The Darra Tribe, The Taliban Arsenal—A Region Close to the Frontier Between Pakistan and Afghanistan Specializes in Copying Foreign Weapons on a Large Scale.' *Le Soir.* 26 September.

Legg, Thomas and Robin Ibbs. 1998. *Report of the Sierra Leone Arms Investigation.* Return to an Address of the Honourable the House of Commons dated 27 July. London: The Stationery Office.

Leppard, David, Chris Hastings, Carey Scott, and Brian Johnson-Thomas. 1999. 'British firms arming Sierra Leone Rebels.' *Sunday Times.* 10 January.

Loeb, Vernon. 2001. 'Jet Purchase, Bin Laden Linked.' *Washington Post.* 15 February, p. 20.

Luft, Gal. 1999. 'The Palestinian Security Services: Between Police and Army.' *MERIA Journal.* June. <http://www.biu.ac.il/SOC/besa/meria/journal/1999/issue2/jv3n2a5.html>

__. 2000. 'Palestinian Military Performance and the 2000 Intifada.' *MERIA Journal*. December. <http://www.biu.ac.il/SOC/besa/meria/journal/2000/issue4/jv4n4a1.html>

Lugo, Leotes Marie T. and Cathy Rose A. Garcia. 1999. 'Insurgents to release one hostage today.' *Business World*. 5 April.

Lumpe, Lora. 1997. 'The US Arms Both Sides of Mexico's Drug War.' *Covert Action Quarterly*. No. 61. Summer, pp. 39–46.

The Manila Times. 2000. 'Philippines: Fifty Communist rebels storm town hall, seize firearms.' 25 December.

Mariani, Bernardo and Angus Urquhart. 2000. *Transparency and Accountability in European Arms Export Controls: Towards Common Standards and Best Practice*. London: Saferworld. December.

Miller, Derek. 2001a. *Small Arms and Light Weapons in the Middle East*. Background paper. Geneva: Small Arms Survey.

__. 2001b. *Understanding Small Arms Transfer Dynamics*. Background paper. Geneva: Small Arms Survey.

Mitevski, Robert. 2000. 'When Are We Going to Say "Farewell to Arms"?' *Skopje Dnevnik*. 8 May.

The Nation (Thailand). 2001a. 'Light thrown on arms scams in the military.' 18 April.

__. 2001b. 'Army officers to be held liable for men's crimes.' 20 April.

__. 2001c. 'Police seize weapons cache.' 11 May.

Netherlands. Ministry of Economic Affairs. 2001. *The Netherlands Arms Export Policy in 2000*. The Hague: Ministry of Economic Affairs. < http://www.ez.nl/beleid/ext_frame.asp?site=handelspolitiek/hpinx01.htm>

Neue Zürcher Zeitung. 2000. 'Swiss Police uncover secret weapons deal.' 30 August.

Norton-Taylor, Richard. 1995. 'Nigeria arms deals "breach EU Embargo".' *Guardian*. 28 August.

Norway. Utenriksdepartementet. 2000. *Eksporten av forsvarsmateriell frå Noreg i 1999*. Oslo: Utenriksdepartementet. <http://www.odin.dep.no/ud/norsk/publ/stmeld/032001-040006/index-hov007-b-n-a.html>

Pakistan News Service. 2001. 'Russia Delivering Weapons to Northern Alliance.' 9 October. <http://www.paknews.com>

Parreño, Disaell Y. 1995. 'Danao City big supplier of Yakuza guns.' *Manila Chronicle*. 21 January.

The Philippine Star. 2001. 'Philippine military chief says talks with rebels continue despite Samar raid.' BBC Worldwide Monitoring. 14 February.

Phongpaichit, Pusak, Sungsidh Piriyarangsan, and Nualnoi Treerat. 1998. *Guns, Girls, Gambling, Ganja: Thailand's Illegal Economy and Public Policy*. Chiang Mai: Silkworm Books.

Portugal. 2000. *Relatorio Estatístico sobre exportações e importações de bens e tecnologies militares realizadas em 1999*. Lisbon: Ministério da Defesa Nacional, Direcção Geral de Armamento e Equipamentos de Defesa.

Pravda. 2001. 'Ukraine to Suspend Arms Supplies to Macedonia.' 31 July. Internet version. <http://english.pravda.ru/cis/2001/07/31/11328.html>

Raghuvanshi, Vivek. 2000. 'Bulgaria to Supply AK-47's to Indian Government.' *Defense News*. 20 November, p. 40.

Rangel Suárez, Alfredo. 2000. 'Parasites and Predators, Guerrillas and the Insurrection Economy of Colombia.' *Journal of International Affairs*. Vol. 53, No. 2. Spring, pp. 577–601.

Rashid, Ahmed. 2001a. 'Afghanistan: Ending the Policy Quagmire.' *Journal of International Affairs*. Vol. 54, No. 2. Spring, pp. 395–410.

__. 2001b. 'The Taliban: First, the War.' *Far Eastern Economic Review*. 2 August.

Rodan, Steve and Arieh O'Sullivan. 1998. 'PA Smuggles Arms Via Sinai Tunnels.' *Jerusalem Post*. 6 June.

Rogers, Marc. 1997. 'Russian Minister Reveals Arms-to-Armenia Secret.' *Jane's Defence Weekly*. 5 March.

Selth, Andrew. 1996. *Transforming the Tatmadaw: The Burmese Armed Forces since 1988*. Canberra Papers on Strategy and Defence No. 113. Canberra: Australian National University.

Slovak Republic. Ministry of Economy. 2000. *Export of Military Production in 1997 to 1999*. Bratislava: Ministry of Economy.

Small Arms Survey. 2001. *Small Arms Survey 2001: Profiling the Problem*. Oxford: Oxford University Press.

Smith, Chris and Domitilla Sagramoso. 1999. 'Small Arms Trafficking May Export Albania's Anarchy.' *Jane's Intelligence Review*. 1 January.

Sobelman, Daniel. 2001. 'Jordan Nabs Arms Smuggler Heading for West Bank.' *Ha'aretz*. 17 July.

South Africa. Department of Defence. Directorate Conventional Arms Control. 2000. *South African Export Statistics for Conventional Arms 1997-1999*. Pretoria: Department of Defence. <http://www.mil.za/SecretaryforDefence/Frame/Frame.htm>

South Korea. Department of Defence. White Paper. 1998. <http://www.mnd.go.kr/mnden/sub_menu/w_book/1998/ref/appendix36.htm>

Spain. Departamento de Aduanas e Impuestos Especiales. Ministerio de Hadienda. 2001. *Exportaciones Realizadas de Material de Defensa for Categorias de Productos*. Madrid: Departamento de Aduanas e Impuestos Especiales.

Steinberg, Gerald and Human Security Research Group. 2001. *Israel and the Palestinian Authority*. Background paper. Geneva: Small Arms Survey.

Stockholm International Peace Research Institute (SIPRI). 2001. *SIPRI Yearbook 2001: Armaments, Disarmament and International Security*. Oxford: Oxford University Press.

Sullivan, Kevin. 1998. 'Kalashnikovs at the Corner.' *The Nation*. 2 August.

The Sunday Times. 2000. 'Malaysia identifies 44 Islamic Cults.' 9 July.

Sventsitskiy, Ivan. 2001. 'Federal Forces Arming Gunmen.' FBIS translated text, CEP20010518000210. 17 May.

Sweden. Ministry for Foreign Affairs. Strategic Export Control Division. 2000. *Swedish Arms Exports in 2000: A Government Report*. Stockholm: Ministry for Foreign Affairs.

Switzerland. Federal Department for Economic Affairs. 2001. *Ausfuhr von Kriegsmaterial im Jahr 2000*. 6 February. <http://www.seco-admin.ch/seco/pm.nsf/ZeigePM_IDString/ESWH_KMAT_Ausfuhr2000?OpenDocument&l=en>

Tang, Edward. 2001. 'Military Rouges.' *The Strait Times*. 16 May.

Tel Aviv Ha'artz. 2001. 'Israeli Businessmen Suspected of Selling arms to Angolan Rebels.' 19 March.

The Toronto Star. 2001. 'Afghan rebels expecting arms from Russia, Iran.' 3 October. <http://www.thestar.com>

United Kingdom. Foreign and Commonwealth Office. 2000. *Focus International, Western Sahara: Moving the Peace Process Forward*. London: Foreign and Commonwealth Office.

__. 2001. *Strategic Export Controls Annual Report 2000.* London: Foreign and Commonwealth Office. July. <http://files.fco.gov.uk/npd/annualreport4/fullreport.pdf>
United Nations. 1999. *United Nations International Study on Firearm Regulation Database Update.* <http://www.ifs.univie.ac.at/uncjin/mosaic/firearms/index.htm>
United Nations Security Council (UNSC). 1996. *Report of the International Commission of Inquiry (Rwanda).* S/1996/1995 of 14 March.
__. 1997. *Third Report of the International Commission of Inquiry (Rwanda).* S/1997/1010 of 24 December.
__. 1998a. *Interim Report of the International Commission of Inquiry (Rwanda).* S/1998/777 of 18 August.
__. 1998b. *Final Report of the International Commission of Inquiry (Rwanda).* S/1998/1096 of 18 November.
__. 1999a. *Report of the Security Council Committee established pursuant to the resolution 985 (1995) concerning Liberia.* S/1999/1301 of 31 December.
__. 1999b. *Report of the Security Council Committee established pursuant to resolution 1160 (1998).* S/1999/216 of 4 March.
__. 2000a. *Report of the Panel of Experts on Violations of Security Council Sanctions Against UNITA.* S/2000/203 of 10 March.
__. 2000b. *Report of the Panel of Experts appointed pursuant to Security Council resolution 1306 (2000), paragraph 19, in relation to Sierra Leone.* S/2000/1195 of 20 December.
__. 2000c. *Final Report of the Monitoring Mechanism on Angola Sanctions.* S/2000/1225 of 21 December.
__. 2001a. *Addendum to the final report of the Monitoring Mechanism on Sanctions against UNITA.* S/2001/363 of 11 April.
__. 2001b. *Report of the Panel of Experts on the Illegal Exploitation of Natural Resources and Other Forms of Wealth of the Democratic Republic of the Congo.* S/2001/357 of 12 April.
__. 2001c. *Report of the Panel of Experts Pursuant to Security Council Resolution 1343 (2001), paragraph 19, Concerning Liberia.* S/2001/1015
United Nations. Statistics Division. 2001. 'COMTRADE/The Personal Computer Trade Analysis System (PC-TAS) CD-ROM.' New York: United Nations Conference on Trade and Development.
United States. Bureau of Alcohol, Tobacco and Firearms (BATF). 2000. *Following the Gun: Enforcing Federal Laws Against Firearms Traffickers.* Washington, DC: Department of the Treasury, June.
__. 2001. 'ATF and the Firearms Industry Partner to Educate the Public as to the Illegality of "Straw Purchasing" Firearms.' *ATF News.* 26 April.
United States. Defense Security Cooperation Agency (DSCA). Excess Defense Articles (EDA). <http://www.dsca.osd.mil/programs/eda/search.asp>
United States. Department of State. 1996. 'Amendment to the International Traffic in Arms Regulations.' *Federal Register.* Vol. 61, No. 135. 12 July, pp. 36625–26.
__. 1998. 'Suspension of Munitions Export Licenses to Indonesia.' *Federal Register.* Vol. 64, No. 198. 14 October, pp. 55805–06.
__. 2001. 'Defence Trade Controls—Embargo Reference Chart.' <http://www.pmdtc.org/country.htm>
United States. General Accounting Office (GAO). 2001. *Information on U.S. Weapons Deliveries to the Middle East.* Washington, DC. September.
Varouhakis, Miron. 2001. 'Ethnic Albanians Have Advantage.' *Kathimerini (Athens, English Edition).* Internet Edition. 13 June.
Venter, Al. 1998. 'African greets gun-runners with open arms.' *International Defence Review.* 1 August.
Villanueva, Ruffy. 2001. 'Aguinaldo slay claimed by NPA.' *Business World.* 14 June.
Viva Rio. 2001. *As Exportações Brasileiras de Armas Leves 1989-2000.* Background paper. Small Arms Survey. Geneva. November.
Watt, Nicolas and Richard Norton-Taylor. 1999. 'Blair Challenged on arms supplies for African rebels.' *The Guardian.* 11 February.
Weiner, Tim and Ginger Thompson. 2001. 'U.S. Guns Smuggled Into Mexico Aid Drug War.' *New York Times.* 19 May.
West, Julian. 2001. 'Taliban Receives Arms Via Pakistan.' *Washington Times.* 6 November.
Wilkinson, Tracy. 2000. 'Palestinians' Gun Culture Poses Political, Social Perils; Mideast: Proliferation of Weapons Frustrates Efforts at Law and Order—and Leaves Civilians Living in Fear.' *The Los Angeles Times.* 18 August.
Wood, Brian. 2000. 'Arms and related support for the rebels in Sierra Leone.' Testimony to the United Nations Committee on Sierra Leone. 31 July–1 August.
Wood, Brian and Johan Peleman. 1999. *The Arms Fixers-Controlling the Brokers and Shipping Agents.* PRIO Report 3/99. Oslo: International Peace Research Institute.
World Forum on the Future of Sport Shooting Activities (WFSA). 2001. *The Facts.* Rome: World Forum on the Future of Sport Shooting Activities.
Xinhua. 1999. 'Weapons Flowing into Somalia.' 21 February.
__. 2001. '14 Rebels Killed in Clash With Troops in N. Philippines.' 15 January.

Acknowledgements

Main contributors

Centre for Analysis of Strategies and Technologies (CAST), Neil Cooper, William Godnick, the Human Security Research Group, Rafique al Islam, Aaron Karp, Derek Miller, Gerald Steinberg, and Viva Rio.

Caught in the Crossfire:
The Humanitarian Impacts of Small Arms

4

Introduction

Amidst all of the debate about controlling the proliferation and misuse of small arms, there is a glaring, fundamental omission—the human face. Research and policy tends to focus on supply-related issues such as production and the mismanagement of stockpiles, inter-state transfers and the illicit trade, technical aspects of weapons tracing, marking, collection and destruction, and on legal or normative regimes designed to stop the flow of weapons. But missing from all of this is the human dimension—a consideration of how people are affected daily by the presence of these weapons—particularly in regions of armed conflict.

It is essential that we recall the human devastation directly attributable to small arms on an annual basis: more than 200,000 deaths from homicide and suicide in the industrialized world, and at least 300,000 killed during armed conflicts occurring in developing countries. Millions more suffer from non-fatal injuries and crippling disabilities, while the anguish of further untold numbers cannot even be documented.

This chapter considers the plight of the hundreds of thousands of people who are fatally and non-fatally wounded by small arms. It recalls the millions deprived of their homes and assets at gunpoint—whose very livelihoods are threatened by the presence of small arms. It forces the humanitarian advocate and policy-maker to reflect upon the damaged wreaked by a single bullet. It challenges all of us to consider, even before the bullet tears through flesh and splinters bone, how to reduce the humanitarian impacts of small arms.

The chapter also highlights those instances where the humanitarian community has mounted a forceful response to the issue of small arms availability and use. Critical entry-points considered in the chapter include supply-side advocacy and the strengthening of export criteria to ensure the adoption of human rights and humanitarian norms, the application and enforcement of international humanitarian law in war zones, and operational reform among humanitarian agencies themselves.

Key findings of this chapter indicate that:

- The humanitarian impacts of small arms—fatal and non-fatal injury, forced displacement, and declining access to basic needs—are severe, especially in regions affected by armed conflict.
- The impacts are difficult to quantify because of poor or non-existent data collection facilities and insufficient international attention and concern.
- Many humanitarian agencies are reconsidering previously sacrosanct principles of neutrality and impartiality as small arms proliferation endangers their activities and personnel.
- Responding to the humanitarian impact of small arms is a growing priority within the humanitarian sector.
- Strengthening current humanitarian approaches, coupled with the innovative use of international humanitarian law and codes of conduct, offers the best hope for alleviating the humanitarian impacts of small arms.

Refugees on the move in the Democratic Republic of Congo. (© Associated Press/Jean-Marc Bouju)

The *Small Arms Survey 2001* identified the direct and indirect effects of small arms on population, health and criminal activity as well as on relief and development. It compared the consequences of small arms availability and use from different perspectives, examining high- and low-income countries including those at war and peace. This chapter focuses exclusively on the consequences of small arms on individuals and societies in those regions affected by chronic violence and armed conflict. All but a handful of today's armed conflicts take place in developing countries. These are the regions emphasized in this chapter, concentrating on the human cost of small arms for civilians and the responses of predominantly European and North American humanitarian actors. The chapter considers the following three questions:

- **What are the humanitarian impacts of small arms?**
- **Are they getting worse?**
- **How is the humanitarian community responding to the humanitarian impacts?**

The first section introduces a conceptual framework to examine the humanitarian impacts of small arms, which include not only hundreds of thousands of deaths and injuries but also millions of international refugees and internally displaced persons (IDPs). It stresses a range of quantitative and qualitative indicators that can be used to measure the growing human suffering associated with small arms. The section shows how humanitarian and relief agencies and organizations can use evidence of the humanitarian impacts to more effectively influence public opinion and to design more appropriate responses.

The widespread insecurity generated by small arms challenges the principles of neutrality and impartiality implicit to humanitarian action.

The chapter goes on to demonstrate that the humanitarian impacts are escalating in magnitude and severity chiefly as a result of new wars. In contrast to the past, civilians constitute a disproportionate number of both combatants and victims in these conflicts, wars whose dynamics are determined primarily by the availability and use of small arms. Recognition of the persistence of these new wars has triggered a revitalized rights-based response from the human rights and humanitarian community. Humanitarianism is shifting from a primary concern with non-politicized, short-term assistance and protection to a growing preoccupation with the rights of so-called beneficiaries and the responsibilities and obligations of the international community toward those affected by systemic armed violence. The widespread insecurity generated by small arms also presents a direct challenge to the sacrosanct principles of neutrality and impartiality implicit to humanitarian action.

As a result of the significant impacts of armed violence and the enormous challenges associated with controlling small arms, humanitarian actors have started to mount a comprehensive response. The third section charts the emergence of three overlapping, but distinctly humanitarian responses to the availability and use of small arms in areas affected by armed conflict. These include supply-side, legal, and operational perspectives. To be sure, the infusion of principles of human rights and international humanitarian law into multilateral instruments such as codes of conduct and export controls indicates a growing awareness of the humanitarian dimensions of small arms control. The chapter examines the issues involved in these critical debates, considers their strengths and deficits, and highlights the implications of small arms-related insecurity on the future of humanitarian action.

The humanitarian impacts: Growing dangers and new evidence

There can be no doubt that the human impacts of small arms availability and use are considerable and far-reaching in all corners of the world. More than half a million people are fatally wounded as a result of small arms use every year, whether through intentional violence, such as homicide and suicide, or unintentionally, such as through accidental shootings. As noted in the *Small Arms Survey 2001* (Small Arms Survey, 2001, p. 236), 'the gross estimate of global deaths from all forms of homicide, war, and suicide in 1998 stood at 2,272,000 ... from war, the number totalled 588,000'. It is estimated that at least 50 per cent of these conflict-related deaths can be attributed to the intentional use of small arms and light weapons. Alarmingly, recent studies of battlefield statistics have indicated that the proportion of people wounded in combat by small-calibre ammunition (as opposed to other types of munitions) frequently rises above 70 per cent (Sellier and Kneubuehl, 2001).

Arguably the most commonly reported, if under-researched, impacts of small arms emerge from armed conflicts and societies struggling to recover from war. The media is saturated with a bewildering array of images depicting the militarization of despair. Headlines concentrate on child soldiers, traumatized refugees, and the apparent senselessness of contemporary war, with accompanying snapshots invariably portraying the familiar silhouette of an AK-47. The United Nations General Assembly (UNGA) has observed how virtually every department of the UN system is exposed to the consequences of armed conflicts, crime, social dislocation, displacement, and human suffering that are directly or indirectly related to the unregulated availability of small arms (UNGA, 2000). In spite of widespread acknowledgement of the problem, humanitarian and relief agencies are only slowly developing an awareness of the specific impacts of small arms.

Recent studies of battlefield statistics indicate that the proportion of people wounded in combat by small-calibre ammunition can rise above 70 per cent.

Humanitarian actors increasingly acknowledge that civilians are more insecure in today's new wars. They also recognize that their staff are regularly exposed to armed violence and criminality and that small arms are frequently a key element in the various safety and security challenges they face. Even so, the scale and dimension of unregulated small arms availability in regions of conflict are consistently under-appreciated (see Box 4.1). This is because the impacts of small arms have until recently been appraised from legal, military, or policy perspectives rather than with a deliberate focus on the human costs.

Efforts by the International Committee of the Red Cross (ICRC) to reorient the small arms debate toward international humanitarian law in the mid-1990s led to wider recognition of the public health perspective on these issues (Coupland and Meddings, 1999; ICRC, 1999; 1997). Some in the medical community, however, are concerned that publicizing medical information from the study of firearm-related injuries in areas of warfare could have the perverse effect of indirectly leading to improvements in the design and development of weapons themselves. While such concerns must be taken seriously, they need to be balanced against experience. As demonstrated by the banning of dum-dum bullets in the late nineteenth century and anti-personnel landmines in the late twentieth, the success of any humanitarian campaign rests on the mobilization of public opinion. Past experience suggests that weapons are more likely to be restrained out of public abhorrence and political expediency than strictly for reasons of public health or military utility.

Introducing the indicators

A first step towards generating public awareness and designing appropriate preventive measures would be the documentation of the humanitarian impacts of small arms on civilians. The humanitarian implications of small arms availability and use are vast, though the research community has yet to appreciate its full dimensions. Recent contributions by public health specialists, criminologists, social scientists, and economists have helped chart a rough path forward. Humanitarian and development personnel exposed to the impacts of small arms in the course of their daily work have also begun to seriously consider the issue and to voice their concerns.

BOX 4.1 Data collection: Never as easy as it seems

Is the post-Cold War world really a more dangerous place for civilians? With the limited data available, can policy-makers, aid agencies, or the public fully appreciate the scale and costs of fatalities and injuries caused by small arms, let alone their long-term consequences on families and communities and the social and economic environment at large?

Central to the development of basic indicators is reliable and verified data on the number, profile and types of injuries and their causes, and on risk factors associated with small arms and their distribution. There are many challenges associated with collecting statistics on the humanitarian impacts of small arms, not the least of which relate to the lack of humanitarian agency and donor engagement in research itself. Other obstacles include the absence of standardized or comparable definitions and methods for collecting and collating data, logistical barriers and threats to data gatherers, limitations in geographic coverage and consistency among existing surveillance systems, and attitudinal and cultural issues that lead to under and over-reporting. Some specific challenges include:

- *Poor data-collection and monitoring capacities.* A recurring problem for relief agencies and donors relates to the inability of governments and NGOs to determine systematically what is taking place at the global level. On the one hand, firearm-related deaths and injuries often appear to be seriously under-reported where they are caused by warfare or repression. On the other hand, many aid institutions indirectly contribute to the problem by awarding low priority to intelligence—data-collection and evaluation—and limiting their focus only to those areas that are easily accessed. Where research is carried out, it is often tainted by selection bias and poor research methods that favour existing programmes and activities. Furthermore, data is rarely effectively processed or properly maintained and institutional memory is concomitantly limited. On top of this, the novelty of the small arms issue represents a barrier in and of itself, often dismissed as someone else's problem or jurisdiction.
- *The tendency of public or private agencies to exaggerate.* The relief and development communities frequently generate inaccurate and inflated numbers, whether out of ignorance or intentionally, to justify programmatic interventions and to mobilize public opinion. Figures are often cited without any empirical foundation. For example, are 90 per cent of small arms-related casualties in conflict really civilians, as many UN agencies and humanitarian NGOs claim? Are 80 per cent of these really women and children? Are there really 300,000 child soldiers fighting in today's wars? Do we have sufficient or verified data to validate these figures or are they politically expedient estimations? Though potentially useful for advocacy purposes, loose approximations can unintentionally and detrimentally reorient debates from substance to credibility.
- *The importance attached to confidentiality and neutrality.* Confidentiality and neutrality are regarded as mainstays of humanitarian legitimacy. Even where agencies accumulate large reservoirs of data relating to small arms, they are frequently left unused or considered too sensitive because they relate to ostensibly political issues such as military action or human rights violations. Though attitudes are rapidly changing, concerns with the politics of information have led many agencies to withhold data for fear of it being manipulated or abused by other actors. Many humanitarian agencies are reluctant to reveal data on patient injuries or deaths. A few, however, including the ICRC and Médecins Sans Frontières (MSF) occasionally release selected findings.

Rigorous data collection and common definitions are a priority for sound evidence-based policy. Appropriate methods need to be developed and epidemiological surveys administered in order to collect data on the type of deaths and injuries and their causes, and the risk factors and distribution of such deaths and injuries within communities. In this way a baseline for trend analysis and future comparisons can be established. Emergency preparedness and conflict monitoring systems can be developed and improved where they consider small arms. Furthermore, researchers and policy-makers need to be properly trained to generate and assess data on the humanitarian impacts of small arms in order to improve interventions designed to reduce armed violence.

This chapter draws its conceptual approach from a study commissioned by the UN Inter-Agency Standing Committee's Reference Group on Small Arms (RGSA), which described the humanitarian impacts of small arms as 'the short and long term effects of armed violence [on people] carried out with small arms and taking place immediately prior to, during and following from complex, human-made emergencies' (Muggah and Berman, 2001, p. vii). This definition suggests that the humanitarian impacts of small arms are inextricably tied to the human costs. It is less concerned with the policy or distinctly economic consequences of small arms proliferation than with the individual victims and widespread fear made possible by small arms availability. These impacts commonly, but not exclusively, take place during armed conflict and extend to all manner of human rights abuses, including death, torture, rape, recruitment of child soldiers, kidnapping, and forced disappearances (see Table 4.1).[1]

BOX 4.2 Accessibility thesis and the humanitarian impacts

Is the unregulated availability of small arms associated with increases in both unintentional and intentional death and injury, forced displacement, or declining access to basic needs? Evidence drawn together in the *Small Arms Survey 2001* (Small Arms Survey, 2001, pp. 200–5) demonstrated that firearms availability and ownership are a proximate determinant influencing the likelihood of homicide, suicide, and accidental death. In addition to other structural causes, intentional death and injury, involuntary displacement, and rapidly declining access to basic needs can be facilitated by the availability of firearms during conflict.

While cultural, social, and economic factors are important in determining the likelihood of whether an individual will resort to violence, the presence of a weapon—whether used in criminal violence or in conflict—conditions the severity and magnitude of the humanitarian impacts. A range of direct and indirect factors, from horizontal inequality to cultures of violence and war, all affect the likelihood of mortality and morbidity. Since making the leap from correlation to causation is notoriously difficult in such matters, many scholars have been cautious about establishing causality where small arms are concerned. But when dealing with epidemics that kill or maim—including firearm injuries—the price of inaction is devastatingly high.

The range of impacts of firearm violence on relief and development is very diverse. There is a virtually unlimited array of objective and subjective impacts related to small arms availability and use, most of which are under-appreciated. These often relate to fear—of death, injury, and long-term insecurity—and are difficult to quantify. A major reason why these impacts remain under-analysed is because they can be described only by affected people themselves. Methods aimed at revealing these vital perspectives, including victimization surveys and participatory research, are considered throughout the chapter.

TABLE 4.1 A framework for assessing the humanitarian impacts of small arms

Humanitarian impact	Primary indicators
Firearm-related death and injury	Firearm homicide rates Firearm suicide rates Unintentional firearm injury rates Intentional firearm injury rates Firearm-related disability rates Psychosocial and psychological trauma associated with armed violence
Violence-induced displacement	Number of refugees and internally displaced persons (IDPs) Incidence of firearm-related death and injury among displaced people Incidence of armed intimidation and assault among displaced people Arms availability in refugee/IDP camps Child mortality rates (CMR) among displaced and relocated populations Social and physical welfare of refugees/IDPs (rape, child soldiers, etc.)
Collapsing access to basic needs and declining social capital	Social and physical welfare of women and children Household access to basic needs Community and customary cohesion Declining access to public goods Sexual violence

Objective indicators of the humanitarian impacts of small arms include the health-related effects on civilian populations such as firearm-related death and injury, as well as long-term disability and psychological trauma. Also considered is the destruction of medical services and the vulnerability associated with deteriorating social welfare provision capacities. However severe the immediate human impacts attributed to small arms, it is important to recall that the larger burden of mortality experienced during episodes of armed violence is attributable to the secondary costs of war, such as death from malnutrition, disease, and preventable illness.

The fact that millions of people die each year, not from the direct acts of violence but because the various functions of armed violence deprive them of access to health services, is not a novel finding (Ghobarah *et al.* 2001; Leaning, Briggs, and Chen, 1999). It should be recalled, however, that small arms can also contribute to an increase in the scale and pace of killing, the likelihood of illness, and the possibility of violations of international humanitarian law. The case of Sierra Leone is indicative. Immediately following the invasion of the country's capital, Freetown, by the Revolutionary United Front (RUF) in 1999, a senior government pathologist reported that more than 7,330 people had been shot and killed in a single month: almost one per cent of the city's entire population. Thousands more suffered lacerations, mutilations, and firearm injuries.[2] Additional surveys carried out in Sierra Leone recorded that almost 60 per cent of all war injuries were gunshot-related, that 11 per cent of all victims were under the age of 15, and that 43 per cent were women (Salama, Laurence, and Nolan, 1999).

Another marker of the humanitarian impacts of small arms on civilians is violence-induced displacement. Depending on whether they have crossed an international border, displaced people can be counted as either internally displaced people (IDPs) or international refugees. It is possible to register the number of people forced to leave their homes at gunpoint. For example, spontaneous internal displacement within Colombia is largely due to massacres involving handguns and assault rifles. Surveys administered in IDP and refugee camps in Albania, Uganda, Sri Lanka, and Georgia record systematic shootings, threats at gunpoint, firearm-related homicides, and other violations of human rights and humanitarian law involving military-style weapons. The militarization of refugee camps is also a growing concern. Finally, violence-induced displacement and relocation is associated with dramatic increases in the risk of illness and communicable disease. The threat and use of small arms, therefore, have long-term health consequences for displaced people that extend well beyond the bullets themselves.

Though more difficult to measure or attribute directly to small arms availability, civilian access to basic needs—including food, water, and shelter—is often eroded by the climate of fear generated by the threat and use of small arms. While men are particularly susceptible to firearm death and injury, women and children are also acutely vulnerable to a range of firearm-related insecurities including recruitment as soldiers, sexual violence, and pathological coping strategies during periods of extreme violence, all of which affects their access to basic needs (WHO, 2001a, p. 10). Invariably, livelihoods are reshaped by arms-related insecurity: customary institutions such as pastoral migration patterns and dowry systems are undermined by threats associated with arms availability. For example, in Kenya, Somalia, Ethiopia, the Sudan, and Uganda, pastoral and agrarian communities are regularly terrorized by the presence of AK-47-wielding bandits and cattle-rustlers. In this context, individual perceptions of armed violence can directly influence day-to-day decision-making. These choices will also have tangible implications for household health and well-being, including the decision to take up weapons in self-defence.

Death and injury in conflict

The WHO World Health Report database estimated in 1998 that 2.3 million people died worldwide as a result of intentional violence. This amounts to an overall death rate of 38.4 per 100,000, compared with 19.9 per 100,000 for all road traffic injuries. Importantly, of these 2.3 million deaths 42 per cent were classified as suicide, 32 per cent were homicide, and 26 per cent—588,000—were war-related.[3] According to a subsequent report released by the WHO (2001a, p. 11), trends involving small arms use and intentional violence were similar in 2001 to those in 1998.

Deaths and injuries from war wounds occur predominantly among young men in the South, the part of the world where armed conflicts are most prolific and precisely where death and injury data is weakest.[4] Since the reliability of health statistics frequently mirrors a country's level of development, many deaths and injuries go unreported. Even where criminal or tribunal investigations are conducted, they frequently underestimate the magnitude of deaths and injuries attributable to small arms. Sometimes, the only deaths and injuries recorded are those that occur in hospital for which a death certificate is completed. But the majority of deaths occur away from hospitals and clinics; injured victims often seek health care from alternative sources and the fatally-injured are frequently buried immediately by their families.

In 1998, the international death rate for intentional violence amounted to 38.4 per 100,000, compared with 19.9 per 100,000 for all road traffic injuries.

BOX 4.3 First question: Where are the injured?

A study by Coupland and Meddings (1999, p. 407) attempted to demonstrate the importance of context in relation to assessing the impacts of small arms and light weapons. The authors reviewed firearm incidents involving combatants, paramilitary groups, and civilians stretching back to the 1920s. Their study confirmed the widespread assumption that the number of people injured in conflicts is usually two to three times the number killed though it can increase to as high as 13.0. They also found situations in which the ratio decreases and fatalities can actually outnumber those injured.

The 'injury to killed' ratio can potentially reveal a number of insights into the circumstances of a particular incident. For example, according to Milas (1999), the total number of Croatian army soldiers fatally injured in the period from the beginning of July 1991 until the end of March 1992 was 487, while the number of severely wounded was 2,301. In this case, the injury to killed ratio was 4.72. By way of contrast, during an incident reported by the BBC involving the Sri Lankan army and the Liberation Tigers of Tamil Eelam (LTTE) in 1997, approximately 80 Tamil guerrilla deaths were reported and only 70 wounded: an injury to killed ratio of 0.87.

Where the number killed is greater than the number wounded, such as in situations where firearms are used against people that are immobilized or unable to run away or defend themselves, the injury to killed ratio will shrink to zero. It can be tentatively assumed that fighting was at close range, may have involved execution-style killings, and potentially involved violations of international humanitarian law. Meddings and Coupland (1999, p. 409) conclude that: 'the increased mortality resulting from the use of firearms in situations other than war requires a complex interaction of factors explicable in terms of wound ballistics and the psychology of the user... understanding these factors has implications for the recognition of war crimes'.

War-related shootings: Kill–injury profile

Incident	Date	Killed	Injured	Ratio
Second World War (Italians)	1944–45	27,944	76,351	2.7
Vietnam (Marines)	1965–73	12,944	512,399	39.6
Northern Ireland (British Army)	1970–80	300	1,700	5.7
Former Yugoslavia (FYU) (Croatians)	1991–92	15	78	5.2
Mogadishu raid (US)	1993	18	70	3.9

Source: Coupland and Meddings (1999)

Mass shootings: Kill-injury profile

Incident	Date	Killed	Injured	Ratio
St Valentine's Day Massacre	1929	6	1	0.17
Wah Hee massacre	1983	13	0	0
Port Arthur, Tasmania	1996	35	18	0.51
Dunblane, Scotland	1996	18	15	0.94
Littleton, Colorado	1999	13	23	1.76

Source: Adapted from Coupland and Meddings (1999) with assorted media reports

International firearm-related mortality rates are difficult to compare because the vehicle leading to death—small arms—is rarely included in the International Classification of Disease (ICD) categories of homicide, suicide, unintentional deaths, or deaths of undetermined intent. As noted by Krug, Powell, and Dahlberg (1998, p. 215), 'even within these categories, the information on the firearm components are not separately provided in the World Health Organization's World Health Statistics Annual, the most common source of data for cross-national comparisons of mortality rates'. Official death reports in many war-affected countries must therefore be interpreted with great caution and awareness of the limitations at each step in the recording process. Even at the local level, hospital statistics are frequently misleading as a record of the overall patterns of serious non-fatal injury if the community does not have access to facilities, if public confidence in the health system is low, or if the referral system for transferring injured patients to hospitals is poor. At this stage, it is more instructive to review trends and patterns of deaths and injuries during conflict rather than rendering longitudinal cross-country comparisons.[5]

A basic epidemiological finding is that for every fatal small arms-related injury in armed conflict there are likely to be many more non-fatal injuries. Estimates from non-conflict affected states, such as the United States, indicate a 1:3 ratio of killed to non-fatally injured victims, though rates fluctuate according to circumstances (WHO, 2001a, p. 13). Ratios of injured to killed tend to be lower in situations of underdevelopment and armed conflict, particularly where human rights violations are taking place (Box 4.3).[6] A detailed appreciation of the injury to killed ratio is often confounded by under-reporting, inconsistent monitoring, and the fact that the number of war-wounded who actually receive treatment in clinics and hospitals are not necessarily representative of the total number of people injured in any particular armed confrontation. Although agencies such as the WHO, the ICRC, and MSF document some injury types in their field operations, many do not differentiate statistically between lesions attributable to fragmenting bombs and artillery fire, weapons other than firearms (such as knives or spears), or secondary effects such as malnutrition or disease in their reporting systems.

Fatal and non-fatal injuries: What are the trends?

The fact remains that firearm-related mortality and injury represent objective markers of the humanitarian impacts of small arms. Though not explored in this chapter, the long-term effects of fatal and non-fatal firearm injuries can also be quantitatively measured through the use of composite econometric indexes such as year's of potential life lost (YPLL) and disability-adjusted life years (DALY).[7] Although some situations and societies present more risk than others, the life-, life-years, and disability- adjusted years, lost rates are often proportionately similar. A seminal study produced by Krug, Powell, and Dahlberg (1998, p. 221) concluded that 'war is probably an important factor contributing to firearm mortality... in 1990, war was estimated to be the sixteenth cause of DALY's lost throughout the world, but by 2020, it is projected to be the eighth leading cause of DALYs lost'. Though data is currently limited, it is increasingly clear that the economic costs associated with the death and disability of young men—often at the most productive time of their lives—are potentially calculable.[8]

While estimating the full extent of the financial costs of firearm casualties is possible, it is confounded by a lack of data on the actual number of fatal and non-fatal gunshot injuries and basic denominator data in the regions of greatest concern[9] (Murray *et al., 2002;* Sethi *et al.*, 2000; Zwi *et al.*, 1996; Cummings, Koepsell, and Mueller, 1995). Nevertheless, the level of professional interest is such that, even where mortality data is inadequate, prominent medical journals regularly publish studies of the number of civilians shot and killed or injured and disabled by firearms. The value of consistent and widely disseminated data on firearm death and injury can extend beyond awareness-building in the public domain to the elaboration of practical preventive health interventions and the protection of civilians more generally.

Who gets hurt?

© Gérard Klijn

Non-fatal small arms injuries leave a legacy of suffering, including disability, psychological trauma, and lost productivity.

Though the impacts of armed combat on soldiers have been studied since the American War of Independence (1775–83), the medical literature on the consequences of small arms on civilians has burgeoned only in the past two decades. The ICRC (1999), for example, has assembled an injury database that consists of systematic data on almost 28,000 patients—combatants and non-combatants—admitted between 1991 and 1998 from its field hospitals in Afghanistan, Rwanda, Chechnya, and the border regions of Kenya and Cambodia. Almost 70 per cent of all admitted patients were wounded by bullets, bombs, shells, mortars, or mines; the rest were treated for burns, for blunt trauma wounds, or for reconstructive surgery. Contrary to popular belief, fewer than half of the weapons-wounded were civilians (see Box 4.4).

BOX 4.4 Are 90 per cent of war-related victims really civilian?

There is a common perception that 90 per cent of all casualties in armed conflict are civilian and that 80 per cent of all war-wounded are women and children. These two statistics were widely cited in the mid-1990s without reference to a methodology describing how the figure was actually determined.

The wound database of the ICRC was installed in January 1991. It includes records on all patients wounded in war who have been admitted to the Red Cross hospitals on the Afghan border of Pakistan, as well as in Kabul and Khandahar, in Khao I Dang on the Cambodian border of Thailand, in Butare in Rwanda, in Novi Atagi in Chechenya, and also in Lokichokkio on the Sudanese border of Kenya. Between January 1991 and July 1998, 18,877 of the 27,825 admitted patients registered sustained injuries relating to bullets, bombs, shells, and mines, while the rest were admitted as a result of burns, blunt trauma wounds, or for reconstructive surgery.

The ICRC does not request information on combatant status but rather on the sex and age of patients. A follow-up analysis of approximately half of the victims reported since 1991 noted that only 35 per cent were female, male and aged under 16, or male aged 50 and above: that is, civilian.

Source: Coupland and Meddings (1999), updated following interviews with ICRC personnel in Geneva, October 2001.

Surprisingly, arms-related death and injury do not necessarily decrease dramatically when wars come to an end. Two seminal studies by teams of epidemiologists and surgeons at the ICRC provide a more textured analysis of the potential humanitarian impacts of arms saturation on civilians.[10] One study reveals that, during some armed conflicts, mines and fragmenting munitions—for example, mortars, bombs, and shells—are more likely than bullets to cause civilian casualties. Analysing patient data from Afghanistan, the research indicates that weapons-related casualties decreased by less than 35 per cent during the 18 months following the cessation of hostilities in war-torn areas where large numbers of arms remained in circulation (Michaels *et al.*, 1999; Coupland and Samnegaard, 1999).

The other study, drawing on field hospital data from Cambodia, notes that civilians accounted for over 40 per cent of combat-related casualties and over 70 per cent of all non-combat weapon casualties (Meddings and O'Connor, 1999). The study reveals that the threat of arms-related death or injury to civilians in non-combat settings can surpass rates experienced during conflict periods if weapons remain diffused in society. As noted below with the example of the Republic of Congo, these trends hold elsewhere (see Box 4.5).

A disturbing feature of contemporary conflict is the ease with which the most basic human rights of combatants and non-combatants, particularly women and children, are violated. In virtually every war-torn country, many of these violations can be traced to the availability and use of small arms. For example, of the estimated 2.5 million civilians believed to have died between 1998 and 2001 in the Democratic Republic of Congo (DRC), more than ten per cent are estimated to have been shot outright. The targeting of pregnant women is reportedly common (Les *et al.*, 2001; MSF, 2001).

The effects of small arms availability on specific groups can vary dramatically. At the height of the fighting in Chechnya, almost one half of civilian casualties were reported to be children (United Nations, 1999; UNSG, 1996). In Colombia's southern Putumayo Department in 1998, the firearm homicide rate among males rose above 1,900 per 100,000, compared to a national rate of 53.99 per 100,000. Local health officials have aptly described Putumayo as 'one of the most violent places on the planet' (Departamento Administrativo de Salud, 1999). By way of contrast, the firearm homicide rate for Brazil was recorded at 25.78 per 100,000, and for comparatively safe Costa Rica it was 2.57 in the mid 1990s.

Often overlooked is the fact that warfare can result in an even greater number of civilian disabilities from firearm injury. These injuries carry a disproportionately high cost, even if they are difficult to quantify using conventional methods (see Box 4.6). If properly funded, the cost of physical rehabilitation and prosthetic limbs alone could easily absorb the entire health budget of a nation (Handicap International, 2000). Firearm injury and disability rates are particularly compelling measurements of the humanitarian impacts of small arms on civilians. According to the WHO (2001b), interpersonal violence, self-inflicted injuries, and war injuries are among the top five largest contributors to global

BOX 4.5 Who gets hurt? Evidence from hospitals in Congo-Brazzaville

In a study commissioned by the International Organization of Migration (IOM) and the United Nations Development Program (UNDP) of small arms use during the conflict and post-conflict period in the Republic of Congo, a Small Arms Survey research team assessed the impacts of firearm violence during 1998–2000 in the capital, Brazzaville. Examination of admittance records from three hospitals with broadly similar numbers of patients revealed that male youth were acutely predisposed to firearm-related injuries both during and following the civil war. Though the study is descriptive—and vital denominator data could not be established due to poor record-keeping (and the absence of data during the conflict periods)—it yielded a number of revealing trends.

Reported firearm casualties in Brazzaville, 1998–2000

Type of weapon	Districts of Brazzaville/Reported cases							Total
	Makélékélé	Bacongo	Poto Poto	Moungali	Ouenze	Talangai	Mfilou	
Small arm	131	75	219	238	220	471	0	1,354
Blade	17	3	0	3	0	6	1	30
Grenade	6	3	0	0	0	33	0	42
Rocket-propelled grenade (RPG)/mortar	0	0	0	0	0	1	0	1
Other	0	0	0	0	0	0	0	0
Total	**154**	**81**	**219**	**241**	**220**	**511**	**1**	**1,427**

Source: Demetriou, Muggah, and Biddle (2001)

Of the estimated 1,354 men who were injured with firearms (95 per cent of all casualties listed above) approximately one-quarter were under 20 years old. More than 40 per cent were between the ages of 21 and 30. Among admitted female patients, the majority of reported firearm injuries were experienced by adolescents and women between the ages of 11 and 30. Overall, the proportion of those injured that were almost certainly civilian casualties—namely, under the age of ten, over the age of 50, or female—was approximately 38 per cent.

health burdens among people aged 15–44. Though only a preliminary estimate, the WHO contends that 0.7 per cent of the global burden of disease in 2000 was attributed to war, including years of life lost and DALYs.

According to Murray *et al.* (2002, p. 348): 'The ratio of years to life lost due to premature mortality caused by conflict to years lived with disability from conflict is 4.75.' Not surprisingly, this burden is distributed across regions in similar proportions to direct deaths from conflict.

Firearm-related disabilities are prevalent in both conflict and post-conflict affected societies. For example, in El Salvador, Honduras, and Guatemala the abundance of leftover small arms coupled with endemic cultures of violence and high levels of inequality are believed to have resulted in a high injury burden (WEAPONS COLLECTION). Rough estimates indicate that for every 1,000 weapons collected from former combatants in the region there are approximately 100,000 unaccounted for in circulation (Godnick, 2001). It is perhaps no surprise, then, that Central America has a firearm homicide rate that fluctuates between 30 and 50 per 100,000 which, in spite of latent crime reporting, is equal to those of some of the most volatile war zones on the planet (Fajnzylber, Lederman, and Loazya, 2000). As illustrated in Figures 4.1 and 4.2, and Appendix 4.1, the firearms injury rate is much higher.

The countries of Central America have firearm homicide rates equal to those in some of the most volatile war zones on the planet.

BOX 4.6 The Intifada: Behind the headlines

Since the second *Intifada* began in September 2000, some 1,500 Palestinians, mostly young men, have been left with permanent crippling injuries. Lethal fire comes from M16 rifles and M24 sniper weapons, as well as from higher-calibre munitions. Some of the worst injuries come from the fragmenting bullets of M16s. Designed for maximum stopping power, the M16 bullet can tumble or break into pieces after penetration, causing multiple internal injuries and massive exit wounds, much like those of the internationally banned dum-dum bullets. Ballistics studies have indicated that the wounding potential of a fragmenting bullet is far higher than that of a regular bullet. Such bullets have a 90 per cent likelihood of leading to a loss of a limb—including by amputation—as opposed to only ten per cent among victims injured by a traditional full metal-jacketed bullet.

But even the more benign ammunition designed for riot control, like so-called rubber bullets—actually rubber-coated steel—can be fatal at short range. According to Dr Jihad Marshal: 'They are the nightmare of the neurosurgeon... Every time the patient moves his head, it's like a marble moving in jelly. There's nothing you can do about it' (Andoni and Tolan, 2001). This situation reveals the impacts of both the discriminate and the indiscriminate usage of small arms to specifically maim, disable, and kill. In a letter written by General Giora Eiland to a prominent Israeli human rights lawyer, it was suggested that '[a] large part of those wounded by live bullets are those we indeed wanted to not only injure but kill... The fact that most of them are wounded in the upper body or head is a positive thing'. After a flurry of international condemnation, the rate of head and chest injuries dropped, replaced by devastating leg and abdomen wounds. But according to Dr Robert Kirschner who works for Physicians for Human Rights: '[s]hooting people with high-velocity bullets to wound them is a form of summary punishment being inflicted in the field' (Andoni and Tolan, 2001).

Limited health services further exacerbate the possibility of permanent injury and death. For example, after being shot, Fouad Mahed had to be taken to Amman for surgery, where he still lost his right leg. Doctors in Gaza gave him 17 pints of blood to replace what was pouring from the wound. Skin graft complications forced doctors to send him to Amman because he could not get adequate treatment in Gaza hospitals. At the age of 17, Ibrahim Mustafa Darwish has also undergone multiple surgeries to remove a metre of intestines after being shot in the abdomen. This occurred during protests at a checkpoint dividing Gaza from Israel, when Israeli soldiers were firing at the group of stone-throwers from a distance of 15 metres.

Source: Andoni and Tolan (2001)

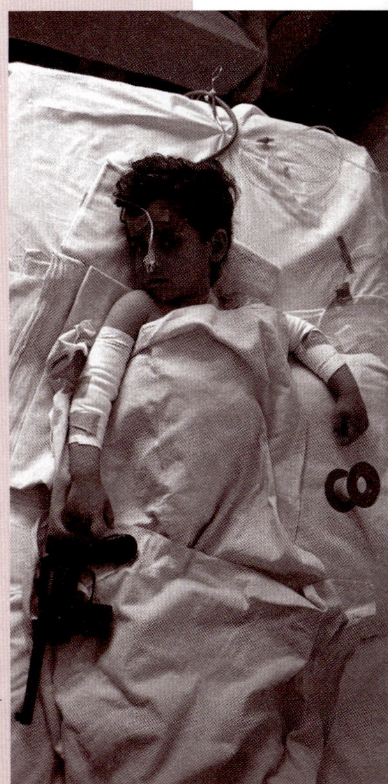

In the war of adults, children are often victims. If violence begets still more violence, what lies ahead for these children?

Similar trends are apparent in both conflict-affected and peaceful regions of east Africa. Studies carried out in clusters of districts in Uganda, Sudan, and Kenya have noted disproportionately high levels of firearm-related injuries in comparison with national averages.[11] An ICRC field hospital in Lokichokkio (Kenya) and a public hospital in Uganda registered between 40 per cent and 65 per cent of all monthly in-patients with firearm injuries from automatic weapons (Kobusingye *et al.*, 2002; Muggah and Berman, 2001).

FIGURE 4.1 Insecurity in Nicaragua

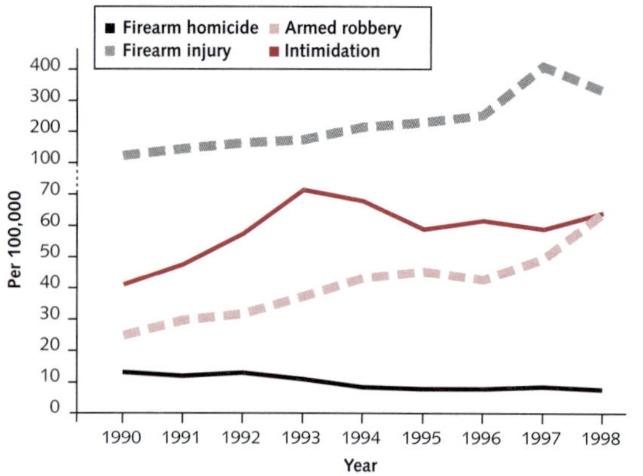

Source: Muggah and Batchelor (2002); Appendix 4.1

Even for those victims able to reach clinics in war-torn countries, the most elementary instruments, medicines, and services are lacking. The physical and human capacities of such facilities are often severely overstretched. They can even be deliberately targeted with small arms during complex humanitarian emergencies. Particularly among afflicted African countries, hospitals and staff are overwhelmed and ill-equipped to treat firearm injuries. One Ugandan surgeon implores: 'How can you triage when faced with a five-year-old with a bullet hole in his chest and a three-year-old with her leg blown off by a landmine... it's simply ridiculous.'[12]

The systemic under-capacities of hospitals and the limited access of civilians to facilities also heighten the risk of infection. A surgeon contracted by Oxfam-GB (2001b, p. 29) to assess the impacts of small arms in the Democratic Republic of Congo (DRC) described the situation thus: 'Bullet wounds, if treated without proper surgical debridement, rapidly develop into deep festering wounds and bone infections (osteomyelitis). Many of the wounded admitted... had been in hiding for weeks before being able to seek refuge and care in the hospital. Others were treated in health posts that had already been looted or burnt.'

Psychological effects

Though armed conflicts lead to a tremendous number of civilian deaths and disabilities, the humanitarian impacts of small arms are not restricted to fatalities and physical lesions. The health-related implications of small arms use on psycho-social and psychological well-being must be considered as well. Even with well-functioning trauma systems that maximize survival from devastating injuries, considerable long-term morbidity and disability persists. As has been extensively discussed by military psychologists and war historians, psychological stress associated with armed conflict ordinarily accounts for the larger share of population health impacts. During the Second World War, for example,

FIGURE 4.2 Insecurity in Honduras

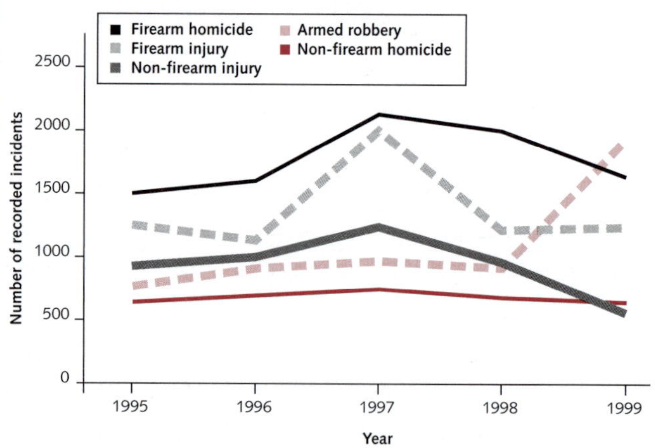

Source: Muggah and Batchelor (2002); Appendix 4.1

'America's armed forces lost... 504,000 men from the fighting effort because of psychiatric collapse... by the 1974 Arab-Israeli war, almost a third of all Israeli casualties were due to psychiatric causes' (Grossman, 1995, p. 43).

Post-traumatic stress disorder (PTSD), for example, is described by the American Psychological Association's (2000) *Diagnostic and Statistical Manual of Mental Disorders* as a reaction to a psychologically traumatic event outside the range of normal experience. It is characterized by, among other things, emotional blunting and detachment, social withdrawal, extreme suspicion of others, and recurrent nightmares.[13] Approximately 15 per cent of the 2.8 million American Vietnam war veterans were diagnosed as suffering from PTSD; and almost 150,000 are affected to the extent that they are still unable to lead a normal life without extensive medication and therapy (Dumas 2001; Slaby 1989). Given that PTSD is a long-term disorder and that the majority of the world's small arms victims do not have access to sustained treatment, medical facilities or counselling, the implications are sobering.

It is important to also consider that *personal* recovery from physical and psychological firearm injuries is to a large extent grounded in *social* recovery. Where societies continue to be plagued with armed violence and cultural, political, and economic livelihoods are systematically undermined, the prospects for recovery are bleak. This is because psychological trauma is unlike physical trauma: 'People do not passively register the impact of external forces (unlike a leg hit by a bullet) but engage with them in an active and social way' (Summerfield, 2000). Community studies in El Salvador have documented increased psychosocial stress that is correlated with increased collective exposure to armed violence. High levels of direct exposure to violence and psychological effects have been reported among Kosovar Albanians and in Sri Lanka and Vietnam (Cardozo *et al.*, 2000; Ugalde *et al.*, 2000). Effective treatment therefore requires, in many cases, a reduction in exposure to prolonged armed violence. In the view of some health specialists, 'the psychological effects of [complex humanitarian emergencies] on victims have not begun to be fathomed' (Leaning, Briggs, and Chen, 1999, p. 10).

Intimidated by a gun: Forced displacement

By the end of the 1990s, up to 40 million people had been violently forced to leave their homes, either crossing a border and officially documented as a refugee or, more likely, internally displaced within their own borders. The United Nations High Commissioner for Refugees (UNHCR) (2001a) estimates the number of refugees at 12.8 million and the Representative of the Secretary-General for Internally Displaced People estimates the number of IDPs at between 20 million and 25 million. Millions more fled who were not counted, eschewing, in some cases, assistance and protection for fear of violent recrimination and social or economic marginalization. To be sure, 'refugees are only the tip of the iceberg of human suffering—at least twice the number of refugees are typically either internally displaced or trapped at home' (Leaning, Briggs, and Chen, 1999, p. xvi) (see Map 4.1 and Appendix 4.2).

One in every 120 people is designated as either internally displaced or as a refugee.

One in every 120 people is designated as either internally displaced or refugee, and small arms-related intimidation represents a critical factor inhibiting sustainable repatriation or resettlement. Most refugees and IDPs appreciate the persuasive power of a single weapon. Agencies such as the UNHCR (2001a, p. 283) have also recently noted that 'armed conflict is now the driving force behind most refugee flows'. The United Nations (1999, p. 2) has repeatedly observed that 'in many recent and current internal armed conflicts, combatants deliberately intimidate, attack, and displace local populations to further their pursuit of economic control over natural resources. In such cases combatants rely on, and indeed profit from, civilian displacement.'

For example, in 1999 widespread violations of human rights occurred against ethnic Albanian Kosovars in the course of mass deportations by the Serbian army, leading to NATO's 78-day air war. A survey carried out by Iacopino and Waldman (1999) notes that more than 30 per cent of all Kosovar households reported at least one of the following abuses among members: shooting, threat at gunpoint, firearm homicide, torture, beating, separation, disappearance, and sexual assault. Similarly, as a result of the sheer scale and lethality of armed violence in Rwanda, between 25 and 40 per cent of the Rwandan population was displaced and more than 500,000 refugees spilled across the country's borders following the genocide (Melvern 2000; Human Rights Watch, 1994). The violence there has not stopped. In Rwanda, testimonial evidence documents how groups of armed men massacre civilian refugees in clinics and makeshift hospitals across the border in (the former) Zaire (Minorisa, 2001).

© Associated Press/David Guttenfelder

When considering the viability of international humanitarian law in protracted wars, such as in Congo-Brazzaville, the question remains: are new warriors fighting by old rules?

Armed danger in the camps

From Srebrenica to Goma, some of the worst small arms-related violations against civilians have taken place against refugees and IDPs during transit or in safe-areas. This is largely because settlements populated by displaced people are highly insecure. They are typically close to an international border or located on desolate or isolated tracts of land. A large part of the humanitarian response to complex emergencies, such as those in Colombia, Sudan, and Sri Lanka, is devoted to ensuring the safety of displaced people who are frequently vulnerable to rampant crime and violence after they have temporarily resettled. In Colombia, estimates of the IDP population range from 400,000 to 2.2 million (CONPES, 1999; NRC, 2001). Disputes over the actual number of IDPs reveal just how inadequate surveillance capacities can be, as well as the state's political reluctance to respond. A large proportion of Colombian displacement can be attributed to massacres; 90 per cent of atrocities committed by the military, paramilitary, or guerrilla actors are reportedly carried out with small arms (Muggah and Berman, 2001). IDPs continue to be targeted long after being displaced, often coerced into precarious and arms-saturated settlements to face destitution, continued persecution, or a life of crime (Crisp, 1999).

It is well-known that locating a refugee or IDP camp near a border conflict zone will never allow the settlement to be permanent and free of refugee warriors or armed actors. The case of Kenya is illustrative. In both of Kenya's 'official' refugee camps—Dadaab and Kakuma—over 200,000 Sudanese, Ethiopian, Somali, and Central African refugees are subjected to armed violence on a daily basis (see Figure 4.3 and Appendix 4.3). According to the Integrated Regional Information Network (IRIN, 2000b), there is a very strong possibility that the camps are being used to traffic arms: 'There have been shooting incidents in the camps... it is easy for people to move around with arms on that border [with Somalia]

MAP 4.1 Far from home: A sample of displaced populations around the world

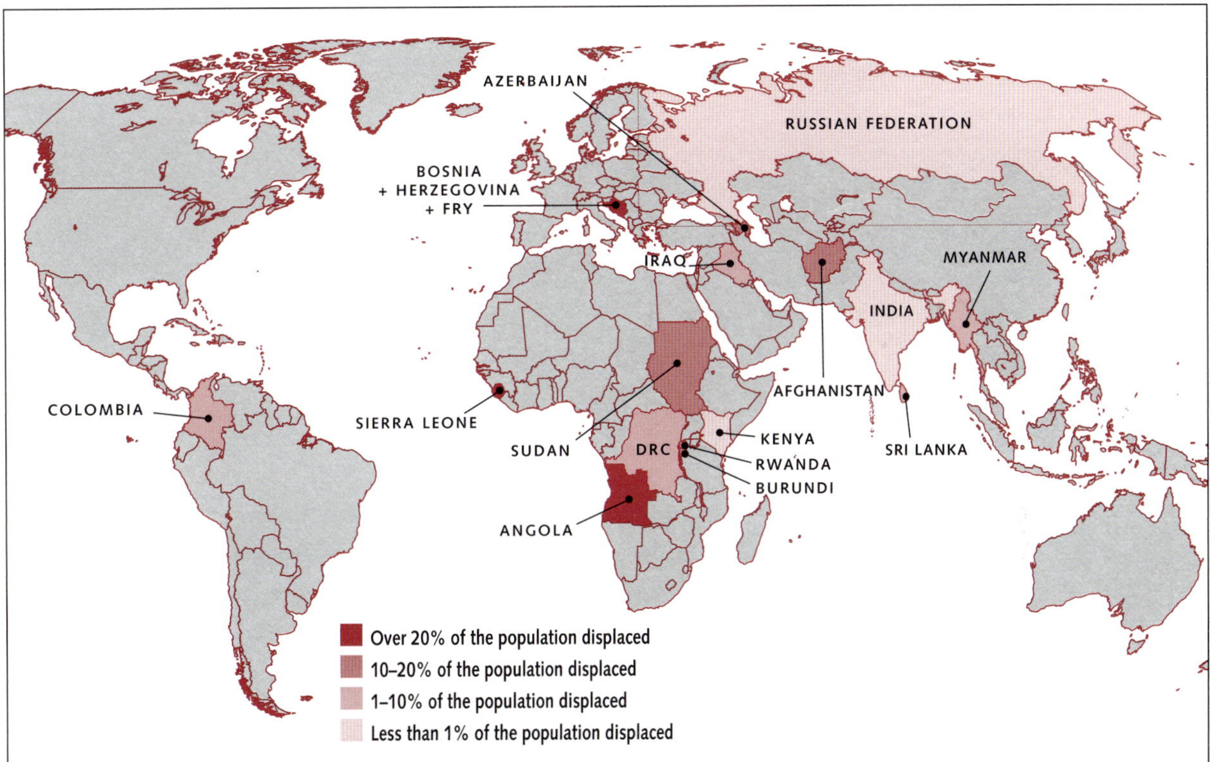

Source: UNHCR (2000b, refugee figures); NRC (2001, IDP figures); UNDP (2001, population figures); Appendix 4.2

because there is no control [in Somalia]'. A study commissioned by UNHCR observed that a large proportion of women collecting firewood out of the Dadaab refugee camp are raped at gunpoint by armed assailants. UNHCR has managed to reduce the number of rapes by having wood trucked into the camps. In 2000, there were an estimated 72 reported rape cases compared with 142 in 1998.[14] But only an estimated 30 per cent of firewood needs are being met. Women remain vulnerable to armed attack and more than 150 informal police reservists have been deployed to police the camp.

These risks have hastened a shift in strategy by humanitarian and development agencies. Though many have refused to engage in military protection, others felt no other solution was viable. For example, in response to increasing camp militarization, key interventions called for by the UNHCR (2000, pp. 1–2) include 'disarming exiled groups who have access to weapons and curtailing any flow of arms into refugee populated areas... [and] disarming exiled soldiers and other armed elements, and ensuring their effective re-absorption into civilian society'. In Tanzania, for example, the UNHCR has begun experimenting with community policing. In areas where it has introduced armed security guards, such as Dadaab (Kenya), it has been criticized for favouring wartime economic activities over peacetime ones and rewarding those in control of a wartime economy.

A large proportion of Colombian displacement can be attributed to massacres; 90 per cent of atrocities committed by the military, paramilitary, or guerrilla actors are reportedly carried out with small arms.

Indirect impact on displaced people

Small arms proliferation indirectly causes suffering from poor conditions in and around the camps. The mobility of refugees and IDPs, and therefore their capacity to trade and collect food or fuel, are often severely constrained. Malnourishment is frequently a consequence of arms related insecurity. As a result, violently displaced people are prone

FIGURE 4.3 Security incidents in Kakuma refugee camp, 1996–2000

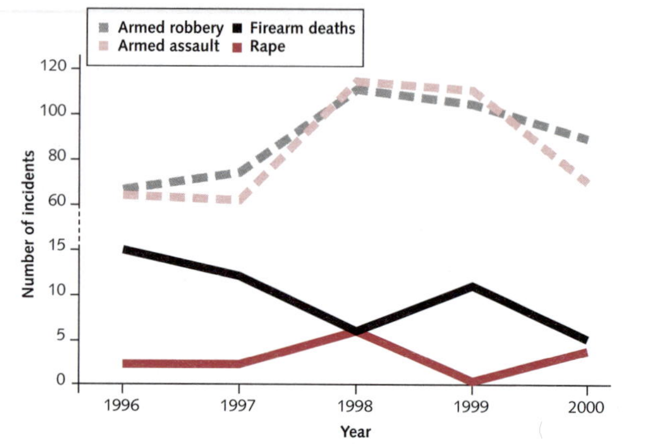

Source: Muggah and Berman (2001); Appendix 4.3

to acute morbidity from disease such as measles, respiratory infections, and malaria. This is most severe among young children, as documented below. Nor are adults immune. A recent study of IDPs in Guinea Bissau discovered that patients with tuberculosis whose treatment was disrupted because of war were three times as likely to die as those who were fully treated in peacetime (Gustafson *et al.*, 2001).

Other diseases have re-emerged as a result of arms-related violence. Sleeping sickness (trypanosomiasis) was thought to have been eliminated in the 1960s, but population displacement and the collapse of health systems as a result of civil wars have caused a resurgence of the disease. It is precisely those countries affected by protracted armed violence—southern Sudan, Sierra Leone, and the DRC—that are reported to be among the worst affected areas: more than 150,591 new cases of sleeping sickness have been detected during the past decade (Dobson, 2001). Internal conflicts, such as those that have taken place in the Balkans and Angola, also threaten the eradication of poliomyelitis (Bush, 2000; Ebersole, 2000).

BOX 4.7 Waving the white flag: Cease-fires and immunization

Despite the overwhelming threats presented to health workers during armed conflict, health interventions such as national immunization days (NIDs) have been successfully implemented and generated creative opportunities for peace-building. One of the first efforts to employ an NID during a conflict situation took place in El Salvador during the 1980s; a three-day cease-fire was agreed among armed combatants so that the United Nations Children's Fund (UNICEF), the Roman Catholic Church, the ICRC, and others could immunize some 400,000 children against polio.

Since then, cease-fires designed to ease the humanitarian impacts of small arms have been carried out in Afghanistan, Angola, DRC, Lebanon, Liberia, Nicaragua, Sierra Leone, and Sudan. In addition to the measurable health impacts of such programmes, the full value of cease-fires extends well beyond the specific health intervention. Cease-fires can be conceived as turning points for generating the trust necessary for underpinning local and national peace. Such initiatives can usefully reduce the 'habit' of fighting and allow the necessary space for combatants to savour the benefits of peace over war. The termination of hostilities among combatants does not have to be a precondition for what have been labelled humanitarian cease-fires, but the humanitarian impacts of small arms can nevertheless be mitigated.

Source: Ebersole (2000)

According to reports by UNICEF (2001), 'polio immunization campaigns have been hampered because of the conflict... Records have been lost and people have been constantly on the move. The whole infrastructure is having to be replaced.' In Somalia, despite being selected according to clan loyalty, 650 vaccinators taking part in a WHO/UNICEF campaign failed to successfully immunize the population: 'In one area, the chief of the subsection said he would shoot any vaccinators who came because we hadn't recruited anyone from his sub-district' (IRIN, 2000b). The introduction of humanitarian cease-fire represents one innovative response undertaken by the UN to allow for increased immunization and vaccination of children in conflict (see Box 4.7).

Basic needs and faltering rights

The humanitarian effects of small arms are not necessarily short-term or even immediately visible. Even in non-conflict affected societies, rampant small arms availability and use can generate a climate of fear among ordinary citizens and in some cases, a culture of violence that can last for generations. Among civilians, perceptions of insecurity influence household decision-making, access to food, water, and shelter, mobility and commerce, and investment of productive resources in labour and land (Muggah and Batchelor, 2002). Pervasive insecurity, both real and perceived, continues even after the formal cessation of hostilities. For example, in Nicaragua, El Salvador, and Honduras, the number of violent deaths in 1998–99 was higher than the number of people known to have died during the wars. It is not surprising, then, that perception surveys from Guatemala to Sri Lanka suggest that many urban residents feel more insecure today than during the war.[15]

In Nicaragua, El Salvador, and Honduras, more people died violent deaths in 1998–99 than during the wars. Many urban residents feel more insecure today than during the war.

The low cost and accessibility of weapons make them particularly attractive commodities to otherwise unemployed, humiliated, or disenfranchised youth. Adolescents are forcibly recruited into the ranks of gangs or informal armies; and many lose access to a range of entitlements, including formal education or access to health care. While the militarization of despair is particularly strong among adolescent boys and young men, it is growing among girls and women. Women are simultaneously recruited at gunpoint into armed factions such as the Revolutionary Armed Forces of Colombia (FARC), the Liberation Tigers of Tamil Eelam (LTTE), or Sierra Leone's Revolutionary United Front (RUF), targeted for sexual abuse and ritual humiliation, and put into situations of extreme insecurity. Women and children feel the humanitarian impacts acutely because they have limited access to basic services. Though the humanitarian impacts of small arms on the basic needs and rights of civilians are manifold, three deserve special attention: child soldiers, sexual violence, and child mortality.

Small arms and child soldiers

The access of children to their basic needs is frequently curtailed where they are recruited at gunpoint into armed groups. Children are often violently seized from elementary schools and orphanages or off the street to serve as combatants on the frontlines, domestic labourers, or sexual slaves to formal and informal armies alike. Though some of these children may benefit from a new range of entitlements made accessible by armed violence, their rights and long-term health and welfare are invariably compromised by a life of informal soldiering.

The Coalition to Stop the Use of Child Soldiers and the Save the Children Fund (SCF) estimate that some 300,000 child soldiers, as young as seven years old, are actively fighting in about 41 countries.[16] An additional 500,000 are reportedly recruited into paramilitary organizations, guerrilla groups, and civil militias. Where there is an abundance of small arms in today's wars, there are armed children: whether suicide bombers in Sri Lanka, soldiers in Myanmar, guerrillas in Colombia, or militia in Sierra Leone. The technology of weaponry only aggravates the problem: 'The availability of modern lightweight weapons is exacerbating the problem because even small children can become killers' (Inocenti, 2001).

In places like Sierra Leone, small arms and ammunition have become integral to the symbolic functions of warfare.

While the number of child soldiers is indicative of the magnitude of the humanitarian impacts of small arms, it is the exposure of children to firearm violence that yields long-term consequences (Machel, 2000). In many countries, small arms and ammunition have become an essential part of the coercive and symbolic functions of warfare. In Congo, for example, a concoction of gunpowder, narcotics, and alcohol was provided to child combatants that was purported to enhance their strength and invulnerability. Virtually all over the world, weapons—from Makarov pistols to RPG-launchers—are openly flaunted as status symbols, much like a cellular phone or a designer label (STOCKPILES). Such relationships are fostered early on and are difficult to reverse, bringing with them long-term generational and developmental challenges.[17]

Small arms and sexual violence

The documented experiences of women in war since Ghengis Khan and the Thirty Years War leave little doubt about the prevalence of rape in warfare. Small arms have facilitated gender-specific atrocities and raised the risk threshold for women (Dybdhal and Pasagic, 2000). For example, Olujic (1998) reported that some 20,000 women were raped at gunpoint, by Bosnian Serb soldiers as part of a deliberate pattern of abuse where sexual violence fulfilled a strategic purpose. Weapons therefore assume a symbolic role in the violent repression of women (Cock, 2000; Fitamant, 1999). Significant under-reporting and concealing of the problem, however, have limited our awareness of it.

Women also suffer from a long-term decline in their access to basic needs through loss of husbands and children to massacres, leading to dramatic transformations in their earning power and social status that extends well beyond the individual to the social and communal realm. These changes can distort the access of household members to basic needs such as adequate nutrition, housing or shelter, health, and education. Humanitarian workers have noted persistent food insecurity among households in arms-saturated areas even in those locations benefiting from substantial agricultural and food aid, indicating a relationship between systematic armed violence and reduced access to basic entitlements (Muggah and Berman, 2001).

BOX 4.8 Measuring terror in Abkhazia

Ritual humiliation and terror carried out on a daily basis with guns may be measured by the number and intensity of acts, but the true extent of its impacts cannot be converted into hard figures. Though a recent study carried out by the Small Arms Survey sought to empirically measure the impacts of small arms availability in the region (Demetriou, 2001), it is worth quoting a Human Rights Watch Report (1995, p. 5) at length to illustrate how small arms were used to terrorize civilians in Abkhazia:

> In a typical scenario, reportedly practised by both Georgian and Abkhaz forces against civilians, a man would be stopped on the street by armed men and asked his identity or place of residence. If he identified himself as from an enemy group, the men would humiliate him, threaten and beat him with fists and rifle butts. Then they would force him to take them to his home, where they would beat and intimidate the family, including children, and sometimes subject one or all to mock executions in front of the others. They would then typically rob the family, and sometimes take the male members, sometimes to terrorize them and their families and sometimes to torture and execute them. Often these visits were repeated. Such ethnically-oriented abuse forced much population displacement.

Small arms and child mortality

Small arms availability disrupts referrals, immunization programmes, and monitoring and surveillance, resulting in some cases, in an increased dependency on foreign aid and personnel (WHO, 2001a; Ugalde *et al.*, 2000). Such disruptions generate a host of secondary consequences including high mortality rates among young children.

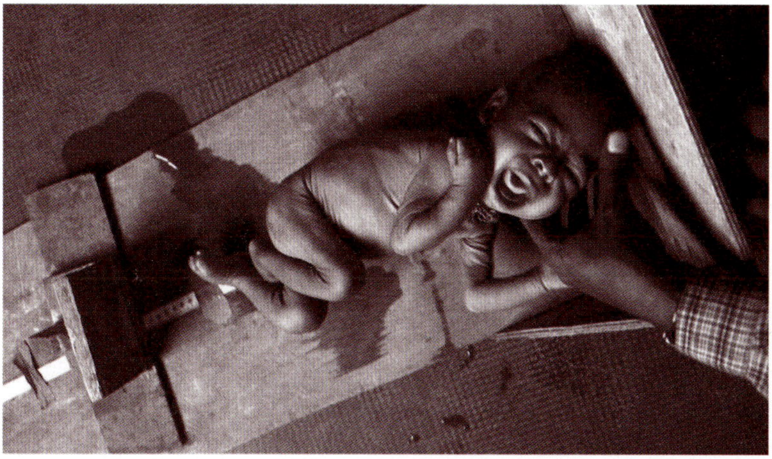

Small arms can have devastating effects on the vulnerable who are unable or too afraid to access essential services.

There is a strong correlation between high rates of armed violence, deteriorating public services, and proportionately higher death rates for children from non-violent causes. Extreme variations can occur within states and even between communities and households. A recent International Rescue Committee (IRC) study (Les *et al.*, 2001) on violence in the DRC confirms this association: 'While only ten per cent of all deaths, or 14 per cent of the excess deaths, were attributed to violence, there is a strong association (across both time and space) between higher violence rates and higher death rates from infectious disease... In Mboa and Kalemie, it is estimated that 75 per cent of children born during this war have died or will die before their second birthday.' Displaced children suffer from malnutrition because of scarcity of food in the area and armed insecurity (MSF, 2001).

A child mortality rate (CMR) of ten per 100,000 per day is regarded as an emergency. In Rwandan refugee camps, for example, child mortality escalated to more than 250 and 300 per 100,000 per day. By way of comparison, in the district of Mindanao (Philippines), UNICEF has reported that the CMR exceeds 310 per 100,000, whereas the CMR for the country as a whole is significantly less (HAIN, 2001). While a high CMR can be attributed to a host of variables including communicable disease, malnutrition, and armed conflict more generally, it is well known that gun violence and armed confrontations have also increased in Mindanao during the period in question, restricting the mobility and access of civilians to health services.

In many violence-affected societies, conventional humanitarian indicators and rates do not reflect the full damages attributable to small arms. For example, in El Salvador's civil war (1980–92), fought primarily with light weaponry, it is estimated that there were approximately 80,000 war-related deaths and up to 19,000 people disabled (Godnick, 2001). But conventional indicators hide the more subtle degradation of El Salvadorian health services. Due to decreases in the health and maintenance budgets, the quality of public services—for example, water supply, sewerage, refuse disposal—deteriorated. Street violence, alcoholism, and drug addition, coupled with the widespread availability of small arms, has resulted in one of the highest firearms homicide rates in the world in the late 1990s (Ugalde *et al.*, 2000).

The source of humanitarian chaos: New wars and small arms

Why are the humanitarian impacts of small arms worsening in particular areas? Much of the change has to do with the dynamics of today's new wars, which take place within rather than between states. Even where other countries have an interest in an internal conflict, virtually all the fighting and social violence is home-grown. Nevertheless, as Nobel Laureate Amartya Sen (2001, p. 50) reminds us: 'Local wars and military conflicts draw not only on regional tensions, but also on the global trade in arms and weapons'.

Though long appreciated by humanitarian field workers, there is growing recognition in the research and policy-making communities that these new conflicts cannot be regarded as temporary aberrations or short-term interruptions of established developmental paths. Analysts are coming to recognize that many of today's wars are protracted affairs not readily amenable to the conventional dose of short-term humanitarian assistance, diplomatic intervention, and reconstruction and development. Particularly worrying for many donors and analysts is the ease with which so-called post-conflict countries are slipping repeatedly back into war or widespread social violence (Muggah and Batchelor, 2002).

After the horrors of the Second World War, the four Geneva Conventions were adopted in 1949 to assure the protection of civilians through the distinction between civilian and military targets. Protocol 1 (1977) additional to the Geneva Conventions extended the narrow definition of international armed conflicts to include armed conflicts in which people are fighting against colonial domination, alien occupation, and racist regimes in the exercise of their right to self-determination whereas Common Article 3 and Additional Protocol 2 apply to non-international armed conflicts. The basic supposition underpinning international humanitarian law is that military interest is best served by attacking other military opponents and not civilians. The effectiveness of humanitarian law is contingent, then, on four key pillars—that:

- military interest is aimed at 'winning the war' and targeting military assets;
- military opponents are clearly recognizable and homogenous groups;
- military opponents have a legitimate monopoly over military-style weapons; and
- civilians are not an explicit target.

Each of these conventional suppositions is undermined by the logic of what have been elsewhere described as the 'new wars' (Duffield, 2001; Kaldor, 1999). The lines between armed conflict and criminality are more ambiguous and unclear than ever. Today's internal conflicts are predicated less on military and political advantage between two warring groups than on the outright collapse of states, the collusion of warring parties, and the access of certain sectors to resources and the free exchange of 'conflict goods' (TRANSFERS).

War is less about a continuation of politics and more about the continuation of economics by other means.

Some theorists have described these new wars as emanating from clusters of grievances left unresolved following the end of the Cold War. Others have described the new wars in rational utility-maximizing terms, with clear 'winners' and 'losers', sustained by self-interested warlords, political elites, and predatory firms (Collier and Hoeffler, 2000; Berdal and Malone, 2000; Reno, 1998). Paraphrasing Karl von Clausewitz, Keen (2001) argues that war is in many cases less about a continuation of politics and more about the continuation of economics by other means. The ends are frequently short-term and profit-oriented, while the means are nasty and brutish (see Box 4.9).

© Associated Press/Enric Marti

Today's warriors include men, women and children, blurring the distinction between combatants and non-combatants.

These new wars are fought not by clearly defined armies but by opportunistic and frequently self-interested armed factions. These factions commonly lack even rudimentary military discipline and are rarely, if ever, trained in the humanitarian laws that inculcate respect for civilian or unarmed populations and the 'proportional' or balanced use of force. As a result, atrocities—the killing of non-combatants such as erstwhile fighters, surrendering combatants or civilians—are more common. Due to the longevity and ferocity of today's conflicts, these factional fighters have few of the reservations that marked soldiers during the First and Second World Wars.

BOX 4.9 Co-operative conflict: Economics by other means

In many countries, armed conflict is more common than peace. Even in those countries where conflict appears to have ended, such as Cambodia, El Salvador, and Mozambique, warlike conditions persist. Their protractedness has defied analysis, as have the absence of political programmes, the proliferation of factions, and ferocious attacks against civilians.

But, protracted internal conflict and social violence can be interpreted as a rational process rather than just chaotic breakdown. Analysts have described today's wars as an extension of economics by other means: the creation of an alternative system of profit, power, and even protection. What may have initially begun with political aspirations mutates into conflicts prioritizing short-term economic benefits. While ideology, identity, and ethnicity remain vital elements, undue focus on them may disguise the real political and economic agendas that sustain war.

At one level, war can be perpetuated and even motivated by the trade in conflict goods. Even more nihilistic, the short-term gains from conflict—such as protection money, the appropriation of labour and land, and the stealing of aid supplies and military benefits—have promoted a new form of co-operative conflict. This involves avoidance of pitched battles, co-ordination of movements in and out of villages, trading arrangements, ransoms for fighters, or even stranger possibilities. In Sudan, government soldiers are known to sell arms and ammunition on the open market, some of which find their way into the hands of the rebel Sudanese Peoples Liberation Army (SPLA). Government troops in Sierra Leone frequently avoided pitched battles with the RUF, opting instead to sell their arms, ammunition, and uniforms (Keen, 1998, p. 20).

The spread of powerful and cheap light weapons makes it particularly important to understand grassroots violence. As societies become better armed, warfare is more likely to involve an arrangement between elites and the civilians they wish to recruit. Violence will increasingly serve a purpose for those at the bottom of a society as well as those at the top. When ordinary people actively embrace violence, they may be seeking revolutionary political change or venting ethnic hatred. They may also be trying to meet other, more immediate or mundane needs, or to preserve their physical security; they may be looking for excitement or for the immediate rectification of a perceived wrong. They may also be following their own economic agendas.

Source: Keen (2001)

This new brand of conflict is waged by life-style warriors: young men who soldier by day and return to their fields and families at night. An extreme example emerges from Congo's capital Brazzaville, where competing militia factions that traded gunfire and mortar fire during weekdays would decamp across the river to Kinshasa for shared weekend festivities. Also common are civilians who actively aid—or are coerced into supporting—insurgents, such as in acutely militarized parts of Colombia or Sri Lanka. The collusion between civilians and insurgents has prompted governments to suspend civilians' basic rights under the guise of internal emergencies and to target entire populations outright. In a number of cases, governments have themselves indirectly recruited paramilitary or militia groups to target suspected conspirators, thereby abdicating their responsibilities under the Geneva Conventions.

The casualties of today's new wars are overwhelmingly a consequence not of heavy conventional weapons, missiles, or even landmines, but rather of lightweight and highly mobile small arms. The virulence of armed conflict and criminal violence is increasing as a result of newly introduced technology and the cascading of older stockpiled weapons (Homer-Dixon, 2001). For example, military-style rifles—mostly AK-47s, G-3s, and the Fusil Automatique Léger (FAL)—are among the most common weapons used by armed criminals in Kenya and within the Greater Horn of Africa and Great Lakes region, while handguns, such as .32s, .38s, and 9mm pistols, revolvers, and grenades, are the most commonly used in atrocities and common crime in Colombia, and throughout Latin America and the Caribbean.

The fact that these weapons are so readily accessed by civilians, criminals, and combatants indicates an alarming failure of the state's theoretical monopoly of violence. In the context of war-affected states, the continued availability of weapons provides a means of resolving differences through armed violence—particularly in the aftermath of armed conflict—contributing to an environment of impunity.

Civilians serve as cover for the operations of rebel movements, as targets for reprisals, as shields against air or artillery attacks, as political tools for international assistance, and as a principal target of ethnic cleansing and genocide. The belief that civilian casualties are not 'collateral damage' but strategic targets finds its antecedents in the treatment of civilians by belligerents in Ethiopia, Mozambique, Angola, and other Cold War theatres. In its extreme form, as in Rwanda in 1994, Bosnia-Herzegovina in 1992–94, and Kosovo in 1998, entire segments of the civilian population have been targeted with small arms. But these conflicts could also be compared to earlier wars in Biafra (Nigeria) in 1967–70, Vietnam in 1965–73, and Afghanistan in 1979–89, where civilian deaths amounted to well over 1.5 million, far more than the estimated military toll (Frohardt, Paul, and Minear, 1999, p. 17).

All of this might seem to paint a pessimistic picture of the prospects for enforcing international humanitarian law and human rights with respect to small arms availability and use in conflict. But systemic violations of humanitarian law, whether in relation to the illegal supply of weapons to regimes abusing human rights, indiscriminate attacks on civilians, or the lack of protection to relief operations, do not mean the obsolescence of humanitarianism. On the contrary, 'international humanitarian law remains highly relevant in contemporary conflicts (such as instances of ethnic cleansing and failed states) and serves to mobilize considerable efforts to further its application' (Bruderlein and Leaning, 1999, p. 432). The development of the International Criminal Court, the *ad hoc* tribunals for the former Yugoslavia and Rwanda, and the banning of anti-personnel landmines are recent examples.

BOX 4.10 A UN response to the humanitarian impacts of small arms?

When planning began for the 2001 UN Small Arms Conference, there were serious misgivings among humanitarian activists. With government disarmament experts clearly in charge, it seemed as if little attention would be given to humanitarian priorities. While these concerns were largely proven prescient, there is some limited language in the *Programme of Action* reaffirming the importance of humanitarian perspectives.

Humanitarian priorities are acknowledged in Preambular paragraph 2, where it is noted that states are 'gravely concerned at... the excessive accumulation and uncontrolled spread [of small arms]... which have a wide range of humanitarian and socio-economic consequences...' (UNGA, 2001). The following paragraph 3 observes that 'small arms and light weapons... sustain conflicts, exacerbate violence, contribute to the displacement of civilians, undermine respect for international humanitarian law [and] impede the provision of humanitarian assistance to victims of armed conflict'. Nevertheless the UN *Programme of Action* still views small arms primarily through a traditional arms control lens.

A few days after the close of the conference, the Security Council met to consider the issue as well. The president of the meeting, Colombia's Ambassador Camillo Reyes, noted that the illicit trade in small arms is 'a challenge that involves security, humanitarian and development dimensions' (United Nations, 2001, p. 1). Although humanitarian perspectives were reaffirmed in the UN process by most operational agencies, in the absence of clear direction from global and regional organizations above, humanitarian agencies are starting to respond with their own approaches from below.

Humanitarian perspectives on arms-related violence

The remainder of the chapter examines the emerging relevance of international humanitarian law and human rights law for dealing with the challenge of small arms in new wars. But it is important first to ask what a humanitarian approach to small arms might actually look like. It is also useful to consider why it has emerged, and to review its strengths and weaknesses.

The international security landscape is shifting. Previously conceived as marginal or soft, issues such as child soldiers, anti-personnel landmines, and small arms are receiving more attention. Though complex and evolving, international humanitarian law and human rights law also are developing to respond more proactively to the small arms issue. Already many countries have self-imposed restrictions on arms exports to countries at war, have supported the evolution of international codes of conduct, and have encouraged NGOs to spread awareness of the Geneva Conventions to warring factions. The initiation of the War Tribunals for Rwanda and the former Yugoslavia, and the International Criminal Court, is another sign of the expansion of humanitarian priorities. Within national civil jurisdictions, the consideration of international humanitarian law with respect to domestic production and trade signals a growing concern with the humanitarian impacts of small arms.

While the climate may be ripening for a humanitarian approach to small arms there are still a great many challenges ahead. Chief among them is the tenacity of traditional notions of security. It has a rival, though, in the growing popularity of the concept of human security: a security strategy that puts human beings' freedom from fear above a more realist approach to security between states.[18] Human rights and humanitarian law are the normative basis for human security, a concept initially designed to expand the notion of security beyond a militarist perspective.

Among an increasing number of governments, there is a growing conviction that the protection of people irrespective of their nationality does not hinge exclusively on the protection of the territorial frontiers of states. National security remains a necessary, albeit insufficient, criterion of ensuring the protection and welfare of people. Just as the nature of war has undergone a perceptible transformation, so too have international responses to war. While the sovereignty and integrity of borders remain vitally important, a number of like-minded states, the United Nations, and the humanitarian community are forcing a new approach to ensuring international security.

Although the UN's primary purpose under Article 1.1 of the Charter is to ensure global peace and security with due respect to the borders of states,[19] a palpable shift was signalled by Boutros Boutros Ghali's observation in 1992 that 'the time of absolute and exclusive sovereignty has passed' (United Nations, 1992). The then-Secretary-General's statement signalled a new age in cross-border interventions ostensibly to preserve humanitarian ideals. Proponents of these principles such as Frohardt, Paul, and Minear (1999, p. 19) optimistically asserted that 'state sovereignty... is now less impenetrable and more infused with humanitarian obligation'. The new spirit of interventionism was reflected in military-style campaigns undertaken in Bosnia, Somalia, Kosovo, and northern Iraq that were labelled—controversially—humanitarian by those who carried them out.

Speaking of the emerging dual sovereignty of the state and individual, UN Secretary-General Kofi Annan noted that today's conflicts 'demand of us a willingness to think anew: about how the UN responds to humanitarian crises affecting so much of the world; and about our willingness to act in some areas of conflict'. He explained that 'intervention must be based on legitimate and universal principles if it is to enjoy the sustained support of the world's peoples. This developing international norm in favor of intervention to protect civilians from wholesale slaughter will no doubt continue to pose profound challenges to the international community' (Annan, 1999). While lofty in ambition, this new spirit of engagement has been half-hearted in practice. Given the indecisive action of the international community in the former Yugoslavia and Rwanda, and the dramatic humiliation of the US Army in Somalia, and its current role in Afghanistan, the future role of humanitarian intervention is unclear.

A humanitarian perspective on small arms

What exactly is a humanitarian perspective on small arms? The word 'humanitarian' is derived from *humanitas*, a Latin equivalent of *paideia*, a concept developed by Greek sophists who debated the place of reason as humankind's key distinguishing feature. In the nineteenth century, the notion of humanity, and to a lesser extent that of humanitarianism, came to be associated with notions of philanthropy and altruism as a moral force (Coupland, 2001a).[20] Seen by the founder of the International Committee of the Red Cross (ICRC) as an inevitable process, humanitarianism was put into practice in the twentieth century as a neutral and impartial exercise in alleviating the suffering of victims of armed conflict, irrespective of their nationality.[21] In the latter part of the last century, however, the meaning of humanitarianism once again expanded to accommodate collective assistance programmes under the mantle of the international community and international humanitarian law.

'The humanitarian community has witnessed a swing away from a sentimental, paternalistic, and privileged discourse of philanthropy and charity to a more political, egalitarian, and empowering ideology of rights and duties in war.'

At the end of the twentieth century, the term 'humanitarian international' was coined to capture the essential failure of philanthropic humanitarianism and its dangerous reliance on technical approaches to ensuring protection. Instead, a new ethos that promoted a rights-based approach to protection and assistance gained currency among scholars, policy-makers and practitioners. Michael Ignatieff (1997, p. 22) called this shift in thinking a revolution that signalled the refurbishment of an Enlightenment heritage of universal human rights and the growth of a 'vast constituency of human rights activists, development workers and aid workers whose moral rationale is the indivisibility of human interests and needs'. Hugo Slim (2001, p. 3) confirms that in the last decade the humanitarian community has witnessed a 'swing away from a sentimental, paternalistic and privileged discourse of philanthropy and charity to a more political, egalitarian and empowering ideology of rights and duties in war'.

Where a rights-based approach is adopted, it is also possible to agree on fundamental notions of what constitutes humane and inhumane. In the context of small arms, it becomes possible to elaborate human rights and health criteria to determine when small arms use reaches a level deemed inhumane or constitutes crimes against humanity. These standards are of direct relevance in international legal and political discourse whenever small arms, their effects, or their regulation are under discussion (Coupland, 2001a).

The spread and application of international humanitarian law is now more important than ever.

But humanitarians committed to the defence of rights in war and armed conflict need to engage much more with international humanitarian law as the main guarantor of human rights in war. Alongside the ICRC, NGOs and United Nations agencies must know the law and share it with civilians in war, with combatants, with politicians, and with civil society. They must take risks to put it into practice and to normalize it (Slim, 2001, p. 10). The rights of civilians and the obligations of all parties in conflict need to be made intelligible, to be brought down from the mountain and made accessible to humanitarians and civilians alike.

International humanitarian law, human rights law, and small arms

The primary purpose of international law, including treaties and customary law, is to lay down rules prescribing the conduct of states, not to regulate the behaviour of individuals. That said, international law is also evolving to account for individuals or non-state actors, as has been recently illustrated by trials for war crimes in the *ad hoc* tribunals established by the United Nations Security Council and the recently adopted statute of the International Criminal Court (ICC). As a subset of international law, the objectives of *international humanitarian law*—also known as the Laws of War—are to impose limits on how weapons are used in conflict and prohibitions on the use of certain weapons in war (see Box 4.11). International humanitarian law encompasses the rules which in times of war aim to protect persons and property affected by armed conflict and to restrict the methods and means of warfare (Levie, 2000; De Mulinen, 1987).

The main sources of international humanitarian law that relate to small arms are the St Petersburg Declaration of 1868, the Hague Conventions of 1899, the four Geneva Conventions adopted by 1949, and the two Additional Protocols adopted in 1966 and entered into force in 1977. The four Geneva conventions relate to: (1) the amelioration of the condition of the sick and wounded armed forces in the field (1864); (2) the sick, wounded and shipwrecked members of the armed forces at sea (1899); (3) the relative treatment of prisoners of war (1929); and (4) the relative protection of civilians during times of war (1949). The Protocols (1977) relate to (1) the protection of victims of international armed conflict, and (2) the protection of victims of non-international armed conflicts. Also included in the corpus of international humanitarian law are customary laws of war and treaties prohibiting certain weapons such as the 1980 UN Convention on Certain Conventional Weapons[22] and the Anti-Personnel Land Mines Convention or so-called 'Ottawa Treaty'.

Though initially designed for inter-state conflict between state parties, international humanitarian law has also adapted to reflect contemporary internal conflicts. According to Emanuela-Chiara Gillard (2000, p. 45) this shift was necessary, since 'legal practice must adapt to the changing dynamics of conflict if it is to have any realistic expectations of protecting civilians'. These changes are also reflected in the findings of the Yugoslavia Tribunal,[23] where it was noted that 'if international law, while of course duly safeguarding the legitimate interests of States, must gradually turn to the protection of human beings, it is only natural that the aforementioned dichotomy should gradually lose its weight' (International Tribunal, 1997, para. 97).

By contrast, *human rights law* seeks to preserve and protect the physical integrity and human dignity of the governed from their governments. It is based principally on the 1948 Universal Declaration of Human Rights. Although the Declaration does not enjoy treaty status, together with the International Covenant on Civil and Political Rights, its Optional Protocol, and the International Covenant on Economic, Social, and Cultural Rights, which both came into force in 1976, it is generally accepted as a means of judging compliance with human rights obligations under the UN Charter.[24] In contrast to international humanitarian law, the case for tying human rights to small arms-related violations during war is currently limited to production and trade—though some have begun to focus on the use of excessive force in the security sector. These limitations exist because, in spite of the unassailable core rights of human beings under the Charter, certain provisions of human rights treaties are regularly suspended during periods of armed conflict.

'Legal practice must adapt to the changing dynamics of conflict if it is to have any realistic expectations of protecting civilians.'

The small arms field is being challenged further by a new convergence between international humanitarian law and human rights law. Commonly referred to as the humanization of humanitarian law, the union is partly attributable to the considerable influence of the human rights movement on the Laws of War, particularly with respect to customary rules of international humanitarian law. This trend is reflected in the jurisprudence of international courts and tribunals as well as normative instruments adopted by international organizations and their acceptance by states

BOX 4.11 International humanitarian law: But can you enforce it?

The consequences of war are mitigated by the application of the rules of international humanitarian law. This is achieved through the regulation of the conduct of hostilities and protecting people who take no part in the conflict. The two major areas of regulation relate to (1) the of type of weapon, and (2) the choice of targets.

First, the use and transfer of certain types of small arms, light weapons, and ammunition are prohibited or restricted by international humanitarian law. A number of the treaties which constitute the cornerstones of international humanitarian law prohibit not only the use of particular weapons but also their transfer. As it is the more recent instruments that include additional prohibitions relating to transfers, they can be seen as reflecting a development of the law and, consequently, can be read into earlier treaties.

A second broader limitation relates to the state's duty to respect international humanitarian law and ensure its respect by other states and parties to conflict. From this broad and customary duty one could infer a prohibition on states themselves transferring or authorizing the transfer of weapons to states or armed opposition groups who are likely to use them in violation of international humanitarian law. These prohibitions are applicable in international and civil armed conflicts, and are binding on both states and armed opposition groups.

International humanitarian law is part of public international law and, accordingly, can be enforced by the imposition of sanctions or by proceedings between states before international tribunals. To date, no such international proceedings have been invoked on the basis of transfers of small arms—for example, in violation of treaty prohibitions—or in situations where it was foreseeable that the weapons would be used to violate international humanitarian law. Ultimately, the transfer of weapons in violation of a treaty or customary prohibition does not necessarily give rise to criminal responsibility; and a person who transfers or authorizes the transfer of weapons used to commit violations of international humanitarian law would be an accomplice only if he or she had the intention to commit the violation himself or herself. The same holds true for individual criminal responsibility for transferring weapons used to perpetrate genocide. It is unlikely that those accused of authorizing small arms transfers would have the intent to 'destroy in whole or in part a national, ethnic, racial or religious group' required by the 1948 Convention on the Prevention and Punishment of the Crime of Genocide for their actions to amount to complicity in genocide.

Human rights law also imposes important constraints on a state's freedom to transfer small arms or to authorize their transfer. A state that does so in circumstances in which the weapons will be used to commit violations of human rights can be considered as participating in the wrongdoing of the recipient state, whether used by its officials to commit extra-legal or arbitrary executions, to torture or to submit individuals to other forms of cruel, inhuman, or degrading treatment or punishment, or to detain individuals in violation of human rights standards.

A prohibition of transfers would be applicable where, rather than committing human rights violations itself through its agents, the recipient state is unable to control the private actors who have control of the weapons. Such a situation would amount to a violation of the state's duty to protect the right to life. In this instance, states should refrain from supplying weapons to states where it is likely that they will fall into the hands of individuals over whom the government is incapable of exercising authority and control. Traditionally, human rights law has been considered as singularly binding on states. The traditional view is being increasingly challenged and the law is developing towards holding organized groups, including non-state actors, responsible for violations of human rights.

The limitations imposed by human rights are wider in scope than those derived from international humanitarian law as they are applicable in times of both peace and war. As human rights law is also part of public international law, it can be enforced by means of inter-state litigation. While this has been fairly limited in practice, there are a number of universal or regional judicial or quasi-judicial bodies with jurisdiction over claims of violations brought by individuals against state parties to the various human rights treaties. Although no complaints have been brought to date against states that have transferred or authorized the transfer of small arms used to commit violations of human rights, such proceedings could be employed to obtain a ruling that such transfers violate international law. Finally, in certain national jurisdictions, most notably the United States, it is possible to bring civil proceedings against individuals or companies for compensation for violations of international law, including international humanitarian law and human rights. Although the US 1789 Alien Tort Claims Act has not been used in the context of transfers of small arms, it may prove a valuable avenue in the future.

Source: Gillard (2001)

(Meron, 2000). As noted by Doswald-Beck and Vite (1993), 'the provisions of international humanitarian law cover the greater part of civil, political, economic, social and cultural rights as represented in UN Human Rights law, but they do so in a way that speaks in terms of the specific duties of fighters'. In this way, 'it is not alien to military thinking and has the advantage of being a realistic code for military behavior as well as protecting human rights to the maximum degree possible in the circumstances'. The confluence between the two streams has instilled a growing realization that people affected by armed violence and war are legally entitled to the very same rights as civilians living in peace.

In spite of attempts by some humanitarian actors to codify key aspects of human rights law with respect to particularly vulnerable groups—for example, IDPs and children—existing laws do not take sufficient account of the rights and needs of the internally displaced, forcibly relocated populations or the right of access for humanitarian organizations. For this reason, a number of key provisions of humanitarian law are potentially more appropriate to protect civilians caught in internal conflict.[25] These laws expressly prohibit killing, summary executions, physical and mental torture, mutilation and corporal punishment, rape, enforced prostitution, indecent assault, pillage, and the taking of hostages.

Even if human rights and international humanitarian law are the cornerstone of a humanitarian approach to small arms, are the new warriors fighting by the old rules?

Even if precepts of human rights and international humanitarian law are accepted as the cornerstone of a humanitarian approach to small arms, a fundamental question remains: Is the existing framework of law suited for today's new wars? As Michael Ignatieff (1997, p. 124) asks, 'Are the new warriors fighting by the old rules?' Many argue that, for the Geneva Conventions to function, there have to be disciplined armies, that distinguish between civilians and combatants. While the debate over the relevance and application of human rights and international humanitarian law to today's new wars is unresolved, the humanitarian community has nonetheless evolved a set of approaches for dealing with the threat of small arms:

- *a supply-side approach* that focuses on constraining the transfer of weapons to regimes that violate human rights and international humanitarian law;
- *a new humanitarian approach* that aims to mitigate the impacts of weapons on civilians through the application of international humanitarian law and demand-related incentives; and,
- *an operational perspective* that stresses the consequences of arms availability on relief workers and peacekeepers, and aims to harden or soften humanitarian targets.

© ICRC/Priska Spoerri

Awareness of the rules of humanitarian law among combatants might reduce the humanitarian impacts of small arms.

Supply-side humanitarianism: Controls and advocacy

The supply-side approach focuses on strengthening supplier controls and end-user conditionality in order to prevent the export of small arms to regimes associated with violations of human rights or humanitarian law. Though Article 51 of the UN Charter recognizes the right of states to self-defence and the acquisition of weapons for military and police forces, Article 1 of the Geneva Conventions (1949) and Protocols (1977) emphasizes the obligation to respect and ensure respect for international humanitarian law. In particular, it states that 'the knowing provision of arms into situations where serious violations of international humanitarian law occur or are likely to occur should be considered a matter of grave concern'. In its strongest form, advocates of this approach contend that countries supplying weapons are accessories to the abuses committed with them, even genocide. The perspective is captured particularly well by the Chair of a recent inquiry in South Africa: 'If [arms supplier] states deliberately or carelessly sell weapons to repressive or aggressive regimes, they bear a measure of culpability for the use to which their weapons are put' (Cameron Commission, 1995).

A persistent concern among advocates of the supply-side approach is whether to focus campaign efforts on black and grey markets or on legally transferred weapons. The question is relevant because 'it is unknown whether, in general, the legal or the illegal small arms trade contributes more directly to ongoing warfare and repression around the world' (Lumpe 2000, p. 2). A humanitarian focus is a reminder that the impacts of small arms on human health are the same regardless of the source. Furthermore, increased availability of both licit and illicit small arms and light weapons are attributed to rising numbers of civilian casualties. Rapid arms build-ups appear to be positively correlated with a surge in violations of humanitarian law and civilian casualties. Human Rights Watch (HRW) has documented the relationship between illicit arms acquisition and civilian impacts in countries such as Angola, Burundi, Cambodia, Colombia, Georgia, India, Israel, Rwanda, Sierra Leone, Sudan, and Turkey.

BOX 4.12 But will they work? Codes of conduct for the small arms trade

The supply-side approach aims to prevent small arms transfers to states likely to commit serious breaches of fundamental human rights and humanitarian law. Violations of humanitarian norms can range from torture and arbitrary executions to violations of international law on the protection of civilians applicable in international and non-international armed conflict. The prohibition of arms transfers might also apply where small arms could fall into the hands of individuals over whom the government is incapable of exercising authority or control. Such a situation would represent a violation of the state's duty to protect the right to life.

Both the European Code of Conduct on Arms Exports and the Organization for Security and Cooperation in Europe's (OSCE) *Document on Small Arms and Light Weapons* initially sought to ensure that recipient states would comply with international humanitarian law and human rights criteria prior to the granting of export licences (CONFERENCE). The European Code was adopted in 1998. The 55 countries of the OSCE (comprising all of Europe, Canada, the United States, and five Central Asian states), have adopted its code, but only after its strongest language was removed.

While the codes demand that human rights and international humanitarian law should be taken into account, there are no specific obligations with respect to the transfer of small arms to abusive regimes. Significant gaps remain regarding coverage of brokering and licensed production, effectively leaving back doors open for evasion. With problems like this in mind, in 1998 Canada proposed a convention prohibiting the international transfer of small arms to non-state actors, but this proposal did not receive broad support either at the time or at the 2001 UN Small Arms Conference.

An international code of conduct has been proposed by a group of Nobel Peace laureates led by former Costa Rican President Oscar Arias, and working with an international coalition of NGOs. Their Framework Convention on International Arms Transfers would establish a legally binding text making arms exports contingent on principles of human rights, humanitarian law, sustainable development, and peace and stability.

Proponents of the supply-side approach call for increased accountability and government scrutiny of small arms from the point of production to end-use certification. Advocacy networks such as the International Action Network on Small Arms (IANSA) and the Humanitarian Coalition on Small Arms, the UN High Commissioner for Human Rights, and Amnesty International have called for stringent codes of conduct for the small arms trade. Such codes would aim to improve transparency, limit the list of permissible recipients based on ethical criteria, and strengthen post-transfer oversight.[26]

Concerted pressure on governments by proponents of the supply-side approach has yielded the adoption of codes at the national level, notably in South Africa and the United States.[27] The US export control system is considered by many to be one of the world's best, and efforts have been made to reinforce it with a formal code of conduct.[28] In Western Europe, both Germany and the United Kingdom have also recently undertaken to impose controls on the transfer of weapons to rights-abusing regimes.[29]

Humanitarian actors have demanded greater commitment to implement and enforce related measures such as UN Security Council arms embargoes. This approach relocates the debate within the wider framework of states' human rights obligations. But among existing export codes and national legislation relating to small arms, only the European Union Code of Conduct explicitly mentions respect for international humanitarian law as a basis for decisions on arms transfers (see Box 4.12).

The first attempt to infuse a humanitarian ethos in a code of conduct for arms exports emerged during independent talks by the five permanent members of the UN Security Council in 1991. Among the many issues considered was a code prohibiting transfers that might prolong or exacerbate armed conflict and the need for member states to ensure compliance with embargoes. Their inability to agree on basic criteria, as well as other underlying concerns such as the Chinese furore over US arms sales to Taiwan, led to the collapse of the talks in 1992. Despite tentative measures by the OSCE in 1993 and the UN Disarmament Commission in 1996, it was not until 1998 that a genuine multilateral effort to improve accountability in export criteria was established in the form of the European Union code. NATO and the OSCE have lauded the principles in the EU code and have noted the need to strengthen provisions for humanitarian law.[30]

From codes of conduct to advocacy

Recognition of the limitations of international humanitarian and human rights law with respect to controlling the legitimate production and trade in small arms has prompted the development of a host of alternative strategies to reduce small arms transfers to abusive regimes and non-state actors. A prominent tool commonly used bilaterally by states and multilaterally by the UN is that of *sanctions*. Needless to say, even after decades of experience in their design and application, sanctions remain extremely controversial. One school prefers targeted or 'smart' sanctions to minimize the disproportionately negative impacts on ordinary civilians rather than the intended targets (BICC, 2001). But critics have insisted that embargoes are a substitute for more resolute action, excusing states from their legal and moral obligations to actively punish abusive regimes.

Other proponents of the supply-side approach have sought to 'name, blame, and shame' actors that are known to violate human rights. Both HRW and Amnesty International (2001) have exhaustively documented illicit small arms transfers to abusive states and armed groups, identifying perpetrators and publicly denouncing those responsible. They highlight the 'morally unconscionable' actions of states that arm rights abusers. By harnessing the media, such approaches have served to embarrass violating parties, tarnishing their images, and encouraging a change in behaviour. The UN (2000) tried the same approach through investigations like the Fowler Report on assistance to UNITA in Angola. By focusing on a 'surrogate villain'—the political sponsor or the arms supplier—human rights organizations seek to call attention to abuses and exert pressure. It is worth recalling, however, that the key policy instrument of these efforts—stigmatization—is useless against perpetrators who are independent of major powers or act outside of customary networks susceptible to economic or political leverage (Leaning, Briggs, and Chen, 1999, p. 199). Often such approaches are rendered impotent if a 'surrogate villain' cannot be identified.

New humanitarianism: Protecting civilians

A second broad approach, adopted by key agencies of the UN, the ICRC, and major international relief agencies, aims to heighten international and domestic consciousness of the impacts of armed violence on non-combatants and vulnerable groups. Referred to here as the 'new humanitarian approach', its proponents acknowledge that countries at war are saturated with small arms and aim to reorient attention downstream, to more operational and demand-driven field activities focused on affected civilians and ill-informed combatants.

Humanitarian agencies are alarmed at the legal and operational implications of civilian small arms possession and the fact that they are often primary targets in armed conflict. They are concerned that the availability of small arms, particularly in the hands of poorly-trained and undisciplined civilians and combatants, acts as a multiplier of violence and threatens the foundations of international humanitarian law (ICRC, 1999). They also recognize that international humanitarian law 'is often the body of law most relevant to the stated purpose for which military arms and ammunition are transferred—to fight an armed conflict' (Herby, 1998, p. 60).

With new wars frequently characterized by a lack of discipline among belligerents and blurred lines between combatants and non-combatants, there seems to be little purchase for the rule of law. Paradoxically, a heightened awareness of the dynamics of these wars has increased consciousness of the need to enforce and reinforce Common Article 3 of the Geneva Conventions and its Additional Protocol 2, which apply to internal armed conflicts requiring all parties to respect those who have laid down their arms and those, such as civilians, who are not taking part in hostilities. Supporters of the new humanitarian approach are adamant that even where a state's justice system is undermined there should not be a legal vacuum with respect to international law. On the contrary, although it is difficult to apply humanitarian law in such contexts, they believe it is precisely in such situations that humanitarian law is most urgently needed.

Application of the rules of law alone will not solve the underlying causes of conflict but it can help minimize the humanitarian tragedy of small arms. To reduce the vulnerability of civilians, humanitarian agencies seek to disseminate information to warring state and paramilitary forces in an attempt to influence their behaviour. The ICRC, for example, has recommended education of armed forces about the basic norms of international humanitarian law as a vital first step to reducing civilian casualties. They have also sought to better apprehend the underlying causes of atrocities against civilians in order to improve their own interventions. Even as they acknowledge the complex motives underpinning small arms use, some development agencies have tried practical interventions, combining awareness and development incentives with voluntary disarmament programmes (WEAPONS COLLECTION). With combatants ignorant or deliberately contemptuous of international humanitarian law, the role of education is potentially far-reaching.

Protecting civilians from small arms-related violence relies in large measure on the ability of the international community to adapt its normative basis for action to the changing reality of conflict. Growth of humanitarian interventions

COMMENTARY 4.1 The limits of humanitarian responses to small arms

Once upon a time, humanitarian action was predicated on the Laws of War. The control of weapons was dominated by the lawful forces of order. Politics was moderated by decisions about insiders and outsiders. Religious or charitable organizations dealt with the ravages of those decisions. Even if this nostalgia for a simpler time isn't entirely accurate, it reflects what was once a clear division of labour. Just as NATO was not established with a mandate to build or protect refugee camps, so UNHCR's mandate was for the legal protection of refugees rather than operational intervention. The lines were drawn; soldiers and lawyers were clearly demarcated.

The old division of labour is gone. Not only are the lines between combatant and non-combatant blurred, so are the differences between war and criminality. As a result, the humanitarian space has become at once confused and expanded. The word 'humanitarian' has two meanings: the first deals explicitly with the laws of war, the second with humanizing and redressing acts of violence. But the twin objectives aimed at ensuring that the laws of war are followed and the rendering of violence more humane have brought humanitarianism and human rights together. Distinctions between human rights and humanitarianism have become minute: the High Commissioner for Human Rights and the High Commissioner for Refugees talk of complementary mandates.

Partially as a result of these changes, humanitarian protection has evolved from being driven by international humanitarian law to being motivated by the needs of specific groups of victims. Humanitarian protection is understood to go beyond the narrow role provided for in the Geneva Conventions, from protection mandated by law to protection based on perceived need. Perhaps prematurely, the humanitarian community has become involved in conflict prevention and small arms regulation, just as UNHCR increasingly become more involved in the provision of physical protection, assistance, and development for refugees and internally displaced people.

These observations give warning that the relationships between small arms availability, humanitarianism, and human rights are not obvious. To say that the availability and use of small arms threatens communities and human security is not the same thing as saying that the humanitarian or human rights community is competent to regulate those weapons. For most of the twentieth century, impartiality and independence have been core to humanitarian action. These principles were relatively easily applicable to treaty-based organizations like the ICRC in traditional contexts, but the current politicization of violence also includes the politicization of those organizations dealing with violence. If we assume that impartiality and independence should remain cornerstones of humanitarian action, one of today's challenges is to preserve their relevance against modern violence. Can humanitarian agencies get fully involved in small arms issues without taking sides? If they refuse to take sides, can they be effective?

Adding the issue of small arms to the humanitarian agenda could catalyse an important re-evaluation. What are the political implications of linking humanitarianism and small arms? More than 100 years ago there was a political decision to split humanitarianism from politics. On what basis was that decision made? Today, we sense a concerted movement toward combining humanitarianism with the use of force: from preventive diplomacy and conflict prevention to the regulation of arms in post-conflict environments. Until we have a better handle on the politics of this shift and the consequences of these decisions, the presumed relationships will remain naive and simplistic. Until then, to say that the humanitarian community should play a central role in small arms regulation is neither necessary nor sufficient.

Source: Warner (2001)

in the 1990s was a product of what has been called an 'Age of Innocence', when optimism peaked following the end of the Cold War with a deepening conviction in the broadening possibilities for humanitarian action. This brief era came to an end with genocide in Rwanda, ethnic cleansing as in Srebrenica (Bosnia), and other horrific tragedies of the 1990s.

But as the nature of human confrontation changed so too have the means by which outside actors respond to inhumanity. At the beginning of the twentieth century, humanitarian action was an impartial exercise epitomized by the ICRC. It started the twenty-first century as a political process in which the effectiveness of short-term humanitarian interventions was measured against the extent to which they contributed to the protection and promotion of human rights in the long term (see Commentary 4.1). These changes were partly a response to accusations of letting war crimes take place and the belief that aid often did more harm than good. Some agencies, however, are less inclined to adopt the new agenda. Michael Ignatieff (1997, p.127) reminds us that 'the ICRC cannot end the flow of arms from Pakistan and Russia [into Afghanistan]; and can't enforce a cease-fire—it can only get factions to observe some basic rules'. These rules include (1) not shooting at the wounded, (2) not attacking ambulances, (3) not targeting hospitals, (4) not attacking civilians, and (5) not torturing prisoners.

Though the ICRC is not in a position to adapt its mandate—it exists to enforce the laws of war and not human rights—other humanitarian actors have taken up a more political agenda. The politicization of humanitarian action has occurred amid considerable controversy: attempts to deliver assistance and publicize atrocities have rendered the act of assistance more difficult. It has become dangerous for many organizations to simultaneously obtain access to victims of small arms abuses while denouncing those who hold the keys to their right of entry.

It is important to stress the radical departure of this new humanitarian agenda from traditional approaches to providing assistance and protection. Notions of aid conditionality and political management mean taking sides, with profound implications for humanitarian assistance generally. The approach took shape in Afghanistan in the 1990s, where assistance was withdrawn as a result of the Taliban's denial of women's rights. It also took place in the Great Lakes region of Africa, where aid agencies withdrew from refugee camps because of the way 'genocidaires' were being assisted to re-invade Rwanda. At the most extreme, it has also resulted in some agencies abandoning a pretence of impartiality, even expediting the arming of non-state actors.

The diversion of small arms and ammunition through relief activities is not new. Accusations have been levied by governments against religious and special interest NGOs from Central America (e.g. the Catholic Church) and west Africa (e.g. World Vision), to the Balkans (e.g. Serbian Red Cross), and the Horn of Africa (e.g. Norwegian Church Aid). Burundian rebels in Tanzania often drew on aid to fund or support their cause. Aid workers in the region admitted that '[their] infrastructure improvements... had become a vector for all kinds of transport... [they were] aware that arms traffickers used [their] UN routes' (Muggah and Berman, 2001). As a result of the threats posed by small arms, a fundamentally new chapter in humanitarian action has been opened.

Coping with humanitarian realities in the field

The third humanitarian approach is described here as the operational perspective. It is a reaction to the humanitarian impacts of arms availability on the quality and effectiveness of international relief and development operations. It stresses the need to improve protection for humanitarian and peacekeeping operations within a deteriorating security environment. It also recognizes that relief work takes place amid unprecedented tension and insecurity. Civilians and those sent to protect and assist them are increasingly viewed as legitimate targets for threats, theft, extortion, and assault at the hands of armed attackers. Instead of trying to reform the entire situation, this approach seeks pragmatic, field-level responses to allow the work to go on.

Risks to humanitarian personnel

Although recognized as a problem for over a decade, many humanitarian organizations have only recently begun documenting and collating information on staff deaths and injuries. The impacts of small arms availability on relief and peacekeeping personnel range from the explicit targeting of staff by warring belligerents and criminals to the opportunity costs attributable to relief aid diversion and unproductive expenditures on security, transport, and logistics. In the early twenty-first century, the UN considered over 50 countries insecure. The number of complex emergencies had increased fivefold over the two decades. Though annual death rates peaked in 1995, approximately 1,500 international and national peacekeepers have been violently killed since 1945 (Seet and Burnham, 2000). A large number of these deaths have been caused by small arms (see Figure 4.4 and Appendix 4.4).

A retrospective study of death and fatal injuries among humanitarian workers over the last decade observed that among NGO, UN, Red Cross, and ICRC personnel almost 70 per cent of those killed were fatally wounded by 'intentional violence'. These deaths should be compared with the 17 per cent killed in vehicle accidents.[31] Available evidence also shows that the firearm homicide rate for UN civilian staff is between 17 and 25 per 100,000 (Muggah and Berman, 2001).[32] Between 1990 and 2001, over 100 ICRC delegates were violently killed with small arms and more than 280 injured. Trends collated from medical journals suggest that violent deaths among humanitarian workers are increasing and that: 'banditry was an important cause of death [with] most victims dying in crossfire or cold blood' (Sheik *et al.*, 2000).

Over the last decade, almost 70 per cent of UN, Red Cross, UN, and ICRC personnel killed were fatally wounded by 'intentional violence'.

FIGURE 4.4 Causes of UN civilian death from hostile actions, 1992–2000

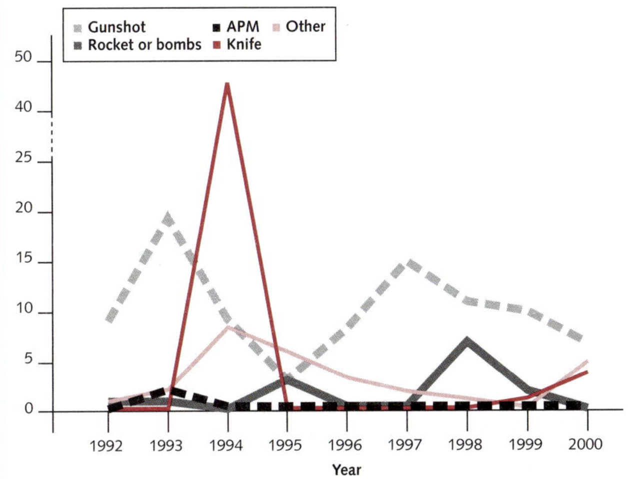

Source: Muggah and Berman (2001); Appendix 4.4

In an influential study, Martin Sheik *et al.*, (2000) examined the causes of death among aid workers worldwide between 1985 and 1998. Overall, deaths attributable to intentional violence increased while those from vehicle accidents fell. Of the 375 deaths analysed, over half occurred in the Great Lakes and Horn of Africa, peaking in 1994 with the Rwandan genocide. Sheik *et al.* (2000, p. 167) record that the 'numbers of deaths among UN peacekeepers and programme staff broadly follows the changes in the number of refugees and asylum seekers worldwide, providing an indirect measure of the prevalence and violence of conflict'.

It is important to note, however, that the sheer volume of humanitarian agencies operating in dangerous environments increased over the same period: in Rwanda and Haiti there were more than 200 and 800 NGOs operating respectively (Seet and Burnham, 2000). What is more, an increasing number of missions are operating in conditions of extreme insecurity and without adequate protection ordinarily afforded under international humanitarian law. While these findings highlight the violent circumstances in which humanitarian workers increasingly find themselves, they do not necessarily indicate a growing trend in homicides.

Between 1948 and 1988, at least 744 'blue helmets' lost their lives in UN missions. Over the next decade the total number of peacekeepers killed almost doubled (Seet and Burnham, 2000). Hostile acts reportedly accounted for more than 36 per cent of all deaths. The risk of death from hostile acts increased while that from accidents decreased. Indeed, 'in missions providing or facilitating humanitarian assistance, both the RR [risk rate] of deaths from all causes and deaths from hostile acts increased' (Seet and Burnham, 2000, p. 598). By early 2000 there were more than 30,000 uniformed peacekeepers and civilian UN personnel deployed in 17 peacekeeping missions: '[t]here was a 1.5 times greater risk of death from hostile acts after the cold war; risk factors included African missions, assertive peacekeeping operations and missions involving humanitarian assistance' (Sheik *et al.*, 2001, p. 300).

Small arms are a day-to-day reality in post-conflict environments.

Personal safety and security is a major preoccupation for expatriate and local staff working in violence-prone areas. The ICRC (1999) estimates that approximately 50 per cent of its international and national staff suffer from emotional and behavioural difficulties during and following their assignment, while an estimated 30 per cent have endured a serious security incident in the field. The stress of working in situations where personal safety is continually jeopardized contributes to the potential for psychological trauma. Whether real or perceived, insecurity adversely influences the productivity of relief and development interventions. Surveys can generate vital data on related risks and vulnerabilities (see Box 4.13).[33]

TABLE 4.2 A sample of violent deaths among peacekeeping and civilian personnel and its distribution

Region	Peacekeeping (1948–98)*	UN civilian (1992–98)
Africa	279	128
East Europe	88	14
Middle East	165	9
Asia	31	20
The Americas	0	5
Total	**563**	**176**

* *A sample of peacekeeping deaths drawn together by Seet and Burnham (2000).*

Sources: Muggah and Berman (2001); Seet and Burnham (2000)

Ensuring the safety of humanitarian personnel

If the safety of personnel cannot be adequately ensured, assistance and protection cannot be effectively provided. According to a UNSG report (2000), there are shared security threats to all humanitarian actors working in the field: 'Threats against NGO staff can... directly affect UN humanitarian and assistance programmes, especially since conflicting parties often do not distinguish between UN and NGO personnel'. This reality has spurred the UN and the ICRC into responding to the humanitarian impacts of small arms at the field level. But the extent to which all relief and development aid agencies are genuinely acting to reduce security risks is debatable.

Some aid organizations are moving quickly to improve their safety and security management (see Box 4.14). There are a variety of internal initiatives and a few inter-agency projects under way, mostly related to incident reporting, research, training, advocacy, and inter-agency co-ordination. The fact remains that there are still far too many aid agencies where senior management fail to see the problem or do not feel responsible (Van Brabant, 2001a, p. 17).

While most of these initiatives are well conceived and executed, too many are *ad hoc* and improvised. In most cases, they do not build on problem analysis, evidence-based management, or sophisticated advocacy planning. As a result, there are critical weaknesses, confusion, and unnecessary duplication in many agency approaches. Many NGOs often confuse security management with the production of a security plan that in practice is little more than an evacuation strategy (Van Brabant, 1997).

BOX 4.13 Opinion surveys among Oxfam field workers

Between December 2000 and February 2001, the Small Arms Survey and Oxfam-GB conducted a pilot survey of small arms-related risks to Oxfam field staff in over 20 countries. Sixty per cent of the 94 respondents described their place of work as 'severely affected by systemic violence'. Field staff were keenly aware of the insecurity generated by small arms, whether by reducing personal mobility or by reversing relief and development interventions in the field.

Small arms were viewed by many as ubiquitous in their areas of operation and correlated with an increased exposure of civilians and personnel to violent threats; automatic rifles and handguns were identified as the single largest contributor to death and injury among civilians. What is more, over 40 per cent of all responding Oxfam-GB staff claimed to have themselves experienced a security incident involving a weapon: whether a non-fatal injury, armed intimidation, banditry, or kidnapping at gunpoint.

The Small Arms Survey, together with the Centre for Humanitarian Dialogue in Geneva and the John Hopkins University's Centre for Refugee and Disaster Studies, will be extending the survey to an internationally representative sample of humanitarian and development agencies during 2002–06.[34]

Threats to Oxfam-GB staff involving small arms

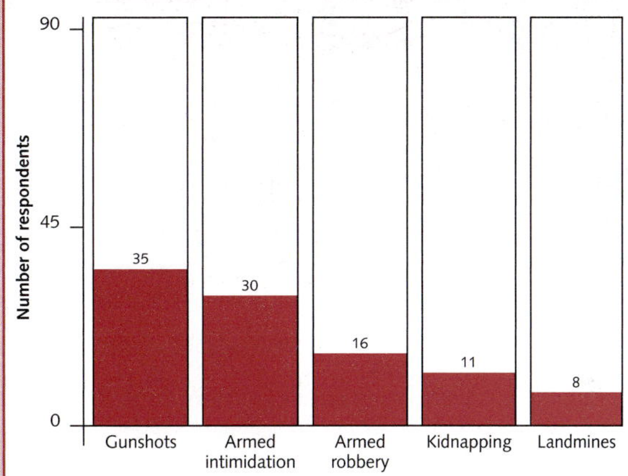

Source: Muggah (2001b); Appendix 4.5

What next: Military protection?

Whereas during the Cold War aid agencies accommodated national militaries or insurgent groups, a shift has taken place, as humanitarian agencies debate their relationships with peace-support operations. These new operations represent an extreme departure from the humanitarian principles of neutrality and impartiality, and have been labelled active humanitarianism and even partisan humanitarianism.

Roberts (1998) identifies three motives for ensuring military protection or military involvement in humanitarian operations: (1) murder and deliberate infliction of suffering on civilians, prisoners, and others; (2) refusal of parties to a conflict to allow access or assist humanitarian relief activities; and (3) violence and threats of violence against humanitarian workers. While this chapter does not explore in great detail the controversial debates over military intervention in humanitarian emergencies, it will briefly review the relationships between the use of military protection and humanitarian mandates.

Aid organizations are divided on whether to use military protection. UN agencies are generally more prepared to co-operate with national or international security organs and operations (see Box 4.15). Peacekeeping is, after all, a political action: deployed on the basis of UN resolutions agreed by governments and mandated to take explicit political action against parties to a conflict. The ICRC, on the other hand, prefers distant engagement with military actors. The organization needs to actively engage with the armed forces as part of its efforts to disseminate and strengthen respect for international humanitarian law. It rarely dwells on the politics that condition military deployments or operations and seeks to maintain a clear and perceivable distance to safeguard its own image of impartiality and neutrality. Though it has been forced to use armed protection in regions such as Somalia, it nevertheless remains committed to the principle that guns draw gunfire (see Box 4.16).

By contrast, international NGOs have been more opportunistic and *ad hoc* in their dealings with the institutions of the military. In some cases, NGOs have called for international military interventions to re-establish security in a bandit-infested country—such as Somalia in 1992—or to separate armed elements from refugee populations—as in eastern

BOX 4.14 Safety, security, and protection

For the purposes of managing field operations, it is useful to distinguish between safety, security, and protection. *Safety* is used here to mean dealing with risks from accidents and disease. *Security* relates to the threat of acts of violence against aid agency staff and assets. *Protection* refers to the threat of acts of violence against civilian populations and non-combatants. Safety and security management is a concern for all organizations, including humanitarian and development agencies. This often is under-appreciated. Development organizations tend to lag in awareness and actions to manage staff safety and security. Multi-mandate organizations that have both relief and development objectives tend to locate the security competence within their emergency or disasters departments. This approach seems outdated now that crime accounts for almost 50 per cent of all violent incidents, and is a risk in many settings where there is no active war-type conflict.

Security management is not an exact science: it is about reducing risk by reducing the threat and/or reducing the impact. There is a tendency to seek guidance for good security management in the form of a security plan, a security checklist, or standard operating procedures. This is partially valid—but also misplaced. There are some relatively straightforward dos and don'ts in security management, such as seeking solid cover when caught in the crossfire—car doors are solid only in Hollywood—or never handling a suspected mine or object that might be booby-trapped.

Security management in violent environments is predominantly about building relationships and accessing information, continuous monitoring, analysis, and interpretation of an evolving scenario, and situational judgement when a threat presents itself. Nothing can be quite as dangerous as failing to scan your environment because you have developed a security plan and are adhering to it.

It is a serious misconception to believe that individuals with police or military experience are always the most qualified to manage the security of aid agencies. It would also be a misconception to hold that they have nothing to offer. People with a background in security forces often have a sense of awareness, discipline, and rigour in how they approach security management that is lacking among civilian aid workers. Aid agency security requires a comprehensive and holistic approach. It stands to reason that a person with a military background may be more effective in dealing with Russian forces in the Caucasus but at a loss when confronted with the dynamics of a tribal segmentary lineage system, as in Somalia.

Source: Van Brabant (2001a)

Zaire, 1994–95. They have called for a more assertive role among military peacekeepers to provide protection to civilian populations under armed attack from brutal militias—as in Bosnia and Sierra Leone. In other cases, some NGOs have used the military, particularly their logistical capabilities, for transport or evacuation, as well as for protection (Van Brabant, 2001b).

The question is not **if** *humanitarian agencies should engage with armed actors, but how and when.*

Many NGO staff are not only critical of but outwardly hostile towards the military. Discomfort with the use of force is frequently a product of latent anti-militarism. Often, however, these criticisms emerge from a legitimate criticism of the political and economic agendas of internationally deployed soldiers. For example, it is no secret that most humanitarian agencies were deeply uncomfortable with NATO's war against Serbia in 1999. Extensive criticism preceded and followed the war for several reasons: it had not been authorized by the UN Security Council; it had provided ineffective and uneven protection for Albanian Kosovars; and it had more to do with avoiding refugee flows into the West and the stability of south-eastern Europe than with the protection of human rights. Other criticisms have addressed the misconduct of national and international military, including their use of torture (in Somalia), rape (Guinea), smuggling of drugs (Afghanistan), hardwoods and antiques (from Cambodia and Myanmar through Thailand), extra-judicial executions (West Africa's Economic Community Monitoring Group, ECOMOG, in Sierra Leone) and their lack of discipline or accountability.

BOX 4.15 Cops and humanitarians: Uneasy bedfellows

Faced with increased threat of armed attack, almost always from small arms, humanitarian agencies have resorted to armed protection, military guards, and private security company services to reduce risk. All of these actors are potentially uneasy bedfellows for aid agencies. What impact does their participation have on how humanitarian agencies or actions are perceived? By hiring these agencies, are humanitarian actors contributing to the proliferation of arms and the privatization of violence?

Not surprisingly, these developments have generated a vociferous debate within the humanitarian aid community about the circumstances in which such strategies would be acceptable and to what degree. No common position can be discerned between aid agencies. Furthermore, the debate is likely to continue because risk patterns on the ground are evolving, as are the politics and doctrines of peace-support operations, codes of conduct, and legislation pertaining to private security companies.

One way of systematically thinking through the decision of whether or not to obtain armed protection would be to consider sequentially: principles, context, and management questions. In the case of armed protection, key principles are raised by the following questions: (1) Can the use of force be justified under certain conditions? (2) Does armed protection run against the basic principles of humanitarian action, or is it only an inevitable response to the conditions in which humanitarian action has to take place? (3) Would armed protection contribute to an arms race? (4) Can paying for armed protection be justified?

While it is possible to defend absolute non-violence, others credibly argue that force used in a principled, targeted, measured, and accountable manner can be justified against a direct violent threat or to obtain compliance with basic standards of human rights and international humanitarian law. Even stable, secure, and democratic societies have police forces, armies, and reserves. If it is acceptable to pay for these services through tax contributions, it might also be acceptable to pay for armed protection where the official law and order forces are non-existent or unable or unwilling to carry out their responsibilities. Force, then, can be for the public good and enhance not only the security of aid workers but also the protection of civilians.

There is little doubt that aid agencies prefer to totally avoid armed protection and need to explore all other means of securing access and providing assistance and protection before resorting to armed protection. But extreme situations may arise where many lives are at risk because insecurity prevents the delivery of assistance. At this stage, the choice is between resorting to armed protection to deliver assistance or preserving the unarmed image of humanitarian intervention. In practice, this can be a difficult judgement call.

These arguments demonstrate that, whereas there is some latitude for decisions on the basis of principle, *contextual* considerations are particularly relevant. A first question to consider is whether one should even remain in a particular environment if one feels threatened to the extent that armed protection is considered. But if one wants to stay, out of a sense of mandate or solidarity, the key questions become: (1) Who provides the armed protection? (2) Does armed protection reduce or increase risk? (3) Who benefits from armed protection?

The question of the provider is very important: Will it be the national police or army, an insurgent group or militia, armed guards hired directly by the agency, tribal warriors, members of an international peacekeeping force, or personnel provided by a private security company? All of these actors have a particular image in any given context and, like the aid agencies, are part of its political economy. How will an association with one or other provider be perceived by other actors in any particular context? In other words, what will it do to one's image and reputation? What is more, armed protection needs to provide an effective deterrent. Being seen to be carrying arms is likely to escalate a situation: bandits shoot before they loot. If you are likely to be outnumbered and out-gunned, armed protection is likely to increase risk rather than provide adequate protection.

Finally, is it only the agency that benefits from armed protection, or is there a significant public good to be gained from it? Does its enabling effect on humanitarian assistance provide benefits to a substantial number of people? Can the armed protection be used to serve a broader public good, for example by being deployed to provide security to a refugee camp rather than just an agency compound or warehouse, or to secure safe passage for all on a given road rather than just an agency convoy?

If the conditional use of force is acceptable in principle—and careful analysis has indicated that armed protection is the only means left to maintain a humanitarian operation on which a significant number of civilians are dependent, and an acceptable provider is available—there remains the question of its management. Whether armed guards come under the direct management control of an aid organization, or whether they are provided by external security forces with their own command and control procedures, fundamental management questions must be resolved. For example: (1) Who has authority and responsibility for what, notably with regard to the selection of armed guards, the choice of weaponry, rules of engagement, supervision, and disciplinary action? (2) What ethical and legal framework will apply to the armed guards and what personal code of conduct? (3) Who provides the armed guards with their equipment and who is responsible for their food, accommodation, and transport?

Source: Van Brabant (2001b)

Box 4.16 Humanitarian security threats in Somalia

> ... Following the firing of AK-47s on a single-engine 9-seater Caravan (with no injuries or damage sustained to the aircraft), plans to resume work in Kismayo anytime soon are severely curtailed... Kismayo is off-limits for all UN operations until it becomes clear who carried out the attack and until the threat has been removed. Elsewhere, in Merka (70 km south of Mogadishu), the UN has also been forced to suspend flights. (IRIN, 2000a).

In September 2000, security officers in Somalia, conducting a survey to determine whether UN agencies could resume humanitarian operations after a six-month suspension, 'were attacked at the UN WHO compound by a group of 30–40 men bearing small arms... UN programmes had ceased after unidentified militiamen sprayed gunfire at a European Community humanitarian plane.' Other incidents in Merka include a bomb discovered on the roof of the WHO compound, a grenade attack at both the WHO compound and the offices of the Coordinating Committee of the Organization for Voluntary Services, an Italian Agency. According to IRIN (2000c), 'attempts to impose curfews, gun control and disarmament have been limited—as "there is no authority here"'.

Conclusion

Many NGO staff are not only critical of but outwardly hostile towards the military.

Small arms availability is not the exclusive cause of conflict. It is, however, the primary mechanism by which combatants and civilians violate international humanitarian law and human rights law on a massive scale. Evidence from many peaceful, conflict-affected, and post-conflict societies suggests that armed violence, criminality, and displacement increase where there is an abundance of small arms. Every year, at least 500,000 men, women, and children are violently killed as a result of small arms; many millions more die from their untreated injuries and secondary illnesses. The devastating toll of recent internal wars has prompted much discussion of the nature of post-Cold War conflict and the relative contribution of small arms. As these types of weapons become more readily available, complex emergencies, internal conflicts, and state collapse have exacerbated the scale and pace of human suffering. The growing availability of such weapons in societies embroiled in or emerging from war poses a long-term challenge to the humanitarian community.

The humanitarian impacts of small arms are much more significant than current data and research indicate. Recording the humanitarian impacts of small arms on death and injury rates, on expanding patterns of forced displacement, and on the eroding access of civilians to their basic needs, is an urgent priority. These impacts are significant for, as Albert Schweitzer reminded us almost 40 years ago, pain and suffering constitute a bigger burden than even death itself (Anderson, 1965).

The unregulated availability of small arms threatens the physical safety and security of humanitarian personnel and agencies working in the field. Due to the increasing perception of risk in the field, already scarce resources have to be diverted to security management, logistics, monitoring, and evaluation. The presence of large quantities of small arms also contributes indirectly to a culture of intimidation, violence and, consequently, humanitarian withdrawal. Pervasive arms-related insecurity hampers a return to stability or human security, let alone an environment conducive for reconstruction or development.

States have primary responsibility for preventing violations of international humanitarian law. To reduce such violations, armies need to be under the command of trained and disciplined leaders. Furthermore, militias must be disarmed, unregulated weapons must be collected, and the violence that has spread through the social fabric must be tamed. The development of norms with respect to international humanitarian law and the formulation of international standards are necessary first steps to make states and non-state actors more accountable in relation to the flow of these weapons. But as Chapter 6 (CONTROLS) demonstrates, legislation to control unrestrained small arms flows is still missing, or in the early stages of development, in many parts of the world.

A humanitarian approach to small arms demands that activities be initiated both at the top of the policy-making spectrum and in the field where aid operations take place. On the one hand, states need to adopt more stringent codes of conduct on small arms exports, especially to governments with poor human rights records and to non-state actors. On the other hand, the new humanitarians must intensify their efforts to protect the rights of civilians whose lives have been disrupted by the spread of small arms. It is also critical for humanitarian actors to collect more evidence of the impacts of small arms from their work in the field. While new humanitarians recognize that their work has shifted from a charitable or philanthropic effort to one that acknowledges the role of politics and the centrality of humanitarian law, they must also balance their approach to small arms with goals that can realistically be achieved.

A humanitarian perspective on small arms provides a critical space for achieving consensus in an otherwise politicized arena, in which disarmament groups, gun lobbyists, and international lawyers argue against each another. Though disagreements over how best to approach the issue of campaigning action on small arms persist within the humanitarian sector, all parties basically agree that a humane approach to small arms is desirable. Such an approach would aim to reduce the risk to civilians of the humanitarian impacts of armed violence and reduce or eliminate their exposure to violations of their basic rights with these weapons. In essence, a humanitarian perspective enables us to focus on the practice of warfare rather than the narrative of war. It compels producer states to account for the legal or illicit transfers of weapons to those regimes violating the basic human rights of civilians or to countries affected by conflict. It also usefully focuses attention on the needs and rights of all people who face arms-related violence on a daily basis, whether in situations of war or of peace.

4. List of Abbreviations

ADOL	Action for the Development of Local Communities
APM	Anti-personnel landmine
CMR	Child mortality rate
DALY	Disability-adjusted life years
DRC	Democratic Republic of Congo
ECOMOG	Economic Community of West African States' Cease-fire Monitoring Group
FAL	Fusil Automatique Leger
FARC	Revolutionary Armed Forces of Colombia
FYU	Former Yugoslavia
HPCR	(Program on) Humanitarian Policy and Conflict Research
HRW	Human Rights Watch
IANSA	International Action Network on Small Arms
ICC	International Criminal Court
ICD	International Classification of Disease
ICECI	Injury Classification of External Causes of Injury
ICRC	International Committee of the Red Cross
IDP	Internally displaced person
IOM	International Organization of Migration
IPIFA	Injury Prevention Initiative for Africa
IRC	International Rescue Committee
ITAR	International Traffic in Arms Regulation
LTTE	Liberation Tigers of Tamil Eelam
MSF	Médecins Sans Frontières
NATO	North Atlantic Treaty Organisation
NID	National immunization day

OSCE	Organization for Security and Co-operation in Europe
PTSD	Post-traumatic stress disorder
RGSA	Reference Group on Small Arms
RUF	Revolutionary United Front
SCF	Save the Children Fund
SPLA	Sudanese Peoples Liberation Army
UNSECOORD	UN Security Coordinator
UNDP	United Nations Development Program
UNGA	United Nations General Assembly
UNHCR	United Nations High Commissioner for Refugees
UNICEF	United Nations Children's Fund
UNITA	União Nacional para a Independência Total de Angola
VIP	Violence and Injury Prevention Programme
WHO	World Health Organisation
YPLL	Years of potential life lost

4. Endnotes

1 The variables highlighted in Table 4.1 were judged priority issues during consultations with literally hundreds of victims of armed violence, relief and development personnel, policy-makers, and researchers. Objective criteria are important in order to provide a basis for their application in relation to humanitarian law and to inform practical and appropriate humanitarian intervention.

2 A US Department of State (2000) report noted that an estimated 200,000 persons in and near Freetown were made homeless in the January fighting. It also claimed that 3,800 children went missing or were abducted and that by November there were 2,400 still registered as unaccounted. The United Nations stated in the same report that rebel forces abducted approximately 20,000 persons throughout Sierra Leone between 1991 and 1999.

3 According to WHO (1999), the mortality rates for suicide, homicide and war were 16.1, 12.5 and 10 per 100,000 respectively. For more detail, consult the WHO *World Health Report* web site <http://www.who.int/whr>

4 The issue of data availability in countries at war is treated by Murray *et al.* (2002). See also the work of SIPRI (2001), Monty Marshall at the Center for Systemic Peace (2001), the International Peace Research Institute (2001), and the Post-Conflict Unit (2001).

5 In order to bridge gaps in national health data, the Small Arms Survey and the World Health Organization's Department of Violence and Injury Prevention Programme (VIP) have obtained support to launch a substantial three-year initiative in 2002 to generate high quality surveillance and survey data on firearm mortality and morbidity in ten developing countries.

6 Kill–injury ratios are low in developing countries because of the absence, limited access to, and poor quality of existing medical care available to those who are wounded by small arms. It is therefore more likely that a greater proportion of those injured by firearms will die as a consequence of their injuries.

7 YPPL relates to the loss of individual and household productivity associated with firearm injuries. It is calculated by multiplying the number of deaths at a given age by the years that persons of that age would be expected to live and summing the total number of years lost. A broader spectrum of morbidity costs can be determined by costing rates and duration of hospitalization, emergency room visits, and temporary and permanent disabilities. See Ordog *et al.* (1995) and Max and Rice (1993) for more on applying different economic methods—for example, human capital or incidence prevalence approaches—to costing the impacts of injury.

8 Beginning in January 2002, the Small Arms Survey will be carrying out two separate projects on fatal and non-fatal injury attributed to small arms. Together with the Injury Prevention Initiative for Africa (IPIFA), the Survey will be assessing the economic costs of firearm injuries in a representative sample of urban and rural clinics in Kenya, Uganda, and Ethiopia. Starting in the middle of 2002 and in partnership with the WHO's VIP, the Survey will also be assessing the social and economic costs of small arms-related injury over a three-year period in ten developing countries.

9 No study has yet sought to systematically review the costs of war on public health through the International Classification of Diseases (ICV) categories of homicide (ICV 960–969) and the operations of war (ICV 990–999). For a review of different approaches to measuring injury, whether through death certificates or in-patient and physician data, consult the Injury Classification of External Causes of Injury (ICECI) <http://www.iceci.org>

10 Garfield (2000: 2) has criticized these two ICRC studies for methodological limitations and also for their inability to provide conclusive evidence of the level of injuries in the population or their causes. He also notes that 'neither study provides any information on people injured who did not reach an ICRC hospital'.

11 A report released by the Action for the Development of Local Communities (ADOL), an NGO that conducted research on small arms between April and June 2001, indicated that Gulu (Uganda) was a major source of guns and ammunition (IRIN, 2001b).

12 Conversation with Dr Olive Kobusingye of the Injury Control Centre in Uganda.

13 Summerfield (2000), however, mounts a substantive critique of the Western-dominated literature on PTSD. He argues that the relationship between traumatic experience and outcomes is not clear-cut. See also Summerfield (1996).

14 CASA, a Canadian firm, was contracted to carry out an evaluation of UNHCR programmes in Dadaab and Kakuma, Kenya, in 2001.

15 See, for example, Muggah (2001d); Moser and McIlwaine (2000); Moser and Holland (1997); Samaranayake (2001).

16 See, for example, the International Coalition to Stop the Use of Child Soldiers web site <http://www.child-soldiers.org>

17 A new treaty banning the use of children in combat took effect from February 2002. See, for example, <http://www.hrw.org> for details.

18 See, for example, the bibliography on human security prepared by Harvard University's Program on Humanitarian Policy and

Conflict Research (HPCR) <http://www.hsph.harvard.edu/hpcr/events/hsworkshop/bibliography.pdf>

19 The UN Security Council authorizes the use of force across borders only where personal insecurity within a given state is of such a magnitude that it poses a threat to 'international security'.

20 According to Bok (1999, p. 188), in nineteenth-century England the 'adjective "humanitarian" was used... in a manner nearly always contemptuous, connoting one who goes to excess in humane principles. The word conveyed deep-rooted suspicion, unlike such words as "humane", "kindly", or "good".'

21 In the context of humanitarian intervention, neutrality requires that a particular actor is not party to an armed conflict and refrains from taking a position on its causes or the motives of actors. Impartiality mandates that assistance be delivered without discrimination and according to the needs of the victims. These are not necessarily complimentary objectives and indeed, can often be contradictory.

22 Including the 1995 Review Conference of the 1980 UN Convention on Conventional Weapons where battlefield electro-optical devices—that is, blinding laser weapons—were banned.

23 See, for example, the Tribunal web site <http://www.un.org/icty/judgement.htm>

24 With respect to the relationship between human rights and small arms, there is also a range of additional instruments and binding treaties whose provisions apply in armed conflict, including the 1951 Convention Relating to the Status of Refugees, the 1989 Convention on the Elimination of all Forms of Discrimination Against Women, the 1984 Convention Against Torture, and the 1989 Convention on the Rights of the Child.

25 Particularly Common Article 3 of the 1949 Geneva Convention and Additional Protocol.

26 See, for example, Lumpe (2000) <http://www.nisat.org/publications/armsfixers/>

27 South Africa, for example, adopted a revised code of conduct in 1996 that sought to screen exports of weapons in light of human rights and security criteria of the receiving state.

28 In the US, the International Traffic in Arms Regulation (ITAR) includes a list of proscribed destinations, including countries under UN Security Council arms embargoes, or countries where there is chronic warfare, serious human rights abuses, or state sponsorship of terrorist activities. By 1995, a US Code of Conduct and Arms Transfers Bill was introduced to Congress, and, following intense debate, a compromise bill was passed in 1999 by the US House of Representatives that called for stringent human rights criteria in their transfers (US House of Representatives, 1999).

29 Germany's Policy Principles for the Export of War Weapons and Other Military Equipment (January 2000) and the UK's Export Control Bill (introduced to Parliament in June 2001) are two such exercises.

30 See, for example, NATO (2000, p. 8) and OSCE (2000, sec. III).

31 Beginning in February 2002, a five-year prospective study of humanitarian deaths will be co-ordinated by the Small Arms Survey, the Centre for Humanitarian Dialogue and the John Hopkins University's Center for Refugees and Disaster Studies. It will be a prospective study of the rates, profiles, and costs associated with death and injury among humanitarian workers.

32 This is a revised estimate from the homicide rate of 35 per 100,000 reported in the *Small Arms Survey 2001*.

33 The Centre for Humanitarian Dialogue, together with the Small Arms Survey, will be undertaking a large-scale survey of the risks of small arms among personnel of more than a dozen humanitarian agencies. The survey was piloted between September and December 2001, implemented in February 2002 and preliminary findings will be published in late 2002.

34 The survey administrators sought to reduce self-selection and ensure a representative sample. It is expected that the large-scale global survey will address these issues more thoroughly (see endnote 33).

4. Appendices

Appendix 4.1 But are they humanitarian emergencies? (Figures 4.1 and 4.2)

Insecurity in Nicaragua (per 100,000)

Year	1990	1991	1992	1993	1994	1995	1996	1997	1998
Firearm homicide	12.9	12.01	12.36	10.32	9.28	8.9	8.05	9.09	8.02
Firearm injury	119.4	142.2	166.4	174.2	209.1	228.6	255.4	404.4	329.4
Armed robbery	25.2	29.9	31.9	37.8	43.2	45.6	42.7	49.5	64
Intimidation	40.9	47	57	72	68	59	62	59	64

Source: Muggah and Batchelor (2002)

Insecurity in Honduras (number of recorded incidents)

Year	1995	1996	1997	1998	1999
Firearm homicide	1,504	1,655	2,182	2,000	1,677
Firearm injury	1,235	1,130	2,009	1,427	1,244
Armed robbery	731	864	944	856	1,920
Non-firearm homicide	610	680	713	667	634
Non-firearm injury	918	1,008	1,210	983	547

Source: Muggah and Batchelor (2002)

Appendix 4.2 Far from home: A sample of displaced populations around the world (Map 4.1)

Country		International refugees by country of origin	Internally displaced persons (IDPs) in country of origin	Percentage of population displaced (%)
1	Afghanistan	3,500,000 (2001)	500,000 (November 2001)	16
2	Angola	350,600 (2000)	3,000,000 (Since 1998)	27
3	Azerbaijan	309,400 (2000)	570,000 (October 2001)	11
4	Bosnia and Herzegovina	448,700 (2000)	518,000 (May 2001)	24
5	Burundi	525,700 (2000)	633,000 (October 2001)	18
6	Colombia	2,500 (2000)	2,200,000 (since 1985)	5
7	Democratic Republic of Congo	248,400 (2000)	2,045,000 (October 2001)	5
8	Former Republic of Yugoslavia	121,100 (2000)	510,000 (May 2001)	31
9	India	110 (2000)	507,090 (November 2001)	<1
10	Iraq	572,500 (2000)	700,000 (June 2001)	6
11	Kenya	5,000 (2000)	100,000 (December 2001)	<1
12	Myanmar	127,800 (2000)	1,000,000 (June 2001)	2
13	Russian Federation	16,300 (2000)	500,000 (October 2001)	<1
14	Rwanda	85,500 (2000)	600,000 (May 2001)	10
15	Sierra Leone	487,200 (2000)	1,300,000 (August 2001)	42
16	Sri Lanka	93,200 (2000)	800,000 (June 2001)	5
17	Sudan	467,700 (2000)	4,000,000 (May 2001)	15

Source: NRC (2001, IDP figures); UNDP (2001, population figures); UNHCR (2000b, refugee figures)

Appendix 4.3 Security incidents in Kakuma refugee camp, 1996–2000* (Figure 4.3)

Year	Firearm deaths	Armed assault	Armed robbery	Rape
1996	15	64	67	2
1997	12	61	73	2
1998	6	114	110	6
1999	11	110	104	0
2000	5	70	90	4

**The refugee population in Kakuma has remained relatively steady: between 70,000 and 72,000 refugees.*

Source: Muggah and Berman (2001)

Appendix 4.4 Causes of UN civilian death from hostile actions: 1992–2000* (Figure 4.4)

Year	1992	1993	1994	1995	1996	1997	1998	1999	2000	Total
Gunshot**	9	19	9	3	8	15	11	10	7	91
Rockets or bombs	1	1	0	3	0	0	7	2	0	14
APM	0	2	0	0	0	0	0	0	0	2
Knife	0	0	46 ***	0	0	0	0	1	4	51
Other	1	2	8	6	3	2	1	0	5	28
Total	**11**	**24**	**63**	**12**	**11**	**17**	**19**	**13**	**16**	**136**

* *UNSECOORD estimates that the UN employed an average of 70,000 staff and dependents per year over the last decade.*

** *UNSECOORD reported in 2001 that since 1992, 107 staff members had died as a result of fatal firearm injuries. The figure above (excerpted from an internal UNSECOORD report) is missing 16 unexplained firearm-related deaths.*

*** *The majority of these deaths can be attributed to Rwanda.*

Source: Muggah and Berman (2001)

Appendix 4.5 Threats to Oxfam-GB staff involving small arms* (Box 4.13)

Gunshots	Armed intimidation	Armed robbery	Kidnapping at gunpoint	Landmines
35	30	16	11	8

**Total sample size consisted of 94 respondents (Muggah, 2001b)*

4. Bibliography

Aaby, Peter, Joaquim Gomes, Manuel Fernandes, Queba Djana, Ida Lisse, and Henrik Jensen. 1999. 'Nutritional Status and Mortality of Refugee and Resident Children in a Non-Camp Setting During Conflict: Follow-up Study in Guinea Bissau.' *British Medical Journal.* Vol. 319. August, pp. 878–81.

Aboutanus, Michael and Susan Baker. 1997. 'Wartime Civilian Injuries: Epidemiology and Intervention Strategies.' *Journal of Trauma.* Vol. 43, pp. 719–26.

American Psychological Association. 2000. *Diagnostic and Statistical Manual of Mental Disorders.* 4th edn. Washington, DC: American Psychiatric Association.

Amnesty International. 2001. *Human Rights Abuses with Small Arms: Illustrative Cases from Amnesty International Reports 2000–2001.* London: AI.

Annan, Kofi. 1999. 'The Legitimacy to Intervene: International Action to Uphold Human Rights Requires a New Understanding of State and Individual Sovereignty.' *Financial Times.* 31 December.

Anderson, Erica. 1965. *The Schweitzer Album.* New York: Harper and Row.

Andoni, Lamis and Sandy Tolan. 2001. 'Shoot to Maim.' *Village Voice.* October. <http://www.villagevoice.com>

Archer, Dane and Rosemary Gartner. 1976. 'Violent Acts and Violent Times: A Comparative Approach to Postwar Homicide Rates.' *American Sociological Review.* Vol. 42, No. 4, pp. 937–63.

Ayoob, Massad. 1980. *In the Gravest Extreme: The Role of the Firearm in Personal Protection.* New York: Police Books.

Berdal, Mats and David Malone, eds. 2000. *Greed and Grievance: Economic Agendas in Civil Wars.* London: Lynne Rienner.

Berger, Dinesh and Lawrence R. Berger. 1996. *Injury Control: A Global View.* Oxford: Oxford University Press.

Bok, Sissela. 1999. 'The New Ethical Boundaries.' In Jennifer Leaning, Susan Briggs, and Lincoln Chen, pp. 179–95.

Bonn International Center for Conversion (BICC). 2001. *Design and Implementation of Arms Embargoes and Travel and Aviation Related Sanctions: Results of the Bonn-Berlin Process.* Bonn: BICC, German Foreign Office, and UN Secretariat.

Boutros-Ghali, Boutros. 1992. *An Agenda for Peace.* New York: United Nations.

Boutwell, Jeffrey and Michael Klare, eds. 1997. *Light Weapons and Civil Conflict: Controlling the Tools of Violence.* New York: Carnegie Commission on Preventing Deadly Conflict and the American Academy of Arts and Sciences.

Bracken, Patrick and Celia Petty, eds. 1998. *Rethinking the Trauma of War.* London: Free Association Books/Save the Children Fund.

Bruderlein, Claude. 2001. 'People's Security as a New Measure of Global Stability.' *International Review of the Red Cross.* No. 842, pp. 353–65.

__. and Jennifer Leaning. 1999. 'New Challenges for Humanitarian Protection.' *British Medical Journal.* Vol. 319, pp. 430–35.

Bush, Kenneth. 2000. 'Polio, War and Peace.' *Bulletin of the World Health Organisation.* Vol. 78, No. 3, pp. 191–92.

Cambridge Dictionary of Philosophy. 1995. Cambridge: Cambridge University Press.

Cameron Commission. 1995. 'Commission of Inquiry into Alleged Arms Transactions between Armscor and One Eli Wazan and Other Related Matters.' *Second Report.* 20 November. Cape Town.

Cardozo, Barbara, Alfredo Vergara, Ferid Agani, and Carol Gotway. 2000. 'Mental Health, Social Functioning, and Attitudes of Kosovar Albanians Following the War in Kosovo'. *Journal of the American Medical Association.* Vol. 284, No. 5. August.

Center for Systemic Peace. 2001. <http://www.members.aol.com/CSPmgm/cspframe.htm>

Chesterman, Simon, ed. 2001. *Civilians in War.* London: Lynne Rienner Publications.

Cock, Jacklyn. 2000. 'Weaponry and Culture of Violence in South Africa.' In Virginia Gamba, ed. *Society Under Siege: Managing Arms in South Africa.* Pretoria: Institute for Security Studies.

Cohen, Roberta and Francies Deng. 1999. *Masses in Flight: The Global Crisis of Internal Displacement.* Washington, DC: Brookings Institute.

Collier, Paul and Anke Hoeffler. 2000. *Greed and Grievance: Economic Agendas in Civil Wars.* Washington, DC: Post-Conflict Unit, World Bank.

Collins, Cindy. 1998. *Humanitarian Implications of Small Arms Proliferation: A Paper Prepared for OCHA* (Draft). New York: Organisation for the Coordination of Humanitarian Affairs (OCHA).

Consejo Nacional de Politica Economica y Social (CONPES) 3057. 1999. *Politica de Prevencion y Atencion a la Poblacion Desplazada.* Bogota: Departamento Nacional de Planeacion.

Consultoria para los Derechos Humanos y el Desplazamiento (CODHES). 2001. 'Prévenir o Curar?' *Bulletin 3. Boletínes y Publicaciones.* Bogota. <http://www.codhes.org.co>

Cooper, Neil. 2001. *Conflict Trade and Arms Acquisition.* Background paper. Geneva: Small Arms Survey.

Coupland, Robin. 1999. 'Clinical and Legal Significance of Fragmentation of Bullets in Relation to the Size of Wounds: A Retrospective Analysis.' *British Medical Journal.* Vol. 319. August, pp. 403–6.

__. 2001a. *Applying an Analytical Framework of Armed Violence to Small Arms.* Background paper. Geneva: Small Arms Survey.

__. 2001b. 'Armed Violence.' *Medicine and Global Survival.* Vol. 7, No. 1. April, pp. 33–37.

__. and David Meddings. 1999. 'Mortality Associated with the Use of Weapons in Armed Conflicts, Wartime Atrocities and Civilian Mass Shootings: Literature Review.' *British Medical Journal.* Vol. 319. August, pp. 407–10.

__. and Hans Samnegaard. 1999. 'Effect of Type and Transfer of Conventional Weapons on Civilian Injuries: Retrospective Analysis of Prospective Data from Red Cross Hospitals.' *British Medical Journal.* Vol. 319. August, pp. 410–12.

Cremer, George. 1998. 'On the Problem of Misuse in Emergency Aid.' *Journal of Humanitarian Assistance.* <http://www.jha.ac/articles/a042.htm> June.

Crisp, Jeff. 1999. 'A State of Insecurity: The Political Economy of Violence in Refugee-Populated Areas of Kenya.' Draft manuscript. Boston: Massachusetts Institute of Technology.

__. and Karen Jacobson. 1998. 'Refugee Camps Reconsidered.' *Forced Migration Review.* Vol. 3, pp. 27–30.

Cummings, Peter, Thomas Koepsell, and Beth Mueller. 1995. 'Methodological Challenges in Injury Epidemiology and Injury Prevention Research.' *Annual Review of Public Health.* Vol. 16, pp. 381–400.

Cutts, Mark. 2000. 'Surviving in Refugee Camps', in *Forum: War, Money and Survival.* Geneva: ICRC.

De Waal, Alex. 2000. *Who Fights? Who Cares? War and Humanitarian Action in Africa.* Asmara: Africa World Press.

Department for International Development (DFID). 1997. *Eliminating World Poverty: A Challenge for the 21st Century.* White Paper on International Development. London: DFID.

Demetriou, Spyros. 2001. *Small Arms Proliferation and Insecurity in Georgia.* Background paper. Geneva: Small Arms Survey.

__. Robert Muggah, and Ian Biddle. 2001. *Small Arms Availability in Congo-Brazzaville. A Report for the IOM/UNDP Small Arms Collection and Ex-Combatant Reintegration Programme.* Brazzaville. IOM/UNPD.

Departamento Administrativo de Salud (DAS). 1999. *Estudio de la Mortalidad por Grupos de Causas.* Mocoa: DAS, Oficina de Planeación, Sección Epidemiología (Putumayo).

__. (with Diva Prevelo Calderon). 1999. *Informes Situación de Violencia en el Putumayo.* Mocoa: DAS.

Dobson, Roger. 2001. 'Sleeping Sickness Re-emerges in Africa After Years of Civil War.' *British Medical Journal.* Vol. 322. June, p. 1382.

Doswald-Beck, Louise and Sylvain Vité. 1993. 'International Humanitarian Law and Human Rights Law.' *International Review of the Red Cross.* Vol. 293. March–April, pp. 94–199.

Duffield, Mark. 2001. *Global Governance and the New Wars: The Merging of Development and Security.* London: Zed Books.

Dumas, Lloyd. April 2001. 'Why Mistakes Happen Even When the Stakes are High: The Many Dimensions of Human Fallibility.' *Medicine and Global Survival.* Vol. 7, No. 1, pp. 1–10.

Dybdhal, Ragnhild and Ivana Pasagic. 2000. 'Traumatic Experiences and Psychological Reactions Among Women in Bosnia During the War.' *Medicine, Conflict and Survival.* Vol. 16, pp. 281–90.

Ebersole, Jon. 2000. 'Health, Peace and Humanitarian Ceasefires'. *Health in Emergencies Newsletter.* No. 8. Geneva: WHO.

Egeland, Jan. 1999. 'Arms Availability and Violations of International Humanitarian Law.' *International Review of the Red Cross.* Vol. 83, pp. 673–77.

Erasmus, Desiderius. 1509. *In Praise of Folly.* Translated by Clarence Miller, 1979. New Haven: Yale University Press.

European Commission (EC). 1998 (December). *Cooperation with ACP Countries involved in Armed Conflict.* Brussels: EC.

Fajnzylber, Pablo, Daniel Lederman, and Norman Loazya. 2000. 'Crime and Victimization: An Economic Perspective.' *Economia.* Vol. 1, No. 1, pp. 219–78.

Fallon, William. 1997. 'Surgical Lessons Learned on the Battlefield.' *Journal of Trauma.* Vol. 43, pp. 209–14.

Federal Bureau of Investigation (FBI). 1998. *Crime in the United States, 1997.* Washington, DC: FBI.

Fitamant, Serrano. 1999. *Assessment Report on Sexual Violence in Kosovo.* New York: United Nations Population Fund (UNFPA).

Fox, Fiona. 2001. 'There Is a New Humanitarianism for a New Millennium.' Draft Presentation for a Seminar on Politics and Humanitarian Aid: Debates, Dilemmas and Dissension. London: Commonwealth Institute.

Freeman, Mark. 1997. 'International Law and Internal Armed Conflicts: Clarifying the Interplay Between Human Rights and Humanitarian Protections.' *Journal of Humanitarian Assistance.* 17 October. <http://www.jha.ac/articles/a059.htm

Frohardt, M., Diane Paul, and Larry Minear. 1999. *Protecting Human Rights: The Challenge to Humanitarian Organisations.* Occasional Paper No. 35. Providence: Thomas J. Watson Institute for International Studies.

Galbraith, Kenneth. 2001. 'Combat Casualties in the First Decade of the 21st Century—New and Emerging Weapon Systems.' *Journal of the Army Medical Corps.* Vol. 147, No. 1, pp. 7–14.

Garfield, Richard. 2000. 'Secondary Data Analysis of the Humanitarian Impact of Light Arms.' Unpublished research proposal. June. New York: Columbia University.

Gellert, George. 1995. 'Humanitarian Responses to Mass Violence Perpetrated Against Vulnerable Populations.' *British Medical Journal.* Vol. 311, pp. 995–1001.

Ghobarah, Hazem, Paul Huth, and Bruce Russet. 2001. 'Civil Wars Kill and Maim People–Long After the Shooting Stops.' Paper Presented to the Conference on Security with a Human Face: Expert Workshop on the Feasibility of a Human Security Report. Cambridge: Harvard University, December.

Gillard, Emanuela-Chiara. 2000. 'What's Legal? What's Illegal?' In Lora Lumpe, pp. 27–54.

__. 2001. *International Humanitarian Law, Human Rights and Small Arms Availability.* Background paper. Geneva: Small Arms Survey.

Gleditsch, Nils. 2001. 'Armed Conflict and Human Security.' Paper presented to the Conference on Security with a Human Face: Expert Workshop in the Feasibility of a Human Security Report. Cambridge: Harvard University, December.

Godnick, Bill. 2001. *Stray Bullets: The Impact of Small Arms and Light Weapons in Central America.* Background paper. Geneva: Small Arms Survey.

Greenwood, Colin. 2000. *Cross-Sectional Study of the Relationship Between Levels of Gun Ownership and Violent Deaths.* West Yorkshire: Firearms Research and Advisory Service.

Grossman, Daniel. 1995. *On Killing: The Psychological Cost of Learning to Kill in War and Society.* Boston: Little, Brown.

Gustafson, Per, Victor Gomes, Cesaltina Vieira, Henrik Jaensen, Remonie Seng, Badara Samb, Anders Naucler, and Peter Aaby. 2001. 'Tuberculosis Mortality During a Civil War in Guinea-Bissau.' *Journal of the American Medical Association.* Vol. 286, No. 5, pp. 599–603.

Handicap International. 2000. *Victim Assistance: Thematic Report.* London: Handicap International.

Harvard Program on Humanitarian Policy and Conflict Research (HPCR). 2001. *Bibliography on Human Security.* Harvard. <http://www.hsph.harvard.edu/hpcr/events/hsworkshop/bibliography.pdf>

Health Action Information Network (HAIN). 2001. <http://www.hain.org>

Herby, Peter. 1998. 'Developing a Humanitarian Law Approach to the Problem of Arms Availability.' Norwegian Red Cross, Conference on Arms Availability and Violations of International Humanitarian Law and the Deterioration of the Situation of Civilians in Armed Conflict. Oslo, 18–20 May.

__. 1999. 'Arms Transfers, Humanitarian Assistance and Humanitarian Law.' In Jeffrey Boutwell and Michael Klare, pp. 197–203.

Herman, Judith L. 1992. *Trauma and Recovery.* New York: Basic Books.

Hiltermann, Joost and Loretta Bondi. 1999. *State Responsibility in the Arms Trade and the Protection of Human Rights.* <http://www.hrw.org/campaigns>

Hobsbawm, Eric. 1994. *Age of Extremes.* New York: Vintage Books.

Homer-Dixon, Thomas. 2001. 'The Virulence of Violence: Small Arms, Many Wars, Large Threat.' *Washington Post.* 4 February.

Horwood, Christopher. 2000. *Humanitarian Mine Action.* Humanitarian Practice Network Paper No. 32. London: Overseas Development Institute.

Human Rights Watch (HRW). 1994. 'Arming Rwanda: The Arms Trade and Human Rights Abuses in the Rwandan War.' *Human Rights Watch Report.* Vol. 6, No. 1. New York: HRW.

__. 1995. 'Georgia/Abkhazia: Violations of the Laws of War and Russia's Role in the Conflict.' *Human Rights Watch Report.* Vol. 11, No. 4. New York: HRW.

__. 1996. 'Angola: Between War and Peace—Arms Trade and Human Rights Abuses Since the Lusaka Protocol.' *Human Rights Watch Report.* Vol. 8, No. 1 (A). New York: HRW.

__. 1999. 'Bulgaria: Money Talks—Arms Dealing with Human Rights Abusers.' *Human Rights Watch Report.* Vol. 7, No. 7. New York: HRW.

__. 2000. 'South Africa: A Question of Principle: Arms Trade and Human Rights.' *Human Rights Watch Report.* Vol. 12, No. 5 (A). New York: HRW.

Huntington, Samuel. 1996. *The Clash of Civilisations and the Remaking of World Order.* New York: Simon and Schuster.

Iacopino, Vincent and Ronald Waldman. 1999. 'From Solferino to Kosovo—the Evolving Role of Physicians.' *Journal of the American Medical Association.* Vol. 282, No. 5. <http://www.jama.amaassn.org/issues/v282n5/ffull/jed90059.html>

Ignatieff, Michael. 1997. *The Warriors' Honor: Ethnic War and the Modern Conscience.* New York: Henry Holt.

Inocenti, Nicol Degli. 2001. 'About 800,000 children "being used as soldiers worldwide".' *Financial Times.* 13 June.

Integrated Regional Information Network Reports (IRIN). 2000a. 'UN Plane Fired Upon in Kismayo, Somalia—Operations Suspended.' 30 March.

__. 2000b. 'Somalia: Security Council Condemns Attacks on Civilians, Aid Workers.' 20 June.
__. 2000c. 'Somalia: UN Security Officers Attacked by Gunmen.' 19 September.
__. 2000d. 'Kenya-Sudan: UNICEF Chief Says Conflict Biggest Obstacle to Aid.' 24 October.
__. 2000e. 'Kenya-Somalia: Somali Refugees Struck by Insecurity.' 24 November.
__. 2001a. 'DRC: Some Hope for Humanitarian Action, Oshima Says.' 10 May.
__. 2001b. 'East Africa: Small Arms Exacerbating Regional Insecurity.' 30 November.
Inter-Agency Standing Committee (IASC). 1998. *Thematic Guidance for the Identification of Agency Data on the Humanitarian Implications of Small Arms and Light Weapons.* <http://www.reliefweb.int>
__. 2000. *Global Humanitarian Assistance: 2000. An Independent Report Commissioned by the IASC from Development Initiatives.* New York: IASC.
International Committee of the Red Cross (ICRC). 1997. *The SIrUS Project: Towards a Determination of Which Weapons Cause 'Superfluous Injury or Unnecessary Suffering'.* Geneva: ICRC.
__. 1998. 'The Medical Profession and the Effects of Weapons.' ICRC Symposium. 8–10 March, Montreux.
__. 1999. *Arms Availability and the Situation of Civilians in Armed Conflict.* Geneva: ICRC.
International Peace Research Institute. 2001. <http://www.prio.no>
International Rescue Committee. 2001. *Health Publications.* <http://www.intrescom.org/health/publications.cfm>
International Tribunal. 1997. *Report of the International Tribunal for the Prosecution of Persons Responsible for Serious Violations of International Humanitarian Law Committed in the Territory of the Former Yugoslavia Since 1991.* A/52/375, S/1997/729, 18 September. <http://www.un.org/icty/rapportan/rapport4-e.htm>
Jean, François and Jean-Christophe Rufin, eds. 1996. *Economie des Guerres Civiles.* Paris: Hachette.
Kaldor, Mary. 1999. *New and Old Wars: Organized Violence in a Global Era.* Stanford: Stanford University Press.
__. ed. 2000. *Global Insecurity.* London: Pinter/UNWIDER.
Kant, Immanuel. 1795. *Perpetual Peace: A Philosophical Sketch.* Translated by H. B. Nisbet, ed. Hans Reiss, 2nd ed. New York: Cambridge University Press.
Kaplan, Robert. 1994. 'The Coming Anarchy.' *The Atlantic Monthly.* Vol. 273, No. 2, pp. 44–76.
Keegan, John. 1976. *The Face of Battle.* Suffolk: Chaucer Press.
Keen, David. (1998). 'The Economic Functions of Violence in Civil Wars.' *Adelphi Paper* 320. Oxford/London: Oxford University Press and the International Institute for Strategic Studies.
__. 2001. *The Functions of Violence.* Background paper. Geneva: Small Arms Survey.
Klugman, Jeni. 1999. *Social and Economic Policies to Prevent Complex Humanitarian Emergencies.* Washington DC: United Nations University and World Institute for Development Economics Research (WIDER).
Kobusingye, Olive C. 2000. Presentation at the IPPNW Medical Conference on Injury Prevention. Helsinki, 27–29 September.
__. Ronn Lett, Paul Ekwaru, and Christopher Orach. (2002 forthcoming). 'The Burden of Injury During Civil Conflict in Northern Uganda.' *Journal of Trauma.*
Koplan, Jeffrey, Mark Rosenburg, and Etienne Krug. 1998. *Violence Prevention: A Public Health Policy.* Atlanta: Centers for Disease Control and Prevention.
Krug, Etienne, Kenneth Powell, and Linda Dahlberg. 1998. 'Firearm-Related Deaths in the United States and 35 Other High- and Upper-Middle-Income Countries.' *International Journal of Epidemiology.* Vol. 27, No. 2, pp. 214–21.
Le Temps. 2000. 'Balles Déformantes pour les Polices Cantonales? Le CICR s'inquiète.' Editorial. 17 June.
Leaning, Jennifer, Susan Briggs, and Lincoln Chen, eds. 1999. *Humanitarian Crises: The Medical and Public Health Response.* Cambridge: Harvard University Press.
Les, Robert, Charles Hale, Fethi Belyakdoumi, Laura Covey, Roselida Ondeko, Michael Despines, and John Keys. 2001. *Mortality in Eastern Democratic Republic of Congo.* Atlanta: International Rescue Committee Health Unit.
Levie, Howard, 2000. 'History of the Law of War on Land.' *International Review of the Red Cross.* No. 828, pp. 339–50.
Levy, Barry and Victor Sidel. 1997 (updated 2000). *War and Public Health.* Oxford: Oxford University Press.
Lindqvist, Sven. 2001. 'Death Comes Flying.' *Harpers Magazine.* Fall, pp. 15–24.
Cardozo, Barbara Lopes, Alfredo Vergara, Ferid Agani, and Carol Gotway. 2000. 'Mental Health, Social Functioning and Attitudes of Kosovar Albanians Following the War in Kosovo.' *Journal of the American Medical Association.* Vol. 284, No. 5, pp. 569–77.
Lumpe, Lora, ed. 2000. *Running Guns: The Global Black Market in Small Arms.* London: Zed Books.
Machel, Graca. 2000. *The Impact of Armed Conflict on Children: International Conference on War-Affected Children.* Winnipeg: UN.
Mali Government. 2001. 'The Humanitarian Challenge of Small Arms Proliferation: The Argument for a Universal, Comprehensive and Non-Discriminatory Small Arms Control Regime.' Paper presented at the first Prepcom for the UN 2001 Conference on Illicit Trade in Small Arms and Light Weapons in all its Aspects. New York, March 2000.
Marshall, Samuel L. A. 1978. *Men Against Fire.* Gloucester, MA: Peter Smith.
Martin, Randolph. 2001. *A More Proactive UN Role in the Security of NGO Staff?* HPG Report 18. March. London: Overseas Development Institute.
Mathews, Robert and Timothy McCormack. 1999. 'The Influence of Humanitarian Principles in the Negotiation of Arms Control Treaties.' *International Review of the Red Cross.* No. 834, pp. 331–52.
Max, Wendy and Dorothy Rice. 1993. 'Shooting in the Dark: Estimating the Cost of Firearm Injuries.' *Journal of Health Affairs.* Vol. 12, No. 4, pp. 171–85.
Meddings, David and Stephanie O'Connor. 1999. 'Circumstances Around Weapon Injury in Cambodia After Departure of a Peacekeeping Force: Prospective Cohort Study.' *British Medical Journal.* Vol. 319, pp. 412–15.
Médecins Sans Frontières (MSF). 2001. *Violence and Access to Health in Congo (DRC): Results of Five Epidemiological Surveys.* December. <http://www.msf.org>
Medicina Legal y Ciencias Forenses (2000a) 'Algunas Consideraciones Acerca de las Muertes por arma de Fuego, su Control y Legislación.' *Centro de Referencia Naciojal Sobre Violencia (CRNV).* Vol. 5, No. 7. July, pp. 25–28.
__. (2000b) *Forensis: Herramienta para la Interpretación, Intervención, y Prevención del Hecho Violento en Colombia.* Bogota: Medicina Legal y GTZ.
Melvern, Linda. 2000. *A People Betrayed: the Role of the West in the Rwanda Genocide.* London: Zed Books.
Meron, Theodor. 2000. 'The Humanisation of Humanitarian Law.' *American Journal of International Law.* Vol. 94, No. 2, pp. 239–78.
Michaels, Markus, David Meddings, Salah Ramez, and Juan Luis Gutiérrez-Fisac. 1999. 'Incidence of Weapon Injuries to Inter-Factional Combat in Afghanistan in 1996: Prospective Cohort Study.' *British Medical Journal.* Vol. 319, pp. 415–17.
Milas, Ropac. 1999. 'Military Casualties in Baranja and East Slavonia during the First 9 Months of the War in Croatia.' *Journal of Military Medicine.* Vol. 164, No. 9. September, pp. 643–47.

Minear, Lary and Thomas Weiss. 1993. *Humanitarian Action in Times of War: A Handbook for Practitioners.* London: Lynne Reinner.

Minorisa. 2001. *Goma/Bukavu: Direct Eye Witness Testimony, January 1997.* <http://www2.minorisa.es/inshuti/gomaa.htm>

Mooney, Erin. 1995. 'Presence ergo Protection? UNPROFOR, UNHCR and the ICRC in Croatia and Bosnia and Hercegovina.' *International Journal of Refugee Law.* Vol. 7, No. 3, pp. 407–35.

Moser, Caroline and James Holland. 1997. *Urban Poverty and Violence in Jamaica.* Washington, DC: World Bank.

__. and Chris McIlwaine. 2000. *Urban Poor Perceptions of Violence and Exclusion in Colombia.* Washington, DC: World Bank.

Muggah, Robert. 2000 'Through the Developmentalist's Looking Glass: Conflict-Induced Displacement and Involuntary Resettlement in Colombia.' *Journal of Refugee Studies.* Vol. 13, No. 2, pp. 133–64.

__. 2001a. 'Globalisation and Insecurity: The Direct and Indirect Effects of Small Arms Availability.' *IDS Bulletin.* Vol. 32, No. 2, pp. 70–79.

__. 2001b. *Perceptions of Small Arms Availability and Use Among Oxfam-GB Field Personnel.* Geneva: Oxfam-GB/Small Arms Survey.

__. 2001c. *Why a Humanitarian Perspective on Small Arms? Humanitarian Practice Network Review.* London: ODI.

__. 2001d. 'Participatory Research on Human Insecurity and Small Arms in South Asia'. Mimeo. Geneva: Small Arms Survey.

__. and Eric Berman. 2001. *Humanitarianism Under Threat: The Humanitarian Impacts of Small Arms and Light Weapons.* Special Report Commissioned for the UN Inter-Agency Standing Committee. Geneva: Small Arms Survey.

__. and Peter Batchelor. 2002. *Development Held Hostage: Assessing the Effects of Small Arms on Human Development. A Study of the Socio-Economic Impacts and Development Linkages of Small Arms Proliferation, Availability and Use.* New York: Emergency Response Division, UNDP.

de Mulinen, Frederic. 1987. *Handbook of the Laws of War for Armed Forces.* Geneva: ICRC.

Murray, Christopher, Gary King, Alan Lopez, Niels Tomijima, and Etienne Krug. 2002. 'Armed Conflict as a Public Health Problem.' *British Medical Journal.* Vol. 324, pp. 346-349.

Murray, Christopher and Alan Lopez, eds. 1996. *The Global Burden of Disease.* Cambridge, MA: Harvard University Press.

Musah, Ali and Fayemi Kayode. 2000. *Mercenaries: An African Security Dilemma.* London: Pluto Press.

NATO. 2000. *Resolution on Small Arms Control.* Resolution 303. November. <http://www.naa.be/publications/resolutions/00-berlin-303.html>

Norwegian Refugee Council (NRC) 2001. *Global IDP Database: 2001.* <http://www.idpproject.org>

Office for the Coordination of Humanitarian Affairs (OCHA). June 2001. *Use of Military or Armed Escorts for Humanitarian Convoys.* Discussion Paper and Non-Binding Generic Guidelines. <http://www.reliefweb.int>

Office of the United Nations High Commissioner for Human Rights (OHCHR). 2001. *The Human Rights Impact of Small Arms and Light Weapons.* Press Release DPI/2195. May. Geneva.

Olujic, Marija. 1998. 'Embodiment of Terror: Gendered Violence in Peacetime and Wartime in Croatia and Bosnia-Herzegovina.' *Journal of Medical Anthropology.* Vol. 12. pp. 31–50.

Ordog, Gary, Jonathan Wasserberger, and Greg Ackroyd. 1995. 'Hospital Costs of Firearm Injuries.' *Journal of Trauma.* Vol. 38, No. 2, pp. 291–298.

Organisation for Economic Cooperation and Development (OECD). 1998. *OECD/DAC Guidelines on Conflict, Peace and Development Cooperation on the Threshold of the 21st Century.* Development Cooperation Guideline Series. Paris: OECD-DAC.

Organisation for Security and Cooperation in Europe (OSCE). 2000. *OSCE Document on Small Arms and Light Weapons.* 308th Plenary Meeting. Vienna, November.

Oxfam-GB. 2001a. 'The Human Cost of Small Arms'. Unpublished paper. Oxford: Oxfam-GB.

__. 2001b. *Study on the Impacts of Small Arms in Mindanao Philippines.* Oxford: Oxfam-GB.

Post-Conflict Unit. 2001. *World Bank Post-Conflict Unit.* <http://www.worldbank.org/research/conflict/>

Project Ploughshares. 2001. 'Improving Human Security Through the Control and Management of Small Arms.' Paper delivered at the Arusha Conference on Small Arms. Arusha, 23–24 March.

Ramsbotham, Oliver and Thomas Woodhouse. 1996. *Humanitarian Intervention in Contemporary Conflict.* Cambridge: Polity Press.

Red Cross. 1999. 'The Humanitarian Challenge of Small Arms Proliferation.' Paper delivered to the 27th International Conference of the Red Cross and Red Crescent. Mali. <http://www.redcross.int/en/conference/ws_reports/ws_report6.asp>

Reno, William. 1998. *Warlord Politics and Africa States.* London and Boulder: Lynne Reinner.

__. 2002. 'Arms, Internal War and the Causes of Peace.' Background Paper for Social Science Research Council Workshop Series on Law and International Relations. Washington. February 6-7.

Richardson, David, Dana Davidson, and Frank Miller. 1996. 'After the Shooting Stops.' *Journal of Trauma.* Vol. 41, pp. 789–93.

Roberts, Adam. 1998. 'The Role of Humanitarian Issues in International Politics in the 1990s.' Paper delivered to the Third International Security Forum. Zurich, 19–21 October.

Robertson, Leon. 1998. *Injury Epidemiology: Research and Control Strategies.* Oxford: Oxford University Press.

Salama, Peter, Bruce Laurence, and Monica Nolan. 1999. 'Health and Human Rights in Contemporary Humanitarian Crises: Is Kosovo More Important than Sierra Leone?' *British Medical Journal.* Vol. 319, pp. 1569–71.

Samaranayake, Mallika. 2001. *Participatory Research on Human Insecurity and Small Arms in Sri Lanka: Field Study Report on Aluthwatta Village.* Background Report for the Small Arms Survey and the Regional Centre for Strategic Studies. Colombo.

Schindler, David and John Toman. 1988. *The Laws of Armed Conflicts.* 3rd edn (reprint). Geneva: Martinus Nijhoff, Henry Dunant Institute.

Seck, Jacqueline. 2000. *West Africa Small Arms Moratorium: High-Level Consultations on the Modalities for the Implementation of PCASED.* Geneva: UNIDIR.

Seet, Benjamine and Gilbert Burham. 2000. 'Fatality Trends in United Nations Peacekeeping Operations: 1948–1998.' *Journal of the American Medical Association.* Vol. 284, No. 5. August, pp. 598–603.

Sellier, Beat and Karl Kneubehl. 2001. *Wundballistik und ihre Ballistischen.* 2nd edn Grundlagen: Springer-Verlag.

Sen, Amartya. 2001. 'Addressing Global Poverty.' *The Economist: The World in 2002*, p. 50.

Sethi, Dinesh, Syad Aljunid, Saperi Sulong, and Anthony Zwi. 2000. 'Injury Care in Low and Middle-Income Countries: Identifying the Potential for Change.' *Journal of Injury Control and Safety Promotion.* Vol. 7, No. 3, pp. 153–64.

Shearer, David. 1998. *Private Armies and Military Intervention.* Adelphi Paper 316. Oxford/London: Oxford University Press and the International Institute for Strategic Studies.

Sheik, Martin, Gutierrez Isabel, Paul Bolton, Paul Spiegel, Michel Thieren, and Gilbert Burham. 2000. 'Deaths Among Humanitarian Workers.' *British Medical Journal.* Vol. 321, pp. 166–68.

Slaby, Andrew. 1989. *Aftershock: Surviving the Delayed Effects of Trauma, Crisis and Loss.* New York: Villard Books.

Slim, Hugo. 1997. 'Relief Agencies and Moral Standing in War: Principles of Humanity, Neutrality, Impartiality and Solidarity.' *Development in Practice.* Vol. 7, No. 4, pp. 342–52.

__. 2001. 'Not Philanthropy But Rights: Rights-Based Humanitarianism and the Proper Politicisation of Humanitarian Philosophy of War.' Draft Paper for the Centre for Development and Emergency Practice. Oxford: Oxford Brookes University.

Sphere Project. 2000. *Humanitarian Charter and Minimum Standards in Disaster Response.* Oxford: Oxfam Publishing.

__. 1998. *SPHERE: Humanitarian Charter and Minimum Standards in Disaster Response.* Geneva: Steering Committee for Humanitarian Response.
Stewart, Frances. 2002. 'Root Causes of Violent Conflict in Developing Countries.' *British Medical Journal.* Vol. 324, pp. 342–345.
Stockholm International Peace Research Institute (SIPRI). 2001. *SIPRI Yearbook 2001: Armaments, Disarmament and International Security.* Oxford: Oxford University Press.
Summerfield, Derek. 1996. *The Impact of War and Atrocity on Civilian Populations: Basic Principles for NGO Interventions and a Critique of Psychosocial Trauma Projects.* HPN Network Paper. London: ODI.
__. 2000. 'War and Mental Health: A Brief Overview.' *British Medical Journal.* Vol. 321, July, pp. 232–35.
Swiss Proposal. 2001. 'Restrictions on Small Calibre Ammunition Which Causes Excessive Wounding and Unnecessary Injuries.' A Proposal for the Second CCW Review Conference (Preparatory Process). Geneva: Swiss Ministry of Foreign Affairs.
Taux, Jean-Daniel. 2000. 'The ICRC and Civil-Military Cooperation in Situations of Conflict.' Rose-Roth Seminar. Montreux.
Trask, Arthur, Franklin Richards, Cary Schwartzbach, and Robert Kurtzke. 1995. 'Massive Orthopedic, Vascular, and Soft Tissue Wounds from Military Type Assault Weapons A Case Report.' *Journal of Trauma.* Vol. 38, No. 5, pp. 428–31.
Tubbs, Norman. 1998. 'Abhorrent Weapons and Superfluous Injury Or Unnecessary Suffering': A Concern for Trauma Surgeons?' *Journal of Injury.* Vol. 29, No. 2. March, pp. 85–6.
Ugalde, Antonio, Ernesto Selva-Sutter, Carolina Castillo, Carolina Paz, and Sergio Canas. 2000. 'The Health Costs of War: Can They be Measured? Lessons from El Salvador.' *British Medical Journal.* Vol. 321. July, pp. 169–72.
United Nations. 1992. *An Agenda for Peace Preventive Diplomacy, Peacemaking and Peace-Keeping.* A/47/277 – S/24111.
__. 1998. *International Study on Firearm Regulation.* New York: UNCCP.
__. 1999. *Report of the Secretary General to the Security Council on the Protection of Civilians in Armed Conflict.* S/1999/957.
__. 2000. *Final Report of the UN Panel of Experts on Violations of Security Council Sanctions Against Unita.* S/2000/203.
__. 2001. *Statement of the President of the Security Council.* S/PRST/2001/21, September.
United Nations Children's Fund (UNICEF). 1999. *The Progress of Nations.* New York: UNICEF.
__. 2001. *Humanitarian Response to Children: January–June 1999. Democratic Republic of Congo.* <http://www.unicef.org/cap/congo.pdf>
United Nations General Assembly (UNGA). 2000. *Safety and Security of United Nations Personnel: Report of the Secretary General.* A/55/494, October.
__. 2001. *Draft Programme of Action to Prevent, Combat and Eradicate the Illicit Trade in Small Arms and Light Weapons in All Its Aspects.* A/Conf.192/L.5/Rev.1.
United Nations High Commissioner for Refugees (UNHCR). 1997. *Staff Stress and Security: A Management Challenge for the UNHCR.* An Evaluation Report Prepared by the Inspection and Evaluation Service. Geneva: EVAL/02/97.
__. 2000a. Introductory Remarks by Nicholas Bwakira to the Open-Ended Ad Hoc Working Group on the Causes of Conflict and the Promotion of Durable Peace and Sustainable Development in Africa, Created by General Assembly Resolution, 53/92. <http://www.un.org/esa/africa/adhocWg/54/unhcr.html>
__. 2000b. *Refugees and Others of Concern to UNHCR: 1999 Statistical Overview.* Geneva: Registration and Statistical Unit, UNHCR.
__. 2001a. *State of the World's Refugees: Fifty Years of Humanitarian Action.* Geneva: Oxford University Press.
__. 2001b. *Le rapport du Représentant sur sa Mission en Azerbaïdjan.* Geneva: UN E/CN.4/1999/79.
United Nations Secretary General (UNSG). 1996. *Impact of Armed Conflict on Children.* A/51/306, September.
__. 1999. *Resolution 1265 Adopted by the Security Council at its 4046th Meeting.* S/RES/1265, September.
__. 2000. *Report of the Panel on United Nations Peace Operations.* A/55/305, November.
United States Commission for Refugees (USCR). 2001. *World Refugee Survey 2001.* <http://www.refugees.org/pub/recent.cfm>
United States. Department of State. 2000. *1999 Country Reports on Human Rights Practices: Sierra Leone.* Released by the Bureau of Democracy, Human Rights, and Labor. February.
United States. House of Representatives. 1999. *International Arms Sales Code of Conduct.* Sec. 1262, para. a.
Van Brabant, Koenraad. 1997. 'Security Guidelines: No Guarantee for Security.' *Humanitarian Practice Network Newsletter* 7. London: Overseas Development Institute, pp. 6–9.
__. 2001a. *Mainstreaming the Organisational Management of Safety and Security: A Review of Aid Agency Practices and a Guide for Management.* Humanitarian Policy Group Report 9. London: Overseas Development Institute.
__. 2001b. *Small Arms and the Safety and Security of Humanitarian and Development Personnel.* Background paper. Geneva: Small Arms Survey.
Van Crevald, Jan Martin. 1995. 'Land of the Free.' *Economist.* 26 March.
Warner, Daniel. 2001. *Small Arms in an Expanding Humanitarian Space? Think Again.* Background paper. Geneva: Small Arms Survey.
Waszink, Camilla. 2001. 'The Use of Small Arms in Internal Armed Conflicts: How to Make Armed Opposition Groups Accountable.' Mimeo. Geneva.
Wintemute, Garren. 1996. 'The Relationship Between Firearm Design and Firearm Violence: Handguns in the 1990s.' *Journal of the American Medical Association.* Vol. 275, No. 22, pp. 1749–53.
World Health Assembly. 1996. *Prevention of Violence: Public Health Priority.* Geneva: WHO (WHA 49.25).
World Health Organisation (WHO). 2000. *World Health Report 2000.* Geneva: WHO.
__. 2001a. *Small Arms and Global Health.* Geneva: Injuries and Violence Prevention Department, WHO.
__. 2001b. *Injury: A Leading Cause of the Global Burden of Disease.* Geneva: WHO.
Zwi, Anthony, Sam Forjuoh, Shiva Mrugusampillay, Wilson Odero, and Charlotte Watts. 1996. 'Injuries in Developing Countries: Policy Response Needed Now.' *Transactions of the Royal Society of Tropical Medicine and Hygiene.* Vol. 90, pp. 591–95.

Acknowledgements

Main contributors

Robin Coupland, Koenraad Van Brabant, Emanuela-Chiara Gillard, David Keen, and Daniel Warner.

small arms
KILL
women
and children
THE UN WORKS FOR
PEACE AND SECURITY
BURKINA FASO
EGYPT

Reaching Consensus in New York: The 2001 UN Small Arms Conference

5

Introduction

At about 6.00 a.m. on Saturday, 21 July 2001, the *United Nations Conference on the Illicit Trade in Small Arms and Light Weapons in All Its Aspects* ('the Conference') concluded. After two weeks of difficult negotiations at UN headquarters in New York and several years of preparation, delegates reached a final consensus agreement on a *Programme of Action to Prevent, Combat and Eradicate the Illicit Trade in Small Arms and Light Weapons in All Its Aspects* (*'Programme'*).[1]

The *Programme* spelled out the security, humanitarian, and socio-economic consequences associated with both the illicit trade in these weapons and their excessive and destabilizing accumulation, and also included a set of norms to guide the actions of states in addressing this problem. By agreeing to this *Programme*, national governments made a commitment to implement a series of measures at the national, regional, and global levels. They also agreed to meet every two years and hold a conference no later than 2006 for the purpose of reviewing the implementation of the *Programme*.

The *Programme of Action* provides a mandate for the United Nations, national governments, donor states, and NGOs to develop and implement practical projects in every critical area of endeavour that will lead to the prevention and reduction of the problems associated with these weapons. UN Secretary-General Kofi Annan warmly welcomed the news that the Conference 'was able to reach a consensus on many important first steps in alleviating this grave threat to international peace and security and to human security'. He praised these steps as 'essential in building norms and in implementing collective measures against this global scourge' (United Nations, 2001b).

Despite these measures, many governments and NGOs were disappointed that critical elements needed for the solution of the problem were either not included in the *Programme* or were watered down to the point of irrelevance. Sally Joss, Co-ordinator of the umbrella group International Action Network on Small Arms (IANSA), stated: 'When this whole process began we were aiming for a clear international plan of action. Instead this has been eroded to the lowest common denominator, and the framework that remains falls short of what is so urgently needed.' While welcoming the consensus *Programme*, IANSA expressed disappointment that certain key elements—including commitments to negotiate international legal instruments on marking and tracing, as well as brokering, or promises of greater transparency—had been left out of the *Programme* (IANSA, 2001). Human Rights Watch termed the outcome a 'Program of Inaction' (HRW, 2001).

The issue areas that left many disappointed were those that had been identified as contentious from the beginning of the process in 1995. The debate was largely over the extent to which the Conference would address legal as well as illicit small arms and light weapons. As the production, import, and export of these weapons is a legal activity critical to their national defence, many governments resisted any attempt by the Conference to limit their sovereign rights in this area. Others insisted that, since legal arms were the source of most illicit transfers, a comprehensive response required consideration of legal activity too.

Debating delegates at the 2001 UN Small Arms Conference. (©Associated Press/Shawn Baldwin)

The debates during the Conference centred on the following key issues:

- the linkage of small arms to human rights violations;
- excessive and destabilizing accumulations of both legal and illicit small arms;
- transparency;
- a commitment to negotiate legally binding instruments on arms brokering and on the marking and tracing of weapons;
- national regulation of civilian possession of military weapons;
- export and re-export control criteria; and
- the supply of arms to governments only.

This chapter provides a description of the events leading up to the Conference and the two weeks of the Conference itself. It also offers an in-depth analysis of the *Programme of Action*, and looks at how the various contentious issues were resolved.

BOX 5.1 Chronology of events

- 12 December 1995: in its Resolution 50/70B, the UN General Assembly requests the Secretary-General to prepare a report on small arms with the assistance of a panel of governmental experts (UNGA, 1995b).
- June 1996–August 1997: the 1997 Panel completes its report, recommending that the UN consider holding 'an international conference on the illicit arms trade in all its aspects, based on the issues identified in the present report' (UNGA, 1997a, para. 80(k)).
- 9 December 1997: in its Resolution 52/38J, the General Assembly requests the Secretary-General to seek the views of Member States on the convening of a conference, and to prepare a second report on small arms with the assistance of a group of governmental experts (UNGA, 1997b).
- May 1998–August 1999: the 1999 Group produces a report addressing the objectives, scope, agenda, dates, and venue of the Conference (UNGA, 1999a).
- 15 December 1999: in its Resolution 54/54V, the General Assembly launches the last stages of the Conference process (UNGA, 1999b).
- February 28–3 March 2000: the first session of the Preparatory Committee (PrepCom) solicits state views on procedures for the Conference and on a programme of action (UNGA, 2000a).
- 6 October 2000: the Chairman of the PrepCom produces an informal paper as a basis for a first draft of a programme of action (dos Santos, 2000).
- 11 December 2000: the first draft of the *Programme of Action* (L.4) is completed, serving as a basis for further drafts (UNGA, 2000b).
- 8–19 January 2001: during the second session of the PrepCom, states submit views on the first draft of the *Programme of Action*.
- 12 February 2001: the second draft of the *Programme of Action* (L.4/Rev.1) is distributed to states (UNGA, 2001a).
- 19–30 March 2001: the third session of the PrepCom begins negotiating the final *Programme of Action*.
- 4 June 2001: the PrepCom Chair issues an Informal Paper containing his assessment of PrepCom III.
- 9 July 2001: the Conference begins.
- 16 July 2001: the third draft of the *Programme of Action* (A/CONF.192/L.5) is distributed to states.
- 21 July 2001: Conference participants agree on a consensus *Programme of Action* (UNGA, 2001d).
- 24 December 2001: the UN General Assembly welcomes the adoption by consensus of the Conference *Programme of Action* and calls upon states to implement it. It also decides to hold the first in a series of biennial meetings in 2003, and to convene a conference no later than 2006, for the purpose of reviewing progress made in implementing the *Programme* (UNGA, 2001f).

The road to the Programme of Action

The initial task of defining the problems associated with small arms and light weapons fell to the UN Panel of Governmental Experts on Small Arms ('1997 Panel'), constituted pursuant to General Assembly Resolution 50/70B of 12 December 1995 (UNGA, 1995b). The 1997 Panel met from June 1996 until issuing its report to the UN Secretary-General in August 1997. This report addressed the following issues:

- the types of small arms and light weapons actually being used in conflicts being dealt with by the United Nations;
- the nature and causes of the excessive and destabilizing accumulation and transfer of small arms and light weapons, including their illicit production and trade; and
- the ways and means to prevent and reduce the excessive and destabilizing accumulation and transfer of small arms and light weapons, in particular as they cause or exacerbate conflict (UNGA, 1995b, para. 1).

As part of its recommendations, the 1997 Panel called for an 'international conference on the illicit arms trade in all its aspects, based on the issues identified in the present report'. The specific objectives of the Conference recommended by the Panel were only gradually delineated and not formally agreed until the Conference itself.

Some states were concerned about putting the small arms issue into the orbit of the UN disarmament system.

While some states favoured holding such a conference, others were opposed, for a variety of reasons. Some states viewed the conference as a way to raise awareness, develop norms of behaviour where none existed, and—they hoped—begin to develop specific actions to which states would formally commit themselves. Additionally, some viewed a conference as the only way to broaden the issue of illicit arms beyond a narrow arms control focus. Others were more cautious, fearing that it was too soon to hold such a conference, which could detract energy from the 'real work' of dealing with the small arms issue, or even be hijacked by the pro-arms lobby. Some states were also concerned about putting the small arms issue into the orbit of the UN disarmament system, given the success achieved on the landmines issue outside the UN system.

Agreeing to hold a conference

In the end a compromise was agreed whereby the views of states would be solicited on convening such a conference and a Group of Governmental Experts on Small Arms ('1999 Group') would be convened to assess the implementation of the Report of the 1997 Panel and further develop the concept of the Conference (UNGA, 1997b).

As of May 1998, the 1999 Group was meeting to develop the recommendations of the 1997 Panel, one of which was an international conference. Many states had in fact indicated their support for a conference on the illicit trade in small arms and light weapons and, as a result, the General Assembly resolution on small arms of 1998 contained the decision 'to convene an international conference on the illicit arms trade in all its aspects' (UNGA, 1998, para. 1).

By late 1998, the move towards a conference had been significantly affected by the signing in 1997 of the *Convention on the Prohibition of the Use, Stockpiling, Production and Transfer of Anti-Personnel Mines and on Their Destruction* (*'Ottawa Treaty'*). The UN bureaucracy and those states which relied on the UN to solve global problems were very concerned that the *Ottawa Treaty* would set a precedent that would see more and more disarmament issues go outside the UN for solution, and were thus intent on keeping the small arms issue within the organization. They were well aware that the International Campaign to Ban Landmines (ICBL) and supportive governments felt that the UN process had failed them. While only in its formative stages, IANSA, an NGO modelled on the ICBL, was beginning to push governments for a solution to the small arms problem in all its dimensions.[2]

The 1999 Group paves the way

The mandate of the 1999 Group was twofold. First, it was to assess progress made on the 1997 Panel's recommendation of measures designed 'to reduce the excessive and destabilizing accumulation and transfer of small arms and light weapons in specific regions of the world where [these] have already taken place' and 'to prevent such accumulations and transfers from occurring in future' (UNGA, 1997a, para. 78). The second task of the 1999 Group was to report 'on further actions recommended to be taken'. This included developing specific aspects of the Conference, such as venue, agenda, objectives, and other procedural matters (UNGA, 1997b).

In the process of addressing 'further actions to be taken' by the United Nations, international and regional organizations, and states (UNGA, 1999a, sec. IV), the 1999 Group provided a set of recommendations that served as a major source of actions for the Conference to consider. It should be noted that the 1999 Group was quite representative of the UN as a whole as it comprised European Union (EU) states, the US, China, and key members of the Non-Aligned Movement (NAM). In fact, the debates that took place within the 1999 Group during 1998–99 served to preview the difficulties that would dominate the Conference itself two years later.

In Section V of its Report, the 1999 Group made specific recommendations on issues relating to the Conference. These had a significant impact on the Conference process—including its Preparatory Committee (PrepCom)—and outcome. Significantly, the Group recommended that the objective of the Conference should be to develop and strengthen *international* efforts to prevent, combat, and eradicate the illicit trade in small arms and light weapons in all its aspects. Prior to this, several key states, especially the US, had played down the need for initiatives of an international nature. Other states were concerned that the focus on international efforts would reduce, or undermine, regional and subregional efforts.

The 1999 Group recommended that the objective of the Conference should be to develop and strengthen international efforts to prevent, combat, and eradicate the illicit trade in small arms and light weapons in all its aspects.

The Group also recommended that the Conference seek to strengthen or develop norms and measures at the global, regional, and national levels. As for the scope of the Conference, the 1999 Group addressed the fine line between illicit and licit in a comprehensive manner, noting that:

> the illicit trade in small arms and light weapons is closely linked to the excessive and destabilizing accumulation and transfer of such arms. The scope of the Conference should therefore not be limited to criminal breaches of existing arms legislation and export/import controls but consideration should be given to all relevant factors leading to the excessive and destabilizing accumulation of small arms and light weapons in the context of the illicit arms trade (UNGA, 1999a, para. 132).

Like its predecessor, the 1997 Panel, the 1999 Group did not, however, attempt to establish a formal agenda. It also deferred decisions on dates and venue to the PrepCom.

A formal mandate for the Conference

By late 1999, states knew that there would be a conference and, despite the misgivings some states had about the initiative, the contentious work of shaping its outcome began with the drafting of the enabling General Assembly Resolution 54/54V. Prior agreement on key assumptions and principles were codified in this resolution, adopted in December 1999. States had little difficulty agreeing on many of the preambular paragraphs. These defined the problem as urgent, and recognized the importance of civil society, the humanitarian and socio-economic consequences stemming from the ready availability of small arms and light weapons, the links to terrorism, organized crime, and drug trafficking, and the need for a comprehensive approach to a resolution of these problems (UNGA, 1999b).

In Resolution 54/54V, states decided to convene the Conference in June/July 2001 and hold three PrepCom meetings, with the date for the first set for 28 February to 3 March 2000. Details on the venue and dates were left to the PrepCom to decide, along with objectives, agenda, and a draft final document that would include a *Programme of Action.* States could not agree on whether to accept an offer by Switzerland to hold the Conference itself in Geneva, so this was also deferred.[3]

© Associated Press/Stephen Chernin

Helen Mirren, actress and anti-gun violence activist, in front of the UN New York headquarters during the Conference.

Getting organized

The work of the Conference began with the first meeting of the PrepCom at UN Headquarters in New York, from 28 February to 3 March 2000. This meeting got bogged down in procedural issues—such as NGO participation—and there was little discussion of substantive issues. Despite this, the proposed *Programme of Action* began to take shape as states laid out their basic positions.

Significantly, there was a fair amount of agreement on a range of measures more strictly related to the illicit trade in small arms. These included the need for effective border controls, better enforcement of UN arms embargoes, some improvement in marking and tracing across state boundaries, the destruction of surplus weapons, the need for safe and secure storage of stockpiles, and the right of sovereign states to acquire, use, and export small arms and light weapons. Predictably, there was no agreement on the more contentious issues.

Given the preliminary and inconclusive results from its first meeting, the PrepCom Chairman, Ambassador Carlos dos Santos of Mozambique, conducted two informal consultations prior to the second PrepCom session. These were held in New York from 17 to 19 July 2000 and on 6 October 2000. At the second consultation, Ambassador dos Santos circulated an Informal Proposal of the Chairman titled *Draft Structure and Elements for a Programme of Action to Prevent, Combat and Eradicate the Illicit Trade in Small Arms and Light Weapons in All Its Aspects.* States generally agreed that this document should serve as the basis for the first draft of a *Programme of Action* that would be debated at the second PrepCom. Based on informal responses to this *Draft Structure*, the PrepCom Chairman revised it and in early November gave it to those assigned the task of drawing up the draft *Programme of Action.*

Agreements reached before the Conference

The second meeting of the PrepCom was held at UN Headquarters in New York from 8 to 19 January 2001. At this meeting, states mainly concerned themselves with starting negotiations on a consensus text of the *Programme of Action.* Crucially, procedural issues were less evident, and proved far less controversial, than in the first PrepCom.

The first draft of the *Programme of Action* (A/CONF.192/PC/L.4) was issued a month before the second PrepCom meeting. This 18-page document attempted to integrate various proposals for a preambular section, measures at the national, regional, and global levels, and for an implementation and follow-up process. The central part of the draft delineated measures to be adopted for each of the following issues:

- illicit manufacturing, acquisition, stockpiling, and transfer;
- measures to prevent diversion from legal manufacture, acquisition, and transfer;
- stockpile management and safe storage;
- collection and destruction of illicit and surplus small arms;
- civilian possession;
- post-conflict situations; and
- transparency, confidence building, and exchange of information (UNGA, 2000b).

The fact that the first draft was issued only a month prior to the second PrepCom meant that most delegations had very little time to study it before the start of the session. Some delegations, including the US, voiced strong objections to many of the elements it contained. Many delegations also criticized its structure, arguing that, while it was extremely comprehensive, it was also repetitive and unduly complicated.

While the first draft, by necessity, incorporated a wide range of views gathered from many sources over a two-year period, the second draft (A/CONF.192/PC/L.4/Rev.1), issued on 12 February 2001, more than a month before the third PrepCom, was much more concise and reflected much of the discussion of the January meeting. In effect, it removed much of the language from the first draft that was considered too controversial for the achievement of consensus. It also narrowed the *Programme of Action* to focus only on those critical issues that had been identified during the debates at the second PrepCom (UNGA, 2001a). Most countries were much more positive about the second draft, and many delegations stated that it provided a good basis for negotiating the final Conference text.

By the end of the third session of the PrepCom, there was a fair amount of agreement on large sections of the second draft—about 60 per cent of a text covering 86 paragraphs. This momentum carried through to the Conference.

Preamble. States agreed that solving the problems associated with the illicit trade in small arms and light weapons was urgent and that their humanitarian consequences were serious. The impacts on women and children were particularly noted. It was also agreed that the illicit trade sustains conflicts, exacerbates violence, fuels terrorism, and is linked to drug trafficking.

There was also agreement in the Preamble on a basic set of norms or principles that should form the basis for the measures agreed to in the *Programme of Action*. These included: references to the UN Charter; the right of states to import, produce, and retain arms; the need for compliance with arms embargoes; and the need for a co-operative international effort to fully address this global issue.

Operational measures. While states continued to differ on specific language, by the end of the third PrepCom a range of measures had been agreed which ultimately found their way into the final *Programme of Action*. At the national level, these were:

- to put in place adequate laws, regulations, and administrative procedures to exercise effective control over the production, export, import, transit, and retransfer of small arms and light weapons (UNGA, 2001d, sec. II, para. 2);

- to establish national co-ordination agencies responsible for policy guidance, research, and monitoring of efforts to prevent, combat, and eradicate the illicit trade in small arms and light weapons in all its aspects (sec. II, para. 4);
- to identify groups and individuals engaged in the illegal manufacture, trade, stockpiling, transfer, possession, and financing for acquisition of illicit small arms and light weapons, and take action against them (sec. II, para. 6);
- to ensure responsibility for all small arms and light weapons held and issued by the state, and effective measures for tracing them (sec. II, para. 10);
- to put in place and implement adequate laws, regulations, and administrative procedures to ensure the effective control over the export and transit of small arms and light weapons, including the use of authenticated end-user certificates (sec. II, para. 12);
- to develop and implement, where possible, effective disarmament, demobilization, and reintegration programmes, including the effective collection, control, storage, and destruction of small arms and light weapons, particularly in post-conflict situations (sec. II, para. 21); and
- to address the special needs of children affected by armed conflict (sec. II, para. 22).

Among the measures agreed at the regional level were the following:

- to encourage negotiations with the aim of concluding relevant legally binding instruments aimed at preventing, combating, and eradicating the illicit trade in small arms and light weapons in all its aspects, and, where they do exist, to ratify and fully implement them (sec. II, para. 25); and
- to establish subregional or regional mechanisms, in particular trans-border customs co-operation and networks for information-sharing among law enforcement, border, and customs control agencies (sec. II, para. 27).

The third PrepCom also saw agreement on a series of global measures, including:

- to co-operate with the United Nations system to ensure the effective implementation of Security Council arms embargoes (sec. II, para. 32);
- to encourage the disarmament and demobilization of ex-combatants, and their subsequent reintegration into civilian life (sec. II, para. 34);
- to encourage states and the World Customs Organization, as well as other relevant organizations, to enhance co-operation with the International Criminal Police Organization (Interpol) to identify those groups and individuals engaged in the illicit trade in small arms and light weapons in all its aspects (sec. II, para. 37);
- to encourage relevant international and regional organizations and states to facilitate the co-operation of civil society, including NGOs, in activities related to the prevention, combat, and eradication of the illicit trade in small arms and light weapons in all its aspects (sec. II, para. 40); and
- to promote a dialogue and a culture of peace by encouraging education and public awareness programmes on the problems of the illicit trade in small arms and light weapons in all its aspects, involving all sectors of society (sec. II, para. 41).

Other processes

The interplay between regional and international processes was an important part of the politics surrounding efforts to develop concerted international action on small arms proliferation issues (Greene, 2001). Regional organizations provided an important focus for wider agenda setting and coalition building, while the Conference process helped to stimulate action at the regional and national levels.[4] It is also important to recognize the important interactions between the Conference process and other international negotiations, particularly those resulting in the legally binding *Firearms Protocol* (UNGA, 2001c).

As international awareness and concern about illicit trafficking and proliferation of small arms grew after 1996, regional initiatives developed alongside the work of UN bodies. In addition to those regional initiatives, discussed below, which fed directly into the *Programme of Action*, several other regional processes contributed to the general momentum that built up before the Conference and helped foster a more positive attitude towards it. In November 1997, members of the Organization of American States (OAS) signed the *Inter-American Convention against the Illicit Manufacturing of and Trafficking in Firearms, Ammunition, Explosives and Other Related Materials* (OAS, 1997), complemented by the 1998 Model Regulations (OAS, 1998). In Africa, the Economic Community of West African States (ECOWAS) Moratorium was signed in October 1998 after some two years of discussions (ECOWAS, 1998). *The Southern Africa Regional Action Programme on Light Arms and Illicit Arms Trafficking* was prepared in May 1998 (SADC, 1998).

Government support for stringent and comprehensive measures at the regional or subregional levels did not always translate into support for similar commitments in the *Programme of Action*.

Such substantial regional initiatives added momentum to the UN process. However, they also highlighted the ways in which different regions and subregions tended to frame the problems and priorities relating to small arms in distinct ways. For example, the OAS addressed the issue primarily as a problem of illicit arms trafficking by criminal groups; the EU addressed it more as a problem of arms export controls and of developing appropriate security assistance to regions emerging from conflict; and Southern African Development Community (SADC) countries addressed it as a problem of enabling post-conflict development and stability and of strengthening co-operation and control over small arms in the subregion.

These frameworks were in many ways complementary, but they did contribute to the controversies in the 1999 Group about the appropriate focus and scope of the Conference. Latin American governments wanted the Conference to concentrate on combating and preventing illicit trafficking in small arms, and thus joined with those in Asia and the Middle East that were suspicious of any broader mandate. Meanwhile the EU—together with Japan, the United States, and virtually all other Western democracies—insisted that the Conference should also address the issues of arms holdings and inter-state transfers, along with norms for assistance in war-torn or vulnerable countries.

The existence of regional agreements reflecting these perspectives tended to consolidate these differences. Diplomats naturally tend to try to internationalize the standards to which their country is already committed, and to avoid new commitments that would require significant changes to domestic policy.

Interestingly, government support for stringent and comprehensive measures at the regional or subregional levels did not always translate into support for similar commitments in the *Programme of Action*. Whereas the EU and Canada, for example, consistently fought for the inclusion of strong international commitments where these were consistent with their regional and national policies, countries such as Australia, New Zealand, Norway, and South Africa tended to adopt a rather more reserved, or minimalist, stance. In part, these different approaches stemmed from a difference in negotiating tactics; yet they also reflected an ongoing debate among various states about the relationship between regional measures and the UN process. There were some fears that the UN negotiations could encourage an inappropriate 'one size fits all' approach which did not adequately recognize regional differences, or that they could detract from the urgent task of developing necessary national and regional measures. Thus, provided the *Programme of Action* was adequately comprehensive in scope and established a reasonable set of principles and norms, Australia, New Zealand, and other South Pacific states, for example, were generally content to focus on developing detailed commitments and programmes at the regional level within the Pacific Island Forum rather than at the international level.

Similarly, South Africa did not support the EU proposals for stringent or fully developed international commitments in several areas (such as national criteria for export controls), even though they were part of the EU-SADC joint action programme on small arms and SADC countries planned to adopt a legally binding Protocol containing such commitments in the near future. These differing priorities caused friction between the EU and some SADC countries during the PrepCom and at the Conference itself.

As the second PrepCom approached, governments began to take the Conference process more seriously. Evidence of this came in the form of a number of regional meetings held towards the end of 2000, which adopted declarations designed to shape the final outcome of the Conference. They were literally being completed as the first draft of the *Programme of Action* was being developed.

The OSCE Document, 24 November 2000

The Organization for Security and Co-operation in Europe (OSCE), after two years of relative neglect of the topic, agreed at its November 1999 Istanbul summit to mandate the OSCE Forum on Security Co-operation to initiate discussions on the problem of small arms proliferation. With 54 members (55 from November 2000) whose national interests cover the gamut of the small arms issue, from states seriously affected by the proliferation and misuse of these weapons to many major suppliers, the organization spoke with unusual authority. It also provided an effective framework within which a coalition of Canada, the EU, Norway, Switzerland, and the United States could co-ordinate positions and aim to persuade other member states—particularly Russia, Ukraine, and other countries of the former USSR—to agree to substantial measures to deal with the small arms issue. Within a year, this was successfully achieved.

On 24 November 2000, the OSCE adopted the *OSCE Document on Small Arms and Light Weapons* (OSCE, 2000). This comprehensive, 29-page document contained many measures that were directly applicable to the global level, with the added relevance of having been negotiated by 54 states. The negotiations of certain parts of the document were precursors for what was to follow at the Conference and its PrepCom sessions:

- The terms 'norms, principles and measures' were used to describe the nature of the contents of the document.

MAP 5.1 OSCE participating states

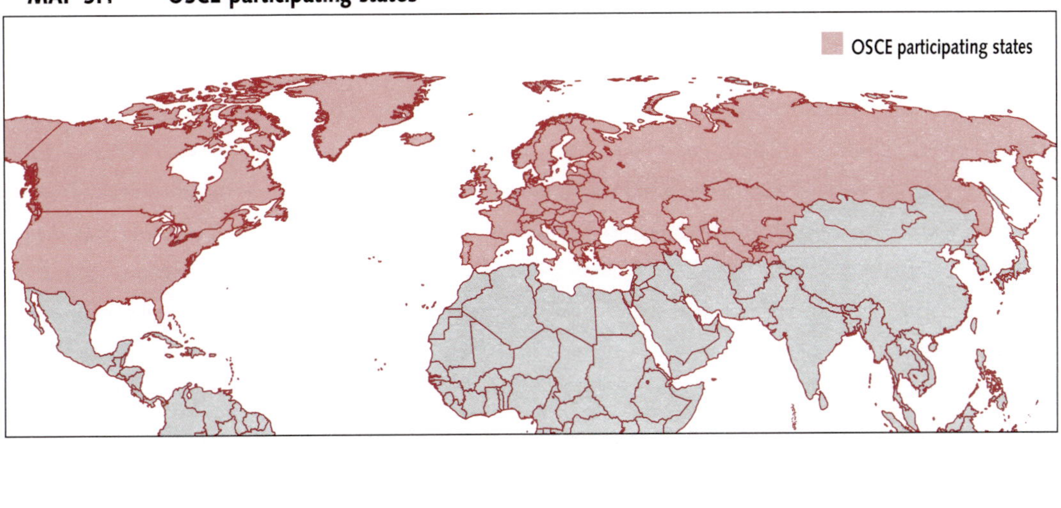

- For the purposes of its document, the OSCE adopted the definition and typology of small arms and light weapons that had been agreed in the report of the 1997 Panel. In so doing, it stated that there was not yet an internationally agreed definition of small arms and light weapons, noting that the OSCE definition did not prejudge any future agreed international definition.
- Until the very last moment, the OSCE draft contained a paragraph in which states agreed to supply arms only to governments and to government-approved organizations. The US had made it clear from the start that this was unacceptable. Meanwhile, the EU pushed for an agreed set of export control criteria reflecting those of the EU, which was also resisted by many OSCE states. In the end, it was agreed to delete the paragraph on supplying arms to non-state actors while retaining the export control criteria.
- The provisions for weapons marking and tracing were very specific but stopped short of a legally binding mechanism that contained standard marking for all countries. The specific markings were left up to individual states.
- No OSCE system was developed to control brokering. National efforts were merely urged.
- Very specific guidelines for identifying weapons surplus to a state's needs were adopted.
- OSCE members agreed that the document was 'politically binding'.

The OSCE document was very specific and comprehensive. Some states viewed it as a source of measures to be adopted at the global level, introducing various elements from the document during the negotiations in the PrepCom sessions. Other states, particularly the US and Russia, used the document as evidence that the basic work to solve the problems stemming from small arms should be conducted primarily at the regional level. Overall, it was the latter view that prevailed during the PrepComs. The same states also pushed this view during the Conference negotiations, making it difficult to reach consensus on measures at the international level.

The Brasilia Declaration of Latin American and Caribbean States, 24 November 2000

With the path-breaking OAS *Inter-American Convention* of 1997 and the *Model Regulations* of 1998 on controlling the international movements of firearms, the states of Latin America had already agreed on some norms and developed measures on small arms that could have been used to shape their positions at the Conference. As a result of the Convention and the *Model Regulations*, many states in the region (e.g. Brazil, Colombia, and Mexico) felt a strong sense of ownership of the small arms issue. However, since the OAS Convention did not address many of the issues to be taken up at the Conference, the states of Latin America and the Caribbean held a special meeting in Brasilia, Brazil, in November 2000 for the purpose of developing a common position to take to the Conference.

While the *OSCE Document*, submitted to the UN for use in developing the *Programme of Action*, was designed as a stand-alone agreement to be implemented by a well-developed institutional structure, the *Brasilia Declaration* was different. It was a short two-page document that simply acknowledged the importance of various existing norms and principles to the development of a UN *Programme of Action*. It also listed a number of goals for the Conference including the need for concrete measures at the global level. While emphasizing the role civil society must play at the Conference, it specifically mentioned 'educational initiatives and publicity campaigns'. Not surprisingly, the meeting gave a strong endorsement to the candidacy of Ambassador Camilo Reyes Rodriguez of Colombia for the presidency of the Conference itself, noting also his endorsement from NAM countries.

The OAU Bamako Declaration, 1 December 2000

MAP 5.2 African Union states

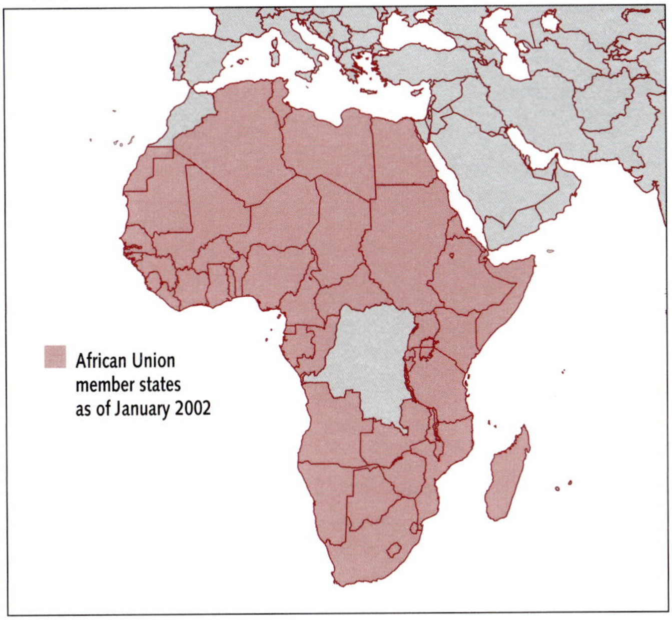

As with the OSCE, the Organization of African Unity (OAU)[5] had been conducting meetings and expert workshops on the issue of small arms and light weapons for some time, serving as an important focus for coalition building amongst African states during 1999–2000. In many ways, the effects of small arms proliferation in Africa were a primary motive for trying to develop an effective UN *Programme of Action* on small arms (Mali, 1996; 1997; South Africa, 1998).[6]

Yet African states rarely spoke with one voice on the small arms issue, and many scarcely followed the UN debates before 1999. Most strikingly, a major difference of perspective had become apparent on small arms issues between the countries of sub-Saharan Africa and those of North Africa. Countries like Algeria and Egypt generally played a leading role in UN negotiating processes, and the general preference for consensus among the NAM meant that they were rarely opposed by other African states in UN negotiations. Yet their positions on small arms were much closer to those of other Arab League countries than to those of sub-Saharan Africa.

A major difference of perspective became apparent between the countries of sub-Saharan Africa and those of North Africa.

After two consultative meetings of national experts in May and June 2000, and against the clear preferences of Egypt and some of its North African neighbours, the *Bamako Declaration* adopted at the OAU Ministerial Conference in Bamako, Mali, in December 2000 largely reflected the comprehensive approach preferred by Sub-Saharan African countries (OAU, 2000). The repeated references to the *Bamako Declaration* during the PrepCom sessions and the Conference itself were as much an effort to maintain North African and wider NAM support for sub-Saharan African preferences as they were aimed at negotiating with other governments. Ultimately the *Bamako Declaration* proved to have an important impact on the final text of the *Programme of Action.*

South Africa had played a leading role on the small arms issue for a number of years, and arrived at the 30 November–1 December 2000 meeting in Bamako, Mali, with a well-developed draft document that formed the basis for the final declaration. Given the severity of the problems related to small arms in Africa, as well as the influence of a co-ordinated African effort in shaping the landmine treaty of 1997, the *Bamako Declaration* was designed, from the beginning, to ensure that Africa had an influence on the final text of the Conference *Programme of Action.*

Unlike the OSCE and Brasilia documents, the eight-page *Bamako Declaration* was in a UN format designed to easily convert into a UN text. The preambular section reaffirmed the importance of regional work and the principles of the UN Charter, and noted the 'devastating consequences' that the 'problem of the illicit proliferation, circulation and trafficking of small arms and light weapons continues to have ... for stability and development in Africa' (OAU, 2000, para. 1). In describing these problems, OAU states presented a comprehensive list of impacts, mentioning international humanitarian law, cultures of violence, security and development, women, children, good governance, and human rights (para. 1). They laid out a general and comprehensive set of measures which emphasized transparency, conflict prevention, strengthening of democracy and respect for human rights, a focus on both supply and demand, capacity-building, and respect for international humanitarian law. The *Bamako Declaration* contained specific recommendations for action at the national and regional levels, setting out a wide range of measures that included practically all of those steps that had long been recommended by many states and NGOs. These included actions that would later be opposed by some OAU members, such as Algeria and Egypt, during the UN deliberations, including transparent controls on manufacturers, suppliers, traders, and brokers (para. 3.A.vii), as well as the 'active involvement of civil society in the formulation and implementation of a national action plan' (para. 3.A.viii).

OAU states held back their enthusiasm for any international mechanisms or commitments that would tie their hands in acquiring or exporting small arms.

It was at the international level that OAU states held back their enthusiasm for any mechanisms or commitments that would tie their hands in acquiring or exporting small arms. The approach was to appeal instead to arms supplier countries to change their behaviour. In a sense this section put much of the blame for the problem on those outside Africa. For example, it asked that non-African states 'Accept that trade in small arms should be limited to governments and authorized registered licensed traders' (para. 4.i), a restriction the OAU states did not place on themselves.

OAU states sought a commitment at the UN Conference to a change of behaviour by supplier states. This theme had been put forward, especially by South Africa, in all subregional, regional, and international forums leading up to the *Bamako Declaration*. The aim, however, was not to create a mechanism for supplier restraint at the global level. In the view of OAU states, before international mechanisms were put in place time had to be given to let regional efforts develop. This approach was clearly seen in the final section of the *Bamako Declaration* where OAU states called for an 'international partnership to curb the illicit proliferation, circulation, and trafficking of small arms and light weapons in Africa' (sec. 5). This involved an appeal for technical and financial support for national and regional initiatives and closer co-operation among regional and international organizations.

During the Bamako meeting, some OAU members disagreed with many of the paragraphs in the final text, particularly Algeria and Egypt, which are also part of the Arab bloc. In the interest of a common African position, many of these disagreements were papered over. However, they soon resurfaced at the Conference itself.

The EU Plan of Action, 11 December 2000

Of all the regional declarations prior to the second PrepCom, only the one agreed to by the EU in Brussels on 11 December 2000 was not completed in time to be fully integrated into the first draft of the *Programme*. However, since the EU had well developed policies on small arms, and had consistently submitted proposals for the *Programme of Action* throughout the PrepCom process, many of its text suggestions were integrated into the first draft, which was also published on 11 December.

As with the *Bamako Declaration*, the EU document was in a format designed to be easily transferred to a draft *Programme of Action*. The EU had already enacted a *Code of Conduct on Arms Exports* (European Union, 1998a) with specific export control criteria, as well as the *Programme for Preventing and Combating Illicit Trafficking in Conventional Arms* (European Union, 1997) and its *Joint Action on Small Arms* (European Union, 1998b). It came as no surprise that the EU *Plan of Action* (2000) contained many elements of these various measures. Key elements of the Plan, that were to surface as the EU position in negotiating the draft Conference *Programme of Action*, included:

- a strong commitment to international humanitarian law and human rights;

- linking security and development;
- restrictive national legislation and procedures that include criminal sanctions for the illicit possession of small arms;
- support for regional and subregional moratoriums;
- transparency of legal arms transfers;
- universal standards for marking of weapons;
- regional and subregional codes of conduct to control and restrain arms exports; and
- capacity-building in support of national and regional actions (European Union, 2000).

In its section on implementation, the EU *Plan* called for increased co-operation, extensive information exchange, effective tracing of weapons through the negotiation of an international convention, linking assistance and co-operation to a state's compliance with human rights and international humanitarian law norms, and the facilitation of the active involvement of civil society. The *Plan* also called for a 'Review Conference' in 2005, reports by states and regional organizations, and the setting up of a committee to prepare the Review Conference and consider additional proposals to prevent, combat, and eradicate the illicit trade in small arms and light weapons in all its aspects.

Though they all agreed to the *Plan*, there were important differences of opinion—and of negotiating strategy—between EU member states in some areas. Most of the EU candidate countries also aligned themselves with the *Plan*, not least for wider foreign policy reasons.[7]

While the regional declarations provided valuable inputs—in terms of both language and content—into the early drafts of the *Programme of Action*, it should also be noted that throughout the deliberations of the Conference many states warned that measures applicable to regions might not be applicable at the global level. The development of the *Programme of Action* was centred on this reality, stressing only those regional measures that did apply at the global level and that all states could agree to by consensus.

Frustrated regional processes: Southeast and South Asia

MAP 5.3 ASEAN and SAARC states

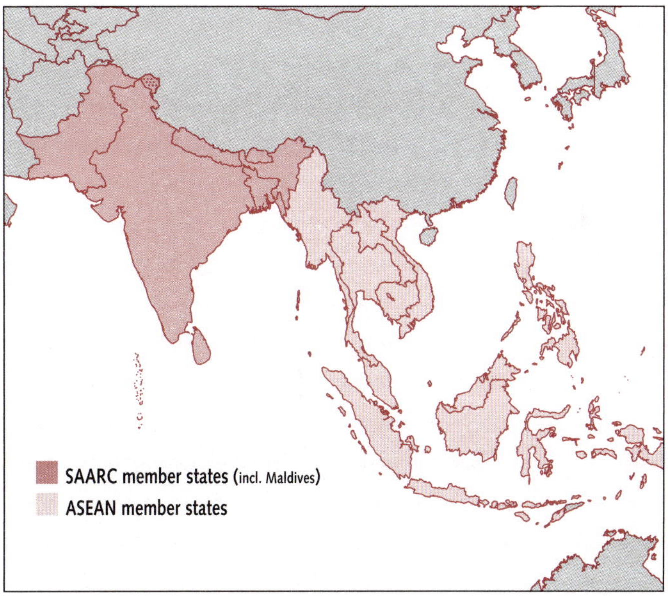

In addition to the relatively successful efforts of those regions previously noted, there were less auspicious efforts to develop regional outlooks in other parts of the world. There were significant attempts to develop a common position amongst the countries of the ASEAN Regional Forum (ARF). Building on the Jakarta Regional Seminar on Illicit Trafficking in Small Arms and Light Weapons of May 2000, the Asian Regional Workshop on Small Arms in Tokyo of June 2000, and elements of the PrepCom draft text, Australia prepared elements of a draft ARF declaration on small arms and light weapons. These were considered at a series of ARF meetings from late 2000 with a view to adopting a statement at the ARF summit in June 2001, just before the UN Conference. Canada and Japan also sponsored a special ARF conference in Cambodia, in February 2001, to raise awareness and promote action on small arms transfers among ASEAN states and their ARF partners.

In practice, these were attempts to use the UN Small Arms Conference process to stimulate action on small arms within ARF rather than the other way around. However, China was unwilling to support this. With the backing of Myanmar and the acquiescence of several other ASEAN states, China was able to use the ARF consensus rule to block the adoption of any declaration. As a result of the failure of these attempts, regional efforts relating to small arms control in Southeast Asia have remained restricted to programmes primarily designed to combat trans-national crime (Small Arms Survey, 2001, pp. 267–68).

Similarly, there were moves to stimulate co-operative action on small arms in South Asia through a series of subregional workshops and seminars during 2000–01. For example, two workshops organized by the Regional Centre for Strategic Studies, based in Colombo, were held in Sri Lanka during 2000. These brought together officials and civil society experts from South Asia and elsewhere. They contributed to raised awareness among South Asian governments, including India and Pakistan, of the relevance of the comprehensive approach to small arms control to their circumstances (up to that point, most South Asian governments tended to believe that their problems with small arms control primarily related to covert and illicit shipments in which their neighbours were implicated). However, the South Asian Association for Regional Cooperation (SAARC)—the primary subregional organization—remains politically paralysed. So efforts to enhance co-operation have had to be developed either through broader international agreements or through bilateral arrangements.

During the PrepCom sessions and the Conference itself, several states from these regions, including India and Sri Lanka, pointed out that they were suffering from the effects of these weapons but had been unable to articulate regional positions. As a result, they constantly reminded the Conference that this situation created a need for global norms, principles, and measures—not just regional action.

The UN Firearms Protocol

The effective implementation and further development of the *Programme of Action* and the Firearms Protocol appear closely linked.

The final *Programme of Action* was shaped not only by the interplay with regional initiatives but also by parallel international processes. The most important of these were the negotiations for the UN *Firearms Protocol*, successfully concluded in March 2001 (MEASURES). The existence of this parallel initiative significantly weakened arguments that the Conference should, like the Protocol, primarily focus on combating and preventing illicit trafficking in small arms. The effective implementation and further development of the *Programme of Action* and the *Firearms Protocol* appear closely linked. Provisions governing such specific areas as marking and tracing, end-use controls, and brokering will be far more effective where they complement and reinforce one other. At a more general level, there are intrinsic links between the issues of crime prevention, security, and disarmament that underpin each instrument (Greene, 2001).

© IANSA/Poster design Bob Boucher

NGO members of IANSA made their presence felt at the UN Conference.

BOX 5.2 NGOs and the 2001 UN Small Arms Conference

No picture of those two weeks in July 2001 would be complete without a view of the variety of roles played by NGOs both before and during the Conference. Just as they influenced the Conference, one clear result of the Conference has been its impact on NGOs.

Throughout the preparatory process and during the Conference itself, there were two distinct NGO communities at work. By far the largest was the broad coalition of some 300 groups working to control the proliferation of small arms and their effects, brought together under the International Action Network on Small Arms (IANSA). The second group represented the firearm community, which saw the Conference as a threat to the rights of gun owners and sport shooters. Chiefly represented by the World Forum on the Future of Sport Shooting Activities (WFSA), this group worked vigorously to minimize Conference outcomes. Its most outspoken member, the National Rifle Association of America (NRA), worked closely with the US delegation throughout the Conference. One issue on which both NGO communities agreed—and on which they worked together during the PrepCom process—was maximizing the official role of NGOs at the Conference.

NGO access to UN processes is typically controversial, especially where these touch on national and international security matters. Yet one of the key functions of holding a UN conference is to draw greater public attention to the relevant issue. NGOs have thus been seen, from the UN's earliest days, to play an important public relations role for the organization. In the case of the Conference, not surprisingly, states differed widely on the permissible degree of NGO participation. During the first two sessions of the PrepCom there were differing views among states over the types of NGOs that should be allowed to be present. For example, states such as Algeria and China wished to restrict NGO participation because they feared they would highlight human rights issues. The modalities agreed at the third PrepCom in March 2001 gave NGOs a recognized role in the remainder of the preparatory process and in the Conference itself (UNGA, 2001b).

A number of NGOs were present during the three PrepCom sessions, and groups of NGOs were allowed to address special meetings of the second and third sessions. These presentations were well received and helped to allay the fears of certain delegations (e.g. Algeria, China) that NGOs would 'hijack' the Conference or cause embarrassment to certain delegations. The NGO presence during the PrepComs also helped to 'normalize' the presence of NGOs at the Conference itself.

A large number of NGOs were present at the Conference—119 organizations registered and 380 representatives attended the Conference (UNDDA, 2001). They represented domestic gun control groups, groups concerned by the humanitarian effects of small arms proliferation or their impacts on development or health, research organizations, church groups, gun collector and sport shooting bodies, and others. While these numbers were modest in comparison with some other UN conferences, they succeeded in making their presence felt.

The plenary session, held on 16 July, was probably the most important expression, at least in symbolic terms, of the NGO role at the Conference. The 42 presentations given by representatives of both NGO communities reflected the full range of their concerns and expertise. A series of side events was held by various NGOs in rooms near the main Conference room on themes like 'The Impact of Small Arms on Children' or particular regions of the world such as 'Small Arms Issues in Africa'. Considerable NGO attention and support was also given to the gun destruction events held by various governments in different parts of the world on 9 July, the opening day of the Conference.

NGOs were also an important source of information on Conference developments, issuing daily reports distributed inside the UN and worldwide. A number of countries such as Canada, Ireland, Netherlands, New Zealand, Norway, Switzerland, and the United Kingdom included NGO representatives on their delegations.

It is difficult to determine with any precision the impact on the Conference outcome of the NGO presence in New York in July. The *Programme of Action* that emerged from the process represented a watering down of previous drafts. In this sense, the firearm community had the greater influence, because of its influence on the negotiating position of one state—the US. It was widely credited as inspiring the uncompromising US position that forced the withdrawal of key provisions from the *Programme of Action* (Stohl, 2001).

However, the real long-term impact of the IANSA-affiliated NGOs on the Conference process is undoubtedly much more substantial. The presence of these NGOs in New York not only underlined the importance of this agenda but also strengthened the hand of those states which struggled to hold the line against further erosions of the language contained in previous drafts of the *Programme of Action*. At the same time, the encounters of delegation members and NGOs in New York will undoubtedly consolidate relationships that will form a key part of post-Conference follow-up.

BOX 5.2 NGOs and the 2001 UN Small Arms Conference (continued)

It is also important to note that the Conference was the result of years of activity by governments and NGOs alike. Arguably more important than the impact of NGOs on the results of any UN conference is the impact the holding of the conference has had on NGOs themselves. In the small arms case, as with so many previous UN conferences, the run-up to the Conference saw NGOs develop their positions, strengthen their networks, engage new constituencies, and raise essential funds. Indeed, it seems clear that July 2001 has strengthened the capacity of NGOs to further engage in the post-Conference work that lies ahead. The experience of NGO involvement in the 2001 UN Conference process can leave no doubt that they will continue to be essential actors in the next stages of international action on small arms—as recognized in the *Programme of Action* itself.[8]

Source: Atwood (2001)

The Conference

The end game: Negotiating a consensus at the Conference

At the end of the third and final PrepCom session in March 2001 it was agreed that the Chairman would produce an Informal Paper containing his assessment of the discussions during the third PrepCom.[9] It was also agreed, largely as a result of pressure from NAM states, that the Chairman would not produce a rolling text of the second draft of the *Programme of Action*, based on the discussions during the third PrepCom, and that L.4/Rev.1 as issued on 12 February would provide the basis for negotiations at the July Conference. As a result, delegations arrived at the Conference at UN Headquarters in New York on 9 July with only L.4/Rev.1 plus a compilation of views—often contradictory—from the discussions at the second and third sessions of the PrepCom.

Actual negotiations on the final *Programme of Action* did not start until the second week of the Conference. At the third PrepCom it had been agreed that the first week would be devoted to statements at the ministerial level. This is standard practice for major international conferences, intended to give added weight to the proceedings.

Ministerial level attendance at the Conference was poor, arguably reflecting a lack of commitment to the issue on the part of many countries. During the first week the Conference heard statements from 143 speakers, including one vice-president, two deputy prime ministers, 38 foreign and other ministers, and 23 deputy ministers. Of those, 131 represented their countries, four spoke for regional groups, and eight spoke for UN agencies and intergovernmental organizations (Batchelor, 2001).

Despite the presence of few ministerial-level officials at the Conference, the ministerial session turned out to be much more than a pro forma event. On the first day, the US representative, Under-Secretary of State for Arms Control and International Security John Bolton, gave a tough speech, which literally woke up the Conference. He called the draft *Programme of Action* defective and identified several aspects, which the US could not, and would not, support (see Box 5.3). Most of the other ministerial-level statements were not nearly as critical or provocative as the US one. Generally they tended to highlight specific dimensions of the illicit trade in small arms related to their own country and/or region, while at the same time reaffirming their commitment to ensuring the success of the Conference.

The Conference President, Ambassador Camilo Reyes of Colombia, also used the first week of the Conference to give states a final opportunity to comment on the draft *Programme of Action* (L.4/Rev.1). He felt that he needed to hear directly from states himself, as it was clear that he would be required to write a new draft prior to the final phase of negotiations. He also appointed various vice-presidents of the Conference to conduct consultations on problematic paragraphs in the hope of developing more acceptable text for his final draft.

BOX 5.3 John Bolton at the UN Conference

The United States was a major player throughout the Conference process. At the Conference itself, however, its primary role was one of obstructionist. This was clearly demonstrated on the opening day of the Conference on 9 July, when the US Under-Secretary of State for Arms Control and International Security, John R. Bolton, gave a blistering speech that literally woke up the Conference. After a few words praising the goals of the Conference, Bolton made it clear that the US would not support a *Programme of Action* which he characterized as defective. Specifically he stated that the United States would not support:

- measures constraining legal trade and legal manufacturing of small arms and light weapons;
- the promotion of international advocacy activity by international or non-governmental organizations;
- measures that prohibit civilian possession of small arms;
- measures limiting trade in small arms and light weapons solely to governments; and
- a mandatory Review Conference.

The media covering the event used various phrases to describe this speech—tough, mean, hard, a warning, prepared to do battle, adding insult to injury, etc. The speech also made news for reasons other than its tone. The extensive citation of the Second Amendment of the US Constitution reflected the NRA's influence on the American position, which was cemented through the presence of some of its members on the US delegation.

US Under-Secretary of State, John Bolton, warns against a Programme of Action *that intrudes on sovereignty.*

The paragraph in L.4/Rev.1 referring to advocacy (UNGA, 2001a, sec. III, para. 18) was so uncontroversial that it had not drawn one intervention during the third PrepCom. Delegations were truly puzzled by the reference and assumed that the US had included it to increase the unpleasantness of its attack on the draft text. Veteran disarmament diplomats at the Conference had a more sober assessment. Their goal was to reach a consensus on a *Programme of Action*, even if imperfect, for the cost of failure was unacceptable. In their view, two of the points made by Bolton on supplying non-state actors and civilian possession were well known. The brusqueness of the attack was not surprising to them, given that the various drafts of the *Programme* had kept these paragraphs in despite strenuous US objections. The speech merely emphasized that the US was indeed serious and something had to be negotiated on these two points. The paragraph on advocacy was not that important. The point on constraining legal trade and manufacturing was somewhat exaggerated, and in any case many other states shared that view. Finally, many felt that the threat to not support a Review Conference was a bargaining chip that the US would give up in exchange for the deletion of the two sensitive paragraphs on supplying non-state actors and civilian possession.

The final push

On the basis of statements heard during the first week of the Conference, on Monday 16 July the President distributed a revised version of the draft *Programme of Action* (A/CONF.192/L.5). Since very little new text had been suggested since the end of the third PrepCom, this draft mainly reflected the review conducted in the first week of the Conference. The draft contained some new paragraphs that had been agreed during the third PrepCom, but, with a few exceptions, it left much of the language on the contentious issues the same as the previous draft (L.4/Rev.1). Some states, including the Arab League, however, were still uncomfortable with the language on humanitarian concerns and transparency that remained in the draft text.

At this point the President began a series of formal consultations on the key contentious issues, such as the use of the phrase 'excessive and destabilizing accumulations', civilian ownership, supplying non-state actors, transparency, self-determination, export controls, and follow-up. Based on his assessment of the debate, he invited the representatives of selected states to meet with him privately on each issue. By Thursday 19 July there were still a number of paragraphs that could not be agreed.

On Friday 20 July, the last scheduled day of the Conference, the President introduced a package of proposals in the form of six Conference Papers (CRP1–CRP6) with new language on all the outstanding contentious issues. One by one, as the end of the Conference approached, agreement was reached on all issues except supplying non-state actors and civilian ownership, the two issues that the US alone had insisted be dropped if it were to join a consensus on the draft *Programme of Action.*

Given the late hour most delegations had conceded that these two paragraphs would have to be removed. In exchange, the US delegation agreed to allow a follow-up conference (not capitalized) to be held in 2006. They were also able to insist that the date and venue would be decided at the 58th session of the General Assembly in 2003. The African bloc, in a display of unity characteristic of its overall behaviour at the Conference, insisted that the two paragraphs remain in the text. This stalemate held until around 6.00 a.m. on Saturday, 21 July, when the African states agreed to drop the two paragraphs in exchange for a strong final statement by the Conference President describing what had happened. Ambassador Reyes did so during the adoption of the final *Programme* later that afternoon, noting:

The struggle to achieve consensus on the draft *Programme of Action* was in the end reduced to negotiations around two issues: civilian possession and transfers to non-state actors.

> While congratulating all participants for their diligence in reaching this new consensus, I must, as President, also express my disappointment over the Conference's inability to agree, due to the concerns of one State, on language recognizing the need to establish and maintain controls over private ownership of these deadly weapons and the need for preventing sales of such arms to non-State groups.
>
> The States of the region most afflicted by this global crisis, Africa, had agreed only with the greatest of reluctance to the deletion of proposed language addressing these vital issues... They did so strictly in the interests of reaching a compromise that would permit the world community as a whole to proceed together with some first steps at the global level to alleviate this common threat (Reyes, 2001).

In the afternoon of Saturday 21 July, the Conference adopted the orally amended draft *Programme of Action* (A/CONF.192/L.5/Rev.1) and its report to the General Assembly by consensus.

The Programme of Action: Reaching consensus

The struggle to achieve consensus on the draft *Programme of Action* was in the end reduced to negotiations around two issues: civilian possession and transfers to non-state actors. In both cases the US delegation wanted these dropped from the final document, while many other delegations strongly supported their inclusion. While it appeared that the US was isolated on the two issues, many delegations were in fact content to hide behind the US stance and remain silent during the negotiations on them. This section provides an analysis of the *Programme of Action,* focusing on how some of the contentious issues were resolved.

The Preamble (Section I)

In all UN conferences, the negotiating of the Preamble is critical. It establishes agreed causes and consequences of the problem, the norms and principles needed to solve it, and the objectives of the conference itself. If certain proposals fall by the wayside in negotiations, at a minimum the Preamble stands to guide further action by the international community after the conference is concluded. Hence the serious debate on many of the provisions of the *Programme* Preamble.

© Associated Press/Shawn Baldwin

Conference President Camilo Reyes of Colombia and Secretary-General João Honwana of the UN Department for Disarmament Affairs co-ordinating political and procedural aspects.

Definition of small arms and light weapons. Defining the subject of a negotiated document is always controversial. States will logically argue that a programme requires a shared understanding of the problem. On the other hand, delegations opposed to any treaty or programme would just as soon tie up the process in definitional arguments. By the final round of negotiations, three different ways of dealing with the definitional issue had been tabled: specific definitions by the US and the EU that limited the weapons to strictly military types, the definition developed by the 1997 Panel, and no reference at all to a definition. The differences arose from the interest of specific countries in tailoring the *Programme of Action* to the types of small arms and light weapons of greatest concern to them. For example, India, Sri Lanka, and some African states insisted that the *Programme* address ammunition and explosives. The US was under pressure from the NRA to ensure that the *Programme* did not cover civilian weapons.

In the first draft, the solution proposed was to simply refer in a footnote to the definition developed by the 1997 Panel, a three-page section of the report. The option adopted at the final session of the Conference was to have no definition. Proponents explained that precision was not necessary in a political document. The issue melted away in the final days of the Conference. Nowhere in the *Programme* is there a definition of small arms and light weapons.

Humanitarian issues and human rights. Despite some early resistance, most states came to see that small arms were in very important ways a humanitarian issue (HUMANITARIAN). Much of the Preamble reflects this and was agreed early in the process. The one exception was the link between the illicit trade and the excessive accumulation of these weapons on the one hand, and violations of human rights and international humanitarian law on the other. Despite efforts by the EU, Canada, Costa Rica, and some African states, this did not make it into the final text. This reflected the position of a number of states, including China, which made it clear they would not agree to any *Programme of Action* that contained any reference to human rights violations. The final *Programme* thus contains no language on human rights violations.

A number of states made it clear they would not agree to any *Programme of Action* that contained any reference to human rights violations.

Multifaceted nature of the problem. A majority of states came to see small arms as more than a narrowly defined arms control problem. These states made a major effort to get as many dimensions of the problem as possible into the Preamble. This was opposed by other states that did not wish to see the subject expanded into other issue areas. One such dimension was the effect on children, suggested by Canada and some African states. The US particularly wished to limit such references. Another example was Norway's fight to keep 'health' in the preambular paragraph listing the dimensions of the problem, in order to provide a mandate for the World Health Organization to continue to develop its work on this issue. In the end the Preamble covered a wide range of consequences and dimensions of the problem (UNGA, 2001d, sec. I, paras 2, 5, 6), which in effect created additional legitimacy for a full range of international and non-governmental organizations—in addition to national governments—to take action to prevent and reduce problems associated with small arms and light weapons.

Self-determination. Veterans of multilateral disarmament diplomacy were very familiar with the debate that took place over the inclusion of a paragraph that recognizes the right of self-determination of all peoples. While this issue is always contentious, it was particularly so in the *Programme of Action*, since the illicit trade in small arms is very closely linked to self-determination and the arming of non-state actors (TRANSFERS). In the Israeli view, 'reference made to this principle is outside the scope of the mandate of the Conference and could be mistaken as legitimization of the illicit traffic in small arms and their use, thus inducing terrorism on the pretext of self-determination' (Israel, 2001). Israel and several countries facing secessionist threats, including Sri Lanka and India, wanted to eliminate this paragraph or at least tone it down by adding a proviso that the rationale is unacceptable if it challenges the territorial integrity of a state.

In the final *Programme of Action* (sec. I, para. 11), the delegates were able to agree on a final text containing both points of view.[10] However, the Israeli delegation in a note verbale dated 20 July 2001 disassociated itself with the language of this paragraph, asserting that it served 'the narrow political interests of certain participants' (Israel, 2001).

Excessive and destabilizing accumulation. Since the small arms issue emerged on the UN agenda in 1995, the excessive and destabilizing accumulation of these weapons had been identified as one of its major manifestations. This implied that the negative consequences of small arms go beyond the 'illicit trade' in them. The phrase had therefore become a code for addressing the legal acquisition, production, and transfer of these weapons, and was an invitation for controversy.

During the third PrepCom session, the Arab group in particular began to call for the elimination of all references to 'excessive and destabilizing accumulation' and tried to substitute the much narrower 'illicit trade'. In one of the more innovative solutions to stalemate at the Conference, the phrase remained in the Preamble (paragraph 22(c)) but was removed from the Measures section of the *Programme*. The definition of the term, as given in the Report of the 1997 Panel, was included as a footnote reference in the Preamble. The removal of the term from the Measures section meant that no (implied) reference was made to restraint in the legal acquisition or export of these weapons—a disappointment to the EU and others who felt that some sort of restraint provision was crucial to an effective solution.

Preventing, Combating, and Eradicating the Illicit Trade in Small Arms and Light Weapons in All Its Aspects (Section II)

National measures

National transparency. Like the term 'excessive and destabilizing accumulation', the concept of transparency was critical to those states who wanted a strong and effective *Programme of Action*. In their view, legal production and transfers first had to become transparent if the illicit trade was to be tackled. The EU and Canada pushed from the beginning to list government transparency as a basic principle of the *Programme of Action*. While such a principle was in the first draft (L.4), Arab League states and China, among others, objected strenuously to it during the Second PrepCom.

The President's draft of 16 July (L.5) did not mention transparency as a norm or principle in the Preamble. It did contain a national measure committing states to provide relevant information on illicit trade (sec. II, para. 23), but did not extend this, as the EU had proposed, to cover all small arms exports and imports. At the regional level, the final *Programme* included 'measures to enhance transparency' (sec. II, para. 31). There are no transparency measures at the global level.

Marking. Some states wished to go beyond what was accomplished in the UN *Firearms Protocol* (MEASURES) and establish a uniform and mandatory marking system for all small arms. But at the Conference there was support only for language calling for a unique, appropriate, and reliable marking on each small arm as part of the production process (sec. II, para. 7). There was no serious attempt to push for a universal, international marking standard at this point, given the resistance of some states such as China.

Many states felt that the *Programme* had to establish that even legal exports could have negative consequences.

Export and re-export controls. Many states, such as the EU and Canada, felt that, if the *Programme* did not establish that even legal exports could have negative consequences, then it would contain a loophole large enough to render ineffective much of the rest of the *Programme.* The key for these states was export and re-export control provisions that would commit states to consider the implications of their exports in advance.

Throughout the deliberations states such as China, Russia, and some NAM states resisted any measures that would restrict their ability to export or re-export small arms. This was consistent with long-standing differences between arms suppliers like the EU and the US and recipient countries that have historically resisted any restrictions on their ability to acquire military equipment. However, the President's draft of 16 July (L.5) attempted to reach a compromise by including language committing states to assess export applications 'according to strict norms and procedures ... consistent with States' existing responsibilities under international law'. This reference to international law was seen as too weak by the European Union. In a further compromise, the President added 'relevant' before 'international law'. When the phrase 'taking into account in particular the risk of diversion of these weapons into the illegal trade' was added, and the word 'norms' replaced with 'national regulations', a consensus was reached (UNGA, 2001d, sec. II, para. 11).

A similar debate took place over states' ability to re-export weapons imported from another state. In this case, both sides got what they wanted. States undertook to make every effort to 'notify the original exporting State in accordance with their bilateral agreements before the retransfer of those weapons', but 'without prejudice to the right of States to re-export small arms and light weapons that they have previously imported' (sec. II, para. 13).

Civilian possession. A majority of national delegations saw the possession of small arms by civilians as a potential problem and believed that the Conference should deal with the issue. No state, however, was prepared to have common rules for civilian possession applied at the national level. That is why the earlier language in L.4/Rev.1 was general, asking states to 'seriously consider the prohibition of unrestricted trade and private ownership of small arms and light weapons specifically designed for military purposes' (UNGA, 2001a, sec. II, para. 20).

The US, not surprisingly, opposed this provision, citing its own constitution. South Africa stated that keeping this paragraph in the document was a 'red line' issue for it. Norway, among others, gave an extensive defence of the paragraph, arguing that civilian possession was directly linked to many of the consequences outlined in the Preamble.

In the clearest case of the NRA's influence on the US position, the latter rejected any mention of the term 'civilian possession' in the text, no matter how general or lacking in commitment. Several states suggested compromise language stressing the prohibition of fully automatic weapons. Others reminded the US that the language of the civilian possession paragraph came from the Report of the 1999 Group, to which the US and the whole General Assembly had agreed (UNGA, 1999a, para. 120). Several options were presented from the floor in this vein, each of which was dismissed by the US.

Reflecting suggestions from the floor, paragraph 22 of the national measures section of the *Programme* was changed to read: 'To establish or maintain, subject to the respective constitutional and legal systems of States, regulations on the ownership of small arms and light weapons and to apply those regulations in a manner consistent with the Programme of Action.' Again, the US dismissed the deal, as did other states that insisted that the issue was important enough to be retained as an operational measure. The US delegation did try to propose language on this issue, but it was unacceptable to the rest of the Conference. Eventually the Conference decided to drop this paragraph altogether.

BOX 5.4 A North–South divide?

On arms control issues generally, states from the South with little capacity for indigenous defence production jealously guard their right to acquire weapons. This norm is enshrined in the Preamble to the *Programme of Action*: 'Reaffirming also the right of each State to manufacture, import and retain small arms and light weapons for its self-defence and security needs'. This explains why, with few exceptions, these states did not support proposed measures that would interfere with this right.

The result was a North–South divide in the Conference on such measures as export and re-export controls, indicators (definitions) of surplus weapons, transparency relating to the legal small arms trade, and linking the *Programme* to development and the prospect of conditionality. Few, if any, southern states supported these measures, but they were the major focus of the European Union, United States, Japan, and other northern industrialized states. For example, the EU's proposals on these measures were systematically excluded from drafts, putting them in a bad bargaining position as time ran short.

One effect of the rejection of the EU proposals on export controls was a hardening of the positions the EU and its supporters held on these issues. At the beginning of the process, the EU had been pushing only for export control criteria at the regional level. Subsequently, along with Canada, it began insisting that specific criteria be developed at the global level. It also pushed for legally binding treaties on marking and tracing, and brokers—sooner rather than later.

The *Bamako Declaration* was a very influential source of proposals by the NAM states for the *Programme of Action*. Significantly, this declaration contains no language that would restrict the right of states to import weapons.

The Conference as a whole, however, did not divide along North–South lines, since on most other contentious issues the coalitions were less clear. For example, the Arab bloc was basically alone among states from the South in opposing the use of the phrase 'excessive and destabilizing accumulation' as it was seen to relate to the Arab–Israeli conflict. Opposition to legally binding instruments was led by the US, China, the Arab group, and a few other key states from the South. And the South certainly did not behave as a bloc on many issues, including marking and tracing, civilian ownership, transparency, destruction of weapons, and the suggestion that Conference follow-up be enhanced with an *ad hoc* committee to monitor progress on the *Programme of Action*. On these and other issues, Canada, the EU, Norway, and Switzerland were supported by a number of affected states from Africa, Asia, and Latin America.

Supplying arms to governments only. Early on in the Conference process, the US flagged as one of its 'red line issues' the paragraph in the draft *Programme* establishing that 'Exporting countries will supply small arms only to Governments, either directly or through entities authorized to procure arms on behalf of Governments' (UNGA, 2000b, sec. II, para. 9(f)). It objected that this would constrain legal commercial sales to collectors and hunters, and would tie government hands more broadly in the future. The speech by the American Under-Secretary of State John Bolton at the ministerial portion of the Conference made this abundantly clear when he stated that 'this proposal would preclude assistance to an oppressed non-state group defending itself from a genocidal government' (Bolton, 2001). Yet many other delegations took the floor in support of the paragraph, suggesting some slight modifications that did not alter its meaning.

The President's compromise package contained new text on this issue, taking it out of the Measures section and making it part of the Preamble: '*Believing* also that Governments bear the responsibility to supply small arms and light weapons only to Governments, or to entities duly authorised by governments.' As in the case of civilian possession, the United States rejected the proposed text, rebuffing the African bloc which had decided to stand firm against the US on this issue.

It is interesting to note that at no time in the debate did any other state side with the US on the non-state actor question, even though many countries regularly export arms to them. Since they knew that the US would not back down, these states could afford to hide behind the US position—letting it appear as the sole obstacle to agreement on this issue.

Destruction of surplus weapons. Prior to the writing of the first draft, there was a relatively strong consensus on the need to destroy surplus weapons. Indeed, it was already happening in many places. So the opposition that emerged to the measures on destruction of surplus weapons came as a surprise to some.

A primary concern was ensuring that national authorities alone had the authority to declare weapons surplus. An EU-Canadian proposal to have universal guidelines for determining a surplus never gained enough support to be included in the final text. Several countries, especially the Arab group, opposed any reference to international standards for destruction, insisting that this should be left to state discretion. There was also opposition to destruction as the preferred means for disposing of collected weapons. Many countries prefer to recycle confiscated weapons for use by their own police and armed forces, or sell them commercially. The prior agreed text of this paragraph was much weaker than many desired. The final text did, however, contain language encouraging public destruction and voluntary surrender of weapons (UNGA, 2001d, sec. II, para. 20).

There was opposition to destruction as the preferred means for disposing of collected weapons.

Regional measures

With few exceptions, points of contention at the regional level involved concepts that some states would find unacceptable if applied in their own region or at the global level. If these could be discouraged or watered down, it would lower the risk that they could be considered at the global level or applied to their regions.

Since several regions have adopted far-reaching measures, such as the EU, the OAS, and the OSCE (Small Arms Survey, 2001, ch. 7), it was more difficult for states from those regions to object. Instead it was countries such as China and India, along with some of the Arab states, that were most active in limiting agreement on regional measures. To ensure that these would be applied only to specific regions and not globally, the terms 'as needed', 'where appropriate', and 'encouraged' were sprinkled liberally throughout this section of the *Programme*. In this way, most objections were met.

The only example of resistance to a regional issue that lingered to the end of the negotiations was objection to the encouragement of moratoriums on small arms acquisition. This came especially from the Arab group, China, and Cuba. This extreme position, despite a text that clearly left it to the countries of the region to develop and implement such moratoriums, arose from their determination to keep legal transfers sacrosanct. Small arms moratoriums were seen as a threat to legal transfers—a bad precedent that ought not to be encouraged. Despite this opposition, the final text encourages moratoriums on the transfer and manufacture of small arms and light weapons in affected regions as well as 'regional action programmes to prevent, combat and eradicate the illicit trade'. It also calls on countries outside such regions to respect these moratoriums and action programmes (UNGA, 2001d, sec. II, para. 26).

Global measures

Marking and tracing. The fate of marking and tracing measures was more critical than most, especially for France and Switzerland, which had tabled a well-developed plan to put a marking and tracing regime into effect. This was their major initiative to significantly help reduce and prosecute the illicit trade in small arms (France and Switzerland, 2000; 2001a; 2001b). The concept itself was not controversial, as indicated by the strong language in support of national marking that had gone unchallenged since the first draft (L.4) of the *Programme of Action* in December 2000.

© Associated Press/Richard Drew

...ultilateral diplomacy produced a consensus Programme of Action, *but no commitment to negotiate legally binding instruments.*

The controversy initially surrounded the goal of a universal marking system, which disappeared early in the process. This was followed by disagreement on negotiation of a legally binding instrument on marking and tracing. The Arab group, China, and the US weighed in heavily against such measures. China had held such a view since negotiations to develop the UN *Firearms Protocol* started in the mid-1990s, and objected to the call for a commitment to negotiate a legally binding marking and tracing instrument at the global level. American opposition reflected a desire to see efforts to eradicate the illicit small arms trade focused on the regional level, and, if global, placed in the criminal context exemplified by the *Firearms Protocol*. The US also opposed such a commitment as it opposed any reference to legally binding UN disarmament action.

The text that did make it into the final document urged the strengthening of 'the ability of States to co-operate in identifying and tracing in a timely and reliable manner illicit small arms and light weapons' (UNGA, 2001d, sec. II, para. 36). Even without a strong global mandate, advocates of a marking and tracing regime, such as France and Switzerland, believed that this paragraph, along with a much stronger commitment to national action (sec. II, para. 7) was enough to begin developing support for an international mechanism, even prior to the completion of the feasibility study called for in the follow-up section of the *Programme* (sec. IV, para. 1(c)).

Brokering. A similar debate took place on the question of small arms brokering. However, discussion of how to tackle illicit arms brokering was much less advanced than in the case of marking and tracing. Whereas marking was accepted in almost all states and needed only to be elevated to the global level, much less was agreed on brokering. As a result there was much less confidence that the regions and the international community were prepared to start negotiating a legal instrument.

The Conference quickly developed a consensus that states needed to regulate the activities of brokers at the national level, especially since very few states actually had such regulations. The United States suggested that model legislation be developed as a precursor to an international instrument. The final *Programme of Action* has a global measure committing states 'To develop common understandings of the basic issues and the scope of the problems related to illicit brokering' (sec. II, para. 39).

The Conference quickly developed a consensus that states needed to regulate the activities of brokers at the national level, especially since very few states actually had such regulations.

Implementation, International Co-operation, and Assistance (Section III)

Section III was the least controversial part of the *Programme of Action*, and many paragraphs from the earlier draft (L.4/Rev.1) were included in the final document. The only new paragraphs are paragraph 15, relating to the provision of 'assistance to combat the illicit trade in small arms and light weapons linked to drug trafficking, transnational organized crime and terrorism', and paragraph 17, concerned with addressing the 'problems related to human and sustainable development' in post-conflict situations.

The majority of paragraphs in this section are concerned with international co-operation, the sharing of information, and the provision of financial and technical assistance with respect to issues such as marking and tracing, stockpile management, the destruction of surplus stocks, and the disarmament, demobilization, and re-integration of ex-combatants. Many delegations, particularly from NAM states, insisted on the inclusion of language such as 'upon request' or 'subject to their national practices' to ensure that such assistance did not undermine national sovereignty. The concerns of donor states and international organizations were also reflected in the use of language such as 'where appropriate', 'in a position to do so', and 'within existing resources'. The US particularly resisted any language that implied any financial obligations or commitments, and in many places the language was diluted to suggest non-financial forms of assistance.

Follow-up to the United Nations Conference on the Illicit Trade in Small Arms and Light Weapons in All Its Aspects (Section IV)

During the PrepComs and throughout the Conference, views had diverged widely on how formal the follow-up to the Conference and its *Programme of Action* should be. This was reflected in the struggle to reach consensus on Section IV.

Time frame for review meetings. Many countries wanted a firm commitment establishing when the Conference review process would start. Many were anxious to see something happen before 2006. Some suggested biennial meetings, starting in 2003. At the insistence of the US, the final wording in the compromise package produced by the Conference President on the last day of the Conference called for a conference (not capitalized) in 2006, the date and venue to be decided by the 58th session of the General Assembly in 2003. This conference will review progress on implementation of the *Programme of Action* (UNGA, 2001d, sec. IV, para. 1(a)).

Section IV also provides for biennial meetings, but does not state when these will begin (sec. IV, para. 1(b)).[11] This was a classic compromise, setting a firm date for a review and establishing a monitoring mechanism (biennial meetings), while giving the US and other sceptics time to resist a review conference if the implementation of the *Programme of Action* is not going well. It also tries to keep the process within the UN, giving those who might be unhappy with the rate of progress reason not to take the issue outside the UN.

Topics for review. The agenda of any review process was equally important to Conference follow-up. This involved a repeat of the debate over determining the agenda for the Conference itself. Colombia submitted a lengthy proposal outlining the information it thought should be submitted for review. Canada and the EU wanted the follow-up to include not just monitoring implementation of the *Programme* but also consideration of new measures. Until the very end of the negotiations, the draft of the Conference President included the mandate to 'examine ways to strengthen and develop measures contained' in the *Programme*. This was resisted and dropped from the final text. It is now an open question whether the work of the biennial meetings will deal exclusively with the *Programme of Action* as written.

***Ad hoc* review mechanisms.** Sharply divergent views developed on how formal the follow-up to the Conference and its *Programme of Action* should be. The first draft of the *Programme of Action* (L.4) suggested a variety of mechanisms, including a specific *ad hoc* mechanism to assess and monitor progress implementing the *Programme of Action* soon after the end of the Conference. A number of states, including South Africa, responded that since the General Assembly called for the Conference, it was up to the General Assembly to develop all aspects of the review process.

Brazil, Canada, and many other states continued to advocate an *ad hoc* mechanism until the very end of the Conference. Some states argued that they could not afford to participate in such *ad hoc* mechanisms, as opposed to formal General Assembly-sanctioned meetings that provided funding. But most of the opposition stemmed from fears that an independent global body might impinge on states' national small arms policies. No reference to an *ad hoc* mechanism is included in the final text.

BOX 5.5 Small arms and terrorism

The association between terrorism and the illicit trade in small arms and light weapons appears in several places in the final *Programme of Action*:

I. Preamble...

5. Recognizing that the illicit trade in small arms and light weapons in all its aspects sustains conflicts, exacerbates violence, contributes to the displacement of civilians, undermines respect for international humanitarian law, impedes the provision of humanitarian assistance to victims of armed conflict and fuels crime and terrorism...
7. Concerned also about the close link between terrorism, organized crime, trafficking in drugs and precious minerals and the illicit trade in small arms and light weapons, and stressing the urgency of international efforts and co-operation aimed at combating this trade simultaneously from both a supply and demand perspective...

II. Preventing, combating, and eradicating the illicit trade in small arms and light weapons in all its aspects...

38. To encourage States to consider ratifying or acceding to international legal instruments against terrorism and transnational organized crime...

III. Implementation, international co-operation, and assistance...

15. Upon request, States and appropriate international or regional organizations in a position to do so should provide assistance to combat the illicit trade in small arms and light weapons linked to drug trafficking, transnational organized crime and terrorism...

Sri Lanka took the lead in attempting to strengthen the provisions relating to terrorism. It was unsuccessful in adding to the Preamble references to 'hostage taking, bombing, human smuggling and money laundering'. However, it was successful in adding financing to paragraphs 4 and 6 of Section II (national measures).

Given the new emphasis on terrorism after the events of 11 September 2001, the *Programme of Action* highlights the importance of addressing the tools of violence as well as terrorism itself.

Reporting. Various proposals called for reporting on the progress of countries in implementing the *Programme of Action.* Until the very last moment, the draft text included provision for biennial country reports to the Secretary-General and another produced by the Secretary-General. As with other specific follow-up mandates, these were dropped in the final text.

International legal instruments. Before and during the Conference, there were efforts to get states to agree to negotiate legally binding global instruments on marking and tracing, as well as brokering. In the end, due to concerns of states such as China and the US, the final text calls only for a UN study on the feasibility of developing an international tracing instrument (UNGA, 2001d, sec. IV, para. 1 (c)). As for brokering, the final document calls on states 'To consider further steps to enhance international co-operation in preventing, combating and eradicating illicit brokering in small arms and light weapons' (para. 1 (d)). This was weaker than the text of the earlier draft (UNGA, 2001a, sec. IV, para. 1(d)). Despite the absence of agreement on international legal instruments, Sections II and III of the final *Programme of Action* contain a range of operational measures at the national and global levels on marking and tracing, as well as brokering.

Action by international and non-governmental organizations in support of the Programme of Action. Paragraph 2 of the follow-up section provides a mandate for all actors—not only national governments—to contribute to the implementation of the *Programme of Action.* It encourages 'the United Nations and other appropriate international and regional organizations to undertake initiatives to promote the implementation of the Programme'. It further encourages 'non-governmental organizations and civil society to engage, as appropriate, in all aspects of international, regional, subregional and national efforts to implement the present Programme' (UNGA, 2001d, sec. IV, para. 2).

These paragraphs were essential to those states that wanted a mandate to proceed vigorously with supporting efforts to deal with small arms. Equally important for this purpose was paragraph 2(b), encouraging 'all initiatives to mobilize resources and expertise to promote the implementation of the Programme of Action and to provide assistance to States in their implementation of the Programme'. In principle, almost everything done to address the small arms problem can now be supported by the UN *Programme of Action.*

Assessing the UN Small Arms Conference

A successful outcome?

After nearly seven months of negotiations, culminating in the final two weeks in New York, the Conference was able to agree by consensus on a *Programme of Action.* This achievement by itself, given the state of the multilateral climate on the small arms issue even a few years ago, means that the Conference can be considered a success.

The Conference itself was also important for a number of reasons. First, it generated an international debate on the issue and forced states to go on record with their views on the causes and consequences of the problem, as well as on the action they supported.[12] Those governments and NGOs preparing to take the issue forward now have a much better idea of which countries might be part of any coalition of 'like-minded states' that could provide leadership in the next five years. The Conference also generated two weeks of high-level international media attention and as such raised awareness of the various dimensions of the small arms issue. This consciousness-raising aspect of the

The various partnerships built by the Conference will be crucial for future efforts to address the different dimensions of the small arms issue.

Conference should not be underestimated, and may prove to be one of its important legacies. Finally, the Conference helped to build partnerships amongst civil society groups, and between NGOs and government delegations, partnerships that will be crucial for future efforts to address the different dimensions of the small arms issue (see Box 5.2).

The *Programme of Action* itself can also be regarded as a positive outcome in so far as it defines some of the norms and principles that will guide the work of the international community on the small arms issue in the coming years. In particular, it confirms that the problems caused by the proliferation and misuse of small arms are multifaceted, going beyond arms control and disarmament to include conflict prevention, development, crime control, public health, and humanitarian dimensions. It also establishes that durable solutions to armed conflicts must focus on the tools of violence as well as root causes. Furthermore, it assigns primary responsibility for solving the problems of small arms to national governments, which have committed themselves to wide-ranging improvements in the way they manage the production, acquisition, export, re-export, storage, collection, and destruction of surplus small arms and light weapons. Finally, the *Programme of Action* highlights both regional and global dimensions of the problem, dimensions that dictate co-operation at all levels.

Despite major efforts by some states to restrict the language of the document to just the illicit trade in small arms and light weapons, the document does agree to treat the problem 'in all its aspects'. Finally, the document commits states to a follow-up process that includes progress reports, biennial meetings, a review conference in 2006, and, most importantly, a mandate for the mobilization of resources and expertise by states, in co-operation with civil society and international organizations, to promote the implementation of the *Programme of Action.*

Falling short?

But while the Conference and its *Programme of Action* can be deemed successful from certain perspectives, a number of factors temper this positive assessment. The language of the *Programme of Action* is non-binding, and leaves wide margins for states to exercise discretion or interpretation through frequent use of such clauses as 'where applicable', 'as appropriate', 'where needed', or 'on a voluntary basis'. More importantly, several crucial issues identified by certain states and NGOs prior to and during the Conference have not been tackled or included in the final *Programme of Action.* These include concrete commitments to:

Documents do not in themselves destroy surplus weapons, secure stockpiles, punish illicit traffickers, or increase the security of vulnerable groups.

- negotiate an international instrument on marking and tracing;
- regulate the civilian possession of weapons;
- increase transparency in the legal production, stockpiling, and trade in small arms;
- control transfers to non-state actors;
- agree upon specific criteria governing arms exports; and
- negotiate an international legal instrument on arms brokering.

Finally, documents, no matter how consensual or well-crafted, do not in themselves destroy surplus weapons, secure stockpiles, punish illicit traffickers, or increase the security of vulnerable groups such as women and children. Ultimately, the success of the Conference and the *Programme of Action* will have to be judged by the effective implementation of its various measures at national, regional, and global levels.

Conclusion

Whether the 2001 UN Small Arms Conference can be considered a success ultimately depends on which of the points made in the preceding section one regards as most significant. While the *Programme of Action*, which was agreed by consensus by all participating states, is not legally binding, it does contain a concrete set of measures at national, regional, and global levels. As a result, it represents the first global framework to guide the work of national governments, regional and international organizations, and civil society in combating the illicit trade in small arms, and has served to raise the level of commitment of states to address the illicit trade in small arms. Finally, it provides the justification for all actors, including NGOs, to monitor, report on, encourage, and if necessary apply political pressure to those states that are not meeting their commitments.

© Reuters/Peter Morgan

The real cost of small arms is symbolized by shoes representing victims of gun violence.

The holding of such a large-scale Conference may also pave the way for more innovative and effective approaches to international disarmament issues in the future. The language of the *Programme of Action* could provide the basis for at least three ways forward. First, the process could remain under the auspices of the UN First Committee, and thus be dominated by an arms control and disarmament perspective. This would represent a minimalist approach, given the well-demonstrated need to broaden the scope beyond arms control perspectives. Second, continued disappointment with omissions in or the implementation of the *Programme of Action* by states may produce a version of the Ottawa Process outside UN auspices and focusing on the humanitarian and economic costs associated with the proliferation and use of small arms. Third, the UN could change the way it tackles the problem by recognizing that the small arms issue is multi-dimensional and by creating a mechanism that allows the consideration of the economic, social, and humanitarian effects of the accumulation, proliferation, and misuse of small arms, and the effective participation of a wide range of stakeholders.

5. List of Abbreviations

1997 Panel	United Nations Panel of Governmental Experts on Small Arms
1999 Group	Group of Governmental Experts on Small Arms
ARF	ASEAN Regional Forum
ASEAN	Association of Southeast Asian Nations
Conference	United Nations Conference on the Illicit Trade in Small Arms and Light Weapons in All Its Aspects
ECOWAS	Economic Community of West African States
EU	European Union
IANSA	International Action Network on Small Arms
ICBL	International Campaign to Ban Landmines
NAM	Non-Aligned Movement

NRA	National Rifle Association
OAS	Organization of American States
OAU	Organization of African Unity
OSCE	Organization for Security and Co-operation in Europe
PrepCom	Preparatory Committee for the United Nations Conference on the Illicit Trade in Small Arms and Light Weapons in All Its Aspects
SAARC	South Asian Association for Regional Cooperation
SADC	Southern African Development Community
WFSA	World Forum on the Future of Sport Shooting Activities

5. Endnotes

1 The full text of the final *Programme of Action* is included in the *Report of the United Nations Conference on the Illicit Trade in Small Arms and Light Weapons in All Its Aspects* (UNGA, 2001e).

2 IANSA was officially launched at the Hague Appeal for Peace conference in The Hague, in May 1999. Its predecessor, the Preparatory Committee for a Global Campaign on Small Arms and Light Weapons, began as a web site, <http://www.prepcom.org>, in January 1998. See Batchelor (2002).

3 In essence, the Swiss government was offering to fund travel and support for representatives from states not already represented in Geneva, most of which were affected by the small arms problem. This would have allowed for the participation of experts from capitals, not just permanent representatives in the New York missions. Nevertheless, a majority of states felt they would be better represented in New York. States not keen on widening the Conference beyond arms control and disarmament also pushed—successfully—to keep it in New York, within the framework of the First Committee process.

4 See the events listed in the Annex to the final report of the Conference (UNGA, 2001e).

5 As of January 2002, all OAU states except the Democratic Republic of Congo had signed the Constitutive Act of the OAU's successor organization, the African Union. In July 2002, South Africa is scheduled to host the inaugural Summit of the Assembly of the African Union.

6 In 1994, Mali asked the UN to assist in the collection of light weapons circulating in the country. This led to the dispatch of a UN advisory mission to Mali in August 1994 and to the General Assembly's adoption of Resolution 49/75 in December 1994, *Assistance to States for Curbing the Illicit Traffic in Small Arms and Collecting Them* (UNGA, 1994). This was the first, formal recognition by the United Nations of the small arms problem, and there has been a similar resolution every year since.

7 These countries were: Bulgaria, Cyprus, Estonia, Hungary, Latvia, Lithuania, Malta, Poland, Romania, Slovakia, and Slovenia.

8 See, for example: UNGA (2001d, sec. I, para. 16; sec. II, para. 40; sec. III, para. 2; sec. IV, para. 2(c)).

9 On 4 June 2001 the Chairman issued his Informal Paper outlining his assessment of the discussions at the third PrepCom. It summarized the differences that remained, by *Programme* section and by issue.

10 The text used in the final draft of the *Programme of Action* was taken from the *Declaration on the Occasion of the Fiftieth Anniversary of the United Nations* (UNGA, 1995a).

11 Note that the UN General Assembly, in its Resolution 56/24V of 24 December 2001, set 2003 as the date of the first biennial meeting (UNGA, 2001f, para. 2).

12 See the Small Arms Survey's UN Conference database for details of country statements and positions: <http://www.smallarmssurvey.org>

5. Bibliography

Atwood, David. 2001. *NGOs and the 2001 UN Conference on Small Arms.* Background paper. Geneva: Small Arms Survey.

Batchelor, Peter. 2001. 'The 2001 UN Conference on Small Arms: A First Step?' *Disarmament Diplomacy.* No. 60. September.

—. 2002. 'NGOs and the Small Arms Issue.' *Disarmament Forum.* No. 1, pp. 37–40.

Bolton, John. 2001. Statement made at the United Nations Conference on the Illicit Trade in Small Arms and Light Weapons in All Its Aspects. 9 July. Available on the Small Arms Survey's UN Conference database: <http://www.smallarmssurvey.org>

Boutros-Ghali, Boutros. 1996. 'Global Leadership After the Cold War.' *Foreign Affairs.* Vol. 75, No. 2. March–April, pp. 86–98.

Brasilia Declaration of the Regional Preparatory Meeting of the Latin American and Caribbean States for the 2001 United Nations Conference on the Illicit Trade in Small Arms and Light Weapons in All Its Aspects ('Brasilia Declaration'). 2000. 24 November. Reproduced in UN document A/CONF.192/PC/19 of 19 December.

Convention on the Prohibition of the Use, Stockpiling, Production and Transfer of Anti-Personnel Mines and on Their Destruction ('Ottawa Treaty'). 1997. Ottawa, 3 December. In force 1 March 1999. <http://domino.un.org/TreatyStatus.nsf>

dos Santos, Carlos. 2000. *Draft Structure and Elements for a Programme of Action to Prevent, Combat and Eradicate the Illicit Trade in Small Arms and Light Weapons in All Its Aspects.* Informal Paper. October.

Economic Community of West African States (ECOWAS). 1998. *Declaration of a Moratorium on Importation, Exportation and Manufacture of Light Weapons in West Africa.* Abuja, 31 October. Reproduced in UN document A/53/763 – S/1998/1194 of 18 December.

European Union (EU). 1997. *European Union Programme for Preventing and Combating Illicit Trafficking in Conventional Arms.* 26 June. Reproduced in UN document A/CONF.192/PC/3 of 13 March 2000.

—. 1998a. *European Union Code of Conduct on Arms Exports.* 8 June. Reproduced in UN document A/CONF.192/PC/3 of 13 March 2000.

—. 1998b. *Joint Action of 17 December 1998 Adopted by the Council on the Basis of Article J.3 of the Treaty on European Union on the European Union's Contribution to Combating the Destabilising Accumulation and Spread of Small Arms and Light Weapons ('Joint Action on Small Arms').* 1999/34/CFSP. 17 December. Reproduced in Official Journal L 9 of 15 January 1999.

—. 2000. *Plan of Action to Prevent, Combat and Eradicate the Illicit Trade in Small Arms and Light Weapons in All Its Aspects.* Reproduced in UN document A/CONF.192/PC/21 of 28 December.

France and Switzerland. 2000. *Food-for-Thought Paper: Contribution to the Realisation of an International Plan of Action in the Context of the 2001 Conference: Marking, Identification and Control of Small Arms and Light Weapons.* Reproduced in UN document A/CONF.192/PC/7, Annex. 17 March.

—. 2001a. *Working Paper by Switzerland and France on Establishing a Tracing Mechanism to Prevent and Reduce Excessive and Destabilizing Accumulation and Transfer of Small Arms and Light Weapons.* Reproduced in UN document A/CONF.192/PC/25. 10 January.

__. 2001b. *Franco-Swiss Workshop on Traceability of Small Arms and Light Weapons: Tracing, Marking and Record-Keeping. Geneva, 12–13 March 2001. Chair's Summary.* Reproduced in UN document A/CONF.192/PC/38, Annex. 23 March.

Greene, Owen. 2001. *The 2001 Conference and Other Initiatives.* Background paper. Geneva: Small Arms Survey.

Human Rights Watch (HRW). 2001. 'U.N.: "Program of Inaction" on Small Arms.' *Human Rights News.* New York, 19 July. <http://www.hrw.org/press/2001/07/smallarms0719.htm>

International Action Network on Small Arms (IANSA). 2001. 'Opportunity Squandered to Introduce Tougher Arms Controls.' Press Release. 21 July. <http://www.iansa.org/calendar/2001UN/confnews/pr_oppor.htm>

Israel. 2001. *Note Verbale Dated 20 July 2001 from the Permanent Mission of Israel to the United Nations Addressed to the Secretariat of the United Nations Conference on the Illicit Trade in Small Arms and Light Weapons in All Its Aspects.* UN document A/CONF.192/14 of 31 July.

Krause, Keith. 2002. 'Multilateral Diplomacy, Norm-Building and UN Conferences: The Case of Small Arms and Light Weapons.' *Global Governance.* Vol. 8, No. 2. Spring.

Mali. 1996. Statement introducing draft resolution A/C.1/51/L.35, 16th meeting of the UNGA First Committee, 51st session. 6 November. UN document A/C.1/51/PV.16.

__. 1997. Statement introducing draft resolution A/C.1/52/L.8, 16th meeting of the UNGA First Committee, 52nd session. 6 November. UN document A/C.1/52/PV.16.

Organization for Security and Co-operation in Europe (OSCE). Forum for Security Co-operation. 2000. *OSCE Document on Small Arms and Light Weapons.* 24 November. FSC.DOC/1/00. <http://www.osce.org/docs/english/fsc/2000/decisions/fscew231.htm>

Organization of African Unity (OAU). 2000. *Bamako Declaration on an African Common Position on the Illicit Proliferation, Circulation and Trafficking of Small Arms and Light Weapons ('Bamako Declaration').* Bamako, Mali. 1 December. SALW/Decl. (I).

Organization of American States (OAS). 1997. *Inter-American Convention against the Illicit Manufacturing of and Trafficking in Firearms, Ammunition, Explosives, and Other Related Materials ('Inter-American Convention').* Washington, DC. 13 November. AG/RES.1 (XXIV-E/97). Reproduced in UN document A/53/78 of 9 March 1998.

__. 1998. *Model Regulations for the Control of the International Movement of Firearms, Their Parts and Components, and Ammunition ('Model Regulations').* 2 June. AG/RES. 1543 (XXVIII-O/98).

Reyes, Camilo. 2001. *Statement by the President of the Conference after the Adoption of the Programme of Action to Prevent, Combat and Eradicate the Illicit Trade in Small Arms and Light Weapons in All Its Aspects.* Reproduced in UN document A/CONF.192/15 (Annex).

Small Arms Survey. 2001. *Small Arms Survey 2001: Profiling the Problem.* Oxford: Oxford University Press.

South Africa. 1998. Statement introducing draft resolution A/C.1/53/L.41/Rev.1, 20th meeting of the UNGA First Committee, 53rd session. 30 October. UN document A/C.1/53/PV.20.

Southern African Development Community (SADC). 1998. *Southern Africa Regional Action Programme on Light Arms and Illicit Arms Trafficking.* November. Reproduced in Andrew McLean and Elizabeth Clegg, eds. 1999. *Towards Implementation of the Southern Africa Regional Action Programme on Light Arms and Illicit Trafficking, 8–9 September 1999, Pretoria, South Africa: Seminar Report.* Pretoria: Institute for Security Studies, pp. 71–85.

Stohl, Rachel, 2001. 'United States Weakens Outcome of UN Small Arms and Light Weapons Conference.' *Arms Control Today.* Vol. 31, No. 7. September, pp. 34–5.

United Nations (UN). 2001a. *Small Arms Conference Concludes with Consensus Adoption of Action Programme.* Press release DC/2795 of 21 July.

__. 2001b. *Secretary-General Congratulates Conference On 'Important First Steps' To Combat Small-Arms Scourge.* Press Release SG/SM/7896–DC/2796 of 21 July.

United Nations Department for Disarmament Affairs (UNDDA). 2001. 'A Programme of Action.' *DDA 2001 Update.* June-July.

United Nations General Assembly (UNGA). 1994. Resolution 49/75G, adopted 15 December. A/RES/49/75 of 15 December.

__. 1995a. *Declaration on the Occasion of the Fiftieth Anniversary of the United Nations.* Resolution 50/6, adopted 24 October. A/RES/50/6 of 24 October.

__. 1995b. Resolution 50/70B, adopted 12 December. A/RES/50/70 of 12 December.

__. 1997a. *Report of the Panel of Governmental Experts on Small Arms.* A/52/298 of 27 August.

__. 1997b. Resolution 52/38J, adopted 9 December. A/RES/52/38 of 8 January 1998.

__. 1998. Resolution 53/77E, adopted 4 December 1998. A/RES/53/77 of 12 January 1999.

__. 1999a. *Report of the Group of Governmental Experts on Small Arms.* A/54/258 of 19 August.

__. 1999b. Resolution 54/54V, adopted 15 December. A/RES/54/54 of 10 January 2000.

__. 2000a. *Report of the Preparatory Committee for the United Nations Conference on the Illicit Trade in Small Arms and Light Weapons in All Its Aspects.* A/CONF.192/PC/9 of 21 March.

__. 2000b. *Draft Programme of Action to Prevent, Combat and Eradicate the Illicit Trade in Small Arms and Light Weapons in All Its Aspects.* UN document A/CONF.192/PC/L.4 of 11 December.

__. 2001a. *Draft Programme of Action to Prevent, Combat and Eradicate the Illicit Trade in Small Arms and Light Weapons in All Its Aspects.* UN document A/CONF.192/PC/L.4/Rev.1 of 12 February.

__. 2001b. *Decision on the Modalities of Attendance of Non-governmental Organizations at the Sessions of the Preparatory Committee as well as the Conference.* A/CONF.192/PC/39 of 23 March.

__. 2001c. *Protocol against the Illicit Manufacturing of and Trafficking in Firearms, Their Parts and Components and Ammunition, Supplementing the United Nations Convention against Transnational Organized Crime ('Firearms Protocol').* Reproduced in UN document A/RES/55/255 of 8 June.

__. 2001d. *Programme of Action to Prevent, Combat and Eradicate the Illicit Trade in Small Arms and Light Weapons in All Its Aspects ('Programme of Action').* 20 July. Reproduced in UN document A/CONF.192/15.

__. 2001e. *Report of the United Nations Conference on the Illicit Trade in Small Arms and Light Weapons in All Its Aspects.* July. A/CONF.192/15. <http://www.un.org/Depts/dda/CAB/smallarms/files/aconf192_15.pdf>

__. 2001f. Resolution 56/24V, adopted 24 December. A/RES/56/24 of 10 January 2002.

Acknowledgements

Main contributors

David Atwood and Owen Greene.

Strengthening Controls:
Small Arms Measures

6

Introduction

Seven centuries after the invention of small arms, governments are beginning to gain control over these weapons. Efforts at regulation are as old as small arms themselves, but until quite recently have had limited reach. The present regulatory framework stands apart from its predecessors in its complexity and geographical scope. At the same time, an increasingly broad spectrum of activities is coming under scrutiny and, in some cases, effective control—for example, production, stockpiling, trade, possession, and use. As this chapter will explain, the road from producer to end-user and end use is often complex. So too are the requirements of effective small arms regulation.

This chapter examines a few, key developments in small arms control. It focuses on select regulatory regimes, which employ various norms and instruments to control small arms at specific points in their life cycles. These studies are supplemented by an examination of a specific instrument (the UN *Firearms Protocol*), region (eastern Africa), and type of actor (NGOs). This chapter has a thematic orientation and does not attempt to systematically update the review of multilateral measures contained in the *Small Arms Survey 2001*, though some updating does occur. In addition, two especially significant new initiatives—the *Firearms Protocol* and the 2001 UN Small Arms Conference—are examined separately, in this chapter and in Chapter 5 (CONFERENCE).

The requirements of effective small arms regulation are complex.

BOX 6.1 UN definitions—a reminder

Small arms: revolvers and self-loading pistols, rifles and carbines, assault rifles, sub-machine guns, and light machine guns.

Light weapons: heavy machine guns, hand-held under-barrel and mounted grenade launchers, portable anti-tank and anti-aircraft guns, recoilless rifles, portable launchers of anti-tank and anti-aircraft missile systems, and mortars of less than 100mm calibre.

Source: UN (1997)

'Control' means ensuring that small arms are held for approved purposes by individuals or groups which, in the judgement of relevant authorities, can be trusted to not misuse them. Control efforts confront a series of problems. One is the remarkable longevity of small arms. If stored carefully, small arms can have a very long life cycle. Usually several decades pass before the original weapon becomes unusable, but its component parts—recycled in newer weapons—may last even longer. Small arms typically pass through many hands before the end of their lives, further complicating control efforts. This life cycle typically begins with production and moves on at some stage to possession, but does not necessarily end with first possession. Transfer, re-transfer, and stockpiling are all part of the normal small arms life cycle. The regulatory regimes examined in this chapter are designed to maintain control over small arms at various stages of this cycle. When efforts to control small arms fail, they frequently end up in the wrong hands for the wrong purposes.

A displaced girl in San Markande, Nicaragua. (© ICRC/Philippe Merchez)

The chapter's key findings are as follows:

- The UN *Firearms Protocol* is a potentially important new instrument for combating the illicit firearms trade and improving control over small arms generally.
- The harmonization of standards and systems between states will be essential to the development of effective small arms tracing.
- Harmonization is also an issue with respect to the end-use practice of exporting countries, currently a soft target for illicit traffickers.
- The regulatory spotlight is slowly shifting to the important, yet still largely uncontrolled, activity of arms brokering.
- While the need for good stockpile management is acknowledged at the international level, this remains a national responsibility, with international action focused on building capacity in newly democratic and developing countries.
- Regulation of civilian ownership at the national level remains a key component of broader international small arms control efforts.

In contrast to Chapter 5 (CONFERENCE), with its emphasis on process, this chapter focuses, wherever possible, on specific control measures. The Conference *Programme of Action* is frequently non-committal in its treatment of the various issues. The emphasis here is on specifics, including actual implementation. The legally binding *Firearms Protocol*, though focusing narrowly on illicit firearms manufacture and trade, is considerably more operational, in conception, than the Conference *Programme of Action*, and is the subject of the chapter's first section. The issues of marking and tracing, export controls, and brokering regulation—all addressed in some form in the *Firearms Protocol*—are examined in subsequent sections. A large part of the chapter is also devoted to an examination of national firearms controls in selected states which, not surprisingly, given their close connection with national sovereignty, figure in neither the Protocol nor the *Programme of Action*.

Below are some of the issues the chapter explores.

- **What control measures are in place or under development at the national, regional, and global levels?**
- **What functions do these measures perform?**
- **Where and when is the multilateral harmonization of national control norms desirable?**
- **What are the relative strengths and weaknesses of the measures examined here?**

Following the analysis of the *Firearms Protocol*, the chapter tends to a more thematic approach, examining different types of control measures. Much of this is necessarily situated at the national level, yet multilateral initiatives, whether existing or in prospect, are also part of the discussion. The review of select regulatory regimes is complemented by two short pieces, the first focusing on the efforts of East African countries to strengthen controls in their subregion, the second discussing the important roles NGOs play in small arms control efforts. We step back in the conclusion to look at the broader picture—assessing the state of efforts to bring small arms under effective control from the various perspectives outlined above.

The UN Firearms Protocol

Just a few short weeks before the opening of the 2001 UN Small Arms Conference in New York, the UN General Assembly opened for signature a major new instrument on small arms and light weapons. Although it is the first legally binding measure on small arms to be adopted at the global level, agreement on the *Firearms Protocol* (UNGA, 2001b) attracted little attention at the time. Negotiations on the Protocol had dragged on far longer than originally intended, lagging well behind agreement on its parent convention on transnational organized crime, because a number of contentious issues held up agreement for some months, including the definition of firearms, the applicability of the Protocol to non-commercial transfers, and the marking of weapons at time of manufacture. Nevertheless, in the end consensus did prove possible on a number of measures to close loopholes in national regulations and criminalize the illicit firearms trade. The Protocol is gathering signatures in New York,[1] but it remains to be seen whether the delay in finalizing the agreement, and the dwindling enthusiasm that accompanied the last stage of negotiations, will have a negative impact on its entry into force and eventual implementation.

The Protocol against the Illicit Manufacturing of and Trafficking in Firearms, Their Parts and Components and Ammunition is one of three supplemental protocols to the *United Nations Convention against Transnational Organized Crime* (UNGA, 2000). The Convention itself and two of the three protocols were agreed and signed in Palermo, Italy, in December 2000, in a blaze of publicity, but the *Firearms Protocol* required an additional negotiating session to secure compromises on various thorny issues. It was finally agreed in March 2001 and opened for signature at the UN General Assembly in New York on 31 May 2001.

Small arms transfers to non-state actors—of prime concern to many African countries—was a hot topic at the Firearms Protocol *negotiations.*

The road to the Protocol

In the world of diplomacy, the *Firearms Protocol* was agreed very quickly. Just over two years after the General Assembly resolution which established the framework for negotiations (UNGA, 1998), the Convention and its three protocols were complete—an outcome that reflected the high priority given the issue by a large number of governments. Unfortunately, some of this momentum appears to have been lost at the end of the negotiations on the *Firearms Protocol*, while the other protocols gained in importance. For a number of governments, combating trafficking in human beings became a noticeably higher priority than tackling gun-running.

However, the idea of a convention on organized crime was mooted much earlier on. It was first proposed at the World Ministerial Conference on Organized Crime in Naples in 1994, and taken up at the Ninth Congress on Crime Prevention in Cairo later the same year. Over the next few years, a number of meetings and consultations, as well as an Inter-governmental Expert Group, moved the concept forward, so that by 1998 Poland was able to propose a first draft of a convention to the General Assembly. In response, the General Assembly established an Ad Hoc Committee to elaborate a new comprehensive international convention against trans-national organized crime along with three additional protocols dealing with specific issues related to this broad problem (UNGA, 1998).

Although all four instruments covered important and sometimes controversial issues, the firearms negotiations were probably the most difficult. Agreement between negotiating states was often elusive, as countries with very different problems proposed widely divergent solutions. Many of the proposals which fell by the wayside would have fundamentally broadened the scope and nature of the agreement.

The process also attracted considerable attention from NGOs, most of which favoured a strong agreement, although some, such as the National Rifle Association (NRA) and the World Forum on the Future of Sport Shooting Activities (WFSA), were opposed to further restrictions on gun ownership. Their participation worried many NGOs and governments, who feared an attempt to undermine agreement. In the event, these organizations did succeed in excluding regulations on domestic possession, which may create problems for the enforcement of measures which remain part of the Protocol. For example, national registries of gun possession could have helped law enforcement agencies to trace weapons after they crossed international borders. Without such registries, the chain of information is more easily broken; it becomes much harder to ascertain where the firearm originally came from. The pro-gun lobby also secured preambular language on the rights of legal gun owners, but did not obstruct other measures, in particular firearms marking. It can be argued that they are tied into the process so that any active opposition to ratification would be hypocritical.

A narrow, enforceable approach

The drafters of the Protocol were determined to keep it narrowly focused on a law-enforcement approach.

The drafters of the Protocol were determined to keep the language narrowly focused on a crime and law-enforcement approach, taking as their model the convention against illicit firearms trafficking agreed by the Organization of American States (OAS) in 1997, the first international convention covering firearms (OAS, 1997).[2] Many of the Protocol's strongest supporters, such as Canada and the United States, both of which were signatories to the OAS convention, wanted to steer the Vienna negotiations clear of issues that were considered arms control rather than crime control measures. They feared that a comprehensive agreement would be weaker and less enforceable. Towards the end of the negotiations, many countries looked to the UN Conference as a more suitable forum for discussion of the more sensitive issues.

As a result, the *Firearms Protocol* is limited in scope and content. It defines the term 'firearms' more narrowly than its OAS predecessor, including the word 'portable' so as to exclude artillery. Also unlike the OAS convention, it excludes explosives and explosive devices, such as grenades, which were deemed to require a separate agreement. In general, the Protocol focuses on just a few areas where it seeks to set standards for national systems and eliminate differences that can be exploited by international traffickers. While the Protocol does not raise prevailing benchmarks, once ratified and implemented it will nevertheless bring about a certain harmonization among states in the areas it does cover, albeit at a low level.

Main provisions

The declared aim of the *Firearms Protocol* is 'to prevent, combat and eradicate the illicit manufacturing of and trafficking in firearms, their parts and components and ammunition' (UNGA, 2001b, art. 2). For this purpose, it deploys two sets of instruments: a system of government authorizations[3] and a system of marking and tracing.[4] Legal transfers of small arms are to be conducted on the basis of reciprocal arrangements between the governments of the states involved. Thus, a licence to export firearms cannot be issued without a corresponding import licence from the importing country. The transaction is also dependent on the approval of any transit countries (art. 10.2). 'Illicit trafficking' is thus defined as a transfer that is not authorized in accordance with these procedures (art. 3.e).

The Protocol requires firearms to be marked at the point of manufacture, import, and transfer from government into private hands, although, as discussed below, it stops short of imposing a new system of universally recognizable markings. Marking was considered to be so vital for combating illicit trafficking that the deliberate removal of markings is also categorized as a criminal offence (art. 5.1.c). Generally, the Protocol requires States Parties to criminalize illicit manufacturing of and trafficking in firearms, their parts and components, and ammunition (art. 5).

Other measures set out in the Protocol include:

- the confiscation, seizure, and disposal of illicit firearms, their parts and components, and ammunition (art. 6);
- the verifiable deactivation of firearms (art. 9); and
- security and preventive measures, covering *inter alia* the security of manufacturers' stockpiles and effective in-transit controls (art. 11).

The Protocol also commits States Parties to co-operate in combating the illicit trade (arts 11.b and 13) and to exchange information for this purpose (arts 12 and 15.2). Provision is also made for training and technical assistance, where needed, to enhance the capacity of states to fulfil their obligations under the Protocol (art. 14).

The issue of brokering is also addressed in the text, although the relevant provisions are of a recommendatory rather than mandatory nature (art. 15). This disappointed many governments and NGOs, conscious of the key role some arms brokers play in the illicit trade (Small Arms Survey, 2001, ch. 3). However, a wide range of countries, including many European Union (EU) members, have yet to adopt national controls on this activity and were reluctant to sign up to international standards first.

Scope

Throughout the negotiations, a debate raged as to whether the Protocol should cover state-to-state transfers as well as the commercial variety. Certain governments, led by Colombia, argued strongly for the inclusion of state-to-state transactions, pointing out that they are just as susceptible to diversion to the illicit market and should thus be subject to the same restrictions as commercial sales. The inclusion of state-to-state transfers is particularly important to ensure the adequate tracing of illicit weapons. However, the view prevailed that their inclusion would broaden the scope of the Protocol too far and risk taking the negotiations into sensitive territory linked to national security concerns. Article 4.2 of the final text thus expressly exempts state-to-state transactions from the Protocol.

The view prevailed that the inclusion of state-to-state transfers would broaden the scope of the Protocol too far.

A second issue relating to scope was whether to apply the Protocol to transfers of firearms from states to non-state actors. Again, negotiating states diverged sharply in their views on this question, with some seeking a clear exemption for such transfers and others wanting them to be covered by the Protocol. In the event, compromise language, making reference to the UN Charter, was proposed by the United States and eventually integrated into the final text:

> This Protocol shall not apply to state-to-state transactions or to state transfers in cases where the application of the Protocol would prejudice the right of a State Party to take action in the interest of national security consistent with the Charter of the United Nations. (UNGA, 2001b, art. 4.2)

This language, consistent with the US refusal in other forums to accept a ban on transfers to non-state actors, will in effect allow States Parties to decide for themselves whether a specific transfer from the state to a non-state actor is covered by the Protocol. The reference to the UN Charter does not appear to significantly limit the scope of the national security exemption given the broad, occasionally divergent, nature of its principles.

Several states expressed reservations with respect to the text of art. 4.2 when the Protocol was agreed. India, for example, indicated that 'the exclusions foreseen in that paragraph would be viewed only in narrow, precisely defined terms'. Other governments, notably Colombia, indicated they rejected the exceptions altogether (UNGA, 2001a, paras 16, 30). In fact, there is little doubt that the subjective language of the national security exemption will make it difficult for any judicial body called upon to adjudicate a dispute under the Protocol to reject a State Party's invocation of this provision (UNGA, 2001b, art. 16).

BOX 6.2

Article 8
Marking of firearms

1. For the purpose of identifying and tracing each firearm, States Parties shall:

   (a) At the time of manufacture of each firearm, either require unique marking providing the name of the manufacturer, the country or place of manufacture and the serial number, or maintain any alternative unique user-friendly marking with simple geometric symbols in combination with a numeric and/or alphanumeric code, permitting ready identification by all States of the country of manufacture;

   (b) Require appropriate simple marking on each imported firearm, permitting identification of the country of import and, where possible, the year of import and enabling the competent authorities of that country to trace the firearm, and a unique marking, if the firearm does not bear such a marking. The requirements of this subparagraph need not be applied to temporary imports of firearms for verifiable lawful purposes;

   (c) Ensure, at the time of transfer of a firearm from government stocks to permanent civilian use, the appropriate unique marking permitting identification by all States Parties of the transferring country.

2. States Parties shall encourage the firearms manufacturing industry to develop measures against the removal or alteration of markings.

Source: UNGA (2001b)

The marking compromise

The second area in which the text of the *Firearms Protocol* was weakened was that of marking. Article 8 of the Protocol allows countries to use 'geometric' as well as 'alphanumeric' symbols to mark their weapons. In the negotiations, China had pushed for the right to use geometric symbols, especially for purposes of identifying manufacturers. Many countries were, however, extremely reluctant to concede to China on this issue. Sole use of alphanumeric markings would have ensured a high degree of transparency, enabling governments to trace weapons back to the manufacturer without the assistance of the original exporting government.

A compromise of sorts was reached. The use of geometric symbols was allowed—but only for those countries which already use them. No others may adopt such a system in the future.[5] Most countries that do use geometric symbols, and are unwilling to change because of cost implications, provide their information to Interpol, so some transparency is assured (*IISS Strategic Comments*, 2001). Nevertheless, several governments expressed reservations in relation to Article 8 in the final report of the negotiating Committee (UNGA, 2001a).

Implementation

There is no doubt that the Protocol suffered when it missed out on the collective pressure that led several countries to sign the main Convention and the two other protocols at the signing ceremony held in Palermo in December 2000. If signature of the *Firearms Protocol* is likely to be slow, ratification will be even slower; it is unlikely to enter into force in the next two to three years.[6] A conference of the States Parties, to be held not later than one year after the instrument's entry into force, might provide an opportunity to review some of the more contentious provisions and clarify such questions as the application of the Protocol's provisions to government transactions. The conference is also intended to boost information exchange on patterns and trends, and promote co-operation with relevant international organizations—such as Interpol and the World Customs Organization (WCO)—and NGOs. Perhaps most importantly, the conference will afford an opportunity to review national implementation.

Implementation of certain provisions will no doubt present challenges to some countries, primarily those in the less developed world. Record-keeping will probably be especially difficult for countries without sophisticated data collection systems, although in some cases they may be able to 'leapfrog' to computerized databases and other technologically advanced systems. However, considerable resources will be needed for this purpose, with assistance from wealthier countries essential. As mentioned earlier, the Protocol recognizes this, committing States Parties to make necessary training and technical assistance available (UNGA, 2001b, art. 14). The draft plans of action of the ECOSOC Crime Commission, which were agreed in September 2001, stress the importance of international co-operation in implementing the *Firearms Protocol*, and mandate the UN Centre for International Crime Prevention (CICP) to develop 'technical co-operation projects'. Subject to available resources, assistance to developing and newly democratic countries in implementation is envisaged (UN ECOSOC, 2001, para. 16). The CICP also has an ongoing pre-ratification assistance project for interested countries.

Implementation of the Protocol will not just depend on resources.

Implementation will not just depend on resources, however. Some countries might find that their national systems prove a barrier to ratification and implementation. The United States is the best example, where legislative distrust of gun control initiatives may hold up ratification in the Senate. Even if ratification is completed smoothly and the Protocol enters into force, it will put a heavy burden of implementation on the US, which does not track the movement of civilian weapons *within* the country and has no national registry of ownership.

As ever, ratification and implementation will depend heavily on political will. While the attention gained by the 2001 UN Small Arms Conference may have a beneficial effect on the ratification prospects of the *Firearms Protocol*, the trafficking of human beings has recently occupied a higher place on the political agenda. In fact, only two countries—Mali and Brazil—took the opportunity at the UN Conference to sign the *Firearms Protocol*. Without a similar high-profile event, signature and ratification may fall down the list of priorities for many countries. Nevertheless, it remains the responsibility of supportive governments to sign and ratify quickly as a sign of their good faith.

BOX 6.3 NGO contributions to small arms control efforts

© Panos Pictures/Robert Knoth

High school students calling for an end to armed violence following the death of a teacher, Freetown, Sierra Leone.

The 2001 UN Small Arms Conference helped legitimize the key roles NGOs play in the evolution of small arms control. These are more important than ever in the wake of the Conference (CONFERENCE).

Among the most critical responsibilities are *public awareness raising and constituency building*. NGOs will be essential in the immediate post-Conference period in broadening public awareness and concern, which will in turn determine the extent to which states follow up on commitments made in the *Programme of Action*. Absent strong national interest, governments shy away from decisive action on disarmament-related matters unless they feel the pressure of public opinion. Present levels of public awareness and mobilization are patchy, with large parts of the world, including many small arms-affected regions, not yet effectively engaged. NGOs will be important not only in engaging new publics but also in sustaining public involvement where already strong. This role involves the use of the kinds of strategies employed so effectively by NGOs in other issue areas for engaging media interest—such as the anti-personnel landmines campaign. The effectiveness of a number of other NGO roles depends, to some extent, on how well NGOs fulfil this public engagement one.

Closely related to this role are *advocacy* and *policy development*. The development of effective control regimes requires action of various kinds by national governments. NGO advocacy and lobbying at the national level, backed by mobilized publics, are often critical to governmental action, whether strictly national in scope or conducted within a regional framework. Concerted campaigning, both national and international, will be necessary, for example, if the good intentions of the *Nairobi Declaration* (2000) are to be translated into effective action in the Horn of Africa and Great Lakes subregion. To be effective, national NGOs and regional NGO networks will need to be strengthened. NGOs can also be expected to continue to press forcefully for the development of global regimes, however tepid the commitment to these in the Conference *Programme of Action*.

NGOs will continue to be instrumental in the development of particular policy initiatives. Where government action is seen to be inadequate or even inimical to the sorts of controls required, NGOs must play an adversarial role to governments. But NGOs can often play a partnership role assisting governments in the development of specific policy initiatives, whether national, regional, or global. At the national level, they can lobby for specific legislation and define regulatory needs. At the regional and global levels, NGOs play a similar role. Prominent examples include the model conventions on tracing, arms brokering, and arms transfers, briefly described elsewhere in this chapter and in Chapter 4 (HUMANITARIAN). 'Biting the Bullet'—a project of three international NGOs, the British American Security Information Council (BASIC), International Alert, and Saferworld—has prepared policy-oriented papers on a whole series of small arms control issues as tools for advocacy and support to governments in national and international policy development. In policy development, NGOs will continue to work alongside sympathetic governments and international organizations.

Closely related to advocacy and policy development, NGOs also play an important *agenda development* role. The UN Conference *Programme of Action* essentially represents the lowest common denominator on what must be done. In countries and regions where there has been little or no action on small arms to date, NGOs can be expected to be at the forefront of agenda setting, stressing the measures set out in the *Programme*, as appropriate to the circumstances. NGOs will also be seeking to deepen and expand the global agenda beyond its narrow emphasis on the illicit trade. Trans-national NGO coalitions are forming under the general umbrella of the International Action Network on Small Arms (IANSA) to bring varying perspectives to bear on small arms work, including the impacts these weapons have on children and women, as well as their humanitarian, development, and human rights implications. In the coming period, NGOs can be expected to increasingly raise issues and, going beyond the established supply-side orientation, help give shape to a demand-side action agenda at all levels (Atwood and Jackman, 2000).

BOX 6.3 NGO contributions to small arms control efforts (continued)

As commitments and obligations by states expand in number and depth, the *monitoring and evaluation* role of NGOs will become increasingly important. The knowledge that there will be independent analyses by NGOs of state performance in enacting and implementing small arms control measures will be critical to ensuring that rhetoric is turned into real action, whether in relation to domestic gun control or to the trans-national effects of weapon flows. The potential importance of NGOs in this area can be illustrated by the crucial role currently being played by the annual *Landmine Monitor Report* in measuring and evaluating the behaviour of States Parties to the anti-personnel landmine convention. Similarly, the evaluation by NGOs of how well enacted policies and regimes are actually working will help provide important measures of progress in the years ahead.

NGOs can also play a useful *implementation* role, especially at the national and local levels. For example, the Christian Council of Churches of Mozambique has for several years conducted a 'Tools for Arms' collection project in which weapons can be exchanged for farm implements. Viva Rio not only campaigns for stronger state and federal gun control laws in Brazil but is also involved in a broad range of social programmes aimed at addressing the social conditions which lead to gun usage (WEAPONS COLLECTION). The Bonn International Center for Conversion (BICC) and the International Resource Group on Disarmament and Security in the Horn of Africa (IRG) are undertaking similar, capacity building work at the local level in the Horn of Africa as part of their Assessing Small Arms Issues and Developing Capacity for Peace in the Horn of Africa (SALIGAD) project. Some NGOs work alongside or as implementing agencies for international agencies engaged in post-conflict programmes aimed at disarming, demobilizing, and re-integrating ex-combatants. Others are actively involved in various aspects of security sector reform.

A major contribution to effective action on small arms issues is the *research and information* role that NGOs play. NGO research has been important to the growing understanding of the causes, patterns, and magnitude of small arms proliferation. Increasingly, the analysis they provide is also important to measuring the effects these weapons have on society. Government support for research NGOs illustrates the extent to which they appreciate and have come to rely on the policy-relevant research these NGOs are undertaking.

A further role NGOs have is that of *interlocutor and facilitator* of encounter and dialogue. A number of groups act as a bridge between governments and the NGO small arms community. Through international and regional seminars and other activities, they foster a greater understanding of participants' perspectives and special contributions along with the requirements for action. In addition, NGOs often facilitate government-to-government exchanges, playing a necessary third-party role in the development and implementation of small arms initiatives. This role, along with many others mentioned above, will probably increase in importance as efforts are made to implement the *Programme of Action* and move beyond it.

Source: Atwood (2001)

Small arms tracing

There is growing recognition of the importance of tracing and its link to other facets of the fight against small arms.

Small arms tracing has attracted considerable attention and interest in recent years. Firearms and ammunition can be marked with simple inscriptions to show an item's type, serial number, manufacturer, and initial purchaser. These elements allow the firearm or piece of ammunition to be registered and subsequently identified. The process of reconstituting the transfer route followed by a specific weapon over the course of its lifetime is known as 'tracing'.

Successful tracing, dependent on adequate marking and record keeping, allows concerned governments and organizations to identify the producer of a particular weapon and the different intermediaries involved in its transfer to, for example, zones of conflict, criminal groups, or other undesirable end-users and end uses. Illicit trafficking networks can be uncovered and uprooted. Actors involved in the deliberate violation of international arms embargoes can be identified and punished—at least in theory. In practice, existing marking, record-keeping, and tracing arrangements are wholly inadequate to the task. These limitations prevented, for example, the Commission of Inquiry examining arms flows to Rwanda from reaching all but the most tentative and general conclusions about the individuals and arms trafficking routes involved in the violation of a UN arms embargo in the subregion (UNSG, 1996; 1998).

The route small arms take to their final end-user cannot be consistently determined on the basis of existing systems.

Technological advances have made the establishment of comprehensive marking, record-keeping, and tracing systems technically feasible. At the same time, as already noted, there is growing recognition of the importance of the issue and its link to other facets of the fight against small arms proliferation, especially the enforcement of UN arms embargoes. Some of the new initiatives which may begin to change this situation are described below, after a brief review of the current situation.

The tracing status quo

Firearms are normally marked at the time of manufacture by stamping, casting, or engraving. These marks typically identify the manufacturer, model, and serial number, though the latter are not standardized between manufacturers and may not be unique. In certain cases, a number or abbreviation indicating quality standard is also marked on the firearm. No international instrument currently prescribes methods of marking. This facilitates the work of those who erase or modify marks when diverting firearms to the illicit market. Some firearms are never marked at all.

Even where marking exists, it is often not checked or recorded by an independent authority. Record keeping occurs, when it occurs, at the national level. Registers for civilian firearms are almost always incomplete and cover ownership—not import, export, or transfer within a country. While national armed forces tend to keep detailed information on their firearms stocks (STOCKPILES), it is not always shared, or promptly shared, with outside agencies—even friendly foreign governments conducting criminal investigations. While a few, new international instruments require national authorities to co-operate in tracing requests by providing relevant information, including information on official firearms stocks, they are not legally binding.

As things stand, when a firearm is inspected somewhere in the world, one can only be sure of determining its model and type—and quite often the manufacturer and country of origin—on the basis of the weapon's initial markings and/or expert examination of its design features. Depending on the extent of national record keeping and, above all, the willingness of national authorities to co-operate, investigators situated abroad may also be able to learn where the weapon went when it left the source country. After this, the firearm enters something of a black hole. The route it takes to final end-user, the identity of intermediate owners, and, crucially, the point at which it moves from legal to illicit markets cannot be consistently determined on the basis of existing systems.

Ammunition is typically marked, at the base of cartridge casings, with the manufacturer's initials and the calibre. However, these elements are often insufficient to ascertain the place of manufacture of an item, and in no case allow a determination of its transfer route. Additional information, especially production batch, is usually recorded on ammunition packaging, but this is obviously lost as soon as the ammunition is taken from its packaging. This may be done deliberately in order to hide all traces of the ammunition's origin. Ammunition can also be diverted from legal to illicit markets by simulating its use: for example, in fictitious 'training exercises'. As with firearms, states do not consistently impose a marking requirement in order to facilitate ammunition tracing. What national practice exists is not harmonized internationally.

The packaging of ammunition for transport is, however, governed by a harmonized system of procedures and regulations. These are inspired by the recommendations of a UN Expert Committee on the transport of dangerous goods (UN, 2001b). Packaging is tested and marked to designate the category and type of ammunition, the country of origin, and exporter. While this system is harmonized at the global level and thus permits the tracing of ammunition packaging throughout the world, it does not apply to its contents—the partially marked ammunition. Once the packaging is lost, so too is the ability to trace the ammunition.

Multilateral initiatives

The *Convention on the Marking of Plastic Explosives* (1991) was negotiated as an anti-terrorism measure in the wake of the bombing of Pan Am Flight 103 over Lockerbie, Scotland, in December 1988. In force since 21 June 1998, the Convention requires States Parties to use a prescribed chemical marking agent in the plastic explosives they manufacture in order to facilitate their detection. Among the first initiatives adopted with respect to *firearms* tracing was the 1992 resolution of the Interpol General Assembly. After '[r]ecognizing that police work is often hampered by the inability of police services of a country to trace the history of firearms', the resolution recommended the establishment of national firearms tracing offices (Interpol General Assembly, 1992). In 1997, in New Delhi, the Interpol General Assembly recommended the permanent marking of all firearms with specified, minimum information, along with the adoption of 'effective legislation and a registering system' (Interpol General Assembly, 1997). Yet very few countries have followed these recommendations.

The precedent-setting *Inter-American Convention against the Illicit Manufacturing of and Trafficking in Firearms, Ammunition, Explosives, and Other Related Materials* (OAS, 1997) was adopted in Washington, DC, on 14 November 1997 and entered into force on 1 July 1998 after ratification by two OAS states. In a first for a multilateral instrument, the Convention establishes a binding obligation for the marking of firearms at the time of manufacture and import, as well as in cases where illicit firearms are confiscated but not destroyed (art. VI). A related provision requires States Parties to retain 'for a reasonable time'[7] necessary tracing and identification information (art. XI). The Convention also mandates co-operation among States Parties in tracing, including 'accurate and prompt responses to trace requests' (art. XIII.3), together with the 'adequate training' of personnel in this field (art. XV.2). As of August

2001, the implementation of the Convention was not far advanced (OAS, 2001). Yet, even when the Convention is more fully implemented, its marking provisions will have limited practical utility so long as there are no common standards at the global level.

The recently concluded UN *Firearms Protocol* is described earlier in this chapter. The Protocol's marking provisions are quite extensive notwithstanding a compromise forced by the Chinese delegation that led to the acceptance, in certain cases, of the use of geometric symbols as well as the alphanumeric variety. The Protocol requires marking at time of manufacture and import, permitting, at a minimum, identification of the manufacturing and importing countries. Imported firearms must also be given a unique identifying mark where they do not already have one (UNGA, 2001b, art. 8.1.a–b). Firearms transferred from government stocks to the civilian market also have to be marked so as to allow for the identification of the transferring country (art. 8.1.c). Article 7 of the Protocol mandates the retention, for at least ten years, of relevant tracing information, but this falls far short of the lifespan of most firearms. In contrast to the *Inter-American Convention*, the Protocol is a global instrument, open to signature and ratification by all states. However, its marking provisions will not have the impact they might have had given the absence of a universal requirement to mark serial numbers (art. 8.1.a).

The marking provisions of the Firearms Protocol will not have the impact they might have had given the absence of a universal requirement to mark serial numbers.

Another important initiative, though not legally binding, is the *OSCE Document on Small Arms and Light Weapons*. Section II of the Document, adopted by the Organization for Security and Co-operation in Europe (OSCE) on 24 November 2000, sets out various provisions for marking and record keeping at the national level in the context of efforts to combat illicit trafficking. Common standards are not established, but OSCE participating States do 'agree to ensure that all small arms manufactured on their territory after 30 June 2001 are marked in such a way as to enable individual small arms to be traced.' (OSCE, 2000, sec. II.B.1) Marks should be permanent and include information allowing identification of the year and country of manufacture, the manufacturer, and the weapon's serial number (sec. II.B.1). Unmarked small arms transferred from government stockpiles are also to be given a unique identifying mark (sec. II.B.2). OSCE participating States also undertake: to keep 'comprehensive and accurate records' of public and private small arms holdings for 'as long as possible' (sec. II.C); to co-operate with one another in the tracing of illicit small arms (sec. III.E.4); and to exchange information on national marking systems (sec. II.D.1).

The importance of marking, record-keeping, and tracing has been acknowledged in a wide range of other regional initiatives. These include the West African Moratorium,[8] the EU Joint Action,[9] and the Organization of African Unity's (OAU) Bamako Declaration.[10] The issue has also been addressed in a series of UN experts reports, including the 1997 Experts Panel report (UN, 1997, para. 80(l)(i)), the 1999 Experts Group report (UN, 1999c, paras 102, 115–17), the Ammunition Study of the same year (UN, 1999b, paras 104–6), and the 2001 *Feasibility Study* on brokering (UN, 2001a, paras 36–40, 51, 53–54, 57, 59, 61).

Marking, record keeping, and tracing feature quite prominently in the *Programme of Action* agreed at the end of the 2001 UN Small Arms Conference. In this document, states commit themselves to ensuring manufacturers mark small arms (UNGA, 2001c, sec. II.7) and to preventing the manufacture, stockpiling, transfer, and possession of unmarked or inadequately marked small arms (sec. II.8). Other provisions of the *Programme of Action* provide for 'comprehensive and accurate' record keeping with respect to manufacture, holding, and transfer (sec. II.9), effective tracing measures (sec. II.10), and the marking and registration of confiscated or collected weapons that are not destroyed (secs II.16 and II.21). Section III of the *Programme* contains potentially important commitments for building capacity in marking, record keeping, and tracing (secs III.6, III.10–12).

Countries such as France and Switzerland,[11] backed by a large number of NGOs, had argued for the inclusion in the *Programme of Action* of an undertaking to negotiate an international convention on small arms tracing. Yet the final *Programme* simply called for a UN group to study the feasibility of negotiating such an instrument (sec. IV.1.c).

Civil society was an active participant in the preparation for the 2001 UN Conference and—though not part of the negotiating sessions—the Conference itself (CONFERENCE). Seminars, workshops, and conferences were held, with tracing frequently at the heart of the discussions. The International Action Network on Small Arms (IANSA, 2001), the Biting the Bullet project (Greene, n.d.), and the Groupe de recherche et d'information sur la paix et la sécurité (GRIP) (Berkol, 2001b) all published documents analysing the issue and making recommendations for the future *Programme of Action*. GRIP has conducted extensive research on techniques for marking and systems for registration and tracing. In October 2001 it was working to develop a model convention which would set out a series of standards for marking and registration along with the establishment of a centralized tracing mechanism (Berkol, 2001c).[12]

The firearms community—firearms users and industry—represented by such groups as the NRA and the WFSA also sought to influence the course of negotiations at the UN. A principal objective of these groups was to exclude handguns and hunting and sporting firearms from the Conference ambit (*Workshop Report*, 2001; CONFERENCE). Parallel to the 2001 UN Small Arms Conference, the Eminent Persons Group announced an initiative which would see the firearms industry adopt a system of voluntary marking and information exchange (Mali, 2001b). Its proposals are based on the recommendations of a seminar hosted by the WFSA in Sardinia, Italy, in June 2000 (WFSA, 2000).

No international instrument currently prescribes methods of marking, facilitating the work of those who erase or modify such marks when diverting firearms to the illict market.

Moving beyond the status quo

To summarize, only two legally binding initiatives currently regulate marking, record keeping, and tracing at the regional and global levels: the *Inter-American Convention* and the UN *Firearms Protocol*. Both instruments, however, focus on the problem of the illicit trade. The OSCE Document is more comprehensive—and covers important, related areas such as stockpile management—but is only a statement of political commitment. While discussions at the 2001 UN Small Arms Conference touched on virtually every issue relating to the small arms problem (CONFERENCE), a commitment to negotiate a legally binding agreement on tracing—as well as brokering—was purged from the final *Programme of Action*.

For the moment, the status quo described earlier prevails. Effective marking, record keeping, and tracing arrangements tend to be the exception. Where they do exist, they are not harmonized between states. To date, there has been little or no national implementation of agreed multilateral initiatives. Nevertheless, many of the countries that have signed on to these measures are taking initial steps to follow up on their commitments, including the preparation of new legislation. Further action is also on the agenda. In October 2001, Belgium, President of the Council of the EU during the second half of 2001, announced it would initiate a discussion with fellow member states on binding EU norms for marking. The Franco-Swiss initiative, described above, will likely play an important role in plans, announced in the *Programme of Action*, to examine the feasibility of developing an international tracing instrument.

It may also serve, more generally, to channel the efforts of states that wish to pursue and intensify efforts to develop international standards for the small arms trade. NGOs and industry representatives will continue to be closely involved in the elaboration of future norms and instruments for small arms tracing.

BOX 6.4 Accountability in export controls

Export controls are an essential point of intervention in the small arms transfer chain. Crucial decisions are taken, at this time, as to what weapons go where. Yet very often arms export policy is entirely concentrated in the hands of the executive branch of government, raising a problem of accountability in democratic nations *vis-à-vis* the legislature and society as a whole.

The US export control system is often cited as something of a model. It is also one of the few countries whose Legislature—the US Congress—has a significant oversight role with respect to arms exports. Pursuant to the Arms Export Control Act, the executive branch of the US government is required to give Congress advance notice of arms sales reaching a prescribed threshold in terms of total value: USD 14 million or USD 50 million, depending on the weapons. The Foreign Assistance Act also directs the Executive to inform legislators of certain US military stocks that are given or sold at deep discount to foreign armed forces. The US House of Representatives and Senate can act together to block or amend any arms transfer notified to them, but short review periods—15 or 30 days, depending on various factors—make this extremely difficult in practice. As of October 2001, Congress was considering draft legislation which would set a new, lower notification threshold for small arms sales—a potentially important change as these deals often fall below existing thresholds. Other formal and informal levers affording Congress a degree of influence over US arms export policy include its power to pass laws and appropriate money, and the ability of individual legislators to request and publicize information from the Executive (Lumpe and Donarski, 1998, ch. 4).[13]

In Sweden, an Export Control Council, composed of current and former representatives of the Swedish Parliament, advises the country's main export licensing authority, the National Inspectorate of Strategic Products (ISP), on applications for the export of war material and dual-use goods which are not easily decided on the basis of existing guidelines and precedents. The role of parliamentarians is limited to the provision of advice. The licensing decision is taken by the ISP (SIPRI, 2001b).[14] Elsewhere in Europe, the role of legislatures in national export systems is more limited. While parliaments at least have responsibility for adopting the primary legislation that underpins these systems, there are instances where the executive basically sets national policy in this area.

Pursuant to laws in force in October 2001, the executive branch of the UK government had broad discretion to determine the country's arms export policies, and could even change these without informing the UK Parliament. Yet new export control laws were being finalized—the last phase of a process beginning in the mid-1990s, with the 'arms to Iraq' scandal and the resulting Scott Report (Scott, 1996). Scott asserted that export policy should be bound by law and proposed setting out the purposes of export controls in legislation. He also recommended making any changes to export policy conditional on parliamentary approval and called for a greater level of disclosure with respect to individual licensing decisions. The draft legislation under consideration in October 2001 went some way towards addressing these criticisms, with Parliament to be given an enhanced role in the system and the purposes of export controls spelled out. Yet the draft also left the executive considerable freedom in a number of areas (Isbister, 2001).

In April 2001, the French Minister of Defence announced that an inter-ministerial working group would be established for the purpose of considering changes to France's legislative framework for arms exports, currently built upon a 1939 law. In the case of France, too, accountability will be a crucial issue in the reform process, with many observers arguing that the current system—controlled by the French government and armed forces—needs to be balanced by some measure of parliamentary oversight and involve civil society in its ongoing evaluation and development (Bouveret, 2001; Bouveret and Elomari, 2001).

End-use assurances

Small arms control is typically weakest where the jurisdiction of one state ends and that of another begins. Export control regimes play a crucial role in preventing the diversion of small arms to unintended end-users and end uses. End-use assurances (EUAs) are vital components of these systems, offering exporting countries a means of retaining some measure of control over their arms exports after they leave their territory. The information provided to the exporting government about where and how the weapons will be used also helps it assess whether or not a transaction is in accordance with export control regulations. End-use assurances usually contain a non-transfer pledge and facilitate the verification of delivery. As such, they are a vital link in the chain of control over military goods.

However, false end-use documentation is often used to divert arms, particularly small arms, to the illicit market. For example, UN bodies set up to assess compliance with sanctions against União Nacional Para a Independência Total de Angola (UNITA) found that large quantities of weapons had been exported from Bulgaria to the Angolan rebel movement on the basis of forged EUAs (UNSC, 2000a; 2000b). Since formats and procedures vary greatly from country to country, EUAs are especially liable to abuse and fraud. Forged statements are used to gain export licences for weapons that were never intended for their stated destination. As few countries physically verify delivery, forgery is a low-risk strategy.

False end-use documentation is often used to divert arms, particularly small arms, to the illicit market.

Recent initiatives by governments and multilateral organizations have responded to this problem by examining end-use standards and recommending minimum levels of information, limited numbers of authorized signatories, and harmonization of documentation. However, because these agreements are recent and non-binding, countries continue to pursue widely divergent end-use practices.

No uniformity in end-use practices

EUAs are usually required for the export of all military and some dual-use equipment. Some governments, such as Denmark's, waive end-use requirements for government-to-government deals, while other countries exempt exports to like-minded or allied governments. Some governments go one step further and require them only for countries of concern. A recent questionnaire distributed by the government of Switzerland to all members of the Wassenaar Arrangement confirmed there is no uniformity in end-use practice among major arms-exporting countries, especially with regard to export destinations and the types of equipment for which EUAs are required (Switzerland, 2000).[15]

Different types of end-use assurances

Governments use a wide variety of terms to describe their end-use documents, including 'undertaking', 'declaration', and 'statement of ultimate consignee'. The nature of the assurance differs depending on the type of equipment being exported; exports of combat equipment (weapons) require more stringent procedures than material designed for military support. Interestingly, some governments, such as Switzerland's, require more detailed information for dual-use items because they have various uses, including the production of weapons of mass destruction, and are often exported to private companies.

In most countries, the type of EUA will vary according to whether the importing entity is a government or a private company. Some governments will accept a copy of the contract as a substitute for an end-use assurance in a government-to-government deal. In these cases, Sweden, which has five different types of EUA, requires a traditional government-to-government declaration (Declaration by End-User) that is printed on banknote paper to reduce the risk of forgery. For exports of small arms and ammunition to private companies for sporting uses, Sweden has a different form (Sweden, National Inspectorate of Strategic Products, 2000). Some countries, such as the US, require the end-user to fill out official forms (US, US Code). Others accept a declaration on the end-user's letterhead. Some, like the UK, provide a model form which should constitute the basis of an end-use declaration on the importer's letterhead (UK, Department of Trade and Industry, 2000). Whatever format is used, most countries require statements of non-governmental end-users to be authenticated by their governments.

Many countries require an International Import Certificate (IIC) as well as an end-use statement. An IIC provides information about the goods, the importer, and the end-user, but not the end use, and serves as a guarantee that the importer will not divert or re-export the imported goods. An IIC is in certain circumstances accepted in lieu of an EUA. Belgium, for example, requires an IIC alone for exports to NATO and some other European or Western countries.

Minimum information

Just as the type of assurance varies according to the deal, the amount of information it contains also differs. For government-to-government deals involving military equipment, most countries require only limited information; the assurance is more a statement of good faith than a detailed description of the goods and their intended use. For example, Bulgaria requires a declaration from a foreign government that goods it imports from Bulgaria will be 'solely for the use' of the relevant ministry. In contrast, a *Statement by Ultimate Consignee*, which is used for Bulgarian exports of military and dual-use equipment—often to private end-users abroad—contains much more information, including: a detailed description of the goods; the names and addresses of the exporter, importer, intermediate consignee, and ultimate consignee or end-user; and the means of transport and route taken (Bulgaria, Ministry of Foreign Affairs).

Signature or stamp

End-use statements or certificates are usually signed by the consignee and the end-user. When an export is made to a non-governmental end-user, most countries require its government to validate the statement. For example, an Australian EUA must be signed by the consignee, the end-user, and the end-user's government, and each signature must be validated by a seal (Australia, Department of Defence).

Non-transfer clauses

EUAs typically serve to commit those importing weapons to not re-export them—at least not without the exporting country's consent. For example, Italy requires an importing company to undertake 'not to divert, not to transfer and not to re-export' the relevant materials, while the US makes any re-export conditional on the prior, written approval of the Department of State (Italy, Ministry of Foreign Affairs, n.d.; US, Department of State, n.d.).

Delivery verification systems

End-use assurances have little practical effect in preventing diversion unless they are verified. Many EUAs include a commitment to provide a Delivery Verification (DV) certificate once the equipment has reached its final destination.[16] Yet, while DVs help ensure delivery occurs as planned, they are just as open to abuse as EUAs. Ultimately, the only way to effectively guard against diversion is to conduct end-use verification checks. Although some governments attempt to do this, usually through their embassies, they are heavily dependent on available resources and diplomatic representation. A few governments, such as the US, ask for an offer of on-site inspection in their EUAs. This can have a potential deterrent effect, though the exporting country may not always have the capacity to take up such an offer.

BOX 6.5

End-Use Assurances Commonly Used
Indicative List

The following is a non-binding list of end-use assurances to be used by Participating States at their discretion.

1. Parties involved in the transaction
1.1 Exporter's full name and address;
1.2 Intermediate consignee's(s') full name and address (if applicable);
1.3 Final consignee's(s')* full name and address;
1.4 End-user's full name and address (if different from the final consignee);

2. Goods
2.1 A detailed description of the goods which discloses their true identity;
2.2 Include quantities and values;

3. End-use
3.1 Describe the specific end-use of the goods;
3.2 Provide assurances that the goods will not be used other than for stated purposes; and/or
3.3 Provide an undertaking that the goods will be used for civil end-uses; and/or
3.4 End-user certification that the goods will not be used for chemical, biological or nuclear weapons, or for missiles capable of delivering such weapons;

4. Location
4.1 Provide certification that the goods will be installed at the premises of the end-user or will be used only by the end-user;
4.2 The final consignee/end-user agrees to allow on-site verification;

5. Re-export/Diversion
5.1 The final consignee's/end-user's undertaking not to tranship or re-export the goods covered by the End-use Certificate/Statement; and/or
5.2 No re-exports without approval from the government of the original exporting country; and/or
5.3 The final consignee's/end-user's assurance that any re-exports will be done under the authority of the final consignee's/end-user's export licensing authorities;
5.4 The final consignee's/end-user's undertaking not to divert the goods covered by the End-use Certificate/Statement to another destination or location in the importing country;

6. Delivery Verification
6.1 Provide a commitment by the final consignee to provide the exporter or the exporting government with proof of importation, upon request (e.g., provide a Delivery Verification Certificate (DV));

7. Documentation
7.1 Signature, name, and title of final consignee's/end-user's representative;
7.2 Signature and end-use certification by the final consignee's/end-user's government or other authority as to the authenticity of the primary details provided in the document;
7.3 If issued by government authority, a unique identifying Certificate/Statement number;
7.4 Original End-user Certificate/Statement or legally certified copies.

** Meaning the last destination of the export (goods) known to the exporter.*

Source: Wassenaar Arrangement (1999)

Developing best practice

A number of international organizations have examined possible standards for end-use certification and verification. The Wassenaar Arrangement has endorsed an 'Indicative List' of end-use assurances (see Box 6.5), though, in keeping with the institution's voluntary nature, the list is non-binding (Wassenaar Arrangement, 1999).

Other organizations have investigated common standards for end-use statements. Under the auspices of the Stability Pact for South-Eastern Europe, a number of countries agreed on the Sofia Statement, with a shorter list of minimum information. This information should include: the name and address of the foreign end-user; end use; country of ultimate destination; description and quantity of the commodity; intermediate consignees; and certification by the relevant governmental institution. The Statement also emphasized the need to minimize the number of officials authorized to issue and sign EUAs, and included a commitment to communicate to other governments the names of these officials (*Statement on Harmonization*, 1999).[17] The Statement, agreed *ad referendum* at the Sofia Conference, was subsequently adopted by 13 governments;[18] however, there seems to have been little or no follow-up.

The OSCE also addressed the issue of EUAs in its *Document on Small Arms and Light Weapons*, but stopped short of common minimum standards for these instruments. OSCE participating States did commit to minimizing the number of authorized signatories and to exploring a common minimum standard of information 'with a view to developing recommendations based on the "best practice" among participating States' (OSCE, 2000, sec. III.C.1).

Although little progress has been made towards the development of common minimum standards in multilateral forums, there does seem to be an interest in developing best practice. The Swiss government has concluded that a common end-use policy should be established, and has asked Wassenaar's Licensing and Enforcement Group (LEOM) to examine whether a single, harmonized end-use assurance for Wassenaar members could be developed.[19]

The Wassenaar Arrangement, with its technical expertise, as well as other, more formal (and powerful) multilateral bodies, have important roles to play in the development of common procedures and documentation for EUAs. Any standards in this area would need to specify minimum information and establish norms with respect to signatures, validity periods, acceptance of copies, the types of exports covered by an EUA requirement, non-transfer, and verification of delivery/end use. Where standardization or harmonization is politically difficult, these organizations could help facilitate the exchange of information on existing national practice, including, crucially, those institutions and officials authorized to sign EUAs.

Controls on brokering

Though an intrinsic part of the small arms trade, arms brokering is largely unregulated. At the national level, relatively few states control this activity, whether directly or indirectly, and there are no multilateral instruments. Independent research has shown, however, that brokers and associated actors play a crucial role in small arms proliferation—especially in some of the globe's most strife-torn regions—exploiting gaps in national laws and focusing their activities on states with weak export controls and enforcement (Small Arms Survey, 2001, ch. 3; Wood and Peleman, 1999). These findings have helped focus attention on the need for regulation (Bondì, 2001; Bondì and Keppler, 2001; Fund for Peace, 2001; UN, 2001a), although any progress that has been achieved—for example, at the 2001 UN Small Arms Conference—has been modest. The trans-national nature of brokering activities makes multilateral co-operation essential, yet common definitions and approaches to jurisdiction and enforcement remain elusive.

The problem of definition

A key difficulty facing national and multilateral efforts to control brokering is the lack of a standard definition. The few existing national laws differ significantly with respect to the range of activities they consider part of brokering. A UN expert study on brokering controls offers its own definitions of key terms in this area. This document and other recent literature highlight the many related activities which underpin the core brokering activity of bringing arms suppliers and recipients together. These include, most notably, the provision of transportation, insurance, and financial services (UN, 2001a, Annex 1; Wood and Peleman, 1999; Small Arms Survey, 2001, ch. 3).

Independent research has shown that brokers and associated actors play a crucial role in small arms proliferation.

Control options

Governments have a variety of options at their disposal for the regulation of brokering. In order to be effective, two minimum requirements must be satisfied:

- the effective exercise by states of control, within the limits of their territorial jurisdiction, on all arms trade activities; and
- a degree of international co-operation allowing states to identify and address violations of relevant laws (UN, 2001a, para. 63).

Specific control options include:

- *Registration and certification*: The registration of brokers may serve simply as a means of gathering, and sharing, information about them. Alternatively, registration can form part of a scheme—which can include formal certification—designed to determine eligibility to act as a broker (UN, 2001a, para. 64(a)).
- *Licensing*: If states seek to ensure proper control over brokering, they need to license transactions in addition to registering and certifying the actors. In applying for a licence, brokers would have to supply certain basic information concerning their business (para. 64(b)).
- *Disclosure requirement*: In order to fill in the picture of a particular small arms deal, governments can require applicants for export licences to disclose any brokering agents involved (para. 64(c)).

The primary purpose of registration, licensing, and disclosure requirements is to achieve transparency and accountability. These measures facilitate the identification of illegal arms transactions and enable buyers to deal with a more reliable market as transactions are open and brokers, once registered, are accountable. A licensing regime has many practical advantages. Most clearly, it enables national authorities to assess prospective brokering deals on the basis of adequate disclosure and to refuse a licence where a transaction is contrary to law or policy. At the same time, a licensing system allows national authorities to prosecute brokers who operate outside it, without them having to prove some other offence.

National laws regulating brokering

The legislative and administrative bases for brokering regulations vary according to the different traditions and situations of each country. Most states regulate the transfer of arms without a specific reference to brokering. Nearly all states maintain laws, regulations, and administrative procedures to control the export, import, transit, and re-transfer of arms and military goods and technologies. However, in many cases, these require strengthening and updating in relation to small arms and light weapons. Many states that are rarely involved in other aspects of the conventional arms trade are suppliers or recipients of used or new small arms. As noted elsewhere in this book (TRANSFERS), these weapons are susceptible to diversion to illicit markets and end-users. In many states, existing controls are manifestly inadequate to prevent this.

The US law is widely considered to be the most comprehensive effort to regulate brokering. It defines a broker as 'any person who acts as an agent for others in negotiating or arranging contracts, purchases, sales or transfers of defense articles or defense services in return for a fee, commission, or other consideration' (US, Department of State, 2001, sec. 129.2(a)). 'Brokering activities' means acting as a 'broker' and 'includes the financing, transportation, freight forwarding, or taking of any other action that facilitates the manufacture, export, or import of a defense article or defense service, irrespective of its origin' (sec. 129.2(b)). The brokering amendment to the US Arms Export Control Act requires all persons engaged in brokering activities to register as such and to obtain a licence for each transaction in which they are involved (US, Congress, 1976, sec. 2778(b)). As part of the registration and licensing process, brokers have to submit detailed information regarding their identity and the relevant transactions. Brokers are also required to provide annual reports regarding their activities (US, Department of State, 2001, sec. 129.8–9). Persons who violate these requirements are subject to punishment, including criminal penalties.

Most states regulate the transfer of arms without a specific reference to brokering.

The US definition of 'brokering activities' is notable for its inclusion of financial services (Coflin, 2000, p. 28),[20] which are very difficult for law enforcement agencies to trace. The Netherlands also has regulations requiring persons financially involved in the trade of military goods to be licensed (SIPRI, 2001a). Imposing a licensing requirement on the financiers of these deals serves to enlist their help in policing them. They run unacceptable risks by participating in transactions which are unlicensed and therefore illegal. Brokers involved in the illicit trade should thus find that funding becomes scarce or much more expensive.

The Polish Law of 29 November 2000 also has a very broad definition of brokering (Poland, 2000, art. 3(8)), encompassing any activity which could be seen as relating to brokering. The Polish law lays down licensing requirements for brokers in Section 2, requiring the establishment of a register of granted licences and of enterprises which have met certain specific conditions

© Associated Press/Bullit Marquez

However good the legislation, it achieves little if no attempt is made to enforce it. Enforcement of existing brokering regulations is often weak.

laid down by the law (art. 21). The German law avoids a broad definition of brokering but, as noted below in the section on jurisdiction, is relatively broad in its application. It sets out a list of prohibited activities for which licences may never be granted, especially where the activity would endanger peace, or compromise German national interests or the fulfilment of its international obligations. A licence will also be refused if the applicant is considered unreliable (Germany, 1961, sec. 6). The Firearms Act of Kenya imposes a registration requirement on anyone dealing in firearms (Kenya, 2000), while Israeli regulation requires a 'negotiation permit' for negotiating the transfer of arms, as well as an 'export licence' for exporting defence equipment and know-how (Israel, 2000). France, South Africa, and Switzerland have varying registration and licensing requirements (France, 1995, arts 20–21; South Africa, 1998; Switzerland, 1996, art. 15.1).

Some brokering laws prohibit a particular brokering activity altogether. Examples include Canadian and UK laws completely banning the brokering of arms to countries under UN embargo. The Swiss Law on War Material forbids the brokering of certain types of arms, including nuclear, biological, and chemical weapons, as well as anti-personnel mines (Switzerland, 1996, arts 7–8). A similar example is the US law's refusal of licences where the export of the relevant article will, *inter alia*, support international terrorism, violate a US or UN Security Council arms embargo, or aid the development of weapons of mass destruction (US, Department of State, 2001, secs 120–30).

Some countries specifically regulate the transportation of arms. As already noted, the US law encompasses transportation services within its definition of arms brokering.[21] Other states have tackled the transportation of arms independently, without assimilating this activity to arms brokering. For example, Germany's War Weapons Control Act (1961) requires persons to be licensed if they are transporting 'war weapons' inside the national territory or if they are transporting arms outside Germany by ship flying the German flag or by aircraft entered in the national registry (Germany, 1961, secs 3–4).

Despite the relative ease with which they may elude customs inspection and tracing efforts, small arms and light weapons are not usually considered separately from other weapons or defence technologies. For example, Sweden's Military Equipment Act applies to military equipment defined broadly (Sweden, 1992; 1993). The Swedish Military List covers most of the material set out in the *List of Dual Use Goods and Technologies and Munitions List* (Wassenaar Arrangement, 2000). The scope of the German War Weapons Control Act is somewhat narrower, yet small arms and light weapons are still lumped together with a wide range of arms—including weapons of mass destruction—under the general rubric 'war weapons' (Germany, 1961, Annex).

Jurisdiction and extra-territoriality

Jurisdictional scope is a key issue in the regulation of brokering given this activity's trans-national nature. One option involves the extra-territorial application of brokering laws, where states require all nationals—including those without any presence in the national territory—to comply with national brokering or foreign trade legislation. Such laws are designed to catch those citizens who engage in illicit brokering activities in third countries with weak national legislation. Yet relatively few states are willing to assert jurisdiction beyond their borders. Some do not believe they are entitled to do so, and there are significant practical problems with enforcement. However, in certain areas of special international concern—including sex tourism and the sexual exploitation of children, the international fight against drugs (UN ECOSOC, 1988), and terrorism (*Convention for the Suppression of Unlawful Acts Against the Safety of Civil Aviation*, 1971)—extra-territorial jurisdiction has been accepted by the international community as necessary and appropriate. UN arms embargoes imposed by binding Security Council resolutions also require states to control their nationals, wherever located. Such cases could, as has been suggested in the UN *Feasibility Study*, act as precedents for extending extra-territorial controls to arms brokering and related activities (UN, 2001a, para. 68).

For now, the jurisdictional scope of brokering regulation tends to be conservative, where laws apply at all. For example, Switzerland's Federal Law on War Material applies only where the broker is located in Switzerland and procures arms in that country for transfer to a recipient in another country (Switzerland, 1996, art. 15.1). France takes a similar position (France, 1939). The German law has somewhat broader scope, applying to brokering transactions which take place entirely abroad provided the broker is physically present in Germany (Germany, 1961, sec. 4a.2).[22]

At the other end of the spectrum, the most expansive brokering laws claim jurisdiction even where the state has no *physical* link with the activity—provided it has some other connection with the act or the broker. Thus, the South African law applies to brokering activities carried out abroad by citizens of South Africa (South Africa, 1998, sec. 9). Sweden has a different approach to extra-territoriality. It forbids 'Swedish authorities, Swedish companies and persons who are resident or permanently domiciled in Sweden' from brokering abroad without a permit (Sweden, 1992, secs 4–5). The US law takes another tack, applying to all brokers linked to the US by virtue of their nationality, residence, business operations, or the fact that they broker American weapons (Bondì and Keppler, 2001, p. 36; US, Department of State, 2001, sec. 129.2(b)). Although states have tended to avoid adopting extra-territorial jurisdiction for reasons of sovereignty, international comity, or constitutional limitations, it is in fact difficult, even impossible, to effectively regulate an activity like arms brokering—often occurring far from the broker's country of residence—without it.

Multilateral initiatives

Efforts to prevent and combat illicit trafficking, including brokering, are typically hampered by inadequate international co-operation and co-ordination. Regional small arms initiatives have made some progress in this area. The 1997 OAS Convention is an important model for international co-operation. It makes 'illicit trafficking'—defined broadly enough to include many brokering activities—without a licence illegal (OAS, 1997, arts I.2, IX). Significantly, the Convention permits, but does not oblige, OAS States Parties to assert extra-territorial jurisdiction (art. V). It contains useful provisions on extradition (art. XIX), the exchange of information between states (art. XIII), and mutual legal assistance (art. XVII).

In late 2000, the OSCE adopted a *Document on Small Arms and Light Weapons* which recommends that governments require the registration and licensing of brokers (OSCE, 2000, sec. III.D). The protocol for the control of firearms, ammunition, and other related materials, approved by the Southern African Development Community (SADC) at its August 2001 Summit, has been praised as one of the most advanced measures in this field. EU countries have also discussed the issue of arms brokering with a view to the eventual development of a set of principles reflecting best practice. The establishment of common controls is not currently on the agenda, however. EU states differ on many of the tough questions that the regulation of brokering raises, such as the application of controls to transport agents and the question of jurisdictional scope.

A wide range of measures and approaches useful in tackling the illicit small arms trade generally and brokering more specifically were discussed in the UN *Feasibility Study*. Yet the report was short on firm recommendations, with most of it devoted to a presentation of various 'options' for regulation (UN, 2001a, sec. III). Nor did the governmental experts hide their scepticism as to the possibility of achieving a legally binding international instrument on brokering, given the patchiness and diversity of existing national controls. They pointed instead to the regional level as a more promising platform for multilateral action in the short term (UN, 2001a, para. 80).

In the event, the 2001 UN Small Arms Conference was largely unsuccessful in securing commitment to effective action at the global level. A number of governments—especially those of China, Russia, and several Arab countries—refused to consider global measures on brokering. Their reasons included an alleged infringement of state sovereignty, practical implementation difficulties, and a stated preference for regional initiatives. Brokering is mentioned in the section of the Conference *Programme of Action* devoted to global measures, but in this context states merely undertook—in extremely general terms—to discuss the problem and try to address it (UNGA, 2001c, secs II.39, IV.1.d). The *Programme* was, however, somewhat more specific concerning action at the national level, with states committing themselves to 'develop adequate national legislation or administrative procedures' to regulate brokering, including such measures as the registration of brokers and licensing of their transactions, backed by 'appropriate penalties' (sec. II.14).

The 2001 UN Small Arms Conference was largely unsuccessful in securing commitment to effective action on brokering at the global level.

Lack of enforcement

When, at the national level, brokering is adequately regulated, enforcement is often weak. However good the legislation, it achieves little if no attempt is made to enforce it. US brokering rules are widely considered the most specific in the world yet, since the US law was passed in 1986, there has not been a single prosecution under its provisions (*New York Times*, 2001). Obstacles to the effective enforcement of this law include difficulties gathering information where the brokering offence has been committed abroad, problems with extradition, and inadequate information exchange between responsible governmental agencies (Bondì and Keppler, 2001, pp. 28–45). Despite this, the law could be enforced if there were sufficient political will to do so. As of September 2001, the US law remained an empty promise.

While insufficient political will remains the rule with respect to brokering, some governments, including Bulgaria and the UK, have recently moved to strengthen and improve enforcement of their laws following public exposure of illicit brokering activity from their territory. Proposals to establish national registers of individuals and companies convicted of violating brokering rules, mentioned in the UN *Feasibility Study*, would also facilitate implementation and enforcement (UN, 2001a, paras 48, 63).

Next steps

Weaknesses in national laws, regulations, licensing procedures, and enforcement are exacerbated by a lack of capacity and inadequate co-ordination and co-operation at the national, bilateral, regional, and global levels. Yet brokering need not remain a gaping hole in the current regulatory framework for small arms. An important first step would be agreement between states on common principles covering such areas as registration procedures, the content of licences, and the definition of offences (UN, 2001a, para. 79). Harmonization in these areas would facilitate resolution of the thornier problem of jurisdiction, and also improve enforcement.

Encouraged by the growing recognition of a need for international control measures, the Washington NGO The Fund for Peace presented a *Model Convention* on arms brokering to the 2001 UN Small Arms Conference (The

Fund for Peace, 2001). The registration and licensing of arms brokers form the core of this instrument. It also contains provisions to encourage businesses involved in brokering transactions to identify illicit activity and applies measures adopted in the *Anti-bribery Convention* (OECD, 1997) to the regulation of arms brokering—such as mutual assistance, extradition, and money-laundering measures.

While the UN Conference's *Programme of Action* does not tackle the problem of brokering with the determination that many countries and NGOs had wanted, it does at least mention the issue and encourage states to begin to address it. The events of 11 September 2001 may sharpen concerns over the lack of controls over illicit arms trafficking. Yet it remains to be seen whether more states will start to regulate brokering activities within their borders, whether those with adequate laws on the books will move to enforce them, and whether increasing recognition of the need for multilateral co-operation in this area will translate into effective action.

© Panos Pictures/Caroline Penn

Children of an Afghan Mujahideen commander play with ammunition stored outside their home.

Stockpile management and security

The secure management of small arms stocks is of pivotal importance in curbing small arms proliferation. In some cases, stocks greatly exceed immediate and likely future needs. This has been a particular problem for many states in the aftermath of the Cold War. Yet small arms surpluses have arisen and continue to arise for a wide variety of reasons: the end of armed conflict, changing perceptions of threat, the reduction or restructuring of armed forces, changed military doctrines, and the acquisition of new weapons. While it is increasingly acknowledged that surplus stocks are best disposed of through destruction (e.g. OSCE, 2000, sec. IV.C.1), this is still the exception rather than the rule. In practice, surplus small arms that are not sold are stockpiled, where they can remain for decades.

While national armed forces, police, and other security forces hold a significant proportion of a country's small arms stocks, in quite a few countries civilians are also allowed to hold substantial numbers of small arms and ammunition (STOCKPILES). In many cases, manufacturers' stocks are also significant. Poor stockpile security is a prime means through which small arms are diverted from the legal to the illicit markets. Lax security makes theft easy. Corrupt officials may sell or otherwise transfer weapons under their care to criminal groups or rebel forces. Stockpile security is especially precarious in states suffering from violent conflict and/or weak governance. To cite one example, corrupt Cambodian military officials are known to have sold large numbers of arms to rebel forces fighting the Sri Lankan government (*Jane's Intelligence Review*, 2000). There are many others (TRANSFERS).

The basic requirements for good stockpile management include:

- properly designed, specially constructed storage facilities situated in secure locations;
- adequate procedures, including strict control of access to stocks and thorough inventory management and accounting;
- well-trained, dependable personnel;
- the protection of the transport of small arms to and from storage locations; and
- adequate resources for the continued operation of the system.

These conditions are difficult to meet in many developing countries, even those which are not at war or otherwise instable. Yet stockpile management is an issue in industrialized countries as well. Significant losses of military equipment have been reported in many such countries over the years. Whether as a result of theft or poor record-keeping, it is estimated that the United States has lost thousands of arms in the past 30 years. In Australia, military armouries were robbed 131 times between 1985 and 1989 (*Jane's Intelligence Review*, 2000). Other recent examples include important thefts in Israel and the UK (TRANSFERS).

Binding international norms in the stockpile management area are few and far between. The recent UN *Firearms Protocol* requires States Parties to ensure only that manufacturers' stockpiles are secure (UNGA, 2001b, art. 11.a), while the 1997 OAS Convention on illicit trafficking merely obliges, in general terms, States Parties to securely store illicit weapons that they seize:

> States Parties shall adopt the necessary measures to ensure that all firearms, ammunition, explosives, and other related materials seized, confiscated, or forfeited as the result of illicit manufacturing or trafficking do not fall into the hands of private individuals or businesses through auction, sale, or other disposal. (OAS, 1997, art. VII.2)

Neither of these instruments regulates normal government arsenals. While a broad range of other agreements and declarations highlight the importance of secure, properly managed stocks, they are non-binding. The November 2000 *OSCE Document on Small Arms and Light Weapons* devotes an entire section to stockpile security and the related question of surplus destruction (OSCE, 2000, sec. IV). These issues are also raised in the *Programme of Action* adopted at the 2001 UN Small Arms Conference—mostly from a national perspective (UNGA, 2001c, sec. II.17–19), though with some provision also made for regional initiatives (sec. II.29). Significantly, the *Programme* also acknowledges the need to assist interested states in building capacity in these areas (sec. III, paras 6, 8, 14).

Of greatest practical significance at the international level are the programmes for assistance in stockpile management.

Probably of greatest practical significance at the international level are the assistance programmes offering some combination of expertise and resources for stockpile management and security. These are conducted under the auspices of various regional organizations and, in some cases, bilaterally. The following examples are especially significant.

In November 1999, the EU, acting within the framework of its December 1998 *Joint Action* (EU, 1998), launched a project to support Cambodian government efforts to combat small arms proliferation in the country. Part of the project, extended in November 2000 for an additional year, focuses on the problem of secure weapons storage and adequate record keeping (EU, 1999, art. 1.2.b; 2000). In this context, the EU has provided funds for the construction of secure storage facilities and for the establishment of a comprehensive registration system (Fawthrop, 2001).

NATO assistance programmes for stockpile management are conducted under the auspices of its Partnership for Peace (PfP), comprising the 19 NATO member states and 26 Partner countries. Section 9.2 of the PfP's Partnership Work Programme chapter on small arms provides for the dispatch of teams of experts to those PfP countries which request assistance in improving stockpile management and security, and also in the destruction of surplus weapons stocks. The tasks envisaged for these teams include needs assessment, technical advice, training, and potentially the provision of equipment (EAPC, 2000, Annex 2, sec. 9.2; 2001, para. 14). As of July 2001, assistance programmes to PfP countries have focused on the problem of anti-personnel landmines and surplus munitions in Albania, Moldova, and Ukraine (EAPC, 2001, paras 14, 16–17; NATO, 2000; 2001). The issue of secure small arms storage has been addressed in seminars, workshops, and training courses organized, within the NATO/PfP framework, by Bulgaria, the Netherlands, and Switzerland.

The Stability Pact serves to co-ordinate projects mounted by member countries and organizations for the purpose of strengthening political and economic reforms in the countries of south-eastern Europe and facilitating the sub-region's integration with the rest of Europe. As part of the Pact's Quick Start Package of projects, a Norway–US Joint Working Group has offered to send an assessment team to any Stability Pact member that requests one in order to evaluate needs and share expertise in stockpile management and security (Special Co-ordinator of the Stability Pact, 2001; Bollenberg, 2001). The Norway–US offer has also been extended beyond the Stability Pact to include all Euro–Atlantic Partnership Council (EAPC) and OSCE countries, as well as countries in Latin America, Asia, and Africa, with a number of visits concluded as of September 2001 (Bollenberg, 2001). Norway and the US have also offered to assist Stability Pact states in the destruction of surplus small arms. Together with Germany, these countries have lent support to the destruction, in Albania, of weapons collected from Albanian civilians, along with surplus military stocks. The US government has also assisted Bulgaria and the Federal Republic of Yugoslavia in the destruction of surplus stocks in those countries. National procedures for stockpile management vary greatly. The US provides one example of an extensive, though by no means perfect, control system.

Small arms inventory control in the US Army

With a total of 1,649,646 firearms, as of 11 April 2001, the US Army owns one of the world's larger inventories, ranging from pistols to heavy machine guns, from antiques to state-of-the-art prototypes (STOCKPILES). Keeping control over this stockpile is an important task. The danger of losing even a few weapons was illustrated in 1997, when six Marines at Camp Lejeune in North Carolina were arrested trying to sell small arms and munitions to private dealers (Johnson, 1997).

Originally the inventory control system evolved to serve the needs of military preparedness, so that the US armed services would know how many small arms they had for operational planning. This straightforward priority was changed by the 1968 Gun Control Act (US, Congress, 1968). The Act, which mostly regulates the domestic trade in firearms, creates legal exceptions for the armed services (sec. 925.a). But its adoption also reflected a growing awareness of the dangers of unregulated firearms, shaping a climate to which military leaders were sensitive. Army officials realized they had to pay more attention to matters of weapons responsibility. This meant knowing exactly where their guns were and being able to trace weapons for criminal investigations. While record keeping for preparedness remains essential, accountability for criminal diversion consumes more time and effort.

Although military small arms officials do not advertise it, the result has been the development of US's most comprehensive system of gun registration. It is ironic to see that the Department of Defense has created a highly developed version of the very system that gun control opponents fear as a threat to individual rights. Each of the armed services is responsible for tracking their own small arms (US, Department of Defense, 2000). The US Army inventory control programme is based at Rock Island Arsenal in Illinois, under its Executive Agent for Small Arms.

Because of the strong performance of the system, independent weapons auditing has not been seriously considered. Indeed, at this juncture, one can imagine acceptance of independent auditing only as part of an international arrangement on small arms security (Karp, 2001).

Inventory control

The US Army employs multiple layers of inventory control. All Army *units* are required to record the number and condition of all their small arms and ammunition every month. Every *base* or *installation* takes inventory every six months, *depots* only annually. To minimize opportunities for false reporting, inspections cannot be conducted by unit armourers, nor can they be carried out by the same personnel twice in succession. Discrepancies must be reported immediately, without exception. Suspected losses are reported to unit and base commanders and civil authorities. Confirmed losses are reported to a central Department of Defense registry, the Unique Item Tracking (UIT), and to national law-enforcement authorities.

Soldiers are drilled into a culture in which responsibility for weapons is taken extremely seriously. Even so, some are lost now and then. Although incidents have occurred—as at Camp Lejeune in 1997—Army officials maintain that theft is difficult and rare. More typical are losses during manoeuvres and foreign deployment, especially peacekeeping and combat missions.

Any loss leads immediately to a formal investigation. Most incidents are resolved at the unit level, the weapons quickly recovered. Army officials report that virtually all of the Army's lost guns are eventually found. If weapons operators are held responsible for the loss, they are billed personally for replacements and subjected to military discipline. The number of small arms lost and not recovered is estimated at roughly ten a year (Karp, 2001).

Weapons marking

At the 2001 UN Small Arms Conference, the US delegation would not commit the country to the negotiation of an international agreement on weapons marking and tracing. Yet, for the US armed forces, marking is critical. Within all Department of Defense agencies, record keeping currently relies on manufacturers' serial numbers. There is some dissatisfaction with this in the armed services—not with the possibility that serial numbers will be removed, but with the inherent inaccuracy of visual inspection of a number at least nine digits long. Inspectors make errors and the process is time consuming. With qualified personnel in short supply, the US Army is under pressure to release individuals from such routine tasks.

Some US Army bases have established their own bar-coding systems for internal purposes, relying on stickers. These are more convenient but less permanent than serial numbers forged on the weapon. More durable solutions are being sought. The current preference is for Contact-Memory Buttons, a commercially available technology developed by the Geneva-based firm Valtac. This permits the automatic reading of serial numbers and additional information on the manufacturer, supplier, age, and history of an item. However, even with automated reading of serial numbers and related data, errors will occur, necessitating a visual reconciliation process. Inventory control will continue to rely, to some degree, on visual inspection (Karp, 2001).

BOX 6.6 Strengthening controls in the Great Lakes region and the Horn of Africa

As described elsewhere, the Great Lakes region and the Horn of Africa have been hit especially hard by the small arms problem (Small Arms Survey, 2001). A turning point in attempts to address the problem came in 2000 with the launching of a series of key initiatives. The overall framework for action in the subregion was adopted by the foreign ministers of ten countries from the subregion, meeting in Nairobi, Kenya, in March 2000 (*Nairobi Declaration*, 2000). Two follow-up instruments, agreed in November of that year, set out in greater detail the steps governments intended to take to tackle small arms proliferation (*Co-ordinated Agenda for Action*, 2000; *Implementation Plan*, 2000). Pursuant to the mandate given to Kenya to co-ordinate implementation of the *Nairobi Declaration*, the Kenyan government established, also in 2000, the Nairobi Secretariat within its Ministry of Foreign Affairs and International Co-operation.

The challenge for 2001 has been to ensure that the above elements are translated into effective action. This task has seen a wide range of organizations, governmental and non-governmental, working in partnership with the Nairobi Secretariat. The subregion's key intergovernmental organizations are the Inter-Governmental Authority on Development (IGAD), the East African Community (EAC), and the Eastern Africa Police Chiefs Committee (EAPCCO). Much of the activity around small arms has been spurred and facilitated by such NGOs as Saferworld, the Security Research and Information Centre (SRIC), and the Institute for Security Studies (ISS). Other important players in the process include the United Nations African Institute for the Prevention of Crime and the Treatment of Offenders (UNAFRI) and the Interpol Subregional Bureau.

Strengthening and harmonizing legislation: The section of the Coordinated Agenda for Action devoted to 'legislative measures' is quite comprehensive, covering most aspects of small arms regulation. Specific measures provide for the criminalization of unauthorized local production (sec. 3.3) and enforcing adherence to international arms embargoes (sec. 3.4). Perhaps most significantly, states undertake to fill existing gaps in national laws (sec. 3.2), and affirm the need for 'legal uniformity and minimum standards' (sec. 3.1). A seminar was held in Nairobi in February 2001 to discuss how states could strengthen and harmonize their small arms controls. The participants, who included senior government officials and police officers, along with representatives of intergovernmental organizations and NGOs, recommended that a legal instrument be drafted for the subregion on small arms control. The Nairobi Secretariat convened an experts group meeting for this purpose in May 2001. A draft was agreed that incorporated the commitments made by states in the Agenda for Action with respect to legislation. It drew heavily on the UN *Firearms Protocol* but included broader disarmament elements as well. As of October 2001, a revised version of the protocol was awaiting ministerial approval.

Enhancing operational capacity and law enforcement: Legislation serves its purpose only when implemented. For this reason, enhanced operational capacity and improved law enforcement have been identified by the countries of the subregion as essential. Important priorities include increasing cross-border co-operation between law enforcement agencies and improving police-community relations (*Nairobi Declaration*, 2000, sec. iv; *Co-ordinated Agenda for Action*, 2000, sec. 4). A seminar held in Kampala in June 2001 resulted in agreement on a series of practical recommendations in this area, including: the identification of training and capacity-building needs; the improvement of practical co-operation between customs and police; and the enhancement of border controls. Training is a key requirement in the subregion and one of the recommendations of the police officers in the seminar was the development of common subregional curricula. This initiative is being taken forward by the UN Regional Centre for Peace and Disarmament (UNREC), based in Lomé, Togo, which organized a workshop on the theme in Nairobi in July 2001 and was planning to hold a 'Training of Trainers' course for the subregion in Nairobi in November 2001.

Establishing national focal points: Effective co-operation between countries and within countries in the fight against small arms depends upon co-ordination between all relevant government agencies. In the *Implementation Plan*, all States Parties undertook to establish national focal points with functions that include: co-ordination with the Nairobi Secretariat and other national focal points, co-ordination and interaction with civil society, information exchange, research, and capacity building. It is clear that the establishment of these national focal points is a prerequisite for the effective implementation of the *Nairobi Declaration* and the *Co-ordinated Agenda for Action*. While, as of October 2001, progress on this had been relatively slow, national focal points had been established in Kenya, Tanzania, and Uganda.

BOX 6.6 Strengthening controls in the Great Lakes region and the Horn of Africa (continued)

The Kenyan national focal point is chaired by the Office of the President and includes 15 government agencies. At a May 2001 workshop, these agencies agreed to develop a co-ordinated national strategy to combat small arms proliferation and illicit trafficking. Tanzania established its national focal point at an inter-agency meeting in August 2001 and planned to establish inter-agency committees at the local level as well. Tanzania is also the first government in the subregion to carry out a comprehensive national study of the small arms problem. Police officers conducted 3,500 household surveys across the country, while workshops were held with local administration officials and community leaders in each major town. The results have been integrated into a national action plan to combat small arms that is being considered by donors. The Ugandan national focal point was created at a meeting of government departments in October 2001.

Securing donor support: The *Co-ordinated Agenda for Action* and *Implementation Plan* have laid out an ambitious programme for tackling small arms proliferation. Yet effective implementation will at least partly depend on financial and technical assistance from donors. The Nairobi Secretariat has prepared a detailed budget and proposed a timetable for activities, which have been packaged, along with the key regional agreements, in a new publication launched at the UN Small Arms Conference in July 2001 (Kenya, 2001). The creation of a new group, 'The Friends of the Nairobi Declaration', has been suggested for purposes of co-ordinating donor support. The UK's Department for International Development has taken the lead on this, with plans to organize a meeting of the 'Friends' in the first half of 2002.

Looking ahead: Progress in implementing the framework agreed in Nairobi in March 2000 has been significant, yet much remains to be done. The draft regional protocol needs to be agreed and ratified. The initial attempts to improve cross-border co-operation and enhance law enforcement capacity must be built upon. And new initiatives are needed to improve weapons collection and stockpile management, ensure the effective demobilization and reintegration of former combatants, raise awareness in communities of the dangers of small arms, and reduce the demand for weapons. This will require improved co-ordination between and within the governments of the subregion through the creation and effective operation of national focal points, increased support from the donor community, and the enhanced involvement of NGOs.

Source: McLean (2001)

National firearms controls

As established in the *Small Arms Survey 2001* (Small Arms Survey, 2001, pp. 88–89), most of the world's firearms are held, not by national armed forces or police, but by civilians. It comes then as no surprise that regulation of civilian gun ownership is—on a daily basis—among the most important components of broader efforts to minimize the misuse of firearms. Domestic firearms regulation is typically seen as a form of *opportunity reduction*, limiting access to these weapons and thus the risks of violence. While the question of a link between access and misuse continues to be fiercely debated, the weight of evidence appears to point to a relationship between higher rates of firearms ownership and higher mortality and morbidity, at least in the industrialized countries (Small Arms Survey, 2001, ch. 6).

The effects of gun laws can be quite indirect. Rosemary Gartner has argued that these not only reflect but also shape prevailing social values. Strict firearms controls may send out a signal about the acceptability of violence and, over the long term, influence behaviour in much the same way as legislation governing drunk driving and drug abuse (Gartner, 1998; Archer, Gartner, and Beittel, 1983). Innovative approaches such as weapons buy-backs, amnesties, and collection and destruction programmes can play a similar role in shaping attitudes (WEAPONS COLLECTION).

There is some evidence that domestic firearms regulation also influences the ease with which legal arms are diverted to illicit markets. Chapter 3 illustrates some of the specific ways such diversion occurs. Measuring the precise effect of domestic regulation on diversion is difficult, yet a recent study indicates that, in US States with strict controls, the majority of guns recovered in crime originate from outside the state. Where regulation is lax, a much higher proportion of crime guns originates locally (Webster, Vernick, and Hepburn, 2001, pp. 184–89).

The state of the art: Current national regulations

Most countries have legislation governing the civilian possession of firearms, although the nature of these laws varies considerably (Cukier, Sarkar, and Quigley, 2000). The UN, through the Commission on Crime Prevention and Criminal Justice (operating under the auspices of ECOSOC), has conducted a study of the extent of such regulation in member countries.

The underlying principles remain the same: regulate possession and use in an effort to prevent misuse and diversion.

The survey results incorporate the responses of 78 nations, representing a little over three-quarters of the world's population and spanning both industrialized and developing countries. Most require licensing of all new firearms purchases, while import and export controls are virtually universal. A significant majority register firearms, and there are storage regulations in close to three-quarters of the responding countries. A significant number of participating states also reported that they had recently amended their legislation and administration of civilian firearms ownership, possession, and use (UN, 1998; 1999a).

While the approaches vary, the underlying principles remain the same: regulate possession and use in an effort to prevent misuse and diversion. The remainder of this section outlines some of the key elements found in a global review of legislation (Cukier, Sarkar, and Quigley, 2000).

Defining lawful purposes for possession

The reasons civilians are allowed to own firearms vary considerably worldwide. Some countries, such as Japan and Great Britain, severely restrict civilian possession and have very low rates of firearm ownership. Most countries allow firearms possession for hunting or for predator control on farms. Some allow possession of certain types of firearms for target shooting. Some permit collection of firearms. Most industrialized countries strictly control civilian access to handguns and allow the carrying of handguns for personal protection only under very restrictive conditions.

The United States is the notable exception. While most firearm regulation is at the State level and is very diverse, permits to carry concealed handguns are easily obtained in many States. Obtaining permission to carry firearms is also relatively easy in Austria, where 8.1 per cent of households have handguns and 26 per cent of firearm owners have their guns for self-protection (*International Crime Victimization Survey*, 1996). Recent polls in Switzerland reveal comparable findings (STOCKPILES). Some developing countries also allow relatively easy access to handguns, though recent trends are towards stricter controls—as in Brazil and South Africa.[23]

Countries responding to the UN Firearms Study reported many different conditions under which civilians may possess firearms. In China, for example, most civilian firearms ownership is strictly prohibited. Possession of 'hunting guns' is permitted for those who live on hunted animals and those who must guard their homes or businesses from animals. Germany allows firearm possession for a variety of purposes, but there are restrictions; a legitimate requirement must be proven and a firearms license usually applies exclusively to the home of the applicant. In Sweden, no licences are issued for the purpose of self-protection, though firearms ownership is permitted for hunting, collecting, and other recreational pursuits. Moreover, properly authorized security personnel may be granted firearms licences for the purpose of protecting persons or property.

As a rule, countries prohibit the civilian possession of firearms where risk is believed to outweigh utility. The way they make this determination is often shaped by history and culture. Worldwide, most nations prohibit civilian possession of fully automatic military-style rifles. Many also prohibit possession of semi-automatic or convertible versions of such weapons. Access to handguns is also frequently outlawed or severely restricted given how easily concealed they are and their consequent utility in crime.

Regulating the sale of weapons

Most countries regulate the sale of weapons to civilians. Some regulate gun dealers and sellers and have significant record-keeping and reporting requirements. However, the extent of regulation over resale varies. In Switzerland, individuals must keep a record of private sales but need not report them to the authorities. While most US States conduct checks on initial sales, secondary sales among individuals escape control even though these account for a substantial proportion of diversions from the legal to illicit markets. For example, as of September 2001 gun shows, where individuals sell firearms, were unregulated.

Screening and licensing

Legally held firearms are often misused, particularly in suicides and domestic violence. In order to prevent this, and for purposes of tracking legal possession, most countries screen and license firearm owners (Culross, 1996, pp. 15–17). Again, however, there are significant differences. Most countries impose age restrictions and check criminal records. Some require formal safety training. Certain categories of firearms may be restricted to those with a proven need for having them—as with handguns in Canada.

Although a few countries—of which the US is the most prominent example—allow buyers without a criminal or psychiatric history to acquire guns essentially on demand, most use some kind of screening mechanism. These include the provision of references, detailed questionnaires, official interviews, and in-depth investigations. Few go so far as Austria, which requires a psychological test as well.

Licensing provisions not only perform a gate-keeping function, they also allow police to take preventive action where risks become apparent. For example, in Canada eligibility can be revoked where a licensed firearm owner commits an offence. A number of countries have specific mechanisms in place, such as spousal notification, to limit the risks of firearm use in domestic violence. Regulating sale and possession increases the effective cost of possession, making the purchase of a gun different from the purchase of other consumer items. This discourages casual ownership which, given the apparent link between accessibility and misuse, also reduces risk.

Controlling access to ammunition

Most countries define the conditions under which ammunition may be held, often making its purchase conditional on possession of the appropriate licence. Others, like the US, leave retail trade in ammunition essentially unregulated.

Marking, registration, and record keeping

Most industrialized countries require privately owned guns to be permanently marked. Some require registration of the firearm and the reporting of every transaction involving it—from initial manufacture through sale and resale. Others require this only at the initial sale, or not at all.

BOX 6.7 The Zug massacre and Swiss gun laws

On Thursday 27 September 2001, Friedrich Leibacher burst into the parliament building of the Swiss canton of Zug, murdered 14 lawmakers and government officials, wounded as many more, and then killed himself. He was armed with a civilian version of the standard Swiss Army assault rifle, the Sturmgewehr 90 (SIG 550, 5.6mm), a Remington pump-action shotgun, a SIG Sauer 232 pistol, and a canister packed with explosives. A former resident of the canton, Leibacher had a criminal record dating back to 1970 and a history of aggressive behaviour. On the basis of a letter he left behind, it appeared the shooting stemmed from minor grievances he had with the local authorities. He had never done military service and had purchased the crime guns legally.

The incident led to the immediate tightening of security at the Swiss Federal Parliament and other public buildings throughout the country. Although these measures clashed with Swiss traditions of open and accessible government, they reflected the general view that the security of public officials had to be better protected. The harder question raised by the Zug shootings was what to do about Switzerland's relatively liberal gun laws. As of November 2001, this debate had not advanced far, yet it appeared that rules allowing Army conscripts and reservists to keep military assault rifles at home were relatively safe. More at issue was the purchase of weapons by civilians such as Leibacher. While the parties and public figures traditionally opposed to the status quo argued for legislative reforms, most commentators instead emphasized the need to improve law enforcement co-operation between cantons, preventing people such as Leibacher from purchasing guns in one jurisdiction when they have clashed with the authorities of another.

© Associated Press/Sigi Tischler

Candle-light procession in Zug, Switzerland, shortly after the shooting rampage that left 15 people dead.

Swiss traditions of civilian gun ownership are deeply rooted and powerful lobby groups resist any potential erosion of existing freedoms. Switzerland is also a favourite example of gun rights advocates outside the country, who cite its relatively low rates of firearm crime and high proportion of firearm owners as proof that guns do not beget violence. The Zug massacre is in fact unprecedented in recent Swiss history. Yet Leibacher's ability to legally acquire several firearms, despite his troubled history, has unsettled many Swiss.

Sources include: Interview with Roland Schwyter, Investigating Magistrate (Untersuchungsrichter), canton of Zug, 3 December 2001; Bolli (2001); Buettner (2001); The Economist (2001); Fleck (2001); Iten (2001a; 2001b); Lezzi (2001); Meier (2001); Merki (2001); Olson (2001a; 2001b; 2001c; 2001d); Osborn (2001); Tages-Anzeiger (2001)

Safe storage

Safe-storage requirements are designed to reduce the risk that firearms will be stolen or used impulsively. Perhaps their most important function is keeping loaded firearms away from children in the home. Safe-storage measures include the disablement of firearms, their separation from ammunition, and the use of locked containers and trigger locks. In many industrialized countries, firearms must be stored unloaded in a secure container separate from ammunition. An American study found that the risk of suicide increased significantly when firearms were present in the home. The risk was greatest where weapons were stored loaded (Kellermann *et al.*, 1992).

Reducing gun access through weapons collection and destruction

Measures have also been taken to reduce demand for firearms by publicizing the risks they pose, particularly in the home, and offering amnesties and buy-backs to encourage people to get rid of guns they do not want or need. The direct impact of these programmes may often seem modest, but, even where the absolute numbers recovered are low, other, more indirect, benefits appear substantial. Educational programmes have focused on increasing awareness of safe practices and compliance with them (Flinn and Allen, 1995, pp. 296–98). Some countries, notably Australia and Britain, have outlawed certain types of weapons and accompanied the new restrictions with buy-back programmes designed to remove banned firearms from circulation. Others, such as Argentina, El Salvador, Mozambique, and the US, have used voluntary programmes to encourage gun owners to reassess the need for firearms in their homes (WEAPONS COLLECTION).

Law enforcement

Much of the criminology literature plays down the value of enforcement—for example, in the 'war on drugs'—as compared with prevention. Interventions aimed at reducing demand, as well as access to such facilitators as firearms, are often regarded as more effective. Nevertheless, penalties for firearms offences—such as the use of a firearm in a crime, illegal possession, or illicit trafficking—have recently been increased in many countries. Enforcement efforts have also focused on problem areas in the illegal trade, such as the small number of firearms dealers who account for the largest percentage of illegal firearms transactions. Increased enforcement may deter some behaviour by increasing the potential cost of illegal firearms possession or trafficking (Lab, 1997). Measures such as licensing and registration also assist enforcement efforts, specifically in criminal investigations.

Case studies: Legislative reform to narrow gun ownership

Countries have initiated changes to firearm regulation with a focus on addressing the misuse of firearms in crime. Often legislative reforms are precipitated by a high-profile shooting incident or mass murder. The best-known examples are found in industrialized countries, illustrated here by the case studies of Australia, Canada, and the UK. But pressure to tighten regulations is global, as shown by the examples of Brazil and South Africa.

Australia

Prior to 1996, all eight of Australia's States licensed owners, but only five registered firearms. However, following the unprecedented murder of 35 people in Port Arthur, Tasmania, on 28 April 1996, representatives of all Australian States adopted the Nationwide Agreement on Firearms, a ten-point plan, which included:

- registration of all firearms;
- stronger licensing, including: proof of genuine reason for owning a firearm; uniform screening; a five-year prohibition on ownership for anyone guilty of domestic violence; a safety course; a minimum age of 18 for firearms purchase; a 28-day waiting period; and strict, uniform storage requirements;
- a ban on ownership of semi-automatic rifles and shotguns, except for farmers who can prove a genuine need, combined with provisions to buy back weapons from their owners; and
- improved controls on the trading of firearms, including a separate permit for each firearm and a ban on private and mail-order sales (WEAPONS COLLECTION).

It is extremely difficult to isolate the impact of a particular piece of legislation. Any serious study would have to take into account possible substitution effects as well as more long-term historical patterns. For example, while there appears to be a slight downward trend in the use of firearms in homicide in Australia in recent years, the same cannot be said of homicides generally. This would suggest that other methods have partially replaced the use of firearms in fatal crimes. It is also too early to determine whether the apparent decline in firearm homicide stems from the legislative changes or is, rather, part of a longer-term trend—as some statistics seem to indicate (Mouzos, 2000, pp. 9–10, 43; 2001a). Initial studies pointing to a decline in the use of firearms in robberies and suicides, as well as a lower incidence of accidents involving firearms, must similarly be treated with caution—as the author herself points out. Clear impact analysis must await data gathered over a longer time-period (Mouzos, 1999; 2001b, p. 5).

Brazil

With more than 40,000 gun murders every year (25.7 per 100,000), Brazil's problem with gun crime is especially severe. Most of the firearms recovered in crime, moreover, are Brazilian-made handguns (Small Arms Survey, 2001, ch. 1). At the national level, Brazil requires licensing of firearms owners, but applicants need establish only that they are at least 21 years old and have no criminal background. Efforts to tighten gun ownership laws have travelled a roller-coaster path in recent years, with no end in sight.

In February 1997, Brazil passed a law criminalizing the possession of unregistered firearms throughout the country. In February 1999, the State of Rio de Janeiro went considerably further, banning the sale of all firearms to civilians in that State. However, in September 1999 this law was declared unconstitutional in Rio's Supreme Court. Following a massive public outcry over spiralling rates of gun violence, the Brazilian Senate Commission on Justice approved, in June 2000, a proposal banning the possession of, and trade in firearms and ammunition nationwide. Exceptions were to be allowed for security forces, sport, collection, and rural areas, where hunting rifles would be allowed. However, a little less than a year later, in May 2001, the bill was rejected by the Senate Commission for External Affairs and National Defence, which voted instead for a substitute bill explicitly protecting the right of civilians to purchase and carry guns. As of December 2001, the Senate had not agreed on a final draft of the proposed law. Once approved by the Senate, the bill will be submitted to the national House of Representatives and voted upon.

In the meantime, the Rio de Janeiro State House of Representatives approved a new law of its own on 19 September 2001, severely restricting the sale of firearms to civilians. It establishes 16 new requirements for those wishing to purchase firearms, including a tax equal to USD 430 (over six times the monthly minimum wage), proof that the buyer requires the use of the weapon, testimony from neighbours that the purchaser is a law-abiding citizen, a psychological assessment, and a demonstration of technical training in the use of firearms. The new law was passed by an overwhelming margin of 61 votes to 9, following intense pressure from such civil society groups as Viva Rio, the women's disarmament campaign, and the Evangelical Church. The Associação Nacional dos Proprietários e Comerciantes de Armas (ANPCA, the Brazilian NRA) declared it would challenge the constitutionality of this new law.[24]

Canada

Canada has historically had relatively strict controls on handguns. Under changes introduced in the 1977 Criminal Law Amendment Act, handgun ownership was 'restricted' while ownership of fully automatic firearms and sawn-off shotguns was banned. Permits for 'restricted weapons' could be issued to individuals if required for 'lawful occupation', target-shooting under the auspices of an approved shooting club, or collection. In exceptional cases permits were issued for 'self-protection'. Restricted weapons were also registered. Special permits were needed to transport or carry restricted weapons, specifying where and when the owner was allowed to have the firearm.

Controls on 'unrestricted' rifles and shotguns were, by contrast, limited. A Firearms Acquisition Certificate (FAC) was needed to acquire such firearms, yet did not have to be renewed for continued possession. In 1991, screening requirements for FACs were strengthened by adding a mandatory training course, a detailed questionnaire, references, a photograph, and a 28-day waiting period. The law also shifted the burden of proof on FAC decisions, requiring denied applicants to prove the refusal was the result of an error. This law also prohibited semi-automatic versions of fully automatic weapons and imposed detailed storage requirements.

Gun control in Canada was catapulted to the top of the public agenda with the December 1989 mass murder of 14 women at the École Polytechnique in Montréal. A broad coalition of more than 350 organizations—including police, legal, and public-health professional organizations, along with advocacy groups—soon emerged to lobby for stronger firearms controls (Falconer, 2001). Studies showed that rifles and shotguns were the firearms most often used to kill and the most often recovered in crime (Canada, Department of Justice, 1995). At the same time, a series of inquests into particular murders—typically involving legally acquired guns—consistently recommended the licensing of all firearm owners and the registration of their weapons. These developments culminated in the Firearms Act (Canada, 1995). Adopted on 5 December 1995, it is still in the process of being implemented. It provides for:

- the ability to prohibit, through Order in Council, firearms not 'reasonably' used in hunting, thus expanding powers to ban semi-automatic military-style weapons;
- a ban on short-barrelled and small calibre (.25 and .32) handguns;
- the licensing of all firearm owners by 1 January 2001, with renewal every five years;
- the registration of all firearms by 1 January 2003; and
- exceptions intended to accommodate the hunting rights of aboriginal peoples.

The most important challenge to the law came from the Province of Alberta, which claimed that it involved an unconstitutional infringement on provincial powers to regulate rifles and shotguns (Canada, Alberta Court of Appeal, 1998). In 2000, the Supreme Court of Canada dismissed this argument, upholding the constitutionality of the new legislation (Canada, Supreme Court, 2000).

As with the Australian, and also British, reforms, it is too early to judge the effects of the Canadian legislation. In 1998, the number and rate of firearm deaths in Canada were the lowest they had been for three decades (Hung, 2002, Table 14). Yet, Canada's largest city, Toronto, saw a surge in gun violence in 2000–01. There were 27 firearm murders in the city in the first nine months of 2001, compared with 26 for all of 2000.[25] Many of the crime guns were thought to have been illegally imported from the US, but it is too soon to say whether this signals a nationwide trend in which US guns replace Canadian ones in crime.

South Africa

Violence is considered by many to represent the greatest threat to human rights in the new democracy of South Africa, where crime dwarfs political conflict as a cause of death. A total of 12,298 people were murdered with guns in 1998 (29.1 per 100,000), accounting for almost 50 per cent of all murders (Chetty, 2000).

As of September 2001, the source of South African firearms legislation remained the Arms and Ammunition Act of 1969. It required the licensing of all gun owners and the registration of individual firearms. It did not restrict the number of firearms an individual could own. Moreover, South Africa has been one of the few countries allowing the widespread carrying of handguns by civilians for 'self-protection'. During the latter half of 2000, the South African Parliament passed the *Firearms Control Bill* (South Africa, 2000). As of September 2001, relevant implementing regulations and systems were being prepared. Some parts of the bill, including the power to create and enforce gun-free zones, had already been proclaimed. Other provisions were expected to be implemented in 2002. The bill strengthens controls over firearms in a number of areas. Regulation of the storage, transport, and carrying of firearms and ammunition is improved. At the same time, stringent penalties will be imposed on firearms owners who, as a result of negligence, lose their weapons. A principal focus of the bill is acquisition. Screening of licence applicants will be strengthened, while the number of firearms individuals can own will be limited, depending on weapon type and purpose of ownership.

© Associated Press/Themba Hadebe

The owner of an arms and ammunition shop, located near Soweto, South Africa.

United Kingdom

Although gun ownership is surprisingly common in the UK, especially in the countryside, controls on gun ownership have a justifiable reputation for being among the most restrictive in the world. The Firearms Act—adopted in 1937 and subsequently amended—required individuals wishing to acquire a firearm to first obtain a Firearms Certificate from the local police. A separate certificate was issued for each gun where the police were satisfied the applicant had a valid reason for owning the weapon—sport shooting, for example, but not self-protection. The certificate had to be renewed every three years. Following the murder of 16 people at Hungerford in August 1987, The Firearms Act of 1988 expanded the class of prohibited weapons to include most semi-automatic rifles as well as self-loading or pump-action shotguns.

UK firearms laws were further strengthened in the wake of public outrage over the murder, on 13 March 1996, of 16 primary-school students and their teacher in Dunblane, Scotland. A new law was passed which banned most handguns and required the remainder (.22 calibre or less) to be stored at gun clubs. When the Labour Party took power in 1997, it banned public ownership of all handguns, subject to narrow exceptions. These reforms were accompanied by amnesty and buy-back programmes, resulting in the handing over of some 185,000 firearms.[26]

It is too early to assess the effects of this ban, especially since gun homicides are rare in the UK and the figures normally fluctuate considerably (see Table 6.1). The most consistently visible trend in the data suggests that the murder of women with firearms is declining. Although this trend pre-dates Dunblane, it has continued since. This may be partially linked to the reduction in the number of households with firearms.

While there is some evidence that gun-related violence has decreased overall, it appears to have risen in some regions. There are significant problems in assessing this phenomenon, yet the sudden rise in gun deaths in London in 2000–01 is real enough, with 25 in the first eight months of 2001 compared with 12 during the same period in 2000 (Mitchell, 2001). Even so, the incidence of firearm-related deaths in Britain is low compared with most other countries (Krug, Powell, and Dahlberg, 1998; UN, 1998).

TABLE 6.1 Homicides by shooting in England and Wales, 1991–2000

	1991	1992	1993	1994	1995	1996	1997	1997–98	1998–99	1999–2000
Total homicides	623	581	566	632	663	585	612	612	663	728
By shooting/rate per 100,000*	50/ 0.10	52/ 0.10	71/ 0.14	63/ 0.12	66/ 0.13	47/ 0.09	58/ 0.11	52/ 0.10	47/ 0.09	62/ 0.12
By shooting, male	32	38	51	49	55	38	52	45	43	53
By shooting, female	18	14	20	14	11	9	6	7	4	9

** using approximate population figures for each year*

Source: UK, Home Office (2000, Tables 4.2, 4.3)

Conclusion

This chapter has examined attempts to control small arms in a few key areas. These efforts extend from the national to the regional and global levels, and from initial production to final possession and storage. Although only selected areas are covered, these are widely recognized as crucial components of a comprehensive control strategy—as largely reflected in the UN Conference *Programme of Action* (CONFERENCE).

Effective tracing of small arms can help identify and ultimately suppress illicit trafficking networks. End-use assurances play a similar role in preventing the diversion of these weapons to unintended end-users and end uses. Brokering controls, when further developed, will fill a crucial gap in the current regulatory framework, as they will limit the freedom certain brokers now exploit when facilitating illicit arms deals. Secure stocks are equally essential in combating diversion, while the national regulation of civilian ownership is a cornerstone of efforts to minimize the misuse of small arms.

These regulatory regimes are at varying stages of development. Controls over civilian ownership are, with important exceptions, quite extensive and in recent years pressure to tighten regulations in both developed and developing countries has mounted. By contrast, arms brokering remains largely unregulated, even at the national level.

In a world of sovereign states, all these regulatory regimes are inevitably rooted in national legislation and institutions. This is where the real test of the recent control initiatives in eastern Africa will come—ensuring that the political commitments made in subregional forums translate into effective action at the national level. While national regulation will remain the foundation of broader control efforts, the inescapably trans-national nature of small arms proliferation makes multilateral co-operation indispensable. As difficult as this often is for states wary of infringements on their sovereignty, the need for the harmonization of small arms controls is increasingly evident. Effective tracing mechanisms cannot be developed without some minimum level of harmonization. End-use assurances will remain open to abuse and fraud so long as state practices remain sharply divergent. The development of significant controls over brokering will require common agreement on definitions and on scope of jurisdiction.

Small arms controls at national and multilateral levels necessarily reinforce one another.

As reflected in the UN Conference *Programme of Action* and existing practice, the two levels of activity—national and multilateral—necessarily reinforce one another. Any multilateral instrument relies on national implementation for its success. Very often, the implementation of such instruments has a significant impact on existing national systems—in both developed and developing countries. At the same time, the efforts of individual states to independently strengthen controls over small arms quickly run up against the limitations imposed by divergent state practices, creating pressure for the development of multilateral measures designed to harmonize and standardize the same.

This chapter describes the wide variety of actors that are involved, at national, regional, and global levels, in attempts to strengthen controls over small arms. It also highlights the crucial roles NGOs play in building support for such efforts and in shaping and implementing specific policy initiatives. Yet the principal role in small arms control belongs to those entities which have ultimate responsibility for the security of their citizens—states.

6. List of Abbreviations

ANPCA	Associação Nacional dos Proprietários e Comerciantes de Armas
BASIC	British American Security Information Council
BICC	Bonn International Center for Conversion
CICP	Centre for International Crime Prevention (United Nations)
DV	Delivery Verification
EAC	East African Community
EAPC	Euro–Atlantic Partnership Council
EAPCCO	Eastern Africa Police Chiefs Committee
ECOSOC	Economic and Social Council (United Nations)
ECOWAS	Economic Community of West African States
EU	European Union
EUAs	End-use assurances
FAC	Firearms Acquisition Certificate
GRIP	Groupe de recherche et d'information sur la paix et la sécurité
IANSA	International Action Network on Small Arms
IGAD	Inter-Governmental Authority on Development
IIC	International Import Certificate
IRG	International Resource Group on Disarmament and Security in the Horn of Africa
ISP	National Inspectorate of Strategic Products
ISS	Institute for Security Studies
LEOM	Licensing and Enforcement Group

NATO	North Atlantic Treaty Organization
NRA	National Rifle Association
OAS	Organization of American States
OAU	Organization of African Unity
OECD	Organisation for Economic Co-operation and Development
OSCE	Organization for Security and Co-operation in Europe
PfP	Partnership for Peace
SADC	Southern African Development Community
SALIGAD	Assessing Small Arms Issues and Developing Capacity for Peace in the Horn of Africa
SIPRI	Stockholm International Peace Research Institute
SRIC	Security Research and Information Centre
UIT	Unique Item Tracking
UNAFRI	United Nations African Institute for the Prevention of Crime and the Treatment of Offenders
UNITA	União Nacional Para a Independência Total de Angola
UNREC	United Nations Regional Centre for Peace and Disarmament
WCO	World Customs Organization
WFSA	World Forum on the Future of Sport Shooting Activities

6. Endnotes

1 As of 1 January 2002, 21 countries had signed the Protocol. For the latest information, see <http://www.undcp.org/crime_cicp_signatures.html>

2 An adapted version of the OAS convention formed the basis of negotiations in Vienna.

3 With respect to illicit manufacture, see art. 3.d.ii. With respect to illicit trafficking, see art. 3.e. With respect to both, see arts 7.b and 10.

4 With respect to illicit manufacture, see art. 3.d.iii. With respect to illicit trafficking, see arts 3.e and 15.2. With respect to both, see arts 3.f, 5.1.c, 6.2, 7, 8, and 12.4.

5 The relevant section of the Protocol reads 'or *maintain* any alternative unique user friendly marking with simple geometric symbols...' (art. 8.1.a, emphasis added).

6 The Protocol enters into force 90 days after the 40th instrument of ratification is deposited (art. 18).

7 Given the long operational life of most small arms, certain commentators argue that tracing information ought to be retained by states for an unlimited period of time—as is the case in Italy (Berkol, 2001a).

8 The principal supporting instruments for the Moratorium provide for the development of a regional arms register and database (ECOWAS, 1999a, sec. IV; 1999b, art. 6). However, as of October 2001, work on the register was suspended, though the Moratorium itself was renewed in July 2001 for a further three-year period (Mali, 2001a).

9 Article 3.e calls for the establishment of regional small arms registers (EU, 1998).

10 Paragraph V.3.B.ii advocates the development of common standards for marking and record keeping.

11 The two countries promoted their initiative for the development of an international tracing mechanism in a series of official proposals and workshops. See France and Switzerland (2000; 2001a; 2001b).

12 For the latest information concerning the model convention, see <http://www.grip.org>

13 Additional source: correspondence from Tamar Gabelnick, Director, Arms Sales Monitoring Project, Federation of American Scientists, November 2001.

14 Additional source: correspondence from Sibylle Bauer, Researcher, Institut d'Etudes européennes, Université Libre de Bruxelles, November 2001.

15 Twenty-seven of the Wassenaar Arrangement's 33 Participating States responded on a confidential basis. Conversation with Mr Roland Vock, official of the Federal Ministry of Foreign Affairs, 30 July 2001.

16 For example, the Italian End-use Statement requires a commitment to provide 'a statement of taking charge of goods, duly legalized... once the goods are received'.

17 Agreed by Albania, Bosnia and Herzegovina, Bulgaria, Croatia, Hungary, Poland, Romania, Slovakia, Slovenia, Turkey, and the United States.

18 These were the countries listed in the preceding note plus the Czech Republic and Macedonia.

19 Conversation with Mr Roland Vock, official of the Federal Ministry of Foreign Affairs, 30 July 2001.

20 An important caveat is that persons engaged exclusively in the business of financing, whose activities do not also include brokering defence articles or defence services, are not required to register as arms brokers (US, Department of State, 2001, sec. 129.3(b)(3)).

21 Though, as in the case of financial services, persons exclusively providing transportation services are exempt from registration (US, Department of State, 2001, sec.129.3(b)(3)).

22 But see Wood and Peleman (1999, ch. 9).

23 See also Kramer (2001).

24 Interview with and correspondence from staff of Viva Rio, October 2001 and January 2002.

25 Interview conducted by Brooks Gray, SAFER-Net, with Rob Knapper, Corporate Communications, Metro Toronto Police Service, 5 October 2001.

26 For more information on the collection programmes, see Faltas, McDonald, and Waszink (2001, p. 18).

6. Bibliography

Archer, Dane, Rosemary Gartner, and Marc Beittel. 1983. 'Homicide and the Death Penalty: a Cross-National Test of a Deterrence Hypothesis.' *Journal of Criminal Law and Criminology*. Vol. 74, No. 3, pp. 991–1013.

Atwood, David. 2001. *The Contributions of NGOs to Small Arms and Light Weapons Control Efforts*. Background paper. Geneva: Small Arms Survey.

__. and David Jackman. 2000. *Curbing the Demand for Small Arms: Lessons in East Africa and the Horn of Africa*. Seminar report jointly sponsored by Quaker UN Offices (Geneva and New York), Project Ploughshares (Canada), and Bonn International Center for Conversion (Germany). Nairobi, Kenya. 12–16 December. <http://www.afsc.org/quno/smallarmsnairobi.pdf>

Australia. Department of Defence. *End-Use and Non-Transfer Certification to the Australian Government for Defence Goods*.

Berkol, Ilhan. 2001a. *Initiatives sur le marquage et le traçage des armes légères*. Background paper. Geneva: Small Arms Survey.

__. 2001b. *Traçabilité des armes légères: Contribution à la Conférence des Nations Unies sur le commerce illicite des armes légères et de petit calibre sous tous ses aspects (New York, 9–20 juillet 2001)*. Note d'analyse. Brussels: GRIP. July.

__. 2001c. 'Les aspects pratiques du marquage et du traçage des armes légères en vue d'un modèle de convention.' Paper prepared for the IANSA Conference 'Proliferation of Small Arms: What European Policy after the UN Conference?' Brussels. 5–6 October.

Bollenberg, Lieutenant Colonel Craig. 2001. 'Destroying Excess Small Arms and Light Weapons and Properly Securing and Managing Stockpiles.' Paper presented at the joint Azerbaijani-Swiss Workshop, *Small Arms and Light Weapons: Practical Challenges for the Implementation of Current Undertakings in the OSCE and the EAPC*. Baku, Azerbaijan. 21–22 June. <http://www.isn.ethz.ch/pfpdc/documents/2001/06-01_baku/bollenberg.htm>

Bolli, Thomas. 2001. 'In Trauer nach Zug marschiert.' *Tages-Anzeiger*. 1 October, p. 7.

Bondì, Loretta. 2001. *Expanding the Net: A Model Convention on Arms Brokering*. Briefing Paper for the United Nations Conference on the Illicit Trade in Small Arms and Light Weapons in All Its Aspects. Washington, DC: The Fund for Peace. 29 June. <http://www.fundforpeace.org/resources/pubs.htm>

__. and Elise Keppler. 2001. *Casting the Net? Implications of the U.S. Law on Arms Brokering*. Washington, DC: The Fund for Peace. January. <http://www.fundforpeace.org/resources/pubs.htm>

Bouveret, Patrice. 2001. *French Export Controls*. Background paper. Geneva: Small Arms Survey.

__. and Belkacem Elomari, eds. 2001. *Ventes d'armes de la France*. Annual Report of the Observatoire des transferts d'armements. 6th ed. Lyon: CDRPC. July.

Buettner, Jean-Martin. 2001. 'Kein FBI für die Schweiz.' *Tages-Anzeiger*. 23 October, p. 8.

Bulgaria. Ministry of Foreign Affairs. *Statement by Ultimate Consignee*. Sofia: MFA.

Canada. 1995. *An Act Respecting Firearms and Other Weapons (Firearms Act)*. Bill C-68. Adopted 5 December. Reproduced in Annual Statutes of Canada, Chapter 39. <http://laws.justice.gc.ca/en/1995/39/text.html>

__. Alberta Court of Appeal. 1998. *Reference re Firearms Act (Canada)*. 1998 ABCA 305. 128 Canadian Criminal Cases (3d), p. 225.

__. Department of Justice. 1995. *Firearms Recovered By Police: A Multi-Site Study*. Daniel Antonowicz consulting. Ottawa. July.

__. Supreme Court. 2000. *Reference re Firearms Act (Canada)*. 2000 SCC 31.

Carver, Jeremy and James Bourke. 2001. *Controls on Brokering*. Background paper. Geneva: Small Arms Survey.

Chetty, Robert, ed. 2000. *Firearm Use and Distribution in South Africa*. Pretoria: The National Crime Prevention Centre (NCPC) Firearm Programme.

Coflin, James. 2000. *Small Arms Brokering: Impact, Options for Controls and Regulation*. Ottawa: International Security Research and Outreach Programme, International Security Bureau, Department of Foreign Affairs and International Trade, Canada. May. <http://www.dfait-maeci.gc.ca/arms/publications2-e.asp>

Convention for the Suppression of Unlawful Acts Against the Safety of Civil Aviation ('Montreal Convention'). 1971. <http://www.undcp.org/terrorism_convention_civil_aviation.html>

Convention on the Marking of Plastic Explosives for the Purpose of Identification. 1991. Montreal. 1 March. <http://www.undcp.org/terrorism_convention_plastic_explosives.html>

Co-ordinated Agenda for Action on the Problem of the Proliferation of Small Arms and Light Weapons in the Great Lakes Region and the Horn of Africa ('Co-ordinated Agenda for Action"). 2000. Adopted 8 November. Reproduced in *Implementing the Nairobi Declaration on the Problem of the Proliferation of Illicit Small Arms and Light Weapons in the Great Lakes Region and the Horn of Africa*. Nairobi: The Nairobi Secretariat, Ministry of Foreign Affairs and International Co-operation, Kenya. July 2001.

Cukier, Wendy. 2001. *National Firearms Controls*. Background paper. Geneva: Small Arms Survey.

__. Tania Sarkar and Tim Quigley. 2000. 'Firearm Regulation: International Law and Jurisprudence.' *Canadian Criminal Law Review*. Vol. 6, No. 1. December, pp. 99–123.

Culross, Patti. 1996. 'Legislative Strategies to Address Firearm Violence and Injury'. *Journal of Family Practice*. Vol. 42, No. 1, pp. 15–17.

Economic Community of West African States (ECOWAS). 1999a. *Plan of Action for the Implementation of the Programme for Co-ordination and Assistance for Security and Development (PCASED)*. Bamako. 25 March. Reproduced in Jacqueline Seck, ed. 2000. *West Africa Small Arms Moratorium: High-Level Consultations on the Modalities for the Implementation of PCASED*. Geneva/Lomé: United Nations Institute for Disarmament Research/United Nations Regional Centre for Peace and Disarmament in Africa, pp. 31–45.

__. 1999b. *Code of Conduct for the Implementation of the Moratorium on the Importation, Exportation and Manufacture of Light Weapons*. Lomé. 10 December. Reproduced in Jacqueline Seck, ed. 2000. *West Africa Small Arms Moratorium: High-Level Consultations on the Modalities for the Implementation of PCASED*. Geneva/Lomé: United Nations Institute for Disarmament Research/United Nations Regional Centre for Peace and Disarmament in Africa, pp. 53–62.

The Economist (US edition). 2001. 'For Defence, for Sport and... ?' 6 October.

Euro-Atlantic Partnership Council (EAPC). 2000. *Guidelines on Implementation of the PWP Chapter on Small Arms and Light Weapons*. Doc. EAPC(C)D(2000)11, Annex 2. 22 March.

__. 2001. *Euro-Atlantic Partnership Council/Partnership for Peace Contribution to the United Nations Conference on the Illicit Trade in Small Arms and Light Weapons in All Its Aspects*. Reproduced in UN General Assembly document. A/CONF.192/5 of 6 July.

European Union (EU). 1998. *Joint Action of 17 December 1998 Adopted by the Council on the Basis of Article J.3 of the Treaty on European Union on the European Union's Contribution to Combating the Destabilising Accumulation and Spread of Small Arms and Light Weapons*. Doc. 1999/34/CFSP. 17 December. Reproduced in Official Journal L 9. 15 January 1999.

__. 1999. *Council Decision of 15 November 1999 Implementing Joint Action 1999/34/CFSP with a View to a European Union Contribution to Combating the Destabilising Accumulation and Spread of Small Arms and Light Weapons in Cambodia*. Doc. 399D0730. Official Journal L 294. 16 November 1999, pp. 5–6.

__. 2000 *Council Decision of 20 November 2000 Extending and Amending Decision 1999/730/CFSP concerning a European Union Contribution to Combating the Destabilising Accumulation and Spread of Small Arms and Light Weapons in Cambodia.* Doc. 300D0724. Official Journal L 292. 21 November 2000, p. 3.

Falbo, Gilliat, Roberto Buzzetti and Adriano Cattaneo. 2001. 'Homicide in Children and Adolescents: A Case-Control Study in Recife, Brazil.' *Bulletin of the World Health Organization.* Vol. 79, No.1, pp. 2–7.

Falconer, Tim. 2001. *Watchdogs and Gadflies: Activism from Marginal to Mainstream.* Toronto: Viking/Penguin.

Faltas, Sami, Glenn McDonald, and Camilla Waszink. 2001. *Removing Small Arms from Society: A Review of Weapons Collection and Destruction* Programmes. Occasional Paper No. 2. Geneva: Small Arms Survey. July. <http://www.smallarmssurvey.org/OccasionalPapers.html>

Fawthrop, Tom. 2001. '900,000 Small Arms Plague Cambodia.' *Jane's Intelligence Review,* Vol. 13, No. 4. 1 April.

Fleck, Fiona. 2001. 'Swiss Massacre Spurs Calls for Gun Law Reform.' *Sunday Telegraph* (London). 30 September, p. 29.

Flinn, Rick and Linda Allen. 1995. 'Trigger Locks and Firearm Safety: One Trauma Center's Prevention Campaign.' *Journal of Emergency Nursing.* Vol. 21, No. 4, pp. 296–98.

Foot, David with Daniel Stoffman. 1999. *Boom, Bust and Echo 2000: Profiting from the Demographic Shift in the New Millennium.* Toronto: Stoddart.

France. 1939. *Decree-Law of 18 April 1939, Creating a Regime for War Material, Arms, and Munitions.* 18 April. Reproduced by the SIPRI Export Control Project. <http://projects.sipri.se/expcon/natexpcon/France/fralaw.htm>

__. 1995. *Decree no. 95-589 of 6 May 1995 regarding the Application of the Decree of 18 April 1939 Creating a Regime for War Material, Arms, and Munitions.* 6 May. Reproduced by the SIPRI Export Control Project. <http://projects.sipri.se/expcon/natexpcon/France/fra_dec95_589.htm>

France and Switzerland. 2000. *Food-for-Thought Paper: Contribution to the Realisation of an International Plan of Action in the Context of the 2001 Conference: Marking, Identification and Control of Small Arms and Light Weapons.* Reproduced in UN document A/CONF.192/PC/7, Annex. 17 March.

__. 2001a. *Working Paper by Switzerland and France on Establishing a Tracing Mechanism to Prevent and Reduce Excessive and Destabilizing Accumulation and Transfer of Small Arms and Light Weapons.* Reproduced in UN document A/CONF.192/PC/25 of 10 January.

__. 2001b. *Franco-Swiss Workshop on Traceability of Small Arms and Light Weapons: Tracing, Marking and Record-Keeping. Geneva, 12–13 March 2001. Chair's Summary.* Reproduced in UN document A/CONF.192/PC/38, Annex. 23 March.

The Fund for Peace. 2001. *Model Convention on the Registration of Arms Brokers and the Suppression of Unlicensed Arms Brokering.* New York: Fund for Peace. July. <http://www.fundforpeace.org/resources/pubs.htm>

Gartner, Rosemary. 1998. 'Affidavit of Rosemary Gartner.' Alberta Court of Appeal, Vol. 39.

Germany. 1961. *Act Implementing Article 26(2) of the Basic Law ('War Weapons Control Act').* 20 April. Reproduced in Federal Law Gazette I-444, 1961. <http://projects.sipri.se/expcon/natexpcon/Germany/kwkg.htm>

Greene, Owen n.d. *Enhancing Traceability of Small Arms and Light Weapons Flows: Developing an International Marking and Tracing Regime.* Biting the Bullet Briefing 5. London and Washington, DC: BASIC, International Alert, and Saferworld.

Hung, Kwing. 2002. *Firearm Statistics: Updated Tables.* Report. Geneva: Small Arms Survey.

IISS Strategic Comments. 2001. 'The UN "Firearms Protocol": Addressing the Trafficking Problem.' Vol. 7, No. 6. July.

Implementation Plan of the Co-ordinated Agenda for Action on the Problem of the Proliferation of Small Arms and Light Weapons in the Great Lakes Region and the Horn of Africa ('Implementation Plan'). 2000. Adopted 8 November. Reproduced in *Implementing the Nairobi Declaration on the Problem of the Proliferation of Illicit Small Arms and Light Weapons in the Great Lakes Region and the Horn of Africa.* Nairobi: The Nairobi Secretariat, Ministry of Foreign Affairs and International Co-operation, Kenya. July 2001.

International Action Network on Small Arms (IANSA). 2001. *Focusing Attention on Small Arms.* 8 January.

International Crime Victimization Survey. 1996. Cited in Richard Block, *Firearms in Canada and Eight Other Western Countries.* Ottawa: Department of Justice. January 1998.

Interpol General Assembly. 1992. Resolution AGN/61/RES/15. Dakar. 10 November.

__. 1997. Resolution AGN/66/RES/6. New Delhi. 21 October.

Isbister, Roy. 2001. *UK Export Control Bill.* Background paper. Geneva: Small Arms Survey.

Israel. Deputy Director General for Foreign Affairs and Arms Control to the United Nations. 2000. *Communication to the Group of Governmental Experts Established Pursuant to General Assembly Resolution 54/54 V of 15 December 1999, Entitled 'Small Arms'.* 28 June.

Italy. Ministry of Foreign Affairs. *End-Use Statement.* Rome: MFA.

Iten, Oswald. 2001a. 'Amok-Massaker im Zuger Kantonsrat; 15 Tote, 15 Verletzte–der Kanton unter Schock.' *Neue Zürcher Zeitung.* 28 September, p. 13.

__. 2001b. 'Querulant plante Wahnsinnstat kaltblütig.' *Neue Zürcher Zeitung.* 29 September, p. 14.

Jane's Intelligence Review. 2000. 'Clandestine Trade in Arms: A Matter of Ways and Means.' Vol. 12, No. 5. 1 May.

Johnson, David. 1997. 'Six Marines are Charged with Selling Stolen Arms.' *New York Times.* 19 October, p. A24.

Joseph, Kate. 2001a. *End-use Certificates.* Background paper. Geneva: Small Arms Survey.

__. 2001b. *Firearms Protocol.* Background paper. Geneva: Small Arms Survey.

Karp, Aaron. 2001. *Firearms Inventory Control in the US Army.* Background paper. Geneva: Small Arms Survey.

Kellermann, Arthur, *et al.* 1992. 'Suicide in the Home in Relation to Gun Ownership.' *New England Journal of Medicine,* Vol. 327, No. 7, pp. 467–72.

Kenya. Ministry of Foreign Affairs and International Co-operation. 2001. *Implementing the Nairobi Declaration on the Problem of the Proliferation of Illicit Small Arms and Light Weapons in the Great Lakes Region and the Horn of Africa.* Nairobi: The Nairobi Secretariat. July.

__. Revenue Authority. 2000. *Communication to the Group of Governmental Experts Established Pursuant to General Assembly Resolution 54/54 V of 15 December 1999, Entitled 'Small Arms'.* Nairobi: Revenue Authority. 31 May.

Kramer, Katherine. 2001. *Legal Controls on Small Arms and Light Weapons in Southeast Asia.* Occasional Paper No. 3. Geneva: Small Arms Survey. July. <http://www.smallarmssurvey.org/OccasionalPapers.html>

Krug, Etienne, Kenneth Powell, and Linda Dahlberg. 1998. 'Firearm-Related Deaths in the United States and 35 Other High- and Upper-Middle-Income Countries.' *International Journal of Epidemiology.* Vol. 27, No. 2, pp. 214–21.

Lab, Steven. 1997. *Crime Prevention: Approaches, Practices and Evaluations.* Cincinnati: Anderson Publishing.

Lezzi, Bruno. 2001. 'Ist eine Verschärfung des Waffengesetzes nötig?' *Neue Zürcher Zeitung.* 29 September, p. 15.

Lumpe, Lora and Jeff Donarski. 1998. *The Arms Trade Revealed: A Guide for Investigators and Activists.* Washington, DC: Federation of American Scientists. August. <http://www.fas.org/asmp/library/publications/revealed.htm>

Mali. 2001a. *Letter Dated 9 July 2001 from the Permanent Representative of Mali to the United Nations Addressed to the President of the Security Council.* UN document S/2001/700 of 9 July.

__. 2001b. *Letter Dated 16 July 2001 from the Permanent Representative of Mali to the United Nations Addressed to the President of the United Nations Conference on the Illicit Trade in Small Arms and Light Weapons in All Its Aspects.* UN document A/CONF.192/13 of 18 July.

McLean, Andrew. 2001. *The Great Lakes Region and the Horn of Africa: Strengthening Controls.* Background paper. Geneva: Small Arms Survey.

Meier, Peter Johannes. 2001. 'Der Amokschuetze von Zug war ein mehrfach verur- teilter Waffenschmuggler.' *Tages-Anzeiger.* 2 October, p. 8.

Merki, Martin. 2001. 'Mehr Sicherheit heisst weniger Freiheit.' *Neue Zürcher Zeitung.* 29 September, p. 14.

Mitchell, Adam. 2001. 'Gun Killings Double As Police Claim Progress.' *The Daily Telegraph.* 17 August.

Mouzos, Jenny. 1999. *Firearm-related Violence: The Impact of the Nationwide Agreement on Firearms.* Trends and Issues in Crime and Criminal Justice, No. 116. Canberra: Australian Institute of Criminology. May. <http://www.aic.gov.au/publications/tandi/tandi116.html>

__. 2000. *Homicidal Encounters: A Study of Homicide in Australia 1989-1999.* Research and Public Policy Series, No. 28. Canberra: Australian Institute of Criminology. <http://www.aic.gov.au/publications/rpp/28/index.html>

__. 2001a. *Homicide in Australia 1999–2000.* Trends and Issues in Crime and Criminal Justice, No. 187. Canberra: Australian Institute of Criminology. February. <http://www.aic.gov.au/publications/tandi/tandi187.html>

__. 2001b. *Firearm-Related Morbidity in Australia 1994–95 to 1998–99.* Trends and Issues in Crime and Criminal Justice, No. 198. Canberra: Australian Institute of Criminology. March. <http://www.aic.gov.au/publications/tandi/tandi198.html>

Nairobi Declaration on the Problem of the Proliferation of Illicit Small Arms and Light Weapons in the Great Lakes Region and the Horn of Africa ('Nairobi Declaration'). 2000. Adopted in Nairobi, 15 March. Reproduced in Andrew McLean, ed. *Tackling Small Arms in the Great Lakes Region and the Horn of Africa: Strengthening the Capacity of Subregional Organisations.* Pretoria: Institute for Security Studies, 2000, pp. 74–77.

New York Times. 2001. 'Curbing Arms Brokers.' 1 July, p. 12.

North Atlantic Treaty Organization (NATO). 2000. 'Helping Albania Manage Munitions.' *NATO Review.* Web edition, Vol. 48, No. 2, Summer–Autumn. <http://www.nato.int/docu/review/2000/0002-09.htm>

__. 2001. 'Safe Destruction of Anti-Personnel Landmines and Munitions in Moldova.' Press Release (2001) 029. 27 February.

Olson, Elizabeth. 2001a. '15 Die in Swiss Shooting Rampage.' *International Herald Tribune.* 28 September, p. 1.

__. 2001b. '14 Killed in Attack on a Swiss Legislature.' *New York Times.* 28 September, p. 6.

__. 2001c. 'Switzerland Feels Chill of Insecurity After Shooting.' *International Herald Tribune.* 29 September, p. 7.

__. 2001d. 'Killings of 14 Lead Swiss to Reconsider Love of Guns.' *New York Times.* 30 September, p. 4.

Organisation for Economic Co-operation and Development (OECD). 1997. *Convention on Combating Bribery of Foreign Public Officials in International Business Transactions ('Anti-bribery Convention').* 21 November. DAFFE/IME/BR(97)20. <http://www.oecd.org>

Organization for Security and Co-operation in Europe (OSCE). Forum for Security Co-operation. 2000. *OSCE Document on Small Arms and Light Weapons.* FSC.DOC/1/00. 24 November. <http://www.osce.org/docs/english/fsc/2000/decisions/fscew231.htm>

Organization of American States (OAS). 1997. *Inter-American Convention against the Illicit Manufacturing of and Trafficking in Firearms, Ammunition, Explosives, and Other Related Materials ('Inter-American Convention').* Washington, DC. AG/RES.1 (XXIV-E/97). 13 November. Reproduced in UN document A/53/78 of 9 March 1998.

__. 2001. *CIFTA: Presentation of the Ambassador Miguel Ruiz Cabanas, Secretary Pro Tépore of the Consultative Committee Established in the Framework of the Inter-American Convention against the Illicit Manufacturing of and Trafficking in Firearms, Ammunition, Explosives, and Other Related Materials (CIFTA), during the United Nations Conference on the Illicit Trade in Small Arms and Light Weapons in All Its Aspects. New York, 11 of July of 2001 [sic].* <http://www.un.org/Depts/dda/CAB/smallarms/statements/consultativeE.html>

Osborn, Andrew. 2001. 'Murderer with Four Guns and a Grudge.' *The Observer.* 30 September, p. 25.

Poland. 2000. *Law of 29 November 2000 Concerning International Trade in Goods, Technologies and Services of Strategic Significance for State Security and Maintenance of International Peace and Security, and Amending Selected Laws.* Reproduced by the SIPRI Export Control Project. <http://projects.sipri.se/expcon/natexpcon/Poland/law001129.htm>

Scott, Sir Richard. 1996. *Report of the Inquiry into the Export of Defence Equipment and Dual-Use Goods to Iraq and Related Prosecutions.* HC115. London: HMSO. 15 February.

Small Arms Survey. 2001. *Small Arms Survey 2001: Profiling the Problem.* Oxford: Oxford University Press.

South Africa. 1998. *Act No. 15 of 1998: Regulation of Foreign Military Assistance Act.* 20 May. Reproduced by the SIPRI Export Control Project. <http://projects.sipri.se/expcon/natexpcon/SAfrica/formilass.htm>

__. 2000. *Bill to Establish a Comprehensive and an Effective System of Firearms Controls; and to Provide for Matters Connected Therewith ('Firearms Control Bill').* B34D–2000. <http://www.polity.org.za/govdocs/bills/2000/b34d-00.pdf>

Special Co-ordinator of the Stability Pact for South Eastern Europe. 2001. *Small Arms and Light Weapons.* <http://www.stabilitypact.org/stabilitypactcgi/catalog/cat_descr.cgi?subcat=1&prod_id=47>

Statement on Harmonization of End-Use/End-User Certificates ('Statement on Harmonization'). 1999. Sofia, Bulgaria. 15 December.

Stockholm International Peace Research Institute (SIPRI). 2001a. *The Netherlands.* SIPRI Export Control Project. <http://projects.sipri.se/expcon/natexpcon/Netherlands/netherlands.htm>

__. 2001b. *The Role of Parliament in Export Control: The Swedish Export Control Council.* SIPRI Export Control Project. <http://projects.sipri.se/expcon/parliament/ekr.htm>

Sweden. 1992. *Military Equipment Act (1992:1300).* Reproduced by the SIPRI Export Control Project. <http://projects.sipri.se/expcon/natexpcon/Sweden/krigmlag.htm>

__. 1993. National lists of military equipment *(Military Equipment for Combat Purposes (MEC); Other Military Equipment (OME)).* Reproduced by the SIPRI Export Control Project. <http://projects.sipri.se/expcon/natexpcon/Sweden/equiplist.htm>

__. National Inspectorate of Strategic Products. 2000. *National Practices for End Use Certificates.* 15 June.

Switzerland. 1996. *Federal Law on War Material.* 13 December. Reproduced by the SIPRI Export Control Project. <http://projects.sipri.se/expcon/natexpcon/Switzerland/swilaw.htm>

__. 2000. *Questionnaire on International Import Certificates, Delivery Verifications and Private End-User Statements.* Issued on 31 March.

Tages-Anzeiger. 2001. 'Vom Verlust der inneren Sicherheit.' 16 October, p. 7.

United Kingdom (UK). Department of Trade and Industry. 2000. *End-User Undertakings—Guidance for Form EUU01.* London: DTI. 27 November. <http://www.dti.gov.uk/export.control/notices/2000/notice111.htm>

__. Home Office. 2000. *Criminal Statistics: England and Wales 1999.* London: The Stationery Office. December. <http://www.archive.official-documents.co.uk/document/cm50/5001/5001.htm>

United Nations (UN). 1997. *Report of the Panel of Governmental Experts on Small Arms.* A/52/298 of 27 August.

__. 1998. *United Nations International Study on Firearm Regulation.* New York: UN.

__. 1999a. *United Nations International Study on Firearm Regulation.* Database update. <http://www.uncjin.org/Statistics/firearms/index.htm>

—. 1999b. *Report of the Group of Experts on the Problem of Ammunition and Explosives.* A/54/155 of 29 June.
—. 1999c. *Report of the Group of Governmental Experts on Small Arms.* A/54/258 of 19 August.
—. 2001a. *Report of the Group of Governmental Experts Established Pursuant to General Assembly Resolution 54/54 V of 15 December 1999, Entitled 'Small Arms' ('Feasibility Study').* A/CONF.192/2 of 11 May.
—. 2001b. *Recommendations on the Transport of Dangerous Goods.* Model Regulations. ST/SG/AC10/1/Rev.12. 12th revised ed. New York and Geneva. August.
United Nations Economic and Social Council (UN ECOSOC). 1988. *United Nations Convention against Illicit Traffic in Narcotic Drugs and Psychotropic Substances.* E/CONF.82/15. <http://www.incb.org/e/conv/1988/index.htm>
—. 2001. *Revised Draft Plans of Action for the Implementation of the Vienna Declaration on Crime and Justice: Meeting the Challenges of the Twenty-first Century.* E/CN.15/2001/14 of 25 June.
United Nations General Assembly (UNGA). 1998. Resolution 53/111, adopted 9 December. A/RES/53/111 of 20 January 1999.
—. 2000. *United Nations Convention against Transnational Organized Crime.* Reproduced in UN document A/RES/55/25 of 8 January 2001.
—. 2001a. *Report of the Ad Hoc Committee on the Elaboration of a Convention against Transnational Organized Crime on the Work of its First to Eleventh Sessions. Addendum. Report of the Ad Hoc Committee on the Elaboration of a Convention against Transnational Organized Crime on the Work of its Twelfth Session.* A/55/383/Add.2 of 20 March.
—. 2001b. *Protocol against the Illicit Manufacturing of and Trafficking in Firearms, Their Parts and Components and Ammunition, Supplementing the United Nations Convention against Transnational Organized Crime ('Firearms Protocol').* Reproduced in UN document A/RES/55/255 of 8 June.
—. 2001c. *Programme of Action to Prevent, Combat and Eradicate the Illicit Trade in Small Arms and Light Weapons in All Its Aspects.* Reproduced in UN document A/CONF.192/15 of 20 July.
United Nations Secretary-General (UNSG). 1996. *Interim Report of the International Commission of Inquiry to Investigate Reports of the Sale or Supply of Arms to Former Rwandan Government Forces in Violation of the Security Council Arms Embargo and Allegations that those Forces are Receiving Training to Destabilize Rwanda.* 17 January. UN document S/1996/67, Annex. 29 January.
—. 1998. *Final Report of the International Commission of Inquiry (Rwanda).* S/1998/1096 of 18 November.
United Nations Security Council (UNSC). 2000a. *Report of the Panel of Experts on Violations of Security Council Sanctions against UNITA.* UN document S/2000/203, Annex I. 10 March.
—. 2000b. *Final Report of the Monitoring Mechanism on Angola Sanctions.* UN document S/2000/1225, Annex. 21 December.
United States (US). US Code, Title 22, Chapter 1, Part 123 *(Licenses for the Export of Defense Articles).*
—. Congress. 1968. *The Gun Control Act of 1968: An Act to Amend Title 18, United States Code, to Provide for Better Control of the Interstate Traffic in Firearms.* Public Law 90-618. <http://www.atf.treas.gov/pub/fire-explo_pub/gca.htm>
—. Congress. 1976. *Arms Export Control Act.* Washington, DC.
—. Department of Defense. 2000. *Physical Security of Sensitive Conventional Arms, Ammunition, and Explosives.* Department of Defense Manual 5100.76-M. 12 August. <http://www.dtic.mil/whs/directives/corres/html/510076m.htm>
—. Department of State. Office of Defense Trade Controls. 2001. *International Traffic in Arms Regulations (ITAR).* <http://www.pmdtc.org/docs/ITAR/ITAR_129.txt>
—. Department of State. Office of Defense Trade Controls. n.d. *Nontransfer and Use Certificate (Form DSP-83).* <http://www.pmdtc.org/docs/dsp83.pdf>
Wassenaar Arrangement on Export Controls for Conventional Arms and Dual-Use Goods and Technologies (Wassenaar Arrangement). 1999. *End-Use Assurances Commonly Used: Indicative List.* December. <http://www.wassenaar.org/docs/oth_end-use_assurances.pdf>
—. 2000. *List of Dual Use Goods and Technologies and Munitions List.* WA LIST (00) 1. 1 December. <http://www.wassenaar.org/docs/docindex.html>
Webster, Daniel, Jon Vernick, and Lisa Hepburn. 2001. 'Relationship Between Licensing, Registration, and Other Gun Sales Laws and the Source State of Crime Guns.' *Injury Prevention.* Vol. 7, pp. 184–89.
Wood, Brian and Johan Peleman. 1999. *The Arms Fixers. Controlling the Brokers and Shipping Agents.* Oslo: International Peace Research Institute (PRIO).
Workshop on Defining 'Small Arms' as they Pertain to 'Firearms' for the 2001 UN Conference on Small Arms ('Workshop Report'). 2001. London. 27 April.
World Forum on the Future of Sport Shooting Activities (WFSA). 2000. *Draft Report of the Informal Workshop on 'Firearms Marking: Model Standards and Common Serial Number Codes.'* Sardinia, Italy. 22–24 June. <http://www.un.org/Depts/dda/CAB/sardinia.pdf>

Acknowledgements

Main contributors

David Atwood, Ilhan Berkol, James Bourke, Patrice Bouveret, Jeremy Carver, Wendy Cukier, Roy Isbister, Kate Joseph, Aaron Karp, and Andrew McLean.

Other contributors

Sibylle Bauer, Silvia Cattaneo, Tamar Gabelnick, Anna Khakee, Alison Riddell, SAFER-Net, Viva Rio, and Herbert Wulf.

FIATALLIS

Changing Attitudes:
Weapons Collection and Destruction

7

Introduction

The largest public destruction of small arms to date took place in Rio de Janeiro, Brazil on 24 June 2001. One hundred thousand pistols, revolvers, machine guns, and other firearms were piled up on the street and destroyed by a bulldozer. The event attracted 20,000 spectators in support of the campaign against gun violence, which is the leading cause of death among Brazilian youth (Viva Rio, 2001; Sullivan, 2001). This was one of several similar weapons collection and destruction ceremonies that took place in connection with the 2001 UN Small Arms Conference.

The event in Rio was also part of a larger process that has been gaining momentum since the mid-1990s. The collection and destruction of small arms, which grew from small-scale and intermittent efforts to address gun violence by removing weapons from local communities, have grown into an established component of conflict prevention and community building. All over the world in different settings, similar programmes have been carried out. Importantly, the *Programme of Action* of the 2001 UN Small Arms Conference (CONFERENCE) firmly supports the disarmament of civilians and former combatants through weapons collection and destruction programmes. These provisions are likely to encourage the further expansion of such efforts.

This chapter examines the issue of weapons collection and destruction efforts, both in the context of crime prevention and peace-building. Based on the information reviewed in this chapter, it is clear that such undertakings enjoy mixed success. In some cases, programmes have not collected large numbers of weapons, although they have contributed significantly to reconciliation and confidence building between parties in conflict. Others have collected significant numbers of weapons, but ensuing confusion and disagreement over their disposal has detracted from the programme's success. Consequently, even though the number and scope of programmes have been expanding, their effectiveness is still debatable. If weapons collection and destruction is to be used as a tool to prevent conflict and crime, it is imperative that we explore the circumstances in which it is successful. Key issues identified in this chapter include:

The collection and destruction of small arms are now an established component of conflict prevention and community-building efforts.

- Small arms collection and destruction is usually intended to support crime prevention or peace-building.
- Disarming paramilitary groups is a decisive factor in advancing and sustaining peace settlements.
- There are still insufficient evaluations of past endeavours.
- There is a continued need for reliable ways to assess weapons collection and destruction programmes.
- Weapons collection is unlikely to have a long-term impact unless it is part of a broader effort addressing the root causes of violence.
- Programmes that link collection of weapons to development strategies are increasingly incorporated into conflict prevention and post-conflict peace-building strategies.
- The total number of weapons collected may often be less important than other objectives, such as building confidence, raising awareness, and forging collaborative relationships between different sectors of society.

Public destruction of weapons, Rio de Janeiro. (©Associated Press/Renzo Gostoli)

This chapter does not attempt to provide definitive answers to the questions arising from weapons collection and destruction. Instead, it explores the potential of such programmes by considering some key questions:

- **What are some of the key elements of weapons collection and destruction programmes?**
- **How have these programmes developed over time?**
- **How can their effectiveness be improved in the future?**

The chapter begins with a review of several important examples of weapons collection and destruction programmes, outlining their aims, key features, accomplishments, and limitations. It emphasizes their potential role within two major frameworks: crime prevention/reduction and peace-building. The selected examples show the particularly strong tradition in North and Latin America of undertaking weapons collection as part of crime prevention strategies. The bulk of weapons collections implemented as part of peace-building exercises have been carried out in Africa and in the Balkans. Examples ranging from purely voluntary initiatives to mandatory hand-ins required by changing firearms laws are included. This chapter does not focus on seizures by law enforcement agencies. Programmes that destroy government stocks of surplus weapons are dealt with in other chapters (especially STOCKPILES and MEASURES).

There is growing awareness of the mutually reinforcing relationship between security and development. This has led to the realization that both are necessary for conflict prevention. Efforts to remove the tools of violence without addressing the underlying causes of conflict are unlikely to have a durable impact on peace and security. The second section provides an in-depth analysis of recent attempts to link disarmament and development strategies through weapons collection, or 'weapons for development'. Two case studies—from Albania and the Republic of Congo—are presented here as illustrations.

The third section outlines tentative lessons to guide future weapons collection and destruction efforts, focusing on programme objectives, incentives for participation, public awareness-raising, and the disposal of weapons. The broad variety of features that different weapons collection programmes display demonstrates that organizers are still trying out different solutions in order to improve results. Failures and shortcomings of earlier undertakings are currently being analysed by a variety of actors and institutions to determine best practices and avoid problems encountered in the past.[1]

Since weapons reduction may be futile unless demand is addressed simultaneously, programmes are generally expanding their objectives and adopting a more comprehensive approach. This is reflected also in the changing role of compensation, where there is a trend away from individual rewards and towards collective incentive schemes, most notably through the weapons for development concept. Another crucial component is their function in shaping public opinion. An analysis of the weapons collection programme in 2001 in the Argentinian province of Mendoza illustrates this point. This section also charts different ways in which collected weapons have been disposed of, reaffirming that destruction is the only method that ensures weapons are permanently removed from circulation.

The final section highlights the need for the development and application of more reliable criteria to evaluate the outcome of weapons collection. The relative absence of well-documented results has resulted in a major gap in the debate about success and failure. The section therefore points to newly developed approaches to improve programme assessment. To gauge the social impact of these initiatives, field agencies and the research community have started to explore participatory methods. The chapter ends with a new approach to assess operational success in terms of the weapons collected and destroyed.

BOX 7.1 2001 UN Small Arms Conference supports weapons collection and destruction: Something everyone could agree on

Weapons collection and destruction schemes received widespread support during the 2001 UN Small Arms Conference. The *Programme of Action* adopted contained several references to such initiatives. Among other things, states undertake to implement 'where appropriate, the public destruction of surplus weapons and the voluntary surrender of small arms and light weapons' (sec. II, art. 20), as well as 'effective disarmament, demobilization and reintegration programmes, including the effective collection, control, storage, and destruction of small arms and light weapons, particularly in post-conflict situations' (sec. II, art. 21). Most governments also expressed their support for such measures in many of the official statements made during the Conference and the Preparatory Committee meetings leading up to it.[2]

On the initiative of Brazil, Mali, the Netherlands, and the United Kingdom, the Conference's Preparatory Committee had proclaimed the opening day of the Conference a global 'Small Arms Destruction Day'. Countries were encouraged to organize symbolic arms destruction events to mark the opening of the Conference in order to attract the media's attention and raise public awareness about the small arms issue. This initiative was inspired by the 1996 'Flame of Peace' in Mali, a weapons destruction event that became known all over the world and served as inspiration for several other countries. On the Conference's first day, 9 July 2001, diplomats, non-governmental organizations, and media gathered outside the UN for a video presentation produced by the Dutch government, showing the impact of small arms as well as their collection and destruction in different parts of the world. Appeals were also held by victims of war and by government representatives.

A number of countries responded to the call for action by conducting public destruction ceremonies, including Brazil, Cambodia, the Republic of Congo, and several west African states. A proposal was later adopted by the conference to make Destruction Day an annual event. The United Nations, with the assistance of the Netherlands, published a handbook on the subject to aid those undertaking small arms destruction.[3] In addition to a general commitment in the *Programme of Action* to support collection and destruction efforts, some governments also made pledges for financial assistance before the conclusion of the conference. Belgium announced that it would contribute USD 1.5 million to a three-year small arms project in the Great Lakes region in Africa. The British government pledged GBP 19.5 million (over USD 28 million) to combat the small arms problem globally, encouraging other donor governments to do the same. A substantial part of these funds will go to establish a long-term Weapons Collection, Management, and Destruction Programme under the aegis of the United Nations Development Programme (United Kingdom, 2001). In his statement to the Conference, UNDP Administrator Mark Malloch Brown outlined the three-pronged approach of this programme, which will consist of activities in the areas of collection and destruction, demand reduction, and control (UNDP, 2001a).

Building on the past

This section reviews formal programmes designed to take small arms and light weapons out of circulation by collecting them and disposing of them safely, often through destruction. It will not attempt to provide a comprehensive account of all programmes that have been undertaken or are currently ongoing, as this would comprise hundreds of programmes. Additionally, there is much more information available regarding some programmes than others. While some programmes are large-scale, widely publicized efforts carried out with the support of the international community, others are small and locally organized, with little recognition outside the area where they take place.

The section will present only a selection of programmes, which have been well-documented and span a variety of settings. This approach will serve to illustrate the central themes and concepts relevant in this field, while also demonstrating the diversity of weapons collection practices.[4] Although weapons collection programmes share many features regardless of the context in which they are carried out, for organization and clarity, they are divided into the two broad categories of programmes supporting either crime prevention or peace-building.

BOX 7.2 From micro-disarmament to practical disarmament

'Micro-disarmament' and 'practical disarmament' are two terms commonly used to refer to programmes that aim to remove small arms from a society through collection and destruction. Then-UN Secretary-General Boutros Boutros-Ghali raised the issue of micro-disarmament in his *Supplement to an Agenda for Peace*, referring to the light weapons typically used in conflicts that the UN was dealing with (UNSG, 1995, paras 60–65). Today the term 'micro-disarmament' has to a large extent been replaced by 'practical disarmament' which as defined by the UN General Assembly includes a broader range of measures 'such as, *inter alia*, arms control, particularly with regard to small arms and light weapons, confidence-building measures, demobilization and reintegration of former combatants, demining and conversion' (UNGA, 1996, preamble). In the same resolution, the General Assembly stresses the importance of certain practical disarmament measures, including 'the collection, control and disposal of arms, especially small arms and light weapons, coupled with restraint over the production and procurement as well as transfers of such arms, the demobilization and reintegration of former combatants, demining and conversion, for the maintenance and consolidation of peace and security in areas that have suffered from conflict' (para. 1). Generally, however, 'practical disarmament' is primarily used to refer to the collection, control, and disposal of small arms and light weapons.

Practical disarmament primarily refers to the collection, control, and disposal of small arms and light weapons.

Weapons collection as crime prevention

Weapons collections have been undertaken in many communities that experience problems with violent crime. The aim of these programmes is to improve public security directly—through removing weapons from circulation—and/or indirectly by raising awareness about the potential dangers of weapons possession.

Usually, they encourage individuals to voluntarily turn in weapons in exchange for some type of compensation. It is common to also include legal inducements. For instance, while an amnesty is usually offered to encourage people to turn in illegal weapons to the programme, this policy is sometimes even more effective if law enforcement authorities declare that they will simultaneously, or after a brief amnesty period, increase their efforts to confiscate illegal weapons.

In order to alert people to the programme's existence, as well as raise awareness about the message that the organizers are intending to convey, the organizers need to reach out to potential participants and the public at large. Therefore, they generally try to attract publicity through available media outlets—print, television, radio, and Internet. In some cases, programmes have been accompanied by a larger-scale public awareness campaign.

The idea of using weapons collection as a crime prevention measure emerged in the United States, with the first significant attempts taking place in Philadelphia in 1968 and in Baltimore six years later.[5] Since then, a large number of communities have carried out gun 'buy-backs' in which citizens turn in weapons they no longer need or want in exchange for monetary compensation. Apart from cash, a wide range of other inducements have been offered, such as vouchers for consumer goods, educational grants, tickets to sporting or cultural events, or even psychotherapy sessions (Paddock, 1994). The collections have generally been privately funded. They have been organized and supported by a variety of actors, including private businesses, religious groups, civic organizations, law enforcement agencies, and media.

The Clinton Administration launched the first US federal weapons collection programme in the autumn of 1999 through the allocation of USD 15,000,000 to buy back firearms in and around public housing projects. Run through a Department of Housing and Urban Development (HUD) drug-fighting grant, this provided up to USD 500,000 to local police departments, with the aim of collecting and destroying as many as 300,000 weapons (Schmitt, 1999).

According to HUD, 20,000 weapons were collected in 80 different cities during the programme's first year. As of July 2001, the Bush Administration decided to discontinue the programme because it was stated that it did not fit with the core mission of the HUD. Gun control activists criticized the decision, maintaining that it formed part of the new administration's strategy to erode gun control as a result of its close links with the National Rifle Association and the pro-gun lobby (Nakashima, 2001).

The many buy-back programmes undertaken in the United States have had varying degrees of success, and have fuelled an ongoing domestic debate regarding the effectiveness of voluntary weapons collections as a means of preventing and reducing crime as well as accidental deaths from firearms. While local programmes may have a significant impact in their target communities, the number of guns collected through these efforts has not led to a general reduction in the vast stocks of weapons in the United States.[6]

Inducements offered include cash, consumer goods, educational grants, tickets to sporting and cultural events, and psychotherapy sessions.

Although the immediate goal of the gun buy-backs carried out in the United States has been to collect weapons in order to ultimately contribute to a reduction in crime, objectives beyond this include developing norms against weapons possession and misuse, improving community collaboration, and supporting broader community programmes. Other times, some weapons collection programmes have been more narrowly focused, intended to prevent violent crime by removing a specific type of weapon from a community. The introduction of tighter legal controls on gun ownership has, in other countries, stimulated some major efforts of this kind. Firearms owners have been offered compensation to turn in their weapons during an amnesty period after which the new legislation will be fully enforced. Such changes in legislation have sometimes come about in response to highly publicized and tragic events involving firearms.

This was the experience of both the United Kingdom and Australia. On 13 March 1996, 16 primary school children and their teacher were shot dead in the Scottish town of Dunblane. Only six weeks later, a man killed 35 people and wounded 19 at a tourist site in Port Arthur, Tasmania. After these incidents, the governments quickly introduced stricter legislation on weapons possession.

In direct response to the Dunblane shooting, the UK government first carried out an amnesty programme to collect illegally held firearms under the old law, retrieving 23,000 firearms during the month of June that year. A prohibition on high-calibre handguns—that is, with a calibre of more than .22—was then established in July 1997 and expanded in February 1998 to include small-calibre handguns (UK, Home Office, 2000). In Australia, the federal, State, and territorial governments harmonized their previously disparate policies through the banning of all semi-automatic firearms and pump-action shotguns, the establishment of a uniform registration and licensing system, and the imposition of minimum standards for firearm security and storage (MEASURES). In both countries, buy-back programmes were conducted to retrieve weapons that became illegal under the new laws. A total of 643,726 firearms were collected in Australia, according to figures issued at the end of August 1998. The compensations amounted to AUD 319,833,727 (USD 168,725,084). In the UK, 162,000 handguns had been surrendered to the police by the end of February 1998. The compensations paid out totalled just under GBP 90,200,000 (USD 129,896,118) (UK, Home Office, 2000; UK, House of Commons, 1999; Australia, 2000; Meek, 1998).

BOX 7.3 Rio de Janeiro hosts world's largest weapons destruction ceremony

RIO DE JANEIRO: On 24 June 2001, nearly 20,000 people gathered along a bayside walkway in Rio de Janeiro to participate in the largest-ever public destruction of firearms. About 100,000 small arms, from police stocks of confiscated weapons, were laid down on sheets of metal and repeatedly run over by a bulldozer before a cheering crowd. As the weapons were crushed, a shower of rose petals fluttered down over the crowd from helicopters that circled above. These guns will be melted down and transformed into a sculpture symbolizing hopes for peace in Rio.

The gun destruction ceremony, organized by the Rio de Janeiro State government, the Brazilian Army and the anti-violence NGO Viva Rio, underscored Brazilian support for a law prohibiting the sale of firearms to civilians (MEASURES). The collaboration of the Brazilian military with Viva Rio in order to educate the public about gun violence was as remarkable as the unprecedented quantity of arms destroyed. This partnership between the military and the NGO community served to reinforce the notion that the state should actively seek the participation of civil society in order to best address public safety concerns. The destruction event emphasized public education and participation, and included contributions by grassroots associations from Rio's *favelas* (shanty towns) students, artists, and musicians.

As part of the gun destruction ceremony, musicians from the *Grupo Cultural Afro Reggae* performed. Afro Reggae empowers children from Rio de Janeiro's *favelas* through workshops in music and dance and uses music as a way to communicate the reality of the favelas to the rest of society. The band's music combines traditional Afro-Brazilian percussion with capoeira, rap, and hip-hop. Its director, Junior, notes how music is 'the key to getting the kids' attention and convincing them that they can create their own future'.

Afro Reggae and Viva Rio were both born in response to the horror of the 1993 police massacre when Rio de Janeiro police killed 21 residents at Vigário Geral, a *favela* infamous for its rampant drug trade. Armed violence is a daily presence in Rio's 500 *favelas* and the leading cause of death for young Brazilians.

The direct victims of gun violence are predominantly male. In Brazil, young men aged 15–29 are 24 times more likely to die by firearms than women in the same age group.[7] Behind each victim of gun violence is the pain of a mother, wife, girlfriend, sister, daughter, or grandmother who has been left behind. These women also bear the brunt of the financial and familial hardships. Among the other legacies of gun violence are the fear, insecurity, and resentment among the affected families and communities.

One of Viva Rio's latest initiatives is therefore to encourage women to play a key role in the prevention of the use of firearms in Brazil by launching a women's campaign. The idea behind the campaign is that if gun violence is a male disease, women can provide the cure. By equipping women with information and convincing arguments, the movement aims to help them disarm their husbands, sons, and communities. The goal of the campaign during 2001, the International Year of the Woman, is to mobilize the female population to actively say 'no!' to guns and to increase their participation in pro-disarmament events, including school educational programmes, publicity campaigns, and the establishment of gun-free zones. During the 24 June gun destruction ceremony, women from the campaign gathered holding banners with their slogan, 'Choose gun free: Your weapon or me!'

While the number of firearms destroyed in Rio is impressive in itself, the holistic approach to solving the problem of urban violence demonstrated at the gun destruction event provides real hope for addressing the complex threat of urban violence. Destroying guns is a powerful symbol, but the reality of so many different sectors of the community coming together against violence provides a true vision of a more peaceful future for Rio.

I had plans for the future; I wanted to travel the world, take a modelling course and continue training in gymnastics for the Olympics. Suddenly, my dreams were shattered. And worse: this happened because of the irresponsibility of supposedly civilized men, who only felt brave when there's a gun in their hand.

Camilla Maglhães Lima, 14, paraplegic since 1998
when a stray bullet hit her while she was on her way home from school.

I speak out to serve as an example to other mothers... Don't people see that Rio is a time bomb about to explode?

Euristéia Santa Ana, mother of William, a 24-year-old army corporal who,
along with his girlfriend, was gunned down after an argument, leaving a two-year-old daughter.

Source: Sullivan (2001)

© Viva Rio/Kita Pedroza

Musicians and other popular local artists participate in events in Rio de Janeiro to raise awareness about gun violence, the leading cause of death amongst Brazilian youth.

Weapons collection as peace-building

Weapons collections undertaken in the context of peace-building share many features of more traditional community programmes aimed at reducing gun violence. Still, there are important differences dictated by the different environments in which they operate.

Since the conclusion of several armed conflicts in the late 1990s, there has been a growing concern about the negative impact of widespread small arms availability in societies that have recently experienced an armed conflict. It has become increasingly apparent that the presence of small arms and light weapons in post-conflict societies can undermine fragile peace agreements, hinder peace-building and reconstruction, and generally increase the likelihood of a return to violence. If peace is to be sustained, it is vital to remove the tools of war in both an effective and a responsible fashion. The utility of practical disarmament is, however, not limited to mopping up weapons after a conflict has occurred, but may also be used as one component of conflict prevention.

The importance of adopting a comprehensive and integrated approach to peace-building that includes practical disarmament measures is becoming widely acknowledged (see for example United Nations Department for Peacekeeping Operations, Lessons Learned Unit, n.d.; UNSC, 1999; UNSG, 1998, paras 63–66; 2000b). In practice, this usually implies incorporating procedures into peace operations for the decommissioning, collection, storage and destruction of weapons, as well as the cantonment, demobilization, and reintegration of ex-combatants.

BOX 7.4 Removal of small arms and light weapons in the context of peace missions

A broad range of international expertise gathered in Stockholm for a seminar devoted to this issue in March 1999. The seminar not only produced substantive input to peace-building initiatives through its findings but also provided the topic with increased legitimacy. A draft of the study *Disarmament, Demobilization and Reintegration in a Peace-keeping Environment*, undertaken by the UN Department of Peace-Keeping Operations, was used as a basis for the discussions. The seminar report includes guidelines for disarmament, demobilization, and reintegration of combatants, collection of weapons from civil society, safe storage and destruction of arms, and the design of peace agreements and mandates of peace missions to promote the effective removal of small arms and light weapons (Sweden, 1999).

> *The collection, safe storage and destruction of small arms and light weapons in the context of peace settlements is vital for the consolidation of peace, for restoring citizen security and as a foundation to rebuild war-torn countries under the rule of law... they have to be dealt with both in the context of conflict prevention, and conflict resolution.*

Opening statement by Anders Bjurner, Swedish Deputy State Secretary for Foreign Affairs

As conflict subsides: Two phases of weapons collection

The collection of weapons often follows immediately after the end of an armed conflict and the conclusion of a peace agreement. In fact, it is increasingly common to incorporate practical disarmament measures in the mandate of peace missions in the form of a formal disarmament component. This was done in Nicaragua, El Salvador, Liberia, Sierra Leone, and Macedonia. Weapons collection programmes are also sometimes implemented after the formal disarmament process has ended, in an effort to reduce the large quantities of weapons that are often left in civilian hands and fuel continued instability. To distinguish between weapons collection schemes implemented as part of a peace agreement and those implemented later but still in a post-conflict environment, they have been referred to as 'Phase I' and 'Phase II' programmes, or as 'disarmament by command' as opposed to 'voluntary weapons collection' (Laurance and Godnick, 2001; Faltas, 2001a). Their main distinguishing features are outlined in Table 7.1. In practice though, the distinction between these two types of programmes is not always clear-cut.

TABLE 7.1 Practical disarmament for peace-building

	Phase I: Disarmament by command	Phase II: Voluntary weapons collection
Timing	Soon after a peace settlement	Later
Objective	Establish political stability	Maintain political stability, enhance public safety
Inducements	Command, penalties, rewards	Penalties, rewards
Scale	Collective	Individual
Visibility	Public	Public or private
Policy framework	Demobilization	Demobilization, crime prevention
Responsible actors	Governmental and political organizations	Governmental, political, or private organizations

Source: Faltas (2001a)

Disarmament by command (Phase I) tends to be conducted in public, often organized and monitored by peacekeeping forces or other international observers. It is regularly undertaken in conjunction with demobilization and reintegration of ex-combatants, commonly referred to as DD&R. This makes possible the use of wartime organizations and military chains of command, and it usually targets a collective, such as a militia group. In addition, there is often a combination of rewards and penalties, for instance by establishing a deadline for the turning in of weapons after which holders will be disarmed by force. Many peace operations are authorized and conducted under the aegis of the UN, but other inter-governmental or non-governmental organizations have also been involved, such as the Economic Community of West African States (ECOWAS) in Mali or NATO in Macedonia.

The objective of disarmament in this phase is to establish political stability and impede a return to violence. In addition to a volatile security situation, the population usually faces a host of other challenges, such as a lack of basic necessities, unemployment, and the collapse of local social and economic structures. These factors are likely to influence the population's desire to hold weapons, and must therefore be taken into account when designing such programmes.

However, even well-designed collection programmes implemented during a formal peace process sometimes fail. Among other obstacles, ex-combatants and other weapons holders might have valid personal reasons to hold on to their weapons. For example, they might not be convinced about the durability of the peace, the authorities' intentions, or the ability of peacekeeping forces to protect them. Furthermore, they might decide to keep them as insurance in case they are unable to find other means to sustain themselves and feel compelled to turn to pillaging or other unlawful activities (Demetriou, Muggah, and Biddle, 2001).

© UNDP, Republic of Congo

Burning weapons in Congo-Brazzaville: Public weapons-destruction ceremonies are a powerful symbol of peace and reconciliation.

Therefore, after the peacekeeping force has moved out and the ex-combatants are demobilized, the tools of war often remain in society. The goal of disarmament at this stage is generally twofold: on the one hand to maintain political stability and avoid the return to violent conflict, and on the other hand to enhance public safety. At this point, however, collecting weapons is usually more difficult since it can no longer be carried out collectively and be influenced by command, group discipline, or pressure. Besides, the immediate momentum inspired by the peace settlement will often have dissipated. In short, a favourable opportunity will have passed. Instead, at this stage the organizers need to reach out to individual weapons holders, convincing them that they will in fact be better off turning in their weapons than keeping them. Voluntary weapons collection carried out in what is still considered a post-conflict environment, but after the formal disarmament process is over (Phase II), will often closely resemble weapons collection as crime prevention in 'peaceful' societies.

After the peacekeeping force moves out and ex-combatants are demobilized, the tools of war often remain in society.

Phase I weapons collection: Successes and failures in Africa

The African continent has been the most prevalent site for internal armed conflicts in the 1990s, and collection and destruction of weapons by command (Phase I) have been undertaken in connection with peace operations following some of these conflicts. The mixed record of these attempts illustrates that there are many obstacles to such endeavours. Above all, however, these undertakings illustrate the importance of efficient disarmament measures to ensure their success.

In 1992, the Government of Mali signed a peace agreement with a coalition of Tuareg rebels fighting for autonomy, which incorporated provisions for the collection and destruction of weapons. Faced with a breakdown of the peace process and the possible renewal of the conflict in 1993, President Alpha Oumar Konaré requested the UN Secretary-General to assist with the reduction of small arms and light weapons within the country. Subsequently, two UN fact-finding missions visited Mali and other countries in the region. They identified the need for 'a proportional and integrated approach to security and development' in Mali, which came to be known as the 'security first' approach (Eketi-Mboumoua, 1996). Unless confidence in the public security forces could be increased, efforts to reduce the problem of arms would be ineffective.

The UN missions proposed several specific measures, including strengthening legislation and judicial processes related to the civilian use of weapons, reform of the internal security services, reinforcing regional co-operation and confidence building, development of information campaigns supporting weapons collection, and the establishment of a national commission for the control of small arms. Economic and social development had become impossible because of the precarious security situation, and it was therefore suggested that some of the available development funds should be allocated to improving civilian security. The missions also encouraged donors to support the security first approach through the provision of financial, legal, and technical assistance to improve the security sector. Mali was the first country to deliberately adopt an integrated approach to development and security by linking weapons collection to the provision of development assistance, directly targeted at measures that would enhance community security.

Between October 1995 and January 1996, around 3,000 weapons were collected during the demobilization process in Mali, which involved approximately 11,000 ex-combatants. The United Nations Development Programme (UNDP) co-ordinated the programme and established a trust fund to finance it, to which several governments, including Mali, contributed. The weapons were subsequently burned in a public ceremony called the Flame of Peace. Unsurprisingly, the collection and destruction of 3,000 weapons has not solved the problem of illegal circulation of weapons in Mali, where, in addition to a traditional arms-bearing culture among certain groups, weapons leak into the country from other conflict areas in the region. However, the Flame of Peace became a powerful symbol of national reconciliation and peace and has since then inspired similar events in other countries. Finally, it stimulated disarmament initiatives for the whole west African subregion, including the ECOWAS Moratorium on small arms and light weapons (see MEASURES; Poulton and ag Youssof, 1998; van der Graaf and Poulton, 2001).

The Abuja II Agreement of 17 August 1996 was the final step in a long and convoluted peace process that put an end to seven years of civil war in Liberia involving a number of armed factions. The Economic Community of West African States' Cease-fire Monitoring Group (ECOMOG) was to supervise the disarmament, monitored and verified by a UN Observer Mission in Liberia (UNOMIL). They carried out a disarmament and demobilization programme between 22 November 1996 and 9 February 1997 in order to create the stability necessary to hold elections.

The turning in of weapons was the first step in the demobilization process. The combatants received demobilization identification cards, which entitled them to a one-month food ration and a promissory note for longer-term benefits such as extended medical care, an agricultural kit, or participation in a food-for-work programme. Weapons collected both through the voluntary collection programme and through confiscation are indicated in Table 7.2.

TABLE 7.2 Estimated number of weapons and ammunition retrieved by ECOMOG, 22 November 1996–13 June 1997

Weapons collected	Serviceable weapons	Unserviceable weapons	Ammunition (rounds)
Formal programme 22 November 1996–9 February 1997	7,797	1,782	1,218,300
After programme 9 February–13 June 1997	458 *	–	25,000
Weapons surrendered outside official demobilization sites	3,750 *	–	–
ECOMOG cordon and search operations	3,500 *	–	150,000
Total	**15,505**	**1,782**	**1,393,300**

**No data on serviceability of weapons*

Sources: UNSG (1997); Fraser (2001)

The immediate goal of the programme was achieved when elections were carried out as planned. Yet little attention was paid to the post-election period, for example with regard to general weapons legislation and management. A lack of funding further undermined the programme result, as the ex-combatants never received the long-term benefits they had been promised. Another major weakness was that no provisions had been made for the final disposition of the weapons, causing this to become a point of contention between the Liberian government and the UN and ECOMOG. Not until June 1999 did President Taylor agree to destroy all the collected weaponry, and the process was completed that October (UN, Department of Public Information, 1997a; Fraser, 2001).

The quality of the weapons handed in was generally poor, indicating a lack of confidence and will to make a serious attempt to disband and disarm.

From November 1999 to May 2000, an attempt at disarming rebel factions was made by the Government of Sierra Leone with the assistance of ECOMOG and the United Nations Mission in Sierra Leone (UNAMSIL). According to the UN, 12,695 weapons and 253,535 rounds of ammunition were retrieved before the collection programme collapsed together with the entire peace process in May 2000, after the Revolutionary United Front (RUF) rebels captured 500 UN peacekeepers, nine of whom were killed (Berman, 2000, p. 25). The quality of the weapons handed in was generally poor, indicating that the warring groups lacked the necessary confidence and will to make a serious attempt to disband and disarm. In addition, most of the collected weapons were not immediately destroyed but simply disabled, facilitating their subsequent recapture by rebels after the peace process had broken down.

As reported in a study carried out by Eric Berman for the Small Arms Survey, the programme was largely ineffective. Although this was partly due to the reluctance of the parties to engage in the disarmament process, weak management also contributed to the poor outcome. This included accepting inoperable weapons, not making sure that people handing in weapons were genuine ex-combatants, and failing to adequately disable or destroy weapons.

In addition, the USD 300 cash payment that was offered to those surrendering weapons at the disarmament sites fuelled new demand, attracting additional weapons to the area. More broadly, the programme was undermined by the intensive rearmament efforts undertaken by the rebels over the 1999–2000 period, who continued to obtain weapons despite international arms embargoes and the ECOWAS Moratorium (Berman, 2000; TRANSFERS).

Still, efforts to achieve peace in Sierra Leone continued, and disarmament resumed on 18 May 2001 after a new agreement had been reached between the government, the RUF, and UNAMSIL. When the programme ended in January 2002, 45,449 former combatants had turned in their weapons, about 20,000 more than had been expected (UN-IRIN, 2002a; 2002b). In addition, during May and June 2001 close to 10,000 weapons already recovered were destroyed and converted into productive tools to be distributed to participants in the reintegration schemes (UNAMSIL, 2001). A community arms collection (CAC) programme was also set up to mop up weapons not covered by the demobilization process (UN-IRIN, 2002a).

Building peace in the Balkans

Weapons collection and destruction has also been an important element in the recent peace-building processes in the Balkans. The United Nations Transitional Administration for Eastern Slavonia, Baranja, and Western Sirmium (UNTAES) conducted a weapons buy-back programme between October 1996 and August 1997. This was part of its mandate to demilitarize the formerly Serb-controlled region and oversee its peaceful reintegration into Croatia, in accordance with an agreement signed between the two parties.

After completing the demilitarization of Serb paramilitaries, UNTAES estimated that there was still a large quantity of weapons left among the civilian population. The objective of the buy-back was to recover as much unauthorized weaponry, ammunition, and explosives as possible. The Croatian government financed the programme, which rewarded the participants with on-the-spot cash payments. Although UNTAES indicated that the programme was voluntary, it also made it clear that it would confiscate any unauthorized or unregistered weaponry it discovered. Table 7.3 charts the numbers of weapons collected.

Only the weapons that were old or in poor condition were destroyed, while the remainder were stored and monitored by UNTAES until the end of its mandate and then transferred to the Croatian authorities. The number of weapons turned in was much higher than had been expected, though significant quantities probably remain in the area. The programme also served to build confidence between the local Serb and Croatian populations (Boothby, 2001; UN, Department of Public Information, 1997b).

In Kosovo, the disarming of the Kosovo Liberation Army (KLA) was of key importance in order to re-establish public order. The KLA had an estimated 8,000–10,000 members before the conflict ended in June 1999, and the group had acquired huge quantities of weapons—mostly small arms and light weapons—during its war against the Serb authorities (BICC, 2001). In an agreement between the NATO-led Kosovo Force (KFOR) and the KLA of June 1999, the rebels agreed to hand over all weapons, except hunting rifles and pistols, within 90 days.

Complete disarmament was problematic, particularly since the KLA did not have clear-cut command structures or strong political leaders. Nevertheless, by the end of September KFOR had received more than 10,000 small arms, 1,200 mines, 27,000 grenades, 1,000kg of explosives, and over five million rounds of ammunition (BICC, 2001). Still, much of the KLA's arsenal—including the majority of the weapons acquired during the war—was believed to be stockpiled across the border in northern Albania. Additionally, tens of thousands of weapons almost certainly

TABLE 7.3 Weapons collected by the UNTAES buy-back programme in Croatia

Time period	Rifles	Anti-tank rocket launchers	Grenades	Ammunition
Formal programme 2 October 1996– 31 August 1997	8,152	742 (reusable) 5,330 (non-reusable)	13,335	1,700,000
After programme 1 September– 22 September 1997	204	11 (rockets)	238	55,000
Total	**8,356**	**6,083**	**13,573**	**1,755,000**

Source: UN, Department of Public Information (1997b)

remained in the hands of individual ex-combatants (*Jane's Intelligence Review*, 2000). The large stocks of weapons under ethnic Albanian control subsequently went on to fuel the conflict in Macedonia (STOCKPILES).

Since the settlement in Kosovo, violence against non-Albanians has persisted, as has organized crime. In response, KFOR has continued to uncover hidden weapons caches, actively seeking out weapons by searching people, vehicles, and houses. Starting in April 2000, KFOR launched a programme to destroy confiscated arms by melting them down and recycling. By August 2001, 5,295 rifles, 976 pistols, 78 support weapons, 31 mortars, 147 anti-tank weapons, and six anti-aircraft weapons had been destroyed (Walla, 2001). To encourage civilians to hand in additional weapons, KFOR declared an amnesty period from 1 May to 4 June 2001. Even with the threat of severe penalties facing illegal weapons holders after the deadline, the results of this effort were modest. In all, 400 rifles, 65 pistols, 75 support weapons, 21 anti-tank weapons, 16 rockets, over 200 hand grenades and anti-personnel mines, and some 31,000 rounds of ammunition were collected (KFOR, 2001).

BOX 7.5 Cattle rustlers give up their guns

In December 2001, the Ugandan government launched a programme to disarm the Karamajong—a pastoralist community in the north-eastern region of the country. This initiative, actively supported and promoted by Ugandan President Yoweri Museveni, followed several other smaller-scale attempts at disarmament by local authorities. Equipped with large quantities of firearms, the Karamajong had for many years been carrying out armed cattle raids against neighbouring districts, causing widespread fear, food insecurity, and displacement. The injection of modern arms into pastoralist communities along the borders between Uganda, Sudan, and Kenya has dramatically changed the dynamics of the traditional cross-border cattle rustling in this area.

During the month of December 2001, some 7,000 firearms were surrendered under the programme. President Museveni had promised oxen and ox-ploughs, as well as building material, in return for the voluntary handing-in of weapons during the established amnesty period. This period was initially planned to end on 2 January 2002, but at this time the participants had not yet received any of the promised benefits. The President therefore announced its extension to the middle of February 2002, after which a more coercive approach was planned. The programme aims to collect about 40,000 illegal weapons, many of which were originally given to the Karamajong by the government for their protection against cross-border raids by Kenyan pastoralist groups. The 18-year civil war in southern Sudan has also contributed to the influx of weapons into the region. In conjunction with the disarmament programme, security forces are being deployed along the border with Kenya and Sudan in order to protect the Karamajong from armed cattle rustlers from the neighbouring communities.

Source: UN-IRIN (2002c)

BOX 7.6 The role of paramilitary weapons stocks in peace settlements: Disarming politics

When the Irish Republican Army (IRA) announced its decision to decommission its weapons on 23 October 2001, it was hailed as a historic breakthrough in the Northern Ireland peace process, hopefully leading to fulfilment of the Good Friday Agreement of April 1998 and finally ending over 30 years of sectarian violence.

The provisions of the Good Friday agreement included the establishment of a local assembly in which the Protestant majority and the Catholic minority would share power. It also affirmed that paramilitary weapons should be put 'completely and verifiably beyond use'. But the arms issue soon brought progress towards peace to a halt. One of the main sticking points was the IRA's reluctance to comply with the decommissioning schemes set up by the Independent International Commission on Decommissioning (IICD), the organization responsible for monitoring, verifying, and reporting on the progress of paramilitary disarmament (STOCKPILES).

The dispute over the issue of disarmament repeatedly threatened the survival of the new Northern Ireland government and intensified sectarian tension, resulting in a number of violent episodes and culminating in the resignation of the head of the power-sharing executive, David Trimble, in July 2001. Apart from allowing occasional international inspections of some of its arms dumps, the IRA was dragging its feet on the weapons issue, refusing to scrap any weapons. It demanded that Britain accelerate military withdrawal and the process to create a new police force in Northern Ireland. Radical forces on both sides were gaining ground, while paramilitary factions were suspected of rearming.

The political process was at the point of collapse when the IRA made its move, believed to have been influenced by the terrorist attacks in the United States on 11 September and the exposure of links between the IRA and the Revolutionary Armed Forces of Colombia (FARC). The British government promptly responded by resuming the dismantling of military installations. As the first occasion on which an Irish republican group has disposed of its weapons in this way, it gives real hope for a durable peace in Northern Ireland. Among other things, this will undoubtedly depend on whether the loyalist paramilitaries respond by putting down their guns.

The confidence-building impact of weapons collection has the potential to make or break a peace settlement.

Cleaning up the remnants of war through voluntary weapons collection (Phase II)

During armed conflicts, weapons are often widely distributed, not only among those actively participating in the hostilities but also among the civilian population (HUMANITARIAN). Re-establishing security thus depends largely on the successful implementation of measures during the peace process to disarm ex-combatants, recover weapons hidden in arms caches, and collect weapons from civilians. Still, when peace operations end, excessive quantities of weapons and ammunition frequently remain in the hands of ex-combatants or other civilians. Thus, high levels of armed violence conducted with easily available small arms characterize many post-conflict environments.

In Central America, for example, the formal disarmament processes failed to recover large parts of the weapons stocks, which were left in civilian hands when its wars ended. The region is still awash with weapons, associated with the drug trade, other types of crime, and a high level of civilian insecurity. In an effort to deal with this problem, several voluntary weapons collection initiatives (Phase II) have been carried out (Laurance and Godnick, 2001).

Nicaragua was the first country in the region to embark upon a post-conflict disarmament process when the guerrilla forces were disarmed in 1990. Nevertheless, by early 1991 weapons that had remained hidden in caches around the country were fuelling rearmament. The Nicaraguan government created the Special Disarmament Brigades (BED) to collect as many of these arms as possible. With the support of the Organization of American States (OAS) and the Italian government, they initiated a buy-back programme in late 1991, offering ex-combatants cash, food, construction materials, and micro-credit in exchange for their weapons.

The Nicaraguan programme went on for two years, and approximately 142,000 weapons were bought back or confiscated. The weapons were later destroyed in public, burnt in an open pit fire. As pointed out by the analyst

BOX 7.6 The role of paramilitary weapons stocks in peace settlements: Disarming politics (continued)

Meanwhile, a similar process has taken place in Macedonia. On 13 August 2001, a European/US-brokered peace accord was signed, which aims to put an end to the insurgency by the ethnic Albanian National Liberation Army (NLA). Since the conflict broke out in February 2001, it had caused many casualties and displaced more than 100,000 people (*International Herald Tribune*, 2001a). However, despite having achieved an agreement, the peace process was still extremely fragile. In NATO's Operation Essential Harvest, more than 3,800 troops were deployed to the area beginning in late August 2001. Their mission was to collect and destroy 3,300 weapons that the ethnic Albanian rebels had promised to turn over in a 30-day period. Their mandate was limited to the voluntary collection of weapons only, and did not provide for any peace-keeping role. The rebels had agreed to disarm and disband in return for parliamentary approval of a number of constitutional and political reforms expanding the rights of ethnic Albanians. In the end, the results of the collection exceeded expectations. They are indicated in Table 7.4.

TABLE 7.4 Weapons surrendered by the National Liberation Army in Operation Essential Harvest

Weapons collected	Assault rifles	Machine guns	Support weapon systems (mortar/anti-tank)	Total weapons	Mines, explosives, and ammunition
27 August–26 September	3,210	483	161	3,875*	397,625 items

**Includes 4 tanks/armoured personnel carriers and 17 air defence weapon systems*
Source: NATO (2001)

Disagreement over the NLA's weapons holdings repeatedly threatened to undermine the programme. This illustrates the importance of trust between the parties, and good accounting in lieu of trust. Stockpile estimates varied greatly, ranging from the NLA's original claim that it held 2,500 weapons to estimates from the Macedonian government as high as 85,000 (Sennott, 2001; MIA, 2001). NATO set the number to be collected at 3,300, based on an official estimate of what would be a sufficiently high number to signal the good intentions of the NLA (STOCKPILES). Many Macedonians were sceptical of the undertaking, claiming that the rebels were not likely to turn in their best weapons and that they could easily rearm. NATO officials acknowledged this risk, but the success of the operation, they maintained, should be measured by its ability to build confidence between the Macedonian government and the rebels to support the negotiated solution to the conflict. At the same time, they were discreetly monitoring how much weaponry the rebels were transporting out of the country for secret storage (*The Economist*, 2001a).

The ongoing efforts to create a stable peace in Northern Ireland and Macedonia have in both cases been jeopardized by the uncertainty surrounding weapons held by paramilitary groups. But as the breakthrough on IRA disarmament shows, giving up weapons also has the potential to bolster and in some cases even save a political process from collapsing. Both cases illustrate that the confidence-building impact of weapons collection has the potential to make or break a peace settlement.

Critics are often correct in claiming that the numbers of weapons retrieved from armed opposition groups is merely symbolic and that rearmament is easy, whether with new supplies from outside sources or from secret arsenals that have not been surrendered. Nevertheless, giving up weapons as a signal of formally ending a war has a significant psychological impact. It can send a signal to combatants that fighting is over, inspire the authorities' confidence, and raise hopes of peace among the civilian population. This makes it an essential building-block in the restoration of a peaceful society.

© PA Photos/Faith Paul Faith

The Irish Republican Army's announced intent to decommission its weapons may be a step towards peace in Northern Ireland.

Sarah Meek (1998), the programme did not remove all unauthorized weapons, but the initiative demonstrated the government's will to tackle the weapons problem while at the same time attempting to improve participants' living conditions in a sustainable manner (Laurance and Meek, 1996, p. 161; O'Connor, 1996).

During the immediate post-conflict disarmament process in El Salvador, the Farabundo Martí National Liberation Front (FMLN) surrendered about 10,200 small arms and light weapons and 9,200 grenades. This left an estimated 360,000 military-style weapons in civilian hands (Laurance and Godnick, 2001). In an effort to reduce violent crime, a coalition of citizens, businesses, non-governmental organizations, and churches called the Patriotic Movement against Crime (MPCD) established a programme named Goods for Guns.

In El Salvador the emphasis was on persuading civilians to hand over military-style weapons as a contribution to a more peaceful society. In return they received compensation consisting of vouchers for consumer goods. The government supported the effort, and representatives from the police and Ministry of Defence provided logistical and technical support. While the police destroys the explosives, the Ministry of Defence is responsible for rendering weapons unusable and storing them for later incorporation into a peace monument (Laurance and Godnick, 2001). Table 7.5 lists the weapons collected.

Although the total number of weapons in El Salvador did not decline during this period, the programme did remove thousands of military weapons from society and raised awareness of the problems associated with their proliferation. Domestically, the collaboration with the national police helped establish a dialogue regarding public security and enabled the MPCD to lobby effectively in favour of a new arms and munitions law passed in June 1999. It also set a precedent for collaboration between civil society and the government in other areas. On a regional and international level, it has served as a model for subsequent collection programmes (Laurance and Godnick, 2001).

In Africa, the most extensive Phase II collection effort has taken place in Mozambique. After 16 years of civil war between the government and the Resistência Nacional Moçambicana (RENAMO) rebel forces ended in 1992, UN peacekeepers collected approximately 190,000 weapons. Most of the weapons were handed over to the new Mozambican Defence Force (FADM), but due to a poorly executed storage plan they were quickly back in circulation. In an effort to address the problem, the Christian Council of Mozambique initiated a campaign called Swords for Ploughshares in October 1995, which was later developed into the Tools for Arms Project (TAE).

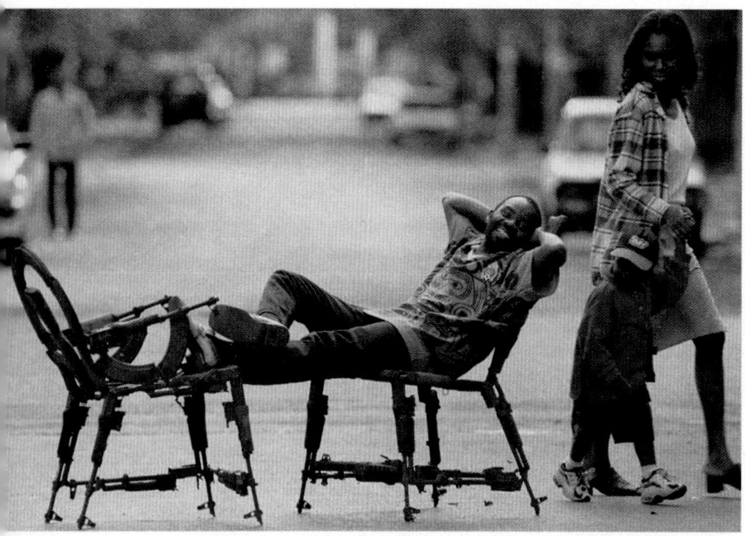

© Christian Aid/Paul Hackett

Mozambican artists put some of the weapons collected by the Tools for Arms Project to creative and practical use.

The Project targets former combatants and illegal arms holders in particular, encouraging them to turn in weapons in exchange for agricultural tools and other equipment. By August 2001, the TAE project had collected and destroyed approximately 200,000 different weapons and related items (Lusa News Agency, 2001). Local artists have turned some of the destroyed weapons into pieces of art, which have been exhibited and sold to raise money for the continuation and expansion of the project.

TABLE 7.5 Weapons collected by the Goods for Guns Programme in El Salvador, 1 January 1996–20 June 1999

Type	Number
Short arms (pistols, revolvers, etc.)	1,354
Long arms (incl. assault rifles)	3,043
Grenades	3,180
Grenade launchers	44
LAW 80 rockets	290
Mines	55
Mortars	4
RPG-7	6
Total weapons*	**9,527**
Ammunition	129,696

**Includes explosives and other items*

Source: Laurance and Godnick (2001)

Conflict and crime: Severing the link

It is often difficult to draw the line between political violence and crime in modern conflicts. Besides, the end of an armed conflict is often followed by an upsurge in crime. Reasons for this include that large numbers of ex-combatants might find themselves without a lawful livelihood in societies already struggling with massive unemployment, that institutions safeguarding law and order have broken down or lack credibility, and that tensions between social groups remain high. This increase in violent crime is exacerbated by the easy availability of firearms in many post-conflict environments. Additionally, the illicit trade and use of arms is associated with other illegal activities, including drug-trafficking, the smuggling of other illegal commodities, and terrorism.

Some weapons collection programmes are specifically tailored to address the increasingly blurred line between the use of arms for political purposes and their use for criminal purposes. One example is the joint weapons retrieval and destruction programme Operation Rachel, launched in 1995 by the South African Police Service (SAPS) and the police of the Republic of Mozambique. Under this programme, the police have destroyed arms caches left in Mozambique following its civil war. The caches are located on the basis of information supplied by local informants rewarded according to the information they provide. South African and Mozambican experts then destroy the weapons on-site. Both countries gain from this co-operation, since weapons from Mozambique are believed to be supplying the illegal arms market in South Africa, contributing to the sharp rise in violent crime rates the country has witnessed during the post-apartheid period (Hennop, 2000).[8] On the other hand, the non-punitive approach is intended to encourage people to reveal weapons caches and to generate support among the local population, thereby contributing to the peace-building process in Mozambique.

South Africa supplies most of the financial and material resources as well as technical expertise. The Mozambicans facilitate contacts with local communities, gather intelligence, and provide the necessary permission for the South African police to operate in Mozambican territory. The programme has helped build confidence between the two countries and their police forces. Between 1995 and 2001, it recovered over 18,000 weapons from more than 500 arms caches hidden throughout the country (Chachiua, 2000; Chachiua and Hennop, 2001; *Maputo TVM Television*, 2001).

BOX 7.7 Cambodia: Steps forward, steps back

Three decades of internal armed conflict have left Cambodia with a huge number of small arms and light weapons: anywhere from 500,000 to one million, according to most estimates, with a very large proportion in civilian hands (STOCKPILES). Although the armed conflict ended in early 1999, the proliferation of small arms throughout the country has contributed to insecurity across Cambodia.

The spread of weapons in Cambodia hinders socio-economic development, affecting families and communities. According to recent research findings by the Working Group for Weapons Reduction in Cambodia (WGWR), women and children are often the indirect victims of weapons misuse and abuse (Yem and Catalla, 2001). They lose loved ones, main income earners, and property. The more intangible consequences can include loss of face, social status, and social support. Children of affected families often drop out of school to help support their family financially or care for younger siblings.

During October 1998 the city of Phnom Penh initiated a weapons confiscation programme. This began with a buy-back strategy. The strategy then changed to roadside confiscation and voluntary turn-in with no cash incentives. The programme was subsequently extended by the Ministry of the Interior to the rest of the country and backed up by a sub-decree (No. 38) limiting legal gun ownership to a thin stratum of upper-level government and military officials.

As of February 2002, this programme had resulted in the collection of 112,562 weapons. Of these, 79,411 weapons were destroyed in a series of 16 public ceremonies in Phnom Penh and across the country up to February 2002. 36,505 weapons were crushed and another 42,906 burned. WGWR has monitored the implementation of the government programme, reporting strong public support for weapons reduction efforts, coupled with a widely-held belief that it has, in fact, contributed towards improving security in the country.

Yet there are serious problems that limit the effectiveness of the programme. First, only a portion of the collected weapons have been destroyed, while remaining weapons are stored in poorly secured and unsafe state depots. Diversion and re-circulation of stored weapons continue to supply a local black market for weapons in Phnom Penh, according to Deputy Prime Minister HE Sar Kheng. Second, current weapons legislation is inadequate and poorly implemented. Finally, the government collection programme has focused more on civilian weapons than the undisciplined use of weapons by Cambodian security forces—police, gendarmerie, and army—a root cause of insecurity and the desire of civilians to remain armed. In response to these problems, the Government of Cambodia has taken several encouraging steps. In co-operation with the European Union Assistance on Curbing Small Arms and Light Weapons in Cambodia (EU-ASAC) created in 1999, it has initiated a weapons storage and registration pilot project and a weapons-for-development pilot project, and drafted a more comprehensive arms law that is currently being reviewed. A National Commission for Weapons Reform and Management was also created in July 2000. The Commission has a policy outlining clear objectives for arms reduction and management, although it has yet to become active and lacks a concrete action plan.

Most people, while broadly supportive of the weapons collection process, remain reluctant to participate in it themselves so long as the rule of law is not fully established in the country and there is a lack of public trust in the security forces. In other words, so long as the state—and its claim to the monopoly of the instruments of violence—does not address root issues and provide security for all people and communities regardless of political affiliation, status, and wealth, Cambodians are likely to continue to arm themselves.

Source: WGWR (2001)

In Pakistan, the government has initiated a large arms control campaign to address its considerable small arms problem. The country has a large domestic illegal arms industry, a strong tribal tradition of carrying and firing guns, several heavily armed religious militant groups, and it suffers from its proximity to conflict-ridden Afghanistan. Because of the extensive spread of illegal small arms and insufficient national controls, it has become a major source of small arms for the whole of South Asia (STOCKPILES; TRANSFERS; Small Arms Survey, 2001, p. 181).

Starting in 2000, the military government in Islamabad set out to crack down on this problem through a multi-phased programme, combining various policy measures. These include undertaking search and confiscation operations, asking people to voluntarily surrender their weapons under an amnesty scheme, regularizing the unauthorized arms manufacturing industry, as well as banning the issuance of new arms licences, arms displays, and the ceremonial firing of weapons at festive occasions. In order to bolster public support for the programme, advertising is raising public awareness and officials are encouraged to hold public meetings to inform and inspire confidence, particularly among people in rural areas not easily reached by other means of communication (Butt, 2000; Cohn, 2001).

The weapons collection component consists of two phases. First, the government announced a general amnesty from 5 June to 20 June 2001 for the voluntary surrender of illicit weapons that would be followed by a countrywide crack-down by law enforcement agencies for the recovery of illicit weapons. According to official figures, in the course of two weeks the authorities acquired a total of 86,757 weapons of various types. Second, the authorities plan to cancel automatic weapons licences that have been issued by previous governments in large numbers, and reassess them using stricter criteria (Pakistan, 2001). Still, the pace of the voluntary arms surrender has fallen short of expectations so far, although the government has stated that it intends to speed up the process (Khan, 2001). Meanwhile, one independent analyst has criticized the campaign for failing to achieve its objectives, pointing to a lack of transparency and political will as some of the main impediments (Siddiqa-Agha, 2001). In certain tribal areas like Baluchistan, public rallies and strikes have been organized in protest against the initiative (*Balochistan Post*, 2001a, 2001b). At the same time, instead of the planned reduction it is believed that there has been an increase in the illegal production of weapons in these areas in connection with the war in Afghanistan.[9]

BOX 7.8 Crack-down on illegal guns in China

Although official data is scarce, the available figures indicate that there has been an explosion in the number of gun-related crimes over recent years in China. This problem is partly associated with a concurrent upsurge in organized crime. Japanese, Taiwanese, and American triads are operating in the country, involved in, among other things, the smuggling of drugs and arms. In response to this, the police are engaged in a tough campaign to confiscate illegal weapons. *The Economist* has reported that between March and June 2001 alone, the police confiscated 600,000 guns, of which 8,800 were military weapons. Over the past five years, 2.4 million weapons have been confiscated countrywide (STOCKPILES).

Source: The Economist (2001b)

Linking disarmament and development

Post-conflict societies face many serious challenges. Disarmament and demobilization of combatants, as well as the removal of weapons, landmines, and other remnants of war from the communities affected, are crucial for the transition to a peaceful society. Yet the affected communities are also confronted with a number of other tasks. In times of war, public and civil institutions are eroded, social services deteriorate, infrastructure and other public assets are damaged or destroyed, socio-economic development and foreign aid are interrupted, and investment and non war-related production slump (Collier, 1999; Colletta, Mendelson, Forman, and Vanheukelom, 1999).

In addition, a great deal of psychological healing is necessary to rebuild trust within and between wounded communities and individuals (HUMANITARIAN). Confidence and trust between previous adversaries must be created and peaceful means of conflict resolution promoted. Researchers have strongly emphasized that collecting weapons is essential but not alone sufficient to sustainably promote peace and security, unless the demand for weapons is simultaneously addressed. This entails addressing the 'root causes of violence' through such methods as promoting development and responsible law enforcement (BICC and SAND, 2000; Faltas, 2001a, 2001b; Lodgaard, 2001).

Different avenues have been explored for tackling small arms disarmament in the broader context of conflict prevention and peace-building. One of the most recent strategies to emerge is 'weapons for development'. This concept refers to programmes that link the removal of weapons from society to development incentives. The approach is based on the assumption that there is a mutually reinforcing relationship between human security and human development. While socio-economic conditions and social inequality may have a detrimental effect on communal relations, thereby raising the likelihood of violent conflict, the insecurity ensuing from conflict will only further impede development. Supporting economic and social development is therefore a crucial factor in preventing conflict as well as in the consolidation of peace in the aftermath of war (UNDP, 1994; 2001b; UNSG, 1998, paras 2, 62). Consequently, efforts to reduce the risk of armed violence by removing weapons from society are increasingly considered within a development framework (Muggah and Batchelor, 2002).

Although this idea emerged during the disarmament process in Mali in 1996 described earlier, the first programme to comprehensively implement this approach was the Gramsh Weapons in Exchange for Development Pilot Programme (GPP) undertaken in Albania in 1999. The situation in the district of Gramsh did not constitute a traditional post-conflict scenario, but it was nonetheless a community flooded with weapons while also facing serious political, social, and economic problems. The UNDP is currently implementing and developing several other projects within this framework in countries or regions where small arms may threaten stability, including in other parts of Albania, El Salvador, the Solomon Islands, and west and east Africa. These programmes are supported by the UNDP's Trust Fund for the Reduction of Small Arms Proliferation, to which several countries have contributed, including Belgium, Norway, the United Kingdom, Switzerland, and South Korea (UNDP, 2001a).

In order to examine this new approach linking disarmament and development, the following analysis written by Spyros Demetriou considers two programmes that have served as path-breakers.[10]

Case 1. The Gramsh Pilot Programme: Exchanging weapons for development

After decades of economic mismanagement under the rule of Enver Hoxha, the population of Albania emerged from behind the Iron Curtain only to fall prey to numerous financial 'pyramid schemes'. Their collapse in 1997 triggered a wave of political and social unrest. As people took the law into their own hands, approximately 650,000 small arms and 1.5 billion rounds of small arms ammunition were looted from army depots. Although a semblance of order was restored by a change in government and the intervention of the Western European Union's Multinational Protection Force, the large number of military weapons still circulating continued to fuel widespread criminality and economic paralysis (Smith and Sagramoso, 1999).

© Reuters/Arben Celi

A new approach to weapons collection was tested in Albania, after 650,000 weapons looted from government stocks ended up in the hands of civilians.

The GPP was designed, following a request by the Albanian government to the UN, to remedy the problems aggravated by widespread weapons availability. It was run by the UNDP in collaboration with the United Nations Office for Project Services and the United Nations Department for Disarmament Affairs. Recognizing that a simple cash buy-back programme might generate inflationary pressures and fuel additional demand for weapons, it was decided to adopt an approach that would rely on collective inducements to maximize weapons collection possibilities. Accordingly, a fivefold strategy was developed (UNDP and the Government of Albania, 1998). Each component will be examined in turn.

TABLE 7.6 Weapons looted from Albanian government armouries in 1997*

Types	Number taken	Number recovered by government	Percentage recovered
Pistols	38,000	170	0.4
AK-47 assault rifles	226,000	17,522	7.7
Ordinary rifles	351,000	66,995	19.0
Machine guns	25,000	11,643	46.0
Grenade launchers	2,450	792	32.3
Mortars	770	242	31.4
7.62mm ammunition	1,560,000,000		
12.7mm ammunition	24,000,000		

**Estimates of the actual numbers of looted weapons vary. According to the Ministry of Public Order, in an internal document provided by the UNDP, Tirana, the numbers looted were 549,775 weapons, 839,310,038 pieces of ammunition, 31,460 hand grenades, and 16,000,000 pieces of explosives. Of these, 176,864 weapons, 107,967,498 pieces of ammunition, and 1,221,429 pieces of explosives had been recovered up to 30 September, 2001.*

Source: Office of the National Security Advisor to the Albanian Prime Minister, quoted in van der Graaf and Faltas (2001, p. 165)

Community participation. At the core of the GPP was a participatory process that involved local communities and authorities in setting the targets for weapons collection, the identification and prioritization of development needs, and a public awareness campaign. The establishment of national, district, and village level co-ordination committees and councils, with representatives from all sectors of society, served to empower communities by giving them a stake in the project. This approach provided communities with a critical discussion and decision-making forum, and a framework for mobilizing society around the objective of improving local security conditions. Most importantly, however, broad participation ensured that programme strategies reflected local realities and needs

Public awareness. The GPP conducted a public awareness campaign on small arms, active at both local and national levels. At the local level, the campaign fostered awareness of the programme itself, educated communities about the dangers posed by small arms and light weapons, and provided a forum for mobilization. It was also a critical confidence-building mechanism, mitigating initial suspicion about the motives of the UNDP. At the national level, a radio and television campaign together with a series of large-scale public events, put the small arms issue on the public agenda, prompting local initiatives in other regions of Albania.

The collection of weapons would reduce the supply of weapons, while development projects would reduce the demand.

Development intervention. In contrast to other voluntary weapons exchange programmes offering cash, food, tools, or other individual benefits, the GPP provided development projects that benefited the entire community. These development projects were intended to complement the collection of weapons as a means to decrease public insecurity. Whereas the collection of weapons would reduce their supply by physically removing them from society, development projects would reduce demand for weapons by offering alternative occupations and improving social and economic welfare.

Capacity building for police. To restore public confidence and trust in local authorities and reduce the sense of insecurity, the police were accorded additional resources to increase their capacity to respond to crimes and accidents. The GPP provided them with vehicles, radio equipment, and training in the democratic application of the law.

Weapons collection. The awareness-raising campaign and the identification and selection of development projects preceded the actual collection of weapons. This sequencing ensured that weapons collection was undertaken in a favourable environment, where the public understood that it was tightly integrated with a larger strategy for improving local conditions. Initially, the scope of development intervention was to be rigidly conditional on the quantities of weapons surrendered per community, but this was later relaxed to accommodate a more equitable approach and forestall unnecessary tensions. When the development projects were launched after the collection, they were perceived not only as collective 'rewards' for the weapons surrendered but also as the next step in a broader strategy to improve community security.

TABLE 7.7 Results of the Gramsh Pilot Programme

Weapons recovered	Ammunition recovered	Target population
5,980	137 metric tons	Approximately 55,000

Source: UNDP (2000a)

Case 2. Consolidating security in the Republic of Congo

In the Republic of Congo (Brazzaville), a series of conflicts during 1993–99 resulted in tens of thousands of casualties and the displacement of more than a third of its population (Pourtier, 1998). Thousands of disaffected youth joined militia groups that, in the absence of centralized control and effective military planning, obtained massive quantities of personal weapons and engaged in widespread pillaging, destruction, and killing of unarmed civilians (Bazenguissa-Ganga, 1996).

The signing of cease-fire accords in late 1999 established a peace process aimed at political reconciliation, institutional reform, and the disarming and disbanding of militia groups. A multi-party body, the Comité de Suivi, was established to oversee the implementation of the cease-fire accords. Despite these efforts, widespread insecurity and social disorder continues to prevail throughout the country. One of the main obstacles to the consolidation of peace is the tens of thousands of ex-combatants who, in the post-conflict period, have found themselves socially marginalized, stigmatized by their communities, and without the skills necessary to play a productive role in social and economic life. In this context, the estimated 40,000 small arms still in possession of ex-combatants constitute both an instrument for criminal activities and a factor sustaining political and social insecurity (Demetriou, Muggah, and Biddle, 2001). In the broader society, this has manifested itself in widespread fear, perpetuating dislocation and impeding a return to normal life. Politically, the easy access to weapons maintains tension and suspicion between the former belligerents, forestalling the reconstruction of trust and confidence.

With these considerations in mind, the International Organization of Migration (IOM) and the UNDP launched a joint Project for the Reintegration of Ex-Combatants and Collection of Weapons in 2000. As opposed to disarmament, demobilization, and reintegration (DD&R) programmes that treat disarmament and reintegration separately, it *links* the provision of reintegration assistance to weapons collection so as to simultaneously tackle both dimensions of the problem. This makes it easier to directly target the social threats that armed ex-combatants pose. In contrast to more rigid DD&R programmes conducted in the formal framework of military operations, it highlights the improvement of social conditions for all as an indispensable component of sustainable peace.

Linking reintegration and weapons collection. In Congo, a formal DD&R process was impossible due to the primarily social, as opposed to military, nature of the militia phenomenon. The lack of organization within the militias meant that, following the end of conflict, a highly armed and socially marginalized group of citizens was left behind. Under such conditions, a formal programme would have been difficult—if not impossible—to implement due to the absence of centralized control over ex-combatants and their weapons.

Although the programme offers reintegration assistance to all ex-combatants, those who turn in a weapon receive priority treatment. In practice, this system has surpassed expectations since the vast majority of beneficiaries have handed in weapons. Although there is no established 'exchange ratio' for the number of weapons needed to be handed in to obtain priority treatment, linking the two still increases the possibility of retrieving large quantities of weapons from former combatants in a short period of time.

Reintegration assistance—in the form of either employment within existing enterprises or the creation of micro-enterprises—directly reduces the ex-combatants' dependence on criminal livelihoods, and consequently their need for weapons. Instead, together with training and technical skills, it provides them with the means to become productive members of society. In this manner, reintegration assistance serves both to 'demilitarize' the minds of ex-combatants and to restore a lost generation of youth to their communities.

Weapons collection. The collection of weapons from ex-combatants is an indispensable and integral component for successful reintegration. First of all, the reduction in supply reduces the possibility of resorting to violence with weapons. In addition, the surrender of a weapon is a particularly significant act which reinforces the process whereby ex-combatants are reintegrated into society. Just as the possession of weapons means something different from one culture to the next, the relinquishing of weapons will have different connotations.

Reintegration assistance reduces ex-combatants' dependence on criminal livelihoods, and consequently their need for weapons.

In the case of Congo, it symbolizes a reorientation from a warrior's life to a civilian life. At the political level, surrendering weapons mitigates the threat to local authorities, bolstering their will to engage in non-violent dialogue and reconciliation. The collection of weapons from ex-combatants not only serves to reduce supply but, if impartial and transparent, may also contribute to rebuilding trust and confidence among ex-combatants and the rest of society.

TABLE 7.8 Results of the IOM/UNDP project in the Republic of Congo as of 31 August 2001

Small arms and light weapons collected	2,800
Grenades and other explosives collected	8,000
Number of ex-combatants receiving reintegration assistance	7,000
Number of micro-projects financed	2,190
Estimated number of ex-combatants not yet receiving assistance	8,000

Source: IOM/UNDP (2001)

Broadening the scope of weapons collection: Implications from Albania and the Republic of Congo

The approaches used in Albania and Congo highlight the merits of linking disarmament and development, and provide important 'roadmaps' for translating these principles into practice. On the one hand, the weapons collection component of both projects fulfilled a core objective of any collection effort: weapons were collected from a target population, thus reducing opportunities for armed violence. In other words, removing the tools of violence from society counts.

At the same time, beyond the physical collection and destruction of weapons, these programmes aimed to embed weapons collection within a broader framework intended to mitigate long-term social and economic threats by merging development and disarmament strategies. Furthermore, they were designed to reinforce collective processes and

TABLE 7.9 Examples of major small arms collection and destruction programmes, 1989–2001

Location	Time period	Framework	Organizer	Weapons collected	Sources
Panama	1989–90	Peace-building	US Army	10,000 weapons	US, GAO (2000)
Nicaragua	1991–93	Peace-building	BED/OAS/ Italy	142,000 weapons	Meek (1998)
El Salvador	1992–93	Peace-building	ONUSAL	10,200 weapons 9,200 grenades 4,000,000 rounds of ammunition	Laurance and Godnick (2001)
Haiti	1994–95	Peace-building	US Army	3,389 weapons	US, GAO (2000)
Mali	1995–96	Peace-building	UNDP	3,000 weapons	van der Graaf and Poulton (2001)
Mozambique 2001	1995–August crime prevention	Peace-building	TAE and related items	200,000 weapons	Lusa News Agency (2001)
Croatia	1996–97	Peace-building	UNTAES	8,356 rifles 6,083 anti-tank rocket launchers 13,573 grenades 1,755,000 rounds of ammunition	Boothby (2001); UN, Department of Public Information (1997b)
Liberia	1996–97	Peace-building	ECOMOG/ UNOMIL	17,287 weapons 1,393,300 rounds of ammunition	UNSG (1997); Fraser (2001)
United Kingdom	1996–97	Legal reform	Government	23,000 firearms (amnesty programme) 162,000 firearms (buy-back)	UK, House of Commons (1999); UK, Home Office (2000)
Australia	1996–98	Legal reform	Government	643,726 firearms	Meek (1998); Australia (2000)
El Salvador	1996–99	Peace-building/ Crime prevention	MPCD	9,527 weapons 129,696 rounds of ammunition	Laurance and Godnick (2001)
Cambodia	1998–June 2001	Peace-building/ Crime prevention	Government	112,562 weapons	WGWR (2001)
Kosovo	June–September 1999	Peace-building	KFOR	10,000 weapons, 27,000 grenades 1,200 mines 5,000,000 rounds of ammunition	BICC (2001)
Gramsh, Albania	1999	Crime prevention/ Conflict prevention	UNDP/ UNDDA/ UNOPS	5,981 weapons 137 metric tons of ammunition	UNDP (2000)
Sierra Leone	1999–2000	Peace-building	UNAMSIL	12,695 weapons 253,535 rounds of ammunition	UN, quoted in Berman (2000)
The Republic of Congo	2000–August 2001	Peace-building	IOM/UNDP	2,800 weapons 8,000 grenades and other explosives	IOM/UNDP (2001)
Bosnia-Herzegovina	1999–November 2001	Peace-building	SFOR	15,169 small arms 57,492 grenades 5,385,130 rounds of ammunition	SFOR (2001)
Macedonia	August–September 2001	Peace-building	NATO	3,875 weapons 397,625 mines, explosives and ammunition	NATO (2001)
Mendoza, Argentina	2000–01	Crime prevention	Government/ NGOs	2,566 weapons	Appiolaza (2001)
Brazil	2001	Crime prevention	Government/ NGOs	100,000 weapons	Viva Rio (2001)

influence attitudes concerning the role of weapons in society. In both cases, the benefits provided (development and reintegration assistance) served not only as an exchange mechanism but also as a road to alternative livelihoods, which in its turn would reduce the demand for weapons. This was an important step towards a 'demilitarization of the mind', a process that was further reinforced by the Gramsh Programme's participatory approach and public awareness campaign.

Both the Albania and Congo programmes attempted to address the extreme social and political fragmentation resulting from deep insecurity. In Gramsh, the collective decision-making process generated co-operation within communities and between citizens and the local authorities. Likewise, in Congo the disarmament of ex-combatants was a critical first step in building confidence between them and the government, increasing possibilities for genuine reconciliation. In both cases, the effects of combining weapons collection and development assistance appear to be similar: the restoration of trust between the general population and the authorities, the improvement of community security, and a reduction in the local demand for weapons. As 'experiments', however, it should be remembered that these projects represent first attempts at operationalizing these principles. Inevitably, they also reveal important lessons learned and additional challenges to be overcome.

Lessons learned

As experience in the field of weapons collection and destruction continues to accumulate, we can learn from this growing body of knowledge in order to improve the effectiveness of future programmes. Although this is still a relatively new activity, which as we have seen includes a broad range of programmes, in some respects practices seem to already be converging.

From the earliest community buy-back programmes in the United States to the most recent and ambitious collection efforts such as the Gramsh Programme in Albania, there have been tangible—though not consistent—changes to the design, implementation, and assessment of collection programmes worldwide. While programmes undertaken in the context of crime prevention and peace-building must differ in some respects (see Table 7.1) so as to be tailored to the need of each situation, their main elements remain the same. They have informed and influenced each other and still continue to do so.

Some of the general trends identified in this chapter include: a broadening of programme objectives; an expansion of the types of incentives used for collecting weapons; and the incorporation of weapons collection as part of a broader strategy to improve human security. It is important to note that observing a general trend does not imply that all programmes contain these features or that the programmes that do are necessarily better. Instead, it means that so far, despite the continued debate about their effectiveness, we are not seeing a scaling back of the number or scope of weapons collections. On the contrary, the number of programmes is increasing and their parameters are generally expanding. Their legitimacy has been further elevated as a result of the support they received at the 2001 UN Small Arms Conference, including in its final document and through pledges of increased funding from donor countries. It therefore seems likely that this trend will continue in the foreseeable future.

It is pertinent to consider the development of some key programme elements in more detail and assess their significance for the implementation and outcome of weapons collection and destruction schemes. What is the potential of such programmes in terms of what they can and should achieve? And what are some of the lessons we have learnt regarding how to best accomplish this potential? Experience allows for tentative conclusions regarding some central components of best practice in this area. These are partly drawn from research already conducted by UN agencies, governments, and NGOs active in this field, but they also incorporate new analysis and research undertaken by the Small Arms Survey which explore some of these issues further.

Establishing appropriate and realistic objectives

Weapons collection and destruction programmes have been carried out in a variety of environments. These include programmes in peace operations to disarm combatants, paramilitary groups, or civilian supporters holding arms, in post-conflict societies to mop up weapons that were not removed during the formal disarmament process, and in other weapons-abundant communities experiencing high incidences of violence and crime. The programme objectives will naturally differ according to the setting. While some programmes limit their goals to collecting a certain type and number of weapons, other programmes aim to improve human security in a more comprehensive manner. We find examples of ongoing programmes that are narrow while others are broader in scope, and as such there is no consistent pattern. This is likely to continue to be the case because organizers will chose the most appropriate objectives according to the specific circumstances. Still, it seems safe to conclude that the current trend is moving towards a broadening of programme objectives rather than the opposite.

There is a growing awareness that efforts to reduce the presence and use of weapons are futile unless the causes of violence and the reasons why people wish to acquire and use weapons in the first place are addressed. Individuals who hold weapons because they feel insecure and seek protection are not likely to be persuaded to give up their means of self-defence before their security is enhanced through other means. Likewise, those who have political or private grievances that they attempt to advance by violent means are not likely to permanently give up their weapons unless some of these underlying factors are addressed. Thus, it is increasingly acknowledged

Individuals are not likely to give up their means of self-defence before their security is enhanced through other means.

BOX 7.9 Confidence is the key

To put an end to the proliferation and misuse of deadly weapons, it is necessary to tackle both supply and demand. This means throttling the supply of such arms and addressing the reasons why people feel a need for them. However, for such practical disarmament measures to be successful, a third ingredient is needed, namely, public confidence.

Take private gun ownership: if weapon laws are tightened, if opportunities to make a living are improved, if more satisfactory channels for the settlement of disputes become available, and if the police do a better job of maintaining law and order, this may or may not have an effect on people's desire to be armed. The outcome will depend on how the public perceive these changes, and how confident they generally feel about themselves, their community, their government, and their future. People will be more inclined to give up their guns if they believe:

- they can provide for their needs without resorting to violence;
- they will be treated with dignity and respect;
- they are safe from robbery, attack, and extortion;
- they can rely on the support of those around them;
- the authorities will protect them;
- the authorities will respect their rights and liberties; and
- the authorities will enforce the law.

Building confidence is a key factor in the pursuit of practical disarmament, just as it is in the quest for international peace and security. Of course, it is not easy. Depending on local conditions, it may require major advances in social and economic development, judicial and police reform, good government, and the growth of strong and independent civil organizations.

Source: Faltas (2001c)

that changing the basic conditions that fuel crime and conflict is an objective that must go hand-in-hand with the collection of weapons (UNDP, 2001b; Faltas, 2001a).

The relationship between what a weapons collection programme is intended to accomplish and what it may realistically achieve is far from straightforward. Generally speaking, there is an assumption that the main purpose of these programmes is to collect weapons and thereby reduce violence. Still, the assumption that one will lead to the other is problematic. Critics object to the claim that weapons collection programmes have an impact on crime rates, and they also question whether they actually reduce the stock of civilian weapons (Kleck, 1996). There is also little evidence to date to suggest that the simple collection of weapons actually leads automatically to the reduction of inter-group conflict if it is not accompanied by sufficient political will and the re-establishment of confidence between the parties. Nevertheless, communities are likely to feel more secure when criminals or belligerents have been disarmed, and this perception might in turn contribute positively to confidence building and the long-term prospect of security and peace. This illustrates the complex link between the purpose of a programme and its actual function. While a programme's stated intent might be to collect guns, even if it does not collect a large number of weapons, it may nevertheless function to reduce public fears of crime and violence or provide hope for peace in an embattled community.

Building confidence is a key factor in the pursuit of practical disarmament.

A collection programme might therefore choose to invert its purpose and function whereby the number of arms collected does not become the purpose or the measure of its success, but instead only a functional side-effect of other objectives. Such objectives could include:

- reducing public perceptions of insecurity with regard to crime or conflict;
- educating citizens about the potential dangers of possessing firearms;
- promoting peaceful means of conflict resolution;
- inspiring confidence in the prospects of a peace agreement and position non-state groups and the government as partners in the process; and
- increasing community cohesion by forging collaboration between different segments of society: government institutions, media, religious organizations, businesses, and others.

Regardless of what the programme objectives are, in order to avoid confusion, misunderstandings, and false expectations, it is essential to clearly convey the intention to everybody taking part: the organizers, the donors, and the participants, as well as the public at large. It is particularly important that the donor community has a realistic appreciation of what the programme attempts to achieve. Enabling donors to easily follow progress towards this goal is essential to ensure their continued support.

Choosing suitable incentives

Offering some form of compensation or reward is helpful in inducing people to disarm, and choosing the most suitable incentive is of crucial importance to the programme outcome. While some incentives can potentially contribute to achieving the programme's goals and bolster its performance, other incentives may undermine programme goals or create additional problems.

It is unlikely that people are going to be willing to hand in their weapons without getting anything in return. Yet this happens occasionally when individuals for some reason see it as a detriment to their safety to hold weapons or when they possess weapons without having a deliberate motivation to be armed. This was the case for many

Albanians after the looting of the government armories, where large numbers of civilians suddenly found themselves in possession of military weapons for which they had no use (Faltas, 2001b, p. 28). These individuals are a much easier target group for a weapons collection effort than those that feel they have legitimate reasons to hold on to their weapons.

There are various reasons why people desire to hold weaponry. A weapons collection programme must attempt to address these in order to be successful. One way of doing this is through the incentives themselves.

Although many of the first buy-back programmes offered participants monetary rewards, and some programmes still do, it has become increasingly apparent that this can backfire. Participants can use the opportunity to turn in old or non-functional weapons, sometimes to use the cash compensation to buy newer ones. Another risk is that the programme will fuel demand for weapons and create a market for illegal trafficking of weapons into the area. One way of counteracting this last possibility is by setting the exchange value below black-market value, so that participants will not make a profit by turning in their weapons (BICC and SAND, 2000, p. 8).

Programmes that target ex-combatants usually offer them some type of reintegration assistance, providing them with the opportunity to create an alternative livelihood. However, rewarding ex-combatants may, in some cases, cause resentment in the target community and present a moral dilemma for the institutions and individuals involved in the collection. The individuals receiving these benefits may have committed abuses and even war crimes during the conflict, and, instead of receiving punishment, they are being rewarded. It is very hard to make a judgement confronted with such a scenario. While, as a principle, perpetrators should be prosecuted for their crimes, and providing victims with a sense of justice is important to the healing of society, pragmatism is sometimes the only option during a period of transition. If prosecution and punishment are not politically viable, the only way to ensure the survival of peace and consequently the long-term well-being of the population may be through compromise and reconciliation.

Partly in response to some of the problems associated with individual compensation schemes, the current trend is to make increasing use of collective incentives as rewards. The merits of collective incentives are best exemplified by the concept of weapons for development outlined earlier. An additional advantage of collective as opposed to individual incentives is the fact that they run a lower risk of commercializing weapons. Observations from the earlier mentioned programme in the Republic of Congo reveal that in some situations owners of weapons may not be aware that their weapons have an economic value. This is partly due to the absence of market structures and significant demand, but also because the arms were acquired through a non-monetary transaction, for example, when arms are looted or issued. Under such conditions, the value of weapons is a function more of their utility in terms of self-defence and the preservation of a warrior identity than of their economic worth. In this context, individual incentives, by establishing a demand for weapons, may spur their commercialization and the creation of a market (Demetriou, 2001).

Still, possible negative repercussions of offering collective incentives must also be considered. In fact, collective incentive schemes may potentially generate community strife, as not all reward systems will be perceived to equitably benefit all community members. This implies a need for analysis of the local impact of the programme, keeping in mind that what is true in one community may have little to do with the reality in another.

Raising public awareness

An ingredient that is increasingly incorporated into weapons collection programmes is public education. In fact, changing attitudes towards the role of weapons in society has been the main objective of some collection and destruction efforts. Supported by these experiences, one might contend that sensitization is a core function of weapons collection programmes and perhaps one of the ways in which they may contribute most effectively to peace, security, and crime prevention. In addition, raising awareness is essential to achieve other key objectives, most importantly to motivate people to turn in their weapons at all. It therefore seems that public education is an element of particular importance to optimizing the outcome of these efforts. The following analysis of a recent initiative undertaken in Argentina, provided by Martín Appiolaza[11] illustrates the potential weapons collection programmes have in this regard, and shows how its organizers went about achieving this goal.

Case 3. Transforming attitudes towards the tools of violence in Mendoza, Argentina

© Diarios Los Andes/Pablo López

A technical expert shows a government official of Mendoza, Argentina, weapons slated for destruction during the Arms Exchange Programme.

The Arms Exchange Programme for Better Living Conditions took place in two phases—during the Christmas week of 2000 and in April and May 2001—in the Argentine province of Mendoza. The weapons collection programme was organized by the province's Ministry of Justice and Security in collaboration with non-governmental organizations. After a series of violent incidents involving firearms created growing insecurity in Mendoza, the Arms Exchange Programme was devised as one part of a larger strategy to improve public security in the region. The upsurge in crime in this province mirrored an exponential increase in armed violence in the country at large. Statistics revealed that the presence of arms in the community had measurable human costs, including:[12]

- firearms were used in 80 per cent of the homicides in Mendoza;
- approximately 900 people were killed with firearms between 1990 and 2000;
- 90 per cent of homicide victims were male; and
- ten per cent of firearms deaths were homicides committed in connection with robberies and rapes, while the rest resulted from fights, suicides, and accidents.

The idea of addressing these problems by removing weapons from society—through voluntary collection rather than more traditional, coercive measures—sparked a heated public debate between those who advocated the active intervention of public security forces as the only way to control crime and those who perceived weapons collection programmes as an additional useful tool to prevent crime as well as accidents. At the outset, the idea was generally not well received. Although there was little opposition to its objectives—weapons reduction, crime prevention, peaceful resolution of conflicts—the criticism was directed at the method.

The collection of 285 weapons and 1,715 rounds of ammunition during the programme's first phase may seem modest, but was actually triple the expectations of the organizers. The second phase, which was extended from the projected two-week period to six weeks in response to popular demand, collected another 2,281 weapons and 6,547 rounds of ammunition. In terms of the weapons collected, the programme therefore far exceeded its goals. Still, the most important achievement was the change in the initially critical attitudes towards the initiative, which gradually turned into strong support, among the general public as well as the media and policy-makers.[13]

Impediments to success. The success of any weapons collection programme depends on raising awareness among the public in order to prompt them to turn in their arms. Their decision, however, is influenced by the information and opinions received through the media. In turn, the media are influenced by the stance of political leaders and other public figures, as well as relevant 'experts'. With regard to the Arms Exchange Programme, the factors that contributed to the initially negative public reactions can be divided into three broad categories:

- **Cultural factors.** Argentina has a strong tradition of civilian arms possession, and legislation has in the past been lax. The common attitude is that weapons are dangerous only in the hands of criminals, and the best way to disarm them is by coercive methods. There is also a growing belief that weapons are necessary for self-protection, partly sustained by a lack of confidence in public institutions and perceptions of personal insecurity, exacerbated by economic and structural developments and the increase in crime.
- **Institutional factors.** There is widespread distrust of public institutions and those in charge of them. Proposals emanating from the authorities are therefore received with scepticism.
- **Communication factors.** The conflicting messages conveyed through the media regarding the Arms Exchange programme generated confusion, uneasiness and fear, all of which reinforced pre-existing distrust due to the above-mentioned factors.

In response to these detrimental factors, the programme organizers needed to come up with a strategy for counteracting them. Their strategy consisted of the following elements:

Demonstrate that weapons collections can be effective. A document was prepared outlining the background, objectives and guidelines for implementation of the programme, describing similar projects around the world (Appiolaza, 2000). The document was widely distributed, and served as the basis for subsequent public debate.

Provide information to defuse apprehension and create confidence. A toll-free phone number was established for information about the programme. These inquiries also served to identify people's concerns and thus provided the organizers with information that they could use to tailor the programme accordingly and adapt their approach to local conditions.

The organizers systematically visited local radio stations all over the province, providing detailed information about their objectives, where and how the collections would be conducted, and what would happen to the collected weapons. To reinforce confidence in the initiative, the organizers declared that the collected weapons would be used to construct a peace monument. Police, public security officers, and social workers aided in spreading the word and providing information about the initiative. Local 'security councils', a forum for community debate where alternative solutions to public security issues were discussed, also played an important role.

To further publicize the programme, sensitize people to its implications, and reach out to potential participants, the public education campaign included a turn-in of violent toys conducted in local elementary schools about a month before the actual weapons collection. About 6,800 pupils turned in toys and games and participated in other activities with an anti-violence theme. These events received a lot of media attention.

Diffuse the idea of weapons collection as a preventive measure. The organizers' ambition to create awareness in the community is underscored in an evaluation by the analyst, William Godnick: 'The organizers admitted from an early stage that the weapons collection was unlikely to bring in the guns in the hands of criminals. The real goal was to influence a change in culture and attitudes towards the role of guns in society. In that context the public education component became equally as important as the proposed weapons turn-in programme.' (Godnick, 2001)

The public education campaign advocated the need for a broader vision concerning the potential problems related to weapons. It emphasized that, in addition to coercive actions undertaken by law enforcement agencies, preventive action is the best protective measure against crime. While disarming communities can positively affect crime rates in the long term, it can have an immediate positive effect on accidents and domestic violence in the short term.

Furthermore, the programme also served to underline that this was an opportunity to disarm, after which the laws would be fully enforced and illegal holders of firearms penalized. It became apparent that many people were actually not aware that by holding unregistered weapons they were breaking the law and could even be considered accomplices to crimes.

Generate publicity through use of the media. The programme received constant coverage by local newspapers, and even some in national media.[14] The general attitude was critical at first, but because the idea was considered original and controversial, the media were never indifferent to the programme. As the organizers disseminated more information about their objectives and achievements, the scepticism subsided. Moreover, this growing support from the media served to mitigate political resistance.

In addition, the organizers advertised the programme on television, focusing on accidents involving firearms and the potential risks they posed to children. A web site was also created with details about the project.[15] This multi-media publicity campaign continued throughout the collection periods.

Evaluation. The total cost of the programme was USD 300,000, of which around 60 per cent was oriented towards public education and publicity. The programme collected a total of 2,566 weapons, which was more than the organizers had anticipated. This constitutes 2.7 per cent of the legal and illegal weapons in the province, or 15 per cent of the estimated illegal weapons—80,000 legally registered weapons and an estimated 15,000 illegal weapons (Zentil, 2000). Regardless of the significance of the quantity of weapons collected, it is equally important to consider the change of attitudes that it aimed to achieve.

Although it is so far difficult to demonstrate what social impact the programme has had, assessments carried out indicate that it did create a public debate about issues related to firearms possession and its potential negative effects, such as accidents and gun violence. It defeated the general apprehension towards using a preventive rather than a coercive strategy to improve public security, and it sensitized the public to the idea of firearms as a risk factor in the community. Surveys carried out in November 2000 and again in March 2001—after the first collection phase was over and the public education campaign had been going on for some time—showed a ten per cent reduction in the number of respondents who thought owning weapons lowered the risk of becoming a crime victim (Appiolaza, 2001). In addition to the public education campaign, the secondary media coverage provided the programme with additional free publicity, which further contributed to raise public awareness.

From an institutional perspective, the programme contributed to increased institutional participation and collaboration between different actors, in addition to providing support for other related endeavours (Appiolaza, 2001; Godnick, 2001). A third phase of the Arms Exchange Programme planned for August 2001 was cancelled due to cutbacks in government expenditures prompted by the economic crisis in Argentina. Still, the project will continue, organized by local NGOs and funded by churches and the private sector (Godnick, 2001). In November 2001, the legislature of Buenos Aires decided to use the same strategy to carry out a weapons collection project in the capital (Ministerio de Justicia y Seguridad de Mendoza, 2001).

© Argentinian Ministry of Justice and Security/Eduardo Antonelli

Weapons collected by the Arms Exchange Programme in Mendoza, Argentina, are crushed in this press.

Disposal of weapons

All programmes need to consider the disposal of weapons collected. Clarity with regard to their disposal is a key factor to ensure a successful outcome. In some of the collection programmes described above, confusion concerning the disposal of weapons detracted from the overall outcome of the programme, undermined confidence in the process, or even added to the political tension after a conflict. Provisions for weapons disposal should therefore be established before initiating the collection of weapons.

Table 7.10 lists different options for the disposal of weapons recovered through collection or confiscation and examples of programmes where these methods have been employed. It also mentions some potential problems or benefits that may follow from each approach.

Destroying or rendering weapons unusable is the only way of ensuring they are permanently taken out of circulation. A number of practical methods have been identified for small arms destruction, and technical guidelines developed by various organizations and experts. On 15 November 2000, the UN Secretary-General issued a report on methods of destruction of small arms, light weapons, ammunition, and explosives (UNSG, 2000a).

The appropriate method must be decided on a case-by-case basis, taking into account, among other things, safety, environmental, infrastructural, and financial considerations. Although it might seem self-evident, practice shows that it is imperative to ensure that personnel with the necessary technical ability and expertise are responsible for the destruction process. Similarly, enhanced cautionary measures are necessary for the destruction of ammunition and explosives.

The act of publicly destroying weapons can have a significant psychological effect. It is therefore increasingly common to conduct destruction ceremonies in public and sometimes also to construct peace monuments or other objects of art using the destroyed weaponry. By demonstrating publicly that the weapons have been destroyed, participants are reassured about the fate of the weapons they have surrendered. In the case of an ongoing effort, this can also encourage more people to participate. Moreover, such events attract publicity, which can further increase participation as well as bolster other programme objectives like public awareness. Finally, it sends a message to the public about the importance of removing the tools of violence and enhancing human security in the target community.

TABLE 7.10 Options for the disposal of collected weapons

Disposal method	Example	Outcome
Hand-over to the government for use by security forces	Haiti, 1994–95 Panama, 1989–90	May undermine confidence and create reluctance to turn in weapons
Hand-over to the government for transfer abroad	Eastern Slavonia, Croatia, 1996–97 Deactivated firearms sold to Austria and Germany to be used for 'decorative purposes'.	Selling deactivated weapons for exhibition can offset programme costs If sold in operable condition, weapons re-enter circulation
Sale on open market to finance expenses of collection agency	Firearms confiscated by police in certain US states.	Weapons re-enter circulation
Storage	Mozambique, 1992 Sierra Leone, 1999–2000	Risk of weapons leaking out if the storage plan is poorly executed Encourages corruption
Weapons verifiably rendered unusable and stored	El Salvador, 1996–99 Northern Ireland, 2001	Signalling good intention of non-state actor (Northern Ireland) Risk of weapons leaking out if the storage plan is poorly executed Risk of weapons re-entering circulation if re-activated
Destruction	Mali, 1995–96 Australia, 1997–98 Brazil, 2001 Macedonia, 2001	Weapons permanently removed Signalling peace and reconciliation

Programme assessment

Weapons collection programmes have been criticized for being ineffective or even counter-productive.

Proponents of weapons collection programmes argue that, by removing weapons from society, one reduces the potential number of accidents, criminal incidents, and violent acts that could otherwise be caused by the use of these weapons. Other supporters emphasize how they raise awareness about these potential dangers, promote peaceful means of conflict resolution, and forge collaborative relationships and confidence between different segments of society (Laurance, 1996; Rosenfeld, 1996). Nevertheless, weapons collection programmes have been criticized for being ineffective or even counterproductive. Since the earliest community collection programmes in the United States, opponents have argued that these programmes do not significantly reduce crime and, by disarming law-abiding citizens, even help make the job easier for criminals (Kleck, 1996). In fact, even after the implementation of hundreds of US buy-back programmes, there is still no conclusive evidence of their impact on violent crime (Godnick, 2001, p. 19). Since weapons collections are expensive and their merits uncertain, it has been asserted that they are a waste of valuable resources.

Critics of more recent collection efforts carried out as part of peace-building exercises have pointed out that both the number and the condition of the weapons collected during such missions tend to be unsatisfactory (US, GAO, 2000, pp. 21–22). After the end of a civil conflict, attempts to disarm factions through weapons collection programmes have sometimes been of limited utility in ensuring future peace and security. The reasons for such failures differ. In Sierra Leone, for example, collected weapons were not destroyed and later re-entered circulation. A persistent and more difficult problem is the temptation among armed groups to keep their best weapons in secret arms caches or to rearm soon after, as in Liberia and in the Balkans. This problem is often exacerbated by a lack of effective controls (Berman, 2000, pp. 24–27; Boothby, 2001, pp. 120–28). Finally, instead of reducing the number of weapons in circulation, programmes may actually increase demand, particularly when they offer cash as compensation (BICC and SAND, 2000, pp. 8–9; Faltas, 2001a, pp. 221–22).

BOX 7.10 Minimizing risks

Many of the activities involved in weapons collection programmes, such as safe handling, transport, storage and destruction of weapons, pose potential risks, to participants as well as organizers. Establishing adequate safety procedures is therefore a major concern.

As a leading implementing agency in this area, the UNDP published the report *Safe and Efficient Small Arms Collection and Destruction Programmes: A Proposal for Practical Technical Measures* in July 2001 (Hughes-Wilson and Wilkinson, 2001). While acknowledging that absolute safety is unattainable, the report stresses the achievement of tolerable risk through risk assessment and the implementation of protective measures. In addition to examining specific recommendations for safe procedures, the report recommends that international technical standards should be adopted to guarantee that weapons collection and destruction programmes are conducted in a safe manner.

The study also points to the contradiction that exists in some cases between different field activities. In areas where landmines and unexploded ordnance (UXO), as well as small arms and light weapons, present a danger, the message conveyed to the population will be confusing. While mine awareness campaigns are intended to educate people about the dangers these artefacts pose, urging the population to avoid contact with them, practical disarmament programmes may simultaneously be urging people to bring in their weapons. Experience shows that people will typically surrender ammunition and explosives in addition to weapons. The responsible organizations must carefully consider the implications of this dichotomy and co-operate to establish a safe environment.

Source: Hughes-Wilson and Wilkinson (2001)

One explanation for the disagreement regarding the effectiveness of weapons collection programmes is the disparate criteria used to assess them.

The divergence of views regarding the effectiveness of weapons collection reflects the fact that the record of such initiatives is indeed mixed. On the other hand, the arguments presented also illustrate that one explanation for disagreement is the disparate criteria used to assess these programmes. A further reason for this problem is the lack of measurable indicators of success.

It is essential for all weapons collection programmes to incorporate an evaluation component. The assessment must include a *process evaluation*, describing and appraising the execution of all programme components, including funds received and expended (BICC and SAND, 2000). The use of performance indicators has been proposed as a way of judging success in financial terms, for example by estimating the cost of each weapon recovered as measured by the total cost of the programme divided by the number of weapons collected (Hughes-Wilson and Wilkinson, 2001).

An *outcome* or *impact evaluation* examines to what extent the programme has achieved what it was intended to do and what its short- and long-term effects were. Goals range from collecting a certain number of weapons to demobilizing and disarming ex-combatants, from reducing the incidence of violent crime to changing perceptions regarding firearms use. In addition, an impact evaluation must consider unexpected or intangible effects that might have come about as a result of the programme. Programmes have sometimes been defined as failures based on too narrow or wrong criteria. If, for example, a programme was intended to raise awareness and promote community co-operation, the fact that it did not collect weapons from criminals is less relevant. Success should be determined in light of the original objectives, but multiple, imprecise, and shifting programme objectives may present a serious challenge to this in practice (Rosenfeld, 1996). This is another reason why it is so important to identify appropriate and realistic goals and to affirm them openly as a way of pre-empting unwarranted or misplaced criticism.

Because social impact studies are generally not conducted in a systematic way, in most cases it is presently not possible to determine the effect these programmes have had in different communities. With this in mind, more attention should be paid to developing measurable indicators of longer-term success, for example in the form of crime and health statistics. Changes in such indicators must then be interpreted taking into account the various factors influencing them. Other key indicators to consider are the attitudes and perceptions among the target population. This is even more important as an increasing number of programmes include awareness-raising as a key objective. This type of evaluation could easily be incorporated by conducting surveys before and after implementation of a programme. For example, the organizers might gauge perceptions regarding insecurity among a representative sample of the target population before the weapons collection programme is announced. When the collection has been carried out, identical surveys should be conducted in order to compare pre- and post-collection results. Nonetheless, there are many caveats to this approach, such as discomfort in the community with being asked certain questions, the complexity of choosing appropriate indicators, and the general difficulty of data collection in environments where such programmes are typically carried out. Another challenge to evaluators is that most programmes operate on principles of anonymity and confidentiality, which precludes them from obtaining the information needed to conduct a reliable assessment (Rosenfeld, 1996).

BOX 7.11 Community involvement: Defining insecurity

Participatory approaches to assessing the impacts of small arms represent an innovative if under-utilized set of tools to evaluate human security. While still insufficiently tested, participatory planning, implementation and evaluation of weapons collection programmes also provide a novel means of strengthening community involvement and the likelihood of programme success. Drawn from development theory and practice, the strength of participatory methods—including participatory rural or urban appraisal (PRA/PUA) and participatory monitoring and evaluation (PM&E)—lie in their recognition that people themselves understand their own risks and have rich insights into their own lives. By providing a voice to vulnerable groups facing violence, participatory research yields locally appropriate intervention strategies. Where properly introduced and facilitated, such approaches can give communities a sense of ownership over programmes undertaken.

Participatory assessments of 'insecurity' and the wide-ranging impacts of small arms have been administered among communities in the Caribbean and Latin America, South Asia, and the Horn of Africa (Ahmed, 2001; Dasgupta, 2001; SALIGAD, 2001). A number of these studies have sought to elaborate relevant criteria and indicators of the effects of small arms on individuals, communities, and regions. As demonstrated by Moser and Holland (1997), who administered a participatory survey of violence in Jamaica, this research can help prioritize small arms-related interventions, identify the unfulfilled needs of arms-related victims, and provide information vital for the effective socio-economic rehabilitation of affected communities.

Recently pioneered by the UNDP in Albania and Kosovo (UNDP, 2001c; Demetriou, 2001), participatory approaches to weapons collection and violence reduction involve beneficiaries in choices about their own priorities and human security needs. This indicates a growing commitment among field agencies to expand donor concerns from a narrow focus on quantitative measurement, results-based management, and cost-effectiveness, to a focus that accommodates qualitative assessments and a recipient-centred perspective.

Ultimately, experiences from the development sector, in which participatory methods are routinely used by multilateral donor agencies and NGOs alike, show that they can improve the quality, effectiveness, and sustainability of a programme. Applying PM&E techniques to verification and monitoring functions—as has been demonstrated in the case of landmines—enables local people to do their own monitoring, data collection evaluation, analysis, and reporting on small arms; to feel they have a stake in the outcome; and to teach donors and implementing agencies by sharing their knowledge (Willet, 2001). By putting local people rather than donors or agencies at the centre of the monitoring and evaluation process, local communities are empowered, ownership is encouraged, and democratic accountability nurtured.

Source: Muggah (2001)

Father and sons transforming weapons into agricultural tools in Afghanistan.

It has been argued that the social impact of a programme might be as important as—and in some cases even more important than—the number of weapons collected. However, this does not imply that the quantity of weapons collected and destroyed is irrelevant. Under certain circumstances, like the efforts to establish a durable peace in Northern Ireland and Macedonia, the actual weapons collected can play a crucial role to the outcome of the process.

A key problem in assessing results in terms of weapons recovered is the general lack of baseline data on small arms possession (Small Arms Survey, 2001, pp. 61–63) This has caused some difficulties and confusion regarding the evaluation of certain collection programmes, like the Gramsh Programme. As long as the initial number of weapons in the community is unknown, it is impossible to determine whether the weapons recovered by a programme is significant. Before initiating a programme, unless reliable information is already available, attempts ought to be made to assemble data on weapons stocks in the target area and ideally produce an estimate of total holdings. This was done for example in connection with the earlier-mentioned project in the Republic of Congo (Demetriou, Muggah, and Biddle, 2001).

The assessment of weapons collected depends not only on their quantity, but also on their quality.

Box 7.12 outlines one proposed approach to assessing the weapons retrieved by collection programmes.

BOX 7.12 Operational assessment: A key to transparency

There are two general approaches to evaluating the outcome of small arms collection and destruction programmes. The first is concerned with the social impact a programme makes on a community. How, for example, does it affect levels of crime or perceptions of safety (Faltas and Di Chiaro III, 2001; Hughes-Wilson and Wilkinson, 2001; Waszink, 2001; BICC and SAND, 2000). The second form of evaluation focuses on *operational success in terms of the weapons collected and destroyed*: did the programme collect a representative sample of the weapons in the target community? In other words, it assesses the number and quality of weapons collected or destroyed, compared to what was anticipated.

This *operational assessment* is important for several reasons. For example, if a programme aims to collect combat weapons from an insurgency group as part of a peace agreement, certain numbers and types of weapons are usually expected. Unless weapons collection is strictly symbolic, it is expected that the weapons turned in, even if they are just a part of the total stock, will be representative of that stock. If a rebel group is in the possession of new and sophisticated weaponry but submits only obsolescent and broken items, its opponents will almost certainly argue that the rebels are not carrying out their end of the bargain or are acting in bad faith. This may even lead to the collapse of a peace agreement. Furthermore, operational assessment is also important when collecting weapons from civilians in order to determine whether project results complied with project aims.

In either case, an assessment of the weapons actually collected is vital to determine success. Such evaluations seldom occur, and when they do it is usually in a subjective and impressionistic manner. In order to undertake this operational assessment, it is necessary to develop an explicit method of appraisal that could be carried out by independent observers or mutually agreed-upon authorities. The remainder of this box focuses on programmes that collect weapons from non-state armed groups as part of a peace settlement. It is proposed that an assessment process work through four basic steps.

Stage 1: Evaluating stockpile liquidity. Stockpiles by non-state actors are liquid, meaning that they are subject to changes in volume due to import—or receipt—loss, breakage, and export. While loss and breakage may be practically impossible to assess, the transfer of weapons to insurgents, paramilitary groups, or other non-state actors (NSA) can in some circumstances be evaluated and quantified. Usually, the objective of a weapons collection programme in this context is the *actual reduction in the volume of NSA weaponry* in order to diminish the group's ability to undertake violent action against the state. Still, in cases where stockpiles can be easily replenished, this objective cannot be achieved through weapons collection alone, since the weapons turned in may easily be replaced. This problem can be alleviated only by taking action to stem the supply of weapons to the NSA before, during, and after the collection programme.

BOX 7.12 Operational assessment: A key to transparency (continued)

Stage 2: Evaluating stockpile lethality. The assessment of weapons collected depends not only on their quantity but also on their quality. A high-calibre machine gun is capable of projecting far greater firepower than a pistol. But in a social context a small pistol may inflict far greater damage if wielded by an individual bent on mayhem, while the machine gun is safely in the hands of responsible authorities. Still, a pile of machine guns, if used to maximum design specifications, can project more firepower than a pile of pistols. This can be quantified and used to provide a verifiable means of comparing stockpiles. A *stockpile lethality measure* can be used to evaluate weapons collection and destruction programmes, as the weapons submitted to the programme should have the same lethality level as the arsenal from which they are taken. Stockpile lethality would be a composite, weighted measure of a sample of weapons, indicating the *average firepower a given set of weapons is able to project as a function of its use to maximum design specification*. Major criteria that can be included are: (1) cyclic rate of fire, (2) ammunition storage capacity, (3) ammunition type, (4) weapon configuration—that is, fully automatic, semi-automatic, single-shot—and (5) condition (i.e. level of functionality).

© Associated Press/Richard Lewis

A British soldier inspects weapons turned in by the ethnic Albanian rebels in Macedonia during Operation Essential Harvest.

An obvious difficulty with this approach is that, when measuring the lethality of a stockpile or cache of weapons in the field, inspectors may be presented with an unrepresentative sample of weapons. By establishing lethality ratings lower than their actual levels, a group can avoid having to turn in much of its best equipment. Solutions to alleviating this problem must be developed in every situation. Factors that can be considered in the overall inspection process include a military assessment of weapons observed in use during a conflict in addition to those claimed by an NSA during peace negotiations, thereby reducing the likelihood of deception.

Stage 3: Evaluating collected stockpile lethality. The lethality rating of weapons collected should be the same as or higher than the known weapons in the NSA's stock. Mechanisms must be established to identify and correct any disparities. During the disarmament of the Farabundo Martí National Liberation Front (FMLN) in El Salvador, for example, the poor quality of the weapons turned in during the first round of demobilization raised doubts about the group's intentions. Only after the monitors from the United Nations Observer Mission (ONUSAL) raised the issue and increased pressure, did they start to receive weapons in acceptable condition (Laurance and Godnick, 2001, p. 20). A lethality scale would enable practitioners to undertake such evaluations in a more routine and systematic fashion, thereby facilitating verification in peace processes like those ongoing in Macedonia and Northern Ireland.

Stage 4: Evaluating stockpile lethality of weapons slated for destruction. The lethality of weapons slated for destruction should be the same rating as those collected. The reason this measurement is necessary is that weapons can be diverted—through looting, robbery, or corruption—between the time they are collected and stored and the time they are moved for destruction. The use of a representative measurement system, combined with publicizing that such an evaluation will take place, will greatly increase the costs of deception or diversion and increase confidence in the programme.

Together, the application of an operational assessment that makes use of a stockpile lethality index to evaluate weapons available for collection, collected, and destroyed, offers the following benefits:

- increased transparency in collection;
- improved confidence in the process and the likelihood of fair results;
- higher costs for deception; and
- lower risk of an agreement crumbling due to contradicting claims over the weapons collected or destroyed.

Source: Miller (2001)

Conclusion

This chapter has reviewed weapons collection and destruction efforts implemented in a variety of settings worldwide. The objectives and features of these efforts differ from place to place, but the core elements are essentially the same. The debate still continues about what purpose these programmes should serve and how effective they actually are. Such debate often arises from misconceptions about the intended goals of the programmes. In some cases, goals are not effectively selected or communicated. Further causes for debate are the insufficient evaluation and analysis of past initiatives and the continued need to develop more reliable methods to assess initiatives. The paucity of baseline data against which results can be measured and the difficulty in obtaining this data make it difficult to evaluate programmes meaningfully. As long as these problems persist, the effectiveness of these measures cannot be ascertained.

Despite this continued disagreement, however, one conclusion can be drawn. Weapons collection programmes continue to receive widespread support from policy-makers, operational agencies, donors, and the general public. The number of programmes seems to be increasing and their parameters expanding. This is even more apparent since the 2001 UN Small Arms Conference, where practical disarmament was one of the few measures that received support from almost all participating states (CONFERENCE).

Given the limitations of such initiatives and their mixed successes, why do they remain so popular? The first reason is probably their intuitive appeal. Removing the tools of violence from a community affected by conflict or crime makes sense. Most people perceive it as a positive initiative, or at the very least, not a negative one. Even individuals opposed to more coercive weapons control might, while not actively supporting such programmes, agree that they do no harm. This applies particularly to programmes that are purely voluntary, allowing individuals to decide for themselves whether or not to hand over their weapons. Programmes that include elements of coercion are inevitably more controversial.

Increasingly, weapons collection programmes form part of a longer-term strategy, addressing the root causes of violence.

Other appealing aspects of weapons collections are their visibility and concrete nature, given that they produce tangible and easily quantifiable results. As we have seen, however, the relationship between the stated objectives of weapons collection programmes and their actual outcomes is not straightforward. While the results of such programmes, in terms of the total numbers of weapons retrieved may be disappointing, certain valuable, indirect effects are often achieved. Some proponents of such initiatives maintain that the removal of every single weapon is nevertheless significant, as it represents a potentially violent incident avoided.

The popularity of the programmes with policy-makers might be the easiest to explain, as most are not considered politically controversial or sensitive. In the area of small arms control, practical disarmament is one of the easier and most visible policy options. Launching a weapons collection may, for example, function as a response to a public outcry for action following a gun-related traumatic event. One exception is disarmament undertaken in the context of certain peace operations, which may present both political and practical risks, particularly when weapons are recovered from paramilitary groups.

The emerging concept of weapons for development shows how weapons collection and destruction programmes are broadening their objectives, integrating new methods and approaches, and lengthening their time-lines. Such programmes increasingly form part of more comprehensive and longer-term strategies which, by promoting development and human security, address the root causes of violence and thus the demand for weapons.

There is a growing realization that, in order to achieve long-term stability and safety, we need to do more than simply remove weapons from a community. Collecting weapons can be crucial in the shorter term. Not only is a tool and symbol of violence removed, but such efforts can also contribute to other security-enhancing outcomes, such as social mobilization and confidence building. However, unless underlying sources of insecurity and violence are addressed, other tools of violence will prevail. This is why the current movement towards addressing the demand for weapons within the framework of practical disarmament holds particular promise.

7. List of Abbreviations

BED	Special Disarmament Brigades
CAC	Community arms collection
DD&R	Disarmament, demobilization, and reintegration
ECOMOG	Economic Community of West African States' Cease-fire Monitoring Group
ECOWAS	Economic Community of West African States
EU-ASAC	European Union Assistance on Curbing Small Arms and Light Weapons in Cambodia
FADM	Mozambican Defence Force
FARC	Revolutionary Armed Forces of Colombia
FMLN	Farabundo Martí National Liberation Front
GPP	Gramsh Weapons in Exchange for Development Pilot Programme
HUD	Department of Housing and Urban Development
IICD	Independent International Commission on Decommissioning
IOM	International Organization of Migration
IRA	Irish Republican Army
ISER	Instituto de Estudos da Religião
KFOR	Kosovo Force
KLA	Kosovo Liberation Army
MPCD	Patriotic Movement against Crime
NATO	North Atlantic Treaty Organisation
NLA	Albanian National Liberation Army
NSA	Non-state actor
OAS	Organization of American States
ONUSAL	United Nations Observer Mission in El Salvador
PM&E	Participatory monitoring and evaluation
PRA	Participatory rural appraisal
PUA	Participatory urban appraisal
RENAMO	Resistência Nacional Moçambicana
RUF	Revolutionary United Front
SAPS	South African Police Service
SFOR	Stabilisation Force
TAE	Tools for Arms Project
UNAMSIL	United Nations Mission in Sierra Leone
UNDDA	United Nations Department for Disarmament Affairs

UNDP	United Nations Development Programme
UNOMIL	United Nations Observer Mission in Liberia
UNOPS	United Nations Office for Project Services
UNTAES	United Nations Transitional Administration for Eastern Slavonia, Baranja, and Western Sirmium
UXO	Unexploded ordnance
WGWR	Working Group for Weapons Reduction in Cambodia

7. Endnotes

1 Key publications on this issue include Sweden (1999); BICC and SAND (2000); Faltas and Di Chiaro III (2001); Faltas, McDonald, and Waszink (2001); and Lodgaard (2001). See also the Help Desk for Practical Disarmament at the Bonn International Center for Conversion, <http://www.disarmament.de>.

2 A full record of all statements can be found at the Small Arms Survey's <http://www.smallarmssurvey.org/Database.html>

3 This handbook can be downloaded in PDF format from <http://www.un.org/Depts/dda/update/jun2001/jun2001.htm>

4 Parts of this section are based on Faltas, McDonald, and Waszink (2001).

5 Kleck (1996) points out that the idea can be traced as far back as the Spanish-American War of 1898, when the US government, in an attempt to disarm guerillas and bandits, offered Cubans a bounty in gold for each firearm they surrendered.

6 For evaluations of US buy-back programmes, see for example Rosenfeld (1996); Callahan, Rivara, and Koepsell (1996); and Romero, Wintemute, *et al.* (1998).

7 Data provided to Viva Rio's disarmament campaign by the Instituto de Estudos da Religião (ISER).

8 This sudden increase can in part be attributed to increased reporting and compilation of data on crime in black neighbourhoods after the end of apartheid.

9 Correspondence from Dr Ayesha Siddiqa-Agha, 4 December 2001.

10 The material used for this section was compiled by Spyros Demetriou during consultancies with UNOPS and IOM on behalf of the Small Arms Survey.

11 This case study summarizes a background paper written by Martín Appiolaza, co-ordinator of the Arms Exchange Programme. Translated from Spanish by main chapter author. For the full document in Spanish, see <http://www.smallarmssurvey.org>

12 These statistics have been developed based on information from public hospitals, law enforcement agencies, opinion polls, and reports from the US Department of Justice.

13 For an evaluation of the programme, see Godnick (2001).

14 The co-ordinator of the programme, Martín Appiolaza, was himself a former journalist, whose contacts ensured that the programme would be covered by the media.

15 This web site can be found at <http://www.seguridad.mendoza.gov.ar/canje/canarm.htm>

7. Bibliography

Ahmed, Naeem. 2001. 'Impacts of Armed Violence on a Community: A Case Study of Karachi.' Background Report for the Small Arms Survey and the Regional Centre for Strategic Studies. Karachi.

Appiolaza, Martín. 2000. *Plan Canje de Armas.* Mendoza, Argentina: Ministerio de Justicia y Seguridad, Gobierno de Mendoza.

__. 2001. *Comunicación para el desarme voluntario: La experiencia del Plan Canje de Armas (Mendoza, Argentina).* Background paper. Geneva: Small Arms Survey.

Australia. 2000. *The Australian Firearms Buyback.* <http://www.gun.law.gov.au/Guns/intromain.htm>

The Balochistan Post. 2001a. 'People Threaten Strike against Arms Recovery Drive.' 20 June.

__. 2001b. 'Strike in Loralai against Deweaponisation Drive.' 26 June.

Bazenguissa-Ganga, Rémy. 1996. 'Milices Politiques et Bandes Armées à Brazzaville: Enquête sur la Violence Politique et Sociale des Jeunes Déclassés.' *Les Études du CERI*, No. 13. Paris: Fondation Nationale des Sciences Politiques.

Berman, Eric. 2000. *Re-Armament in Sierra Leone: One Year After the Lomé Peace Agreement.* Occasional Paper No. 1. Geneva: Small Arms Survey.

Bonn International Center for Conversion (BICC). 2001. *Conversion Survey.* Baden-Baden: Nomos Verlag.

__. and Program on Security and Development (SAND). 2000. *Tackling Small Arms and Light Weapons: A Practical Guide for Collection and Destruction.* Bonn and Monterey: BICC and SAND.

Boothby, Derek. 2001. 'The UNTAES Experience: Weapons Buy-back in Eastern Slavonia, Baranja and Western Sirmium (Croatia).' In Sami Faltas and Joseph Di Chiaro III, pp. 131–56.

Butt, Tariq. 2000. 'Pakistan Government Wants to Regularize Arms Trade in Tribal Area.' *The News Islamabad.* 27 September.

Callahan, Charles M., Frederick P. Rivara, and Thomas D. Koepsell. 1996. 'Money for Guns: Evaluation of the Seattle Gun Buy-Back Program.' In Martha Plotkin, pp. 81–95.

Chachiua, Martinho. 2000. 'The Evolution of Operations Rachel, 1996-1999.' In Virginia Gamba, pp. 195–222.

__. and Ettienne Hennop. 2001. 'Operation Rachel: South Africa-Mozambique.' In Sami Faltas, Glenn McDonald, and Camilla Waszink, pp. 15–17.

Cohn, Martin Regg. 2001. 'Kalashnikov Culture, Talking Tough on Guns in Pakistan.' *The Toronto Star.* 11 April.

Colleta, Nat, Johanna Mendelson Forman, and Jan Vanheukelom. 1999. *Security, Poverty Reduction and Sustainable Development: Challenges for the New Millennium.* The World Bank and Belgian Ministry for Foreign Affairs, Foreign Trade, and International Cooperation.

Collier, Paul. 1999. 'On the Economic Consequences of Civil War.' *Oxford Economic Papers.* Vol. 51, pp. 168–83.

Dasgupta, Anindita. 2001. *Human Insecurity and the Threat of Firearms: Perspectives from South Asia: NorthEast India.* Background Report for the Small Arms Survey and the Regional Centre for Strategic Studies. Assam.

Demetriou, Spyros. 2001. *Securing Development.* Background paper. Geneva: Small Arms Survey.

__. Robert Muggah, and Ian Biddle. 2001. *Small Arms Availability in Congo-Brazzaville.* A Report for the IOM/UNDP Small Arms Collection and Ex-Combatant Reintegration Programme. Brazzaville: IOM/UNPD.

The Economist. 2001a. 'How Many Groups, How Many Guns?' 25 August, pp. 26–27.

__. 2001b. 'The Wild East.' 10 November, pp. 57–58.

Eteki-Mboumoua, W. 1996. *UN Sahara-Sahel Advisory Mission Report.* In-house report. 20 September.

Faltas, Sami. 2001a. 'In Search of a 'Best Practice' of Micro-disarmament.' In Sami Faltas and Joseph Di Chiaro III, pp. 211–30.

__. 2001b. 'Practical Disarmament: Lessons Learned.' In Sverre Lodgaard, pp. 25–33.

__. 2001c. *Lessons Learned in Practical Disarmament.* Geneva: Small Arms Survey.

__. and Joseph Di Chiaro III, eds. 2001. *Managing the Remnants of War: Microdisarmament as an Element of Peace-Building.* Baden-Baden: Nomos Verlag.

__. Glenn McDonald, and Camilla Waszink. 2001. *Removing Small Arms from Society: A Review of Weapons Collection and Destruction Programmes.* Occasional Paper No. 2. Geneva: Small Arms Survey.

Fraser, Douglas. 2001. 'Weapons Collection and Disposal as an Element of Post-Settlement Peace-Building: Liberia Case Study.' In Sami Faltas and Joseph Di Chiaro III, pp. 97–129.

Gamba, Virginia, ed. 2000. *Governing Arms: The Southern African Experience.* Pretoria: Institute for Security Studies.

Godnick, William. 2001. *Transforming Attitudes Towards the Tools of Violence: The Arms Exchange Program in Mendoza, Argentina.* Peace Studies Papers, Working Paper 3. October. Bradford: Department of Peace Studies, University of Bradford.

van der Graaf, Henny and Sami Faltas. 2001. 'Weapons in Exchange for Development: An Innovative Approach to the Collection of Weapons in Albania.' In Sami Faltas and Joseph Di Chiaro III, pp. 159–82.

__. and Robin Edward Poulton. 2001. 'Flames of Peace: Disarmament and Post-Conflict Peace-Building in Mali.' In Sami Faltas and Joseph Di Chiaro III, pp. 49–76.

Hennop, Ettienne. 2000. 'Illegal Firearms in Circulation in South Africa.' In Virginia Gamba, pp. 61–76.

Hughes-Wilson, John and Adrian Wilkinson. 2001. *Safe and Efficient Small Arms Collection and Destruction Programmes: A Proposal for Practical Technical Measures.* New York: UNDP.

Institute of Development Studies (IDS). 1998. *Participatory Monitoring and Evaluation: Learning From Change.* IDS Policy Briefing, Issue 12. November. Brighton: IDS.

International Herald Tribune. 2001a. 'Macedonia Leader Warns of War if Pact is Rejected.' 1–2 September.

__. 2001b. 'NATO Ponders a 'Security Vacuum': Long-Term Stabilization of Macedonia is Likely to be Needed.' 4 September, pp. 1, 5.

International Organization for Migration (IOM) and the United Nations Development Programme (UNDP). 2001. *Programme de Réinsertion des Ex-miliciens et de Ramassage des Armes Légères République du Congo/Brazzaville.* Report. Brazzaville. August.

Jane's Intelligence Review. 2000. 'The UCK's Arsenal.' 1 November.

Khan, Muhammed Ejaz. 2001. 'Deweaponization Drive to Continue, says Moin.' *The News International Internet Edition.* 28 August.

Kleck, Gary. 1996. 'Gun Buy-Back Programs: Nothing Succeeds Like Failure.' In Martha Plotkin, pp. 29–49.

Kosovo Force (KFOR). 2001. *KFOR News Release: Statement to Press Conference.* 4 June. <http://www.nato.int/kfor/news/releases/2001/nrel_010604.htm>

Laurance, Edward. 1996. 'Gun Reduction Programs as a Policy Alternative: Issues for Research, Planning and Evaluation'. In Martha Plotkin, pp. 55–71.

Laurance, Edward assisted by Sarah Meek. 1996. *The New Field of Micro-Disarmament: Addressing the Proliferation and Buildup of Small Arms and Light Weapons.* Brief 7. Bonn: Bonn International Center for Conversion.

Laurance, Edward and William Godnick. 2001. 'Weapons Collection in Central America: El Salvador and Guatemala.' In Sami Faltas and Joseph Di Chiaro III, pp. 15–48.

Lodgaard, Sverre, ed. 2001. *Practical Disarmament.* Oslo: Norwegian Institute of International Affairs.

Lusa News Agency. 2001. 'Mozambique: 8,000 Weapons Collected; Exchanged for Agricultural Equipment.' Maputo. 8 August.

Macedonian Information Agency (MIA). 2001. 'Fyrom Police List "85,000 Pieces of Weaponry" in Possession of NLA.' 21 August. EUP20010822000038.

Maputo TVM Television. 2001. '18,000 Weapons Recovered in Joint RSA-Mozambican operation between 1995–2001.' 6 July. FBIS APF-20010706000124.

McCarthy, Rory. 2001. 'Tackling the Culture of Jihad.' *Guardian Unlimited.* 16 February.

Meek, Sarah. 1998. *Buy or Barter: The History and Prospects of Voluntary Weapons Collections Programmes.* ISS Monograph Series No. 22. Pretoria: ISS.

Miller, Derek. 2001. *Operational Assessments in the Evaluation of Collection and Destruction Programs.* Background paper. Geneva: Small Arms Survey.

Ministerio de Justicia y Seguridad de Mendoza. Dirección de Prensa y Difusión. 2001. *Legislatura Porteña Vota la Aplicación del Plan Canje de Armas Mendocino.* Mendoza. 6 November.

Moser, Caroline and Jeremy Holland. 1997. *Urban Poverty and Violence in Jamaica.* Washington, DC: World Bank.

Muggah, Robert. 2001. *Participatory Approaches to Assessing Impacts and Responses to Small Arms.* Background paper. Geneva: Small Arms Survey.

__. and Peter Batchelor. 2002. *Development Held Hostage: Assessing the Effects of Small Arms on Human Development. A Study of the Socio-Economic Impacts and Development Linkages of Small Arms Proliferation, Availability and Use.* New York: Emergency Response Division, UNDP.

Nakashima, Ellen. 2001. 'Administration Calls Halt to Gun Buybacks.' *Washington Post.* 26 July, p. A23.

NATO. 2001. *NATO's Role in the former Yugoslav Republic of Macedonia.* Press Briefing. 26 September. <http://www.nato.int/fyrom/tfh/2001/t010926a.htm>

O'Connor, Neil M. 1996. '*Buy-Back Programs as an Instrument of Micro-Disarmament: Observation from Haiti, Nicaragua and the Dominican Republic.*' Monterey: Program for Arms Control, Disarmament and Conversion, Monterey Institute of International Studies.

Paddock, Richard C. 1994. 'Therapists to Trade Treatment for Guns.' *Los Angeles Times.* 4 October, p. A3.

Pakistan. Pakistan Mission to the United Nations, New York. 2001. *Statement by Mr. Moin-ud-din Haider, Minister for Interior for Pakistan at the UN Conference on the Illicit Trade in Small Arms and Light Weapons in All its Aspects, New York, 10 July 2001.* New York: Pakistan Mission to the United Nations.

Pakistan News Service. 2001. 'Arms Culture to be Rooted Out, Moin Reiterates.' 3 August. <http://www.paknews.com/archive.php?archive=main1&mon=8&year1=2001>

Plotkin, Martha, ed. 1996. *Under Fire: Gun Buy-Backs, Exchanges, and Amnesty Programs.* Washington, DC: Police Executive Research Forum.

Poulton, Robin Edward and Ibrahim ag Youssouf. 1998. *A Peace of Timbuktu: Democratic Governance, Development and African Peacemaking.* New York and Geneva: United Nations Institute for Disarmament Research.

Pourtier, Roland. 1998. 'Les Raisons d'une Guerre "Incivile".' *Afrique Contemporaine.* No. 186, pp. 7–32.

Romero, Michael P., Garen J. Wintemute, *et al.* 1998. 'Characteristics of a Gun Exchange Program and an Assessment of Potential Benefits.' *Injury Prevention.* Vol. 4, pp. 206–10.

Rosenfeld, Richard. 1996. 'Gun Buy-Backs: Crime Control or Community Mobilization?' In Martha Plotkin, pp. 1–23.

SALIGAD. 2001. *Assessing Small Arms Issues and Developing Capacity for Peace in the Horn of Africa.* <http://www.saligad.org>

Schmitt, Eric. 1999. 'Clinton Unveils Program to Buy Back Guns from Private Owners.' *The New York Times.* 10 September.

Sennott, Charles M. 2001. 'NATO Enters a Well-Armed Land.' *Boston Globe.* 26 August, p. 1.

Siddiqa-Agha, Ayesha. 2001. 'De-Weaponization Campaign is Non-Starter.' *The Friday Times.* Vol. XIII, No. 43. 21–27 December.

Small Arms Survey. 2001. *Small Arms Survey 2001: Profiling the Problem.* Oxford: Oxford University Press.

Smith, Chris and Domitilla Sagramoso. 1999. 'Small arms trafficking may export Albania's anarchy.' *Jane's Intelligence Review.* Vol. 11, No 1. 1 January.

Stabilisation Force (SFOR). 2001. Factsheets on Project Harvest. SFOR Informer Online. <http://www.nato.int/sfor/home.htm>

Sullivan, Jacqueline. 2001. *Rio de Janeiro Hosts World's Largest Weapons Destruction.* Background paper. Geneva: Small Arms Survey.

Sweden. Ministry for Foreign Affairs and National Defence College. 1999. *Removal of Small Arms and Light Weapons in the Context of Peace Missions.* Seminar Report. Stockholm, 11–12 March.

United Kingdom. 2001. *Statement by the United Kingdom during the UN Conference on the Illicit Trade of Small Arms and Light Weapons in All Its Aspects.* New York, 12 July.

__. Home Office. 2000. Correspondence from various officials, August–November.

__. House of Commons. 1999. Select Committee on Public Accounts. *Public Accounts: Twentieth Report.* HC 354, 7 July. <http://www.parliament.uk/commons/selcom/cmsel.htm>

United Nations. 2001. *Report of the United Nations Conference on the Illicit Trade in Small Arms and Light Weapons in All Its Aspects.* A/CONF.192/15. New York, 9-20 July.

__. Department of Peacekeeping Operations. Lessons Learned Unit. n.d. *Multidisciplinary Peacekeeping: Lessons from Recent Past.* <http://www.un.org/Depts/dpko/lessons/>

__. Department of Public Information. 1997a. *United Nations Observer Mission in Liberia.* <http://www.un.org/Depts/DPKO/Missions/unomil.htm>

__. 1997b. *Croatia-UNTAES: Recent Developments.* 22 December. <http://www.un.org/Depts/DPKO/Missions/untaes_r.htm>

United Nations Development Programme (UNDP). 1994. *Human Development Report.* New York: Oxford University Press.

__. 1998. *Weapons for Development: Report of the UNDP mission for an Arms Collection Pilot Programme in the Gramsh District.* Albania, Tirana. 4 September.

__. 2000a. *Weapons in Exchange for Development Update.* UNDP, Albania. 12 May. <http://www.undp.org.al/projects/weapons_upd12may2000.htm>

__. 2001a. *Statement by the United Nations Development Programme during the UN Conference on the Illicit Trade of Small Arms and Light Weapons in All Its Aspects.* New York, 13 July.

__. 2001b. *Sharing UNDP Experiences in Addressing Small Arms and Light Weapons Proliferation from the Development Perspective.* Findings of a Workshop for UNDP Country Staff. New York, 11 July.

__. 2001c. *UNDP Small Arms Reduction Programme: A Growing Portfolio.* Fact Sheet. New York: UNDP.

United Nations Development Programme and the Government of Albania. 1998. *Pilot Programme for Weapons in Exchange for*

Development in the District of Gramsh. Tirana, Albania: UNDP.

United Nations General Assembly (UNGA). 1996. *General and Complete Disarmament.* A/RES/51/45 of 10 December.

United Nations Mission in Sierra Leone (UNAMSIL). 2001. *Sierra Leone: Recovered Weapons to be Destroyed and Recycled.* Press Release. Freetown. 13 May.

United Nations. Office for the Coordination of Humanitarian Affairs. Integrated Regional Information Network (UN-IRIN). 2002a. *Sierra Leone: Over 45,000 Ex-Fighters Hand Over Their Weapons.* Abidjan. 7 January.

__. 2002b. *IRIN-WA Weekly Roundup 105.* 5–11 January.

__. 2002c. *Uganda: Focus on Karamoja Disarmament.* Nairobi. 10 January.

United Nations Secretary-General (UNSG). 1995. *Supplement to an Agenda for Peace: Position Paper of the Secretary-General on the Occasion of the Fiftieth Anniversary of the United Nations.* A/50/60/S/1995/1 of 25 January.

__. 1997. *Twenty-third Progress Report of the Secretary-General on the United Nations Observer Mission in Liberia.* S/1997/478 of 19 June.

__. 1998. *Report of the Secretary-General on the Causes of Conflict and the Promotion of Durable Peace and Sustainable Development in Africa.* S/1998/318 of 13 April.

__. 2000a. *Report of the Secretary-General on Methods of Destruction of Small Arms, Light Weapons, Ammunition and Explosives.* S/2000/1092 of 15 November.

__. 2000b. *The Role of United Nations Peacekeeping in Disarmament, Demobilization and Reintegration.* S/2000/101 of 11 February.

United Nations Security Council (UNSC). 1999. *Statement by the President of the Security Council on the item entitled 'Maintenance of Peace and Security and Post-Conflict Peace-Building.'* S/PRST/1999/21 of 8 July.

United States. General Accounting Office (GAO). 2000. *Conventional Arms Transfers: U.S. Efforts to Control the Availability of Small Arms and Light Weapons.* Report to the Honorable Dianne Feinstein, US Senate. Washington, DC: GAO.

Viva Rio. 2001. *Destruction of 100,000 Fire Arms in Rio de Janeiro.* Press Release. Rio de Janeiro. June.

Walla, Håvard. 2001. 'The Spoils of Peace.' *KFOR Online.* 6 August. <http://www.nato.int/kfor/news/reports/2001/nr_010806.htm>

Waszink, Camilla. 2001. 'Practical Disarmament Through Voluntary Weapons Collection Programmes.' In Sverre Lodgaard, pp. 35–45.

Willet, Susan. 2001. *Citizens Monitoring and Evaluation of Humanitarian Landmine Assistance.* Mimeo from ICBL meeting in Managua.

Working Group for Weapons Reduction (WGWR). 2001. *Weapons Collection and Management in Cambodia.* Background paper. Geneva: Small Arms Survey.

Yem, Sam Oeun, and Rebecca F. Catalla. 2001. 'I Live in Fear: Consequences of Small Arms and Light Weapons on Women and Children in Cambodia.' Ph.D. June. Phnom Penh, Cambodia.

Zentil, Marcelo. 2000. 'Plan Canje de Armas está en marcha.' *Diario los Andes.* Mendoza. 7 November.

Acknowledgements

Main contributors

Martín Appiolaza, Spyros Demetriou, Sami Faltas, Derek Miller, Robert Muggah, Jacqueline Sullivan, and the Working Group for Weapons Reduction in Cambodia.

Other contributors

William Godnick, Ayesha Siddiqa-Agha, and the United Nations Development Programme.

Index